BASIC SPACE PLASMA PHYSICS

Third Edition

BASIC SPACE
PLASMA PHYSICS

Third Edition

Wolfgang Baumjohann

Austrian Academy of Sciences, Austria

Rudolf A. Treumann

Munich University, Germany

World Scientific

NEW JERSEY · LONDON · SINGAPORE · BEIJING · SHANGHAI · HONG KONG · TAIPEI · CHENNAI · TOKYO

Published by

World Scientific Publishing Co. Pte. Ltd.

5 Toh Tuck Link, Singapore 596224

USA office: 27 Warren Street, Suite 401-402, Hackensack, NJ 07601

UK office: 57 Shelton Street, Covent Garden, London WC2H 9HE

Library of Congress Control Number: 2022002168

British Library Cataloguing-in-Publication Data
A catalogue record for this book is available from the British Library.

BASIC SPACE PLASMA PHYSICS
Third Edition

ISBN 978-981-125-405-5 (hardcover)
ISBN 978-981-125-440-6 (paperback)
ISBN 978-981-125-406-2 (ebook for institutions)
ISBN 978-981-125-407-9 (ebook for individuals)

For any available supplementary material, please visit
https://www.worldscientific.com/worldscibooks/10.1142/12771#t=suppl

Contents

Preface
to the Third Edition

Any natural (exact) science is in continuous progression. This was not known until the seventeenth century. However, only since the nineteenth century it became clear that science, in contrast to history and the humanities, shares this fundamental property with evolution. It is thus no surprise that it has to be continuously rewritten in order to adjust for new insights and elements which have been added to the bulk of scientific knowledge, Sir Raymond Popper's Third World, whereby it is important to note that all those scientific insight and laws in the exact natural sciences are based on observation and experimental results, and for this reason, they are removed only in the rarest cases. Instead of removing, they are continuously made more precise, improved, and confirmed while the former versions remain as approximations to the more exact laws to be valid and applicable in restricted domains. The only case when the old laws were permanently removed happened in the transition from medieval and purely descriptive to the modern experimental science in the Seventeenth Century starting with Kepler and Galilei. Descriptive science is subject to opinion and thus changes with time together with the changes of interpretation as has been formulated by Michel Foucault. However, due to the continuous improvement of knowledge, any textbook in any scientific field will for this reason age to become replaced by more modern texts. This happens because, firstly an increasing amount of experimental knowledge is added to the bulk, however secondly, also because science progresses to unify the accumulated knowledge and complete its theoretical understanding while simultaneously opening up new fields and igniting new activities. In this process, methodological advances throw their shadows on any scientific field to allow for simplifications, accelerate the accumulation, making older approaches methodologically but not scientifically obsolete. This methodological shift is particularly strongly felt in fields which evolve very quickly. Space science clearly belongs to those fields even though the phase of its great acceleration is to be traced back to the sixties and seventies of the past century. Meanwhile it has found its place in the mainstream of the scientific domains as an established and well founded scientific branch.

In such a situation it is quite satisfactory for us to recognise that extended parts of the present revised text have not yet lost their actuality such that it makes sense to publish a Third Edition without the need to completely rewrite the basic chapters of the former revision. Since the first ten chapters are intended to serve as an undergraduate course in space plasma physics, they can essentially remain as they are. There is no need for changing their systematics. Of course, all sections based on observations like the paragraphs on magnetic storms in Chapter 3 should be substantially extended in a more elaborate text. Lots of observations should be added here, in particular when referring to the observations which have been provided in the past three decades by the highly sophisticated multi-spacecraft-missions starting with Ampte, Ulysses, the

Cluster spacecraft, Themis, Swarm, MMS and many others including rocket observations and ground-based research like the ionospheric heating facility EISCAT in Scandinavia and stations in Antarctica in the polar regions, beyond the Voyager missions and others which have travelled to the distant inner and outer planets, the outskirts of the Heliosphere which they have left behind to fly away into the distant interstellar and galactic space. On the other hand, the most recent launch of the Solar Parker Probe intends to investigate the innermost parts of our solar system close to the sun inside the orbit of Mercury. All of them contributed and contribute substantially to our knowledge about near-Earth's space and its plasma physics. It remains a task for the future to collect the key new discoveries and all the knowledge which has been accumulated by them, to distill its essentials and compress it into a textbook covering the entire field of Space Physics in digestible form. In thinking over this new edition we were pondering of undertaking such an effort. We refrained from it because of the difficulty of distilling the real essentials. In spite of the large activities of the space physics community, the unification has not yet advanced to a stage where we could do this in any responsible way without neglecting important insight, otherwise we would include too many details which would divert the interested student.

For all these reasons we decided to stay with the current content of this book which has to our pleasure served as what it was intended for: providing the basic information about plasma physics in Earth's environment and beyond. We have, however, corrected a few typos which the eager student will probably have found him/herself. Nevertheless, a new edition always provides the opportunity of getting closer to the ideal print which almost lacks any errors. Other changes to the text and chapters we have not done except for a few rectifications. The book is essentially the same as before, and we hope that this is so for the beginning student. Once he/she has learned the basics of space plasma physics from it, she/he should be in a position to understand observations and theory and contribute to the progress in this field.

We feel, however, that we should provide the student with a list not of the many achievements in space plasma physics but hints on the main contemporary fields of most intense research. In looking back, they have surprisingly remained the same with the addition of only a few. In the ionosphere, intense research has been devoted to the equatorial electrojet physics and its effects on the atmosphere exciting a number of instabilities. Similarly, intense research has concentrated on the auroral zone, the structure of field-aligned currents and their stability, the generation of large localised electric fields during geomagnetic disturbances and emission of surprisingly intense radiation in the kilometer wave band, so-called AKR radiation, first discovered in the sixties and early seventies. Traditionally, radio propagation and the physics of lightnings have been in focus. An important new development is the coupling between plasma and thermosphere which plays an essential role in the so-called physics of Space Weather and the connection to the main disturbance of the geomagnetic field, the magnetic storm with its building element, the magnetospheric substorm. In all these more modern approaches, the neutral atmospheric composition now occupies its place which it didn't do before. Clearly, the ionospheric density, composition, and conductivity, and of course the ionospheric current system and its closure in the magnetosphere in all spatial and temporal variations have been calculated and mapped to high precision today already. Just to note, it has also been found that the upper ionosphere during storms emits not only radio waves but also X rays which indicates strong collisions and high particle energies generated either in lightnings, their upper ionospheric equivalent, or in the outer magnetosphere, preferentially by particle acceleration in strong reconnection in the central near-Earth plasma sheet and cross-tail current. All these processes occurring there are of extraordinary complexity. For this reason, they cannot be covered in a textbook like this one which, however, provides the basics to understand them before entering those fields of current research.

In this spirit, the present book retains its gross structure. We have, however, taken the opportunity of the present Third Edition of the earlier revised version to add a new Appendix in order to extend two of the sections in this book where some more recent theoretical results have been obtained. One of the sections deals with stationary skewed kinetic particle distributions which have been observed continuously in space for long time while lacking their understanding. Skewed distributions indicate several different properties like the presence of higher order nonlinear interactions among particles and fields but should then depend on time resembling diffusive processes which can be treated by a quasilinear Fokker-Planck approach. Recently, however, it was found that the heuristic Kappa distribution invented in the mid-sixties by Stan Olbert could be given its statistical mechanical kinetic foundation. It justifies such distributions as statistical mechanical or thermodynamic states far from thermal equilibrium. A brief account of this approach is included now in the Appendix.

Another addition contained in the Appendix deals with a most interesting zero-frequency plasma instability, the mirror mode. This instability is of basic interest because it indicates the start of a phase transition in high temperature collisionless plasma which resembles a superconducting state in solid state physics. The Appendix briefly refers to the physics, in addition to also providing a small number of references.

Finally a short section on Space Weather as a new applicational field in space plasma physics has been added in order to alert the reader interested in applications to the possibility of starting an applicational career in space plasma. Space weather or better space meteorology, a subfield of the more extended field of Space Climatology, collects the global scientific effort to monitor the impact of space on Earth and civilisation in order to raise the awareness of possible hazards of space and the possibilities to either forecast and predict them in order to prevent or mitigate their dangerous effects. This is of substantial practical importance in view of our technologically advanced societies which have become increasingly vulnerable not just to the positive as negative anthropogenic effects, its impact on nature, but also to the natural threats that Earth and civilisation are exposed to from the so-far yet barely predictable event provided occasionally by the space climate events. We hope the future reader and student of this introductory treatise will acknowledge these efforts of modernising the introductory text to Space Plasma Physics.

                                                                                    The authors

Preface
to the Revised Edition

After a bit more than one decade where, to our surprise and not less also to our delight, this treatise has received widespread interest and use in basic courses in space plasma physics, and after all the positive reactions to its publication and the encouraging comments we received as an echo to our initial efforts in writing it, we feel obliged to modernise its content in order to bring it up to the current state of knowledge in space plasma physics and to meet the needs of students and teachers in this field. We have been approached several times in the meantime by various colleagues to invest some effort into a new edition but have resisted for long, not only because of the heavy working load we both had but also because we knew that the interested students and colleagues had already detected the misprints and had corrected them in their copies not transmitting them farther. We were also thinking of possibly writing quite another book which would have consisted solely of problems such that the interested student could have learned from solving the problems instead of reading a text. However, as things evolve, though we have indeed started it, we never found the time to proceed with that project, and so ultimately decided to take the bull by the horns and produce a new revised and slightly extended edition of our original text now including some problems at the end of each tutorial chapter and adding some more literature. The result is what the reader has in his or her hands.

Most extensions have been seamlessly integrated in the attempt of not to distort the flow when something new is included or the interpretation has changed. We have taken great care in detecting the misprints. Hopefully their remaining number has become vanishingly small. In addition to these quite many changes in paraphrasing here and there, we also added a few new sections like the one on magnetopause reconstruction in Chapter 8. These are based on new developments in space physics which we felt the reader should be informed about in passing.

In the same spirit, we decided to add a chapters on instability and thermal fluctuations. This chapters is needed in order to complete the discussion of the kinetic dispersion relation which in general has complex solutions. In many cases one of these solutions runs unstable. Hence, the alert reader has probably asked what happened to those solutions. The new chapter on instability provides the important answer to that question. We, however, felt that it was then also necessary to raise and answer another question about how a wave could grow if it is not already initially present. Fortunately, quantum theory offers an answer to this question in terms of thermal fluctuation theory which in the classical case simplifies considerably. This simplified theory has been briefed in this volume and illustrated with a few cases of application.

We finally also felt that one should then provide at least a few applications. As the first the important problem of collisionless reconnection has been selected. Reconnection is at the heart of the physics of the magnetosphere and thus directly related to near-Earth space plasma physics. It is, by today's belief, responsible for the general convective motion of plasma in the magnetosphere, for filling the magnetosphere with plasma in the first place and for the global instability of the magnetosphere, the substorm and ultimately the magnetospheric storm. Reconnection is a difficult problem which unites global and microscopic aspects. Its theory is still in active and fast evolution. We therefore give only a cursory account of it adding two brief observational applications: reconnection at the magnetopause and magnetotail respectively which from the space physics point of view are closest to our interest. However, reconnection has much wider application to solar physics, in the universe, and last not least in the laboratory and fusion research.

For the second application we chose the problem of collisionless shocks. This choice was guided by the importance of shocks in the universe and the plasma physical interest in this very complicated and fascinating subject. The theory of shocks is a nice example of lucidity, and in the non-relativistic case it has achieved a state where it makes sense to attempt a concise textbook review. We do not touch the much more complicated question of formation of relativistic shocks however in this book as in near-Earth space no examples of such shocks are known except, possibly, shocks artificially produced in Laser fusion experiments.

This shock chapter and the former chapter on collisionless reconnection are thus less tutorial than all other chapters in this volume, but they provide some meat to the dry theory developed there. We consider this being of some value for the reader to become also confronted with some new developments which in addition go beyond what is contained in our companion volume. For reasons of limited time we are not in the position to revise also that book in particular as so far it is not yet out of stock. Including new material on these matters in a way which simply continues the present book makes sense. That this policy led to an extension over its original size might be considered disadvantageous. It is, however, our feeling that the advantages compensate for the disadvantages.

What concerns the problems added to the tutorial chapters so their degree of difficulty spans a wide range. Most of them can be answered by intelligent guessing, some of them require simple calculations by using the formulae given, some are possibly difficult for the beginner as they require insight and inventive thinking. We have, however, intentionally not given any answers or solutions as is usually done in textbooks. The idea is that, from our own experience, an appendix containing the solutions and answers is not very helpful. Rather it seduces the reader to look for the solution instead of investing some effort himself. Instead he will find just a few hints in order to put him on the right track.

The revised version we present herewith is hopefully avoid of the misprints and also of several misconceptions the first edition contained. We have taken great care in detecting and correcting them. Any comments are requested to be sent electronically by electronic mail to *wolfgang.baumjohann@oeaw.ac.at*. We hope the readers, students or instructors, will enjoy the new edition or, at least, will find it useful as a textbook.

Wolfgang Baumjohann and Rudolf Treumann

Preface
to the First Edition

One more textbook on plasma physics? Indeed, there are a number of excellent textbooks on the market, like the incomparable book *Introduction to Plasma Physics and Controlled Fusion* by Francis F. Chen. It is impossible to compete with a book of this clarity, or some of the other texts which have been around for longer or shorter. However, we found most of the books not well-suited for a course on space plasma physics. Some are directed more toward the interests of laboratory plasma physics, like Chen's book, others are highly mathematical, such that it would have required an additional course in applied mathematics to make them accessible to the students. The vast majority of books in the field of space plasma physics, however, are collections of review articles, like the recent *Introduction to Space Physics* edited by Margaret G. Kivelson and Christopher T. Russell. These books require that the reader already has quite some knowledge of the field.

The only textbook specifically addressed to the needs of space plasma physics is *Physics of Space Plasmas* by George K. Parks. This book covers many aspects of space plasma physics, but is ordered in terms of phenomena rather than with respect to plasma theory. To give the students a feeling for the coherency of our field, we felt the need to find a compromise between classical plasma physics textbooks and the books by Parks and Kivelson & Russell. We tried to achieve this goal during a third-year space plasma physics course, which we gave regularly at the University of Munich since 1988 for undergraduate and graduate students of geophysics, who had an average knowledge of fluid dynamics and electromagnetism.

This textbook collects and expands lecture notes from these two-semester courses. However, the first part can also be used for a one-semester undergraduate course and research scientists may find the later chapters of the second part helpful. The book is written in a self-contained way and most of the material is presented including the basic steps of derivation so that the reader can follow without need to consult original sources. Some of the more involved mathematical derivations are given in the Appendix. Special emphasis has been placed on providing instructive figures. Even figures containing original measurements are mostly redrawn in a more schematic way.

The first five chapters provide an introduction into space physics, based on a mixture of simple theory and a description of the wealth of space plasma phenomena. A concise description of the Earth's plasma environment is followed by a derivation of single particle motion in electromagnetic fields, adiabatic invariants, and applications to the Earth's magnetosphere and ring current. Then the origin and effects of collisions and conductivities and the formation of the ionosphere are discussed. Ohm's law and the frozen-in concept are introduced on a somewhat heuristic basis. The first part ends with an introduction into magnetospheric dynamics, including convection electric fields, current systems, substorms, and other macroscopic aspects of solar wind-magnetosphere and magnetosphere-ionosphere coupling.

The second part of the book presents a more rigorous theoretical foundation of space plasma physics, yet still contains many applications to space physics. It starts from kinetic theory, which is built on the Klimontovich approach. Introducing moments of the distribution function allows the derivation of the single and multi-fluid equations, followed by a discussion of fluid boundaries and shocks, with the Earth's magnetopause and bow shock as examples. Both, fluid and kinetic theory are then applied to derive the relevant wave modes in a plasma, again with applications from space physics.

The material presented in the present book is extended in *Advanced Space Plasma Physics*, written by the same authors. This companion textbook gives a representative selection of the many macro- and microinstabilities in a plasma, from the Rayleigh-Taylor and Kelvin-Helmholtz to the electrostatic and electromagnetic instabilities, and a comprehensive overview on the nonlinear aspects relevant for space plasma physics, e.g., wave-particle interaction, solitons, and anomalous transport.

We are grateful to Rosmarie Mayr-Ihbe for turning our often rough sketches into the figures contained in this book. It is also a pleasure to thank Jim LaBelle for valuable contributions, Anja Czaykowska and Thomas Bauer for careful reading of the manuscript and many suggestions, and Karl-Heinz Mühlhäuser and Patrick Daly for helping us with LaTeX. We gratefully acknowledge the support of Heinrich Soffel, Gerhard Haerendel and Gregor Morfill, and acknowledge the patience of our colleagues at MPE, when we worked on this book instead of finishing other projects in time. Both of us owe deep respect to our teachers who introduced us into geophysics, Jürgen Untiedt and the late Gerhard Fanselau.

Last but not least, we would like to mention that we have profited from many books and reviews on plasma and space physics. References to most of them have been included into the suggestions for further reading at the ends of the chapters. These suggestions, however, do not include the very large number of original papers, which we made use of and are indebted to.

Needless to say, we have made all efforts to make the text error-free. However, this is an unsurmountable task. We hope that the readers of this book will kindly inform us about misprints and errors they may find in here, preferentially by electronic mail to *bj@mpe-garching.mpg.de*. We will be grateful for any hints and post them with other errors on *http://www.mpe-garching.mpg.de/bj/bspp.html*.

— 1 —

Introduction

The context of the term 'geophysics' has changed considerably during the second half of the past century. Well into the fifties the key interest of non-exploratory non-applied geophysics focussed on the interior of our planet, i.e., solely on the geophysics of the solid Earth, covering seismology, rock physics, magnetic and electric properties of crust and mantle, the structure of the outer and inner cores of Earth etc. Clearly, interest in these fields has not diminished; at the contrary some of them have become mature and have even attracted public interest. This is due to the development of methods of data analysis and interpretation in seismology, new methods in high pressure physics and a better understanding of the processes of wave propagation in matter under high pressure and in fluids with the latter providing the key to the understanding of the origin of the geomagnetic field. It is also due to the growing public interest in the resources, in the environment, and in the history and the future of our planet.

On the other hand, with the advent of the spaceflight era, the interests of geophysicists broadened and extended into the external neighborhood of our planet. It was realized that the extraterrestrial matter is in an ionized state, very different from the state of known matter near the ground of the Earth. Gradually, it was also realised that this environment of Earth plays its own and not completely unimportant role. It is a strongly dynamical environment on time scales much shorter than those of the dynamical processes in the Earth's interior. And it even affects, to sometimes non-negligible extent, the processes close to the Earth's surface which are the direct life-space of mankind. This happened in the history of Earth when the geomagnetic field arose in the interior of the planet and provided a screen preventing Earth's surface from being hit by medium energy Cosmic Rays and thus enabled the evolution of sophisticated biological molecules and higher forms of life on Earth. It happened at later times when the geomagnetic field changed polarity or switched from dipolar to quadrupolar geometry and back again in the unstable phases of the geo-magnetic dynamo. It also happens from time to time today when violent events in the immediate vicinity of the Earth, so-called magnetic storms, occur and affect the propagation of electromagnetic waves and other technical installations, having led to the invention of the new research field of Space Weather which was coined in order to awake a feeling of conscience for a relation between the weather and climate in space and the weather and climate on Earth.

Matter in the state of high ionisation behaves unexpectedly because of its high sensitivity to electric and magnetic fields and because of its ability to carry electric currents, which sometimes can be very strong, and to provide a medium in that electric field may occasionally exist, drive motion of the matter and cause violent electric discharges. Within this context, the *concept of a plasma* became introduced and space plasma physics became a new and important branch of geophysics. Nowadays, methods of plasma physics are not only used in external geophysics, but have also been recognised of being essential to the understanding of the dynamics of the Earth's fluid outer core and the generation of the terrestrial magnetic field.

1.1 Definition of a Plasma

A *plasma* is a gas of charged particles, which consists of about equal numbers of free positive and negative charge carriers. Having roughly the same number of charges with different signs in the same volume element guarantees that the plasma behaves *quasi-neutral* in the stationary state. On average a plasma looks electrically neutral to the outside, since the randomly distributed particle electric charge fields mutually cancel.

For a particle to be considered a free particle, its typical potential energy due to its nearest neighbor must be much smaller than its random kinetic (thermal) energy. Only then the particle's motion is practically free from the influence by other charged particles in its neighborhood as long as no direct collisions take place.

Since the particles in a plasma have to overcome the coupling with their neighbors, they must have thermal energies above some electronvolts. Thus a typical plasma is a *hot* and *highly ionized gas*. While only a few natural plasmas, such as flames or lightning strokes, can be found near the Earth's surface, plasmas are abundant in the universe. More than 99% of all known ordinary baryonic matter in the 'visible' universe after light decoupled from matter at about 300 000 years after the Big Bang is in the plasma state.

1.1.1 Debye Shielding

For the plasma to behave quasineutral in the stationary state, it is necessary to have about equal numbers of positive and negative charges per volume element. Such a volume element must be large enough to contain a sufficient number of particles, yet small enough compared with the characteristic lengths for variations of macroscopic parameters such as density n and temperature T. In each volume element the microscopic space charge fields of the individual charge carriers must cancel each other to provide the required macroscopic charge neutrality.

To let the plasma appear electrically neutral, the electric *Coulomb potential* field ϕ_C of every individual point charge, q,

$$\phi_C = \frac{q}{4\pi\varepsilon_0 r} \tag{1.1}$$

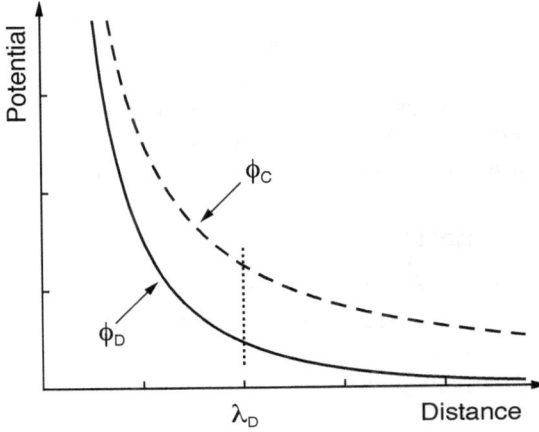

Figure 1.1: Comparison of Debye and Coulomb potential demonstrating the shielding above λ_D.

with ε_0 being the free space permittivity, is shielded by other charges in the plasma and assumes the *Debye potential* form

$$\phi_D = \frac{q}{4\pi\varepsilon_0 r} \exp\left(-\frac{r}{\lambda_D}\right) \tag{1.2}$$

in which the exponential function cuts off the potential at distances $r > \lambda_D$. The characteristic length scale, λ_D, is called *Debye length* and is the distance, over which a balance is obtained between the thermal particle energy, which tends to perturb the electrical neutrality, and the electrostatic potential energy resulting from any charge separation, which tends to restore charge neutrality. Figure 1.1 shows this shielding effect.

In Section 9.1 we will show that the Debye length is a function of the electron and ion temperatures, T_e and T_i, and the plasma density, $n_e \simeq n_i$ (assuming singly charged ions)

$$\lambda_D = \left(\frac{\varepsilon_0 k_B T_e}{n_e e^2}\right)^{\frac{1}{2}} \tag{1.3}$$

where we have assumed $T_e \simeq T_i$ and where k_B is the Boltzmann constant and e the electron charge. We will give the exact definition of the temperature in Section 6.5. Until then we will use the terms temperature and average energy, $\langle W \rangle = k_B T$, as synonyms.

In order for a plasma to be quasineutral, the physical dimension of the system, L, must be large compared to the Debye (charge field shielding) length λ_D

$$\lambda_D \ll L \tag{1.4}$$

This requirement is often called the *first plasma criterion*. Below it will be completed by a more quantitative and thus easier to handle criterion. Otherwise, when this condition is violated, there is not enough space available for the collective shielding effects to occur, and we have a simple neutral gas containing just a small number of ionised particles (ions) insufficiently many to allow to speak of a plasma. The latter is typical for most part of Earth's atmosphere below roughly 80 km altitude.

1.1.2 The Plasma Parameter

Since the shielding effect is the result of the collective behavior inside a Debye sphere of radius λ_D, it is necessary that this sphere contains a sufficiently large particles. The number of particles N_D inside a Debye sphere is $N_D = \frac{4\pi}{3} n_e \lambda_D^3$. The term $n_e \lambda_D^3$ is often called the *plasma parameter*, Λ. With its help one has the *second plasma criterion*

$$\Lambda \equiv n_e \lambda_D^3 \gg 1 \tag{1.5}$$

By substituting λ_D from the expression given in Eq. (1.3) and raising each side of Eq. (1.5) to the 2/3 power, it becomes apparent that the second criterion quantifies what is meant by free particles. The mean potential energy, $q\phi_C = q^2/4\pi\varepsilon_0\lambda_{ip} \ll k_B T_e$, of a particle due to its nearest neighbor, which is inversely proportional to the mean interparticle distance, $\lambda_{ip} = n_e^{-1/3}$, and thus is proportional to $n_e^{1/3}$, must be much smaller than its mean energy, $k_B T_e$, for an ensemble of charged particles of both signs to constitute a plasma.

1.1.3 The Plasma Frequency

The typical oscillation frequency in a fully ionized plasma is the electron *plasma frequency*, ω_{pe}. If the overall quasi-neutrality of the plasma is locally disturbed by some external force or simply by fluctuations in the spatial particle distribution, the light electrons, being more mobile than the much heavier ions, become accelerated in an attempt to restore the violated charge neutrality. Due to their inertia they will move back and forth around their original equilibrium position, which results in fast collective oscillations around the more massive ions and thus changes in electron density n_e. In Section 9.1 it will be shown that the plasma frequency depends on the square root of the plasma density. With m_e the electron mass, the frequency ω_{pe} can be written as

$$\omega_{pe} = \left(\frac{n_e e^2}{m_e \varepsilon_0} \right)^{\frac{1}{2}} \tag{1.6}$$

Some plasmas, like the Earth's ionosphere, are not fully ionized. Here we have a substantial number of neutral particles and if the charged particles collide too often with neutrals, the electrons will be forced into equilibrium with the neutrals. Then the medium does not behave as a plasma anymore, but behaves similar to an electrically

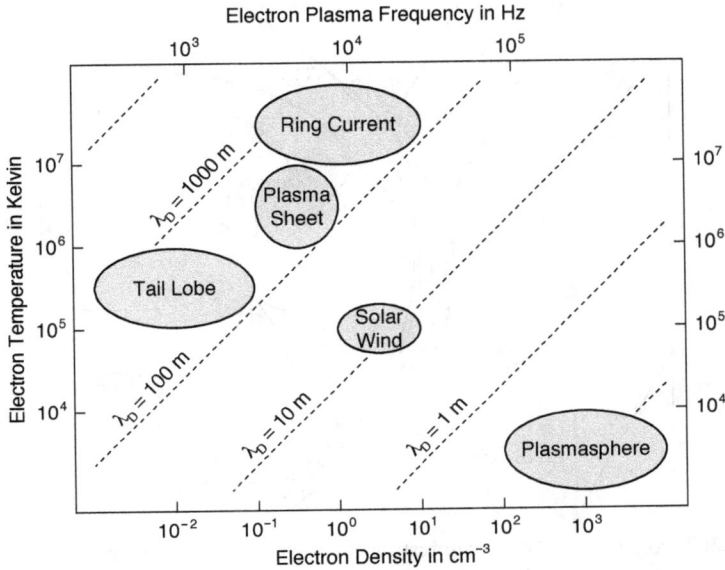

Figure 1.2: Ranges of the typical parameters density, temperature and Debye shielding length for several geophysical plasmas. The plasma parameters differ by orders of magnitude. The increase in the Debye length toward the upper left corner of the figure is of particular interest. It visualises the increasing importance of the first plasma criterion with decreasing plasma density and increasing temperature. Plasmas in that corner can be called plasma only if they are sufficiently extended. This is, for instance, not necessarily the case anymore for the radiation belt particles which therefore are usually treated as single charged particles moving in the geomagnetic field.

conducting neutral gas. For the electrons to remain unaffected by collisions with neutrals, the average time between two electron-neutral collisions, τ_n, must be larger than the reciprocal of the plasma frequency

$$\omega_{pe}\tau_n \gg 1 \tag{1.7}$$

This is the *third plasma criterion* for an ionized medium to behave as a plasma. The same criterion for the ions (replacing ω_{pe} with ω_{pi}) puts a somewhat more stringent condition on the plasma.

1.2 The Geophysical Plasmas

Plasmas are not only abundant in the universe, but also in our solar system. Even in the immediate neighborhood of the Earth, all matter above about 100 km altitude, within and above the ionosphere, must be treated using plasmaphysical methods.

There are quite a number of different geophysical plasmas, with a wide spread in their characteristic parameters like density and temperature. Figure. 1.2 gives a

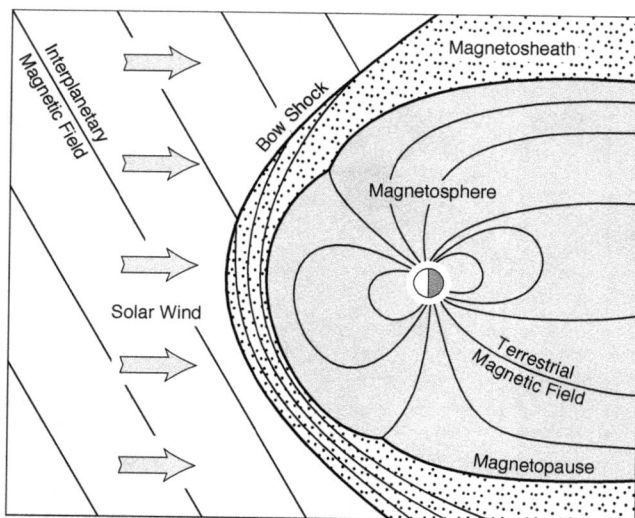

Figure 1.3: Topography of the solar-terrestrial environment with the sun assumed to be located in the far left, blowing the solar wind against the terrestrial (geomagnetic) field and confining the field to the magnetosphere inside its outer boundary, the magnetopause. The solar wind carries with it the interplanetary magnetic field. Because of the super-sonic nature of the solar wind flow, a bow shock wave is generated in front of the magnetosphere. It compresses the plasma and interplanetary magnetic field into the magnetosheath between the bow shock and the magnetopause.

synopsis of the ranges of the plasma parameters density n_e and temperature T_e which the plasmas cover in the environment of Earth. These ranges vary by many orders of magnitude between the different plasmas. Accordingly their properties are very different.

1.2.1   The Solar Wind

The sun emits a highly conducting plasma which streams radially out of the sun at supersonic speeds of about 500 km/s into the interplanetary space. This outflow of plasma is the result of the supersonic expansion of the solar corona which is caused by the heating of the solar corona by some even today barely specified process which creates such high temperatures that a large number of the fast particles, electrons and also ions, in the solar corona can overcome the solar gravitational attraction and escape into interplanetary space being blown out of the sun like a solar breath.

Because of the solar variability the velocity of this stream may vary between 300 km/s and 1500 km/s though both extremes occur only relatively rarely during particular solar eruption events. This streaming and expanding plasma is called the *solar wind* and consists mainly of electrons and protons, with an admixture of 5% Helium ions. Expanding radially out the density of the solar wind decrease with the inverse square of the distance from the sun while its temperature decreases adiabatically. At distance of 1 AU at the orbit of Earth the solar wind is already quite dilute and thus electrically highly conducting. Typical values for the electron density and temperature in the solar wind near the Earth are $n_e \approx 5$ cm^{-3} and $T_e \approx 10^5$ K (1 eV = 11,600 K; see Appendix B.2). The solar wind ions have lower and therefore difficult to measure temperature roughly one order of magnitude less than the electrons.

Because of the very high solar wind conductivity, which for all practical purposes can be assumed to be infinitely high (see Chapter 4 for a discussion of plasma conductivities), the solar coronal magnetic field is completely 'frozen' into the streaming solar wind plasma (very similar to a superconductor, see Section 5.1) and drawn outward by the expanding solar wind flow, gradually becoming the interplanetary magnetic field (which usually is abbreviated as IMF). The *interplanetary magnetic field* at 1 AU near Earth's orbit is of the order of $\sim$5 nT but may vary between 3 nT and occasionally even 12 nT. With radial distance from the sun it decays like the inverse of the radius. Later we will show that with distance it readily assumes an about azimuthal spiral field configuration.

When the solar wind hits on the Earth's dipolar magnetic field, it cannot simply penetrate it but rather is slowed down and, to a large extent, deflected around it as shown schematically in Fig. 1.3. The solar wind hits the obstacle with supersonic speed, and a *bow shock* wave is generated in front of the Earth in the direction of the sun. At this bow shock the solar wind plasma is slowed down and a substantial fraction of the solar wind particles' kinetic energy is converted into thermal energy. The extended region of thermalized subsonic plasma behind the bow shock between it and the outer boundary of the geomagnetic field is called the *magnetosheath*. Its plasma is compressed and thus denser and hotter than the proper solar wind plasma and the magnetic field strength has higher values in the magnetosheath region. Both magnetic field and plasma density undergo quite large fluctuations here in magnitude, and the magnetic field and flow also in direction.

1.2.2 The Magnetosphere

The shocked solar wind plasma in the magnetosheath cannot easily penetrate the terrestrial magnetic field but is mostly deflected around it. This is a consequence of the fact that the interplanetary magnetic field lines cannot penetrate the terrestrial field lines and that the solar wind particles cannot leave the interplanetary field lines due to the aforementioned frozen-in characteristic of a highly conducting plasma.

The boundary separating the two different regions is called *magnetopause* and the cavity generated by the terrestrial field has been named *magnetosphere* (see Figs. 1.3 and 1.4). The kinetic pressure of the solar wind plasma distorts the outer part of the terrestrial dipolar field. At the frontside it compresses the field, while the nightside magnetic field is stretched out into a long *magnetotail* which reaches far beyond lunar orbit.

The plasma in the magnetosphere consists mainly of electrons and protons. The sources of these particles are the solar wind and the terrestrial ionosphere. In addition

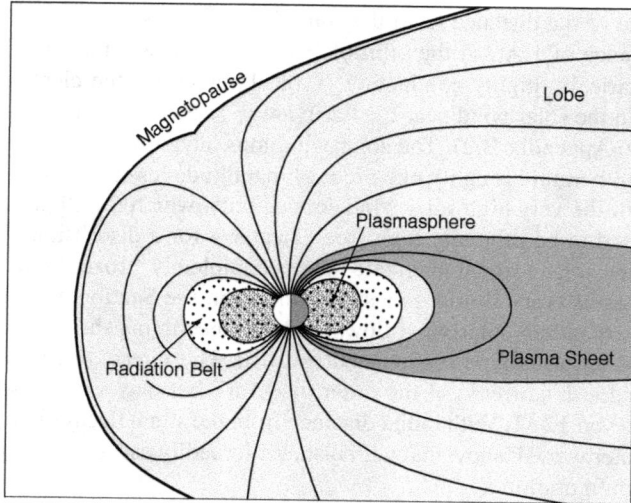

Figure 1.4: The plasma structure of the Earth's magnetosphere showing the four main magnetospheric plasma popultions: a dilute plasma in the lobes, a hot plasma in the plasma sheet, a dense cold plasma in the plasmasphere, and the very dilute and hot radiation belt plasma.

there are small fractions of He^+ and O^+ ions of ionospheric origin and some He^{++} ions originating from the solar wind. However, the plasma inside the magnetosphere is not evenly distributed, but is grouped into different regions with quite different densities and temperatures. Figure 1.4 depicts the topography of some of these regions.

The *radiation belt* lies on dipolar field lines between about 2 and 6 R_E (1 Earth radius = 6371 km). It consists of energetic electrons and ions which move along the field lines and oscillate back and forth between the two hemispheres (see Section 3.2). Typical electron densities and temperatures in plasma background in the radiation belt region are $n_e \approx 1$ cm^{-3} and $T_e \approx 5 \times 10^7$ K. The high energy radiation belt particles which give this region the name have much lower density while being of much higher kinetic energy, of the order of > 30 keV reaching up to several MeV. The magnetic field strength ranges between about 100 and 1000 nT.

Most of the magnetotail plasma is concentrated around the tail midplane in an about 10 R_E thick *plasma sheet* that extends far out into the nightside. Near the Earth, it reaches down to the high-latitude *auroral ionosphere* along the field lines. Average electron densities and temperatures in the plasma sheet are $n_e \approx 0.5$ cm^{-3} and $T_e \approx 5 \times 10^6$ K, with $B \approx 10$ nT.

The outer part of the magnetotail is called the *magnetotail lobe*. It contains a highly rarified plasma with typical values for the electron density and temperature and the magnetic field strength of $n_e \approx 10^{-2}$ cm^{-3}, $T_e \approx 5 \times 10^5$ K, and $B \approx 30$ nT, respectively.

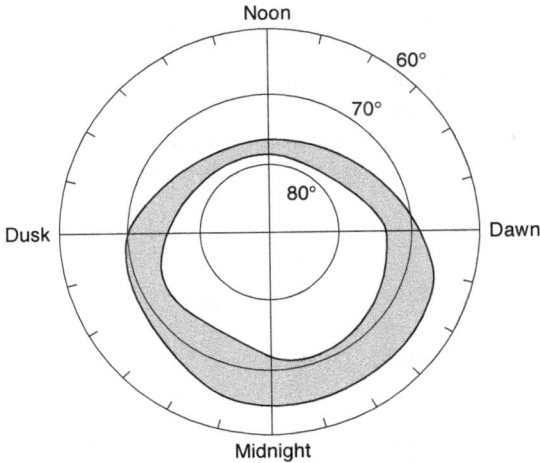

Figure 1.5: Average auroral oval (shaded region) and polar cap. The auroral oval is defined as the region of occurrence of visible auroae. This is centred around the pole regions both on the North over the Arctic and the South over Antarctica. This region is not a perfect oval. On the nightside it extends down to latitudes of $\sim 40°$ while on the dayside it is quite close to the pole and much narrower than during night which accounts for the more frequent occurrence of aurorae during night time. The closed region that is left free around the pole is the polar cap region.

1.2.3 The Ionosphere

The solar ultraviolet light impinging on the Earth's atmosphere ionizes a fraction of the neutral atmosphere. This ionisation effect is highly altitude dependent and thus not solely determined by the solar irradiation of the atmosphere but to a large extent by the constitution of the atmosphere, its chemical composition and its density. The latter is a function of altitude which is determined by the gravitational attraction of Earth and by the temperature of the atmosphere. Thus, the ionisation is also a function of the geographical latitude and local time. To some extent the ionisation is also affected by temporal variations in the atmosphere which may reach altitudes of the ionosphere and to solar variability. At altitudes above 80 km collisions between neutral and freshly ionised ions are too infrequent to result in rapid recombination and a permanent ionized population called the *ionosphere* is formed. Typical electron densities and temperatures in the mid-latitude ionosphere are $n_e \approx 10^5$ cm^{-3} and $T_e \approx 10^3$ K. The magnetic field strength is of the order of 10^4 nT.

The ionosphere extends to rather high altitudes and, at low- and mid-latitudes, gradually merges into the *plasmasphere*. As depicted in Fig. 1.4, the plasmasphere is a torus-shaped volume inside the radiation belt. It contains a cool but dense plasma of ionospheric origin ($n_e \approx 5 \times 10^2$ cm^{-3}, $T_e \approx 5 \times 10^3$ K), which corotates with the Earth. In the equatorial plane, the plasmasphere extends out to about 4 R_E, where the density drops down sharply to about 1 cm^{-3}. This boundary is called the *plasmapause*.

At high latitudes plasma sheet electrons can precipitate along magnetic field lines down to ionospheric altitudes, where they collide with and ionize neutral atmosphere particles. As a by-product, photons emitted by this process create the polar light, the *aurora*. These auroras are typically observed inside the *auroral oval* (see Fig. 1.5), which contains the footprints of those field lines which thread the plasma sheet. Inside of the auroral oval lies the *polar cap*, which is threaded by field lines connected to the tail lobe.

1.3   Magnetospheric Currents

The plasmas discussed in the last section are usually not stationary but move under the influence of external forces. Sometimes ions and electrons move together, like in the solar wind. But at other occasions and in other plasma regions, ions and electrons move into different directions, creating electric currents. Such currents play a very important role in the dynamics of the Earth's plasma environment. They transport charge, mass, momentum and energy. Moreover, the currents create magnetic fields, which may severely alter or distort any pre-existing fields. Actually, the distortion of the terrestrial dipole field into the typical shape of the magnetosphere is also accompanied by electrical currents. As schematically shown in Fig. 1.6, the compression of the terrestrial magnetic field on the dayside is associated with current flow along the magnetopause surface, the *magnetopause current*. The tail-like field of the nightside magnetosphere is accompanied by the *tail current* flowing on the tail surface and the *neutral sheet current* in the central plasma sheet, flowing across the magnetosphere from dawn to dusk; both currents are connected and form a Θ-like current system, if seen from along the Earth-Sun line.

Another large-scale current system, which influences the configuration of the inner magnetosphere, is the *ring current*. The ring current flows around the Earth in a westward direction at radial distances of several Earth radii and is carried by the ionic component and also the low-energy electronic component of the radiation belt particles mentioned above. In addition to their bounce motion, these particles drift slowly around the Earth. Since the protons drift westward while the electrons move in the eastward direction, this constitutes a net charge transport.

A number of current systems exist in the conducting layers of the Earth's ionosphere, at altitudes of 100–150 km. Most notable are the *auroral electrojets* inside the auroral oval, the *Sq currents* in the dayside mid-latitude ionosphere, and the *equatorial electrojet* above the magnetic equator. The *Sq* current is a permanently present about stationary current system caused by the solar illumination of the ionosphere which restricts it to the dayside region. The two jets are concentrations of current flows at about constant geomagnetic latitude.

In addition to these currents which all flow perpendicular to the geomagnetic and magnetospheric field, currents also flow along the electrically highly conducting magnetic field lines. One may distinguish between large-scale and small-scale field-aligned currents. As shown in Fig. 1.6, the large-scale *field-aligned currents* connect

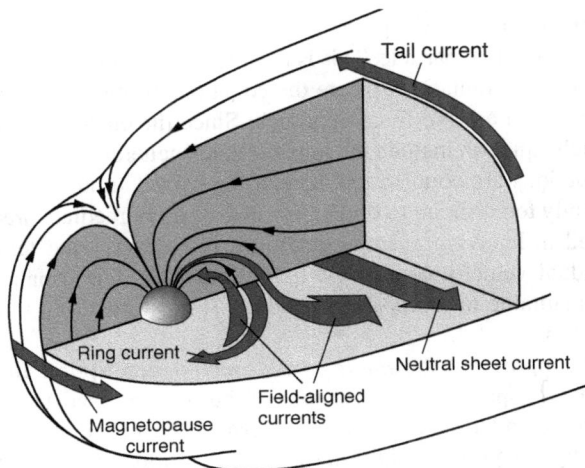

Figure 1.6: Synopsis of the large-scale currents (shaded arrows) flowing in the magnetosphere. The dominant magnetospheric current system consists of currents flowing perpendicular to the magnetic field. These are the magnetopause currents and tail currents along the boundary of the magnetosphere, the neutral sheet current across the magnetospheric tail which connects to the tail current, and the ring current flowing around Earth in the equatorial plane of the magnetosphere. Shown are also the field-aligned currents which connect the magnetospheric (neutral sheet and ring) current system to the ionosphere. The ionospheric currents flowing perpendicular to the magnetic field which close the current system are not shown.

the large-scale magnetospheric current systems in the magnetosphere to those currents flowing in the polar ionosphere. Small-scale field-aligned currents, on the other hand, are generated by temporal processes in the magnetospheric system and do not necessarily connect the entire large-scale magnetospheric current system to the ionosphere. They mostly connect only selected parts of the magnetosphere which are temporarily affected. This happens, for instance, during so-called magnetospheric disturbances during substorms. The field-aligned currents are almost exclusively carried by electrons flowing along the magnetic field and are essential for the exchange of energy and momentum between the magnetospheric and ionospheric regions which are connected through the geomagnetic field.

1.4 Theoretical Approaches

The dynamics of a plasma is governed by the interaction of the charge carriers with the electric and magnetic fields. If all the fields were of external origin, the physics would be relatively simple. However, as the particles move around, they may create local space charge concentrations and thus electric fields. Moreover, their motion can also generate electric currents and thus magnetic fields. These internal fields and their feedback onto the motion of the plasma particles make plasma physics difficult.

In general the dynamics of a plasma can be described by solving the equations of motion for each individual particle. This is the approach applied in so-called numerical particle-in-cell (PIC) simulations where the equations of motion of large numbers of particles are solved in a brute-force approach. Since the electric and magnetic fields appearing in each equation include the internal fields generated by every other moving particle, all equations are coupled and have to be solved simultaneously. Such a full solution is not only too difficult to obtain, but also of no immediate practical use, since one is interested in knowing average quantities like density and temperature rather than the individual velocity of each particle. Therefore, one usually makes certain approximations suitable to the problem studied. It has turned out that four different approaches are most useful.

The simplest approach is the *single particle motion* description. In principle this is not a plasma physics approach because none of the criteria is satisfied. It describes the motion of a particle under the influence of external electric and magnetic fields which are not affected by the presence of the particle(s). This approach neglects the collective behavior of a plasma, but is useful when studying a very low density plasma, like found in the Earth's ring current and in the radiation belts. To zero order approximation this approach provides also very rough information of the particle behaviour in plasma, in particular about the particle parameters.

The *magnetohydrodynamic* approach is the other extreme. It neglects all single particle aspects. The plasma is treated as a single electrically conducting fluid with only macroscopic variables of interest, like average density, velocity, and temperature. The approach assumes that the plasma is able to maintain local equilibria and is suitable to study very-low-frequency wave phenomena in highly conducting fluids immersed into sufficiently strong magnetic fields. We will at a later place more precisely quantify the limitations of this approach. We note, however, that magnetohydrodynamic is very well applicable to the processes in Earth's deep interior, the outer core, where the matter is in the state of a highly conducting metallic fluid but not in the plasma state.

The *multi-fluid* approach, which is much better suited for treating real plasmas than magnetohydrodynamics, is similar to the magnetohydrodynamic approach, but accounts for different particle species (electrons, protons, and possibly heavier ions). It assumes that each species behaves like a separate fluid. In this sense it is the magnetohydrodynamics of each particle fluid separately, coupling all fluids through the common external magnetic field. It has the advantage that differences in the fluid behaviour of the light electrons and the heavier ions can all be taken into account. This can lead to charge separation fields, internal current flow and also the possibility of wave propagation in the low and high frequency ranges and of electromagnetic and electrostatic nature.

The *kinetic theory* is the most developed plasma theory. It adopts a statistical approach. Instead of solving the equations of motion for each individual particle, it considers the development of the distribution function for the system of particles under consideration in phase and configuration space. Yet even in kinetic theory certain simplifying assumptions have to be made and there are different flavors of kinetic theory, depending on the kind of simplification made.

In the present book we will describe all these approaches and apply them to suitable geophysical plasma phenomena. We will start with the single particle approach. Subsequently, we will derive the basic equations of the kinetic theory, but then first turn to the fluid theories and their applications, before we finally go into the details of the kinetic approach.

References

[1] F. F. Chen, *Introduction to Plasma Physics and Controlled Fusion* (Plenum Press, New York, 1974).
[2] D. A. Gurnett and A. Bhattacharjee, *Introduction to Plasma Physics* (Cambridge University Press, 2005).
[3] M. G. Kivelson and C. T. Russell, *Introduction to Space Physics* (Cambridge University Press, Cambridge, UK 1995).
[4] G. K. Parks, *Physics of Space Plasmas: An Introduction* (Westview Press, Advanced Book Program, Boulder CO, 2004).

Problems

Problem 1.1 *What is the physical meaning of the plasma parameter Λ?*
Hint: *Think of the relation between the inter-particle distance and the screening length.*

Problem 1.2 *Figure 1.1 shows the screening of the electric field of an ion. Inspection of this figure suggests, however, that the screening is not complete. Calculate the stray field outside the Debye sphere. What is its value in the solar wind at density $n_e = 5$ cm^{-3} and temperature $k_B T_e = 50$ eV?*

Problem 1.3 *Give a physically intuitive interpretation of the plasma frequency.*

Problem 1.4 *Can one think of Langmuir waves with their high-frequency electric field oscillation as of electromagnetic waves? What are the difference to waves in free space?*

Problem 1.5 *We have listed three regions of plasmas in space: the Solar Wind, the Magnetosphere and the Ionosphere. Why distinguishing in this manner? What are the main differences between these regions?*

Problem 1.6 *Could you think of other regions in space outside the direct environment of Earth (the solar system and possibly even beyond) where plasmas of similar kind would exist? What would be the differences to the Earth's environment?*

Problem 1.7 *Apply the first and second plasma criteria to Earth's ionosphere, solar wind and also to cosmic rays. Which of those regions deserve the term plasma?*
Hint: *Cosmic ray densities depend on energy but are of the order of* $10^{-4} - 10^{-6}$ cm^{-3} *while having kinetic energies ('temperatures') of* $k_B T \sim 10$ MeV *to* 10^3 GeV.

Problem 1.8 *What is the reason for the solar wind density to decay like the inverse square of the radial distance from the sun? Assuming an adiabatic index of* $\gamma = \frac{5}{3}$ *what would be the electron temperature of the solar wind at its start at coronal distance of* 0.01 AU?

Problem 1.9 *Why does the interplanetary magnetic field which is carried radially away from the sun not also decay as the inverse square of the radial distance? Why does it turn into azimuthal direction at large radial distances?*

— 2 —

Single Particle Motion

Plasmas are collections of very large numbers of electrically charged particles. It is the charged state of the particles that distinguishes plasmas from other particle collections such as normal gases or fluids. The electric charge couples the particles to the electromagnetic field, which affects their motions.

In a situation where the charged particles do not directly interact with each other and where they do not affect the external magnetic field significantly, the motion of each individual particle can be treated independently. This *single particle approach* is only valid in very rarified plasmas where collective effects are negligible. Furthermore, the external magnetic field must be rather strong, much greater than the additional magnetic field produced by the electric currents due to the motion of the entire collection of charged particles in the volume under consideration.

We will see later that the single particle approach can be applied only in some plasmas of geophysical interest. However, in order to understand the collective behavior of the plasma, i.e., the motion of the charge carriers under the influence of electric and magnetic fields generated by the motion itself, it is very instructive to study first the motion of charged particles in prescribed external electric and magnetic fields.

2.1 Field Equations

Before describing the particle motion in external electric and magnetic fields, we introduce the electromagnetic field equations. There is a twofold coupling between electric charges and electromagnetic fields. Charged particles at rest are the sources of the electrostatic field, $\mathbf{E}$, which is the origin of the *Coulomb force*

$$\mathbf{F}_C = q\mathbf{E} \tag{2.1}$$

they feel in the combined electrostatic field of all the other particles. On the other hand, charged particles moving with a velocity, $\mathbf{v}$, are current elements generating a magnetic field, $\mathbf{B}$, which is the origin of the *Lorentz force*

$$\mathbf{F}_L = q\,(\mathbf{v} \times \mathbf{B}) \tag{2.2}$$

The motion of charged particles is strongly influenced by the presence of the electromagnetic field, while at the same time it is also the source of the fields. The relation between fields and particles is described by *Maxwell's equations* (see Appendix B.5)

$$\nabla \times \mathbf{B} = \mu_0 \mathbf{j} + \varepsilon_0 \mu_0 \frac{\partial \mathbf{E}}{\partial t} \tag{2.3}$$

$$\nabla \times \mathbf{E} = -\frac{\partial \mathbf{B}}{\partial t} \tag{2.4}$$

15

where $\mathbf{j}$ is the electric current density in the plasma, and ε_0 and μ_0 are the vacuum permittivity and susceptibility, respectively.

These equations show that the electric and magnetic fields are not independent, but are coupled by their spatial and temporal variations. Moreover, the electric current density turns out to be the source of the magnetic field and of fast fluctuations of the electric field. Since $\varepsilon_0\mu_0 = c^{-2}$ is equal to the inverse square of the light velocity, the latter will be negligible in a plasma as long as we do not consider propagation of electromagnetic waves. Hence, the second term on the right-hand side of Eq. (2.3) is small as long as no fast oscillations appear in the electric field.

In order to close the system, the first two equations have to be supplemented by two more equations, namely the conditions

$$\nabla \cdot \mathbf{B} = 0 \qquad (2.5)$$
$$\nabla \cdot \mathbf{E} = \rho/\varepsilon_0 \qquad (2.6)$$

The first of these expressions indicates that there are no sources of the magnetic field and thus the magnetic field lines are always closed. The second condition shows that the source of the electric field is the electric space charge density, $\rho = e(n_i - n_e)$, which is the difference between the charge densities of the ion and the electrons. Similarly, the electric current is defined as the difference between the electron and ion fluxes as $\mathbf{j} = e(n_i\mathbf{v}_i - n_e\mathbf{v}_e)$ where for simplicity we have assumed that the ions are singly charged.

2.2   Gyration

The equation of motion for a particle of charge q under the action of the Coulomb and Lorentz forces given in Eqs. (2.1) and (2.2) can be written as

$$m\frac{d\mathbf{v}}{dt} = q(\mathbf{E} + \mathbf{v} \times \mathbf{B}) \qquad (2.7)$$

where m represents the particle mass and $\mathbf{v}$ the particle velocity. Under the absence of an electric field this equation reduces to

$$m\frac{d\mathbf{v}}{dt} = q(\mathbf{v} \times \mathbf{B}) \qquad (2.8)$$

Taking the dot product of Eq. (2.8) with $\mathbf{v}$ and noting that $\mathbf{v} \cdot (\mathbf{v} \times \mathbf{B}) = 0$ (useful vector relations are found in Appendix B.4), we obtain

$$m\frac{d\mathbf{v}}{dt} \cdot \mathbf{v} = \frac{d}{dt}\left(\frac{mv^2}{2}\right) = 0 \qquad (2.9)$$

which shows that the particle kinetic energy as well as the magnitude of its velocity are constants. A static magnetic field, whatever its spatial variance is, does not change the particle kinetic energy.

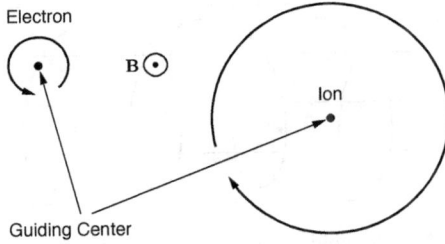

Figure 2.1: Gyration of charged particles around a guiding centre. In addition to the different sizes of their gyroradii (at same velocity), electrons gyrate right-handedly, ions left-handedly.

In a uniform magnetostatic field along the z axis, $\mathbf{B} = B\hat{e}_z$, we get the components

$$
\begin{aligned}
m\dot{v}_x &= qBv_y \\
m\dot{v}_y &= -qBv_x \\
m\dot{v}_z &= 0
\end{aligned}
\tag{2.10}
$$

The velocity component parallel to the magnetic field, $v_\parallel = v_z$, is constant. Taking the second derivative, we get

$$
\begin{aligned}
\ddot{v}_x &= -\omega_g^2 v_x \\
\ddot{v}_y &= -\omega_g^2 v_y
\end{aligned}
\tag{2.11}
$$

where ω_g is the *gyrofrequency* or *cyclotron frequency*, which has opposite signs for positive and negative charges (note that ω_g is often defined as a positive number, independent of the sign of the charge; we will also use it this way in later chapters, starting on p. 269)

$$
\boxed{\omega_g = \frac{qB}{m}}
\tag{2.12}
$$

Equation (2.11) is a harmonic oscillator equation with solutions of the form

$$
\begin{aligned}
x - x_0 &= r_g \sin \omega_g t \\
y - y_0 &= r_g \cos \omega_g t
\end{aligned}
\tag{2.13}
$$

Since ω_g carries the sign of the charge, the x component has opposite signs for electrons and ions. r_g is the *gyroradius* defined as

$$
\boxed{r_g = \frac{v_\perp}{|\omega_g|} = \frac{mv_\perp}{|q|B}}
\tag{2.14}
$$

where $v_\perp = (v_x^2 + v_y^2)^{\frac{1}{2}}$ is the constant speed in the plane perpendicular to $\mathbf{B}$.

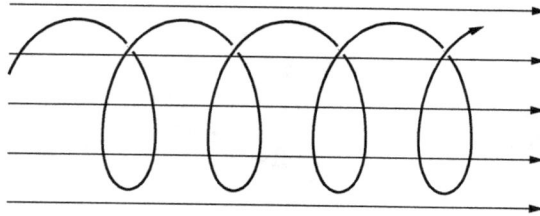

Figure 2.2: Helicoidal ion orbit in a uniform magnetic field.

The components of Eq. (2.13) describe a circular orbit of the particle around the magnetic field, with the sense of rotation depending of the sign of the charge (see Fig. 2.1). The center of the orbit (x_0, y_0) is called the *guiding center*. The circular orbit of the charged particle represents a circular current and the direction of the gyration is such that the magnetic field generated by the circular current is opposite to the externally imposed field. This behavior is called *diamagnetic effect*.

In Fig. 2.1 we have neglected a possible constant velocity of the particle parallel to the magnetic field, $v_\parallel$. Whenever $v_\parallel \neq 0$, the actual trajectory of the particle is three-dimensional and looks like a helix. Such a helicoidal trajectory is shown in Fig. 2.2. The *pitch angle*, α, of the helix is defined as

$$\alpha = \tan^{-1}\left(\frac{v_\perp}{v_\parallel}\right)$$

(2.15)

and depends on the ratio between the perpendicular and parallel velocity components.

2.3 Electric Drifts

Taking the electric field into consideration will result in a drift of the particle superimposed onto its gyratory motion. The exact nature of this *electric drift* depends on whether the field is electrostatic or time-varying and whether it is spatially uniform or not.

2.3.1 E × B – Drift

Let us now assume that an electrostatic field, **E**, is present. Looking for solutions of Eq. (2.7), we can again treat the perpendicular components and the component parallel to **B** separately. The parallel component

$$m\dot{v}_\parallel = qE_\parallel$$

(2.16)

describes a straightforward acceleration along the magnetic field. However, in geophysical plasmas most parallel electric fields cannot be sustained, since they are immediately canceled out by electrons, which along the magnetic field lines are extremely mobile under most circumstances.

Assuming that the perpendicular electric field component is parallel to the x axis, $\mathbf{E}_\perp = E_x \hat{\mathbf{e}}_x$, the perpendicular components of Eq. (2.7) are

$$
\begin{aligned}
\dot{v}_x &= \omega_g v_y + \frac{q}{m} E_x \\
\dot{v}_y &= -\omega_g v_x
\end{aligned}
\tag{2.17}
$$

Taking the second derivative, we obtain

$$
\begin{aligned}
\ddot{v}_x &= -\omega_g^2 v_x \\
\ddot{v}_y &= -\omega_g^2 \left(v_y + \frac{E_x}{B} \right)
\end{aligned}
\tag{2.18}
$$

If we substitute $v_y' = v_y + E_x/B$, we get back to Eq. (2.11), where the particle is gyrating about the guiding center. Thus Eq. (2.18) describes a gyration, but with a superimposed drift of the guiding center in the $-y$ direction. This drift of the guiding center is usually called $E \times B$ *drift* and has the general form

$$
\boxed{\mathbf{v}_E = \frac{\mathbf{E} \times \mathbf{B}}{B^2}}
\tag{2.19}
$$

The E×B drift is independent of the sign of the charge and thus electrons and ions move into the same direction.

Figure 2.3 shows the acceleration and deceleration effect of a perpendicular electric field. An ion is accelerated into the direction of the electric field, thereby increasing its gyroradius. But it is decelerated during the second half of its gyratory orbit, now with decreasing gyroradius. The different gyroradii shift the position of the guiding center in the $\mathbf{E} \times \mathbf{B}$ direction. The electrons are accelerated when moving antiparallel to the electric field and decelerated when moving parallel. But since their sense of gyration is opposite, too, their guiding centers drift into the same direction.

It is instructive to note that the E×B drift has a fundamental physical root in the *Lorentz transformation* of the electric field into the moving system of the particle. In this system the electric field is given by

$$
\mathbf{E}' = \mathbf{E} + \mathbf{v} \times \mathbf{B}
\tag{2.20}
$$

For a free particle this field must vanish, $\mathbf{E}' = 0$, which yields for the electric field

$$
\mathbf{E} = -\mathbf{v} \times \mathbf{B}
\tag{2.21}
$$

Solving for the velocity immediately yields the expression for the electric drift in Eq. (2.19). Because the Lorentz transformation does not depend on the charge of the particles, the E×B drift is also independent of the sign of the charge. The physical meaning of the E×B drift is thus simply to eliminate the transverse electric field $\mathbf{E}_\perp$ by transformation into a frame which is moving with velocity $\mathbf{v}_E$.

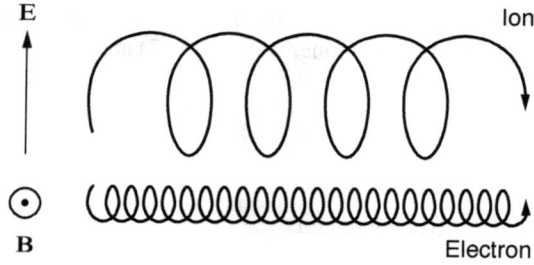

Figure 2.3: Particle drifts in crossed electric and magnetic fields. Even though the sense of gyration of electrons and ions is opposite, they drift into the same direction if an external electric field **E** is applied perpendicular to the magnetic field **B**.

2.3.2 Polarization Drift

We could have derived Eq. (2.19) directly from Eq. (2.7). Taking the cross-product of both sides of Eq. (2.7) with $\mathbf{B}/B^2$, we obtain

$$\mathbf{v} - \frac{\mathbf{B}(\mathbf{v}\cdot\mathbf{B})}{B^2} = \frac{\mathbf{E}\times\mathbf{B}}{B^2} - \frac{m}{q}\frac{d\mathbf{v}}{dt}\times\frac{\mathbf{B}}{B^2} \tag{2.22}$$

We can recognize the left-hand side as a perpendicular velocity vector and the first term on the right-hand side as the E×B drift. Averaging over the gyroperiod and thus neglecting temporal changes of the order of the gyroperiod or faster allows us to take the perpendicular velocity as the perpendicular drift velocity, $\mathbf{v}_d$. Remembering that the magnetic field is assumed time independent, we rewrite

$$\mathbf{v}_d = \mathbf{v}_E - \frac{m}{qB^2}\frac{d}{dt}(\mathbf{v}\times\mathbf{B}) \tag{2.23}$$

which yields with Eqs. (2.12) and (2.21)

$$\mathbf{v}_d = \mathbf{v}_E + \frac{1}{\omega_g B}\frac{d\mathbf{E}_\perp}{dt} \tag{2.24}$$

Equation (2.24) describes the drift of a charged particle in crossed homogeneous magnetic and electric fields, where the electric field is allowed to vary slowly with time t. The last term in this equation is called *polarization drift*.

$$\boxed{\mathbf{v}_P = \frac{1}{\omega_g B}\frac{d\mathbf{E}_\perp}{dt}} \tag{2.25}$$

There is an important qualitative difference between the polarization drift and the E×B drift. The E×B drift does neither depend on the charge nor on the mass of the particle, since it can be viewed as a result of the Lorentz transformation into a system with vanishing transverse electric field $\mathbf{E}_\perp$. Thus electrons, protons, and heavier ions all move into the same direction perpendicular to **B** and **E** with the same velocity. The polarization drift, on the other hand, increases proportional to the mass of the particle. It is directed along the transverse electric field, but oppositely for electrons and ions. Accordingly, it creates a current

$$\mathbf{j}_P = n_e e\left(\mathbf{v}_{Pi} - \mathbf{v}_{Pe}\right) = \frac{n_e(m_i + m_e)}{B^2}\frac{d\mathbf{E}_\perp}{dt} \tag{2.26}$$

which carries electrons and ions into opposite directions and polarizes the plasma. Since $m_i \gg m_e$, this current, which is called *polarization current*, is mainly carried by the ions.

2.3.3 Electric Drift Corrections

One may notice that Eq. (2.24) contains the total time derivative of the perpendicular electric field which is defined as the sum of the partial and convective derivatives

$$\frac{d\mathbf{E}_\perp}{dt} = \left[\frac{\partial}{\partial t} + (\mathbf{v}\cdot\nabla)\right]\mathbf{E}_\perp \tag{2.27}$$

Thus, it is not only the slow time variation of the electric field which contributes to polarisation drift but also the gradient of the perpendicular electric field along the particle path **v**, where **v** can now be any drift velocity which is imposed on the particles, including a drift $v_\parallel$ along the magnetic field **B**. Equation (2.24) thus also describes the drift due to inhomogeneities of the electric field the particle encounters during its motion.

In first approximation the velocity vector **v** can be replaced by the E×B-drift velocity. The convective term in this case becomes proportional to E^2. It is a nonlinear contribution and in spatially very weakly variable electric convection fields is usually much smaller than the partial time derivative and therefore is neglected in most cases. However, if other drift forms of the kind considered below have to be taken into account, then the convective contribution can become very important or even dominant and substantially contributing to the generation of polarisation currents.

The convective derivative takes into account spatial variations of the electric field in the direction of the E×B drift under the assumption that the electric field changes only gradually. When this is not the case and the electric field changes considerably over one gyroradius, there is a further correction on the electric field drift, known as *finite Larmor radius effect*. This correction is a second order effect in r_g and leads to the following more complete expression for the electric field drift

$$\mathbf{v}_E = \left(1 + \tfrac{1}{4}r_g^2\nabla^2\right)\frac{\mathbf{E}\times\mathbf{B}}{B^2} \tag{2.28}$$

The second spatial derivative takes account of the spatial variation of the electric field, averaged over the gyration orbit.

Finite Larmor radius effects are normally neglected in macroscopic applications of particle motion but may become important in the vicinity of plasma boundaries, plasma transitions and small scale structures in a plasma.

2.4 Magnetic Drifts

When analyzing Eq. (2.8), we have assumed that the magnetic field is homogeneous. This is often not the case. A typical magnetic field has gradients and often field lines are curved. This inhomogeneity of the magnetic field leads to a *magnetic drift* of charged particles. Time variations of the magnetic field itself cannot impart energy to a particle, since the Lorentz force is always perpendicular to the velocity of the particle. However, since $\partial \mathbf{B}/\partial t = -\nabla \times \mathbf{E}$, the associated inhomogeneous electric field may accelerate the particles in the way described in the previous section. What concerns the contribution of spatial inhomogeneities of the magnetic field, so it can be twofold. Inhomogeneities of a vector field can either be spatial changes of its modulus or spatial changes of its direction. In the first case the direction remains constant. In the second the field is bent possessing a local curvature. In the following we treat both these cases in view of their contribution to the particle drift.

2.4.1 Gradient Drift

Let us assume that the magnetic field is weakly inhomogeneous, for example increasing in the upward direction. As visualized in Fig. 2.4, the gyroradius of a particle decreases in the upward direction and thus the gyroradius of a particle will be larger at the bottom half of the orbit than at the top half. As a result, ions and electrons drift into opposite directions, perpendicular to both $\mathbf{B}$ and ∇B.

Since we assume that the typical scale length of a magnetic field gradient is much larger than the particle gyroradius, we can Taylor expand the magnetic field vector about the guiding center of the particle

$$\mathbf{B} = \mathbf{B}_0 + (\mathbf{r} \cdot \nabla)\mathbf{B}_0 \tag{2.29}$$

where $\mathbf{B}_0$ is measured at the guiding center and $\mathbf{r}$ is the distance from the guiding center. Inserting this relation into Eq. (2.8) we obtain

$$m\frac{d\mathbf{v}}{dt} = q(\mathbf{v} \times \mathbf{B}_0) + q[\mathbf{v} \times (\mathbf{r} \cdot \nabla)\mathbf{B}_0] \tag{2.30}$$

Expanding the velocity term into a gyration and a drift motion, $\mathbf{v} = \mathbf{v}_g + \mathbf{v}_\nabla$, and noting $v_\nabla \ll v_g$ yields

$$m\frac{d\mathbf{v}_\nabla}{dt} = q(\mathbf{v}_\nabla \times \mathbf{B}_0) + q[\mathbf{v}_g \times (\mathbf{r} \cdot \nabla)\mathbf{B}_0] \tag{2.31}$$

where we have omitted the terms describing gyration in a homogeneous field and neglected $\mathbf{v}_\nabla \times (\mathbf{r} \cdot \nabla)\mathbf{B}_0$ as a small quantity.

Since we are interested in time scales much larger than the gyroperiod, we can average over one gyration. Upon this the left-hand side vanishes since any acceleration a particle experiences when moving into the weak field region is balanced by a deceleration when moving into the strong field region in the other half of its gyratory

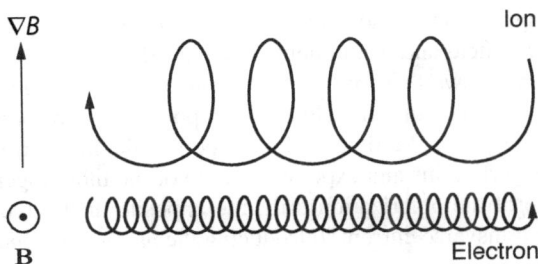

Figure 2.4: Particle drifts due to a gradient ∇B in the strength of a magnetic field **B**. The magnetic field gradient shifts the centre of the gyration circle of a particle in the direction perpendicular to the field and gradient. The shift depends on the charge sign. Particles of opposite signs drift in opposite directions thus causing a gradient drift current to flow across the magnetic field.

orbit. Since we know that $\mathbf{v}_\nabla$ lies in the plane perpendicular to the magnetic field, we can follow the same line as on p. 20 and take the cross-product with $\mathbf{B}_0/B_0^2$. Then we find

$$\mathbf{v}_\nabla = \frac{1}{B_0^2} \left\langle [\mathbf{v}_g \times (\mathbf{r} \cdot \nabla)\mathbf{B}_0] \times \mathbf{B}_0 \right\rangle \tag{2.32}$$

where the angle brackets denote averaging over one gyroperiod. Assuming **B** to vary only in the x direction, $\mathbf{B}_0 = B_0(x)\hat{\mathbf{e}}_z$, we obtain

$$\mathbf{v}_\nabla = -\frac{1}{B_0} \left\langle \mathbf{v}_g x \frac{dB_0}{dx} \right\rangle \tag{2.33}$$

Replacing $\mathbf{v}_g$ and x by the expressions given in Eq. (2.13), we get

$$v_{\nabla x} = -\frac{v_\perp r_g}{B_0} \left\langle \sin \omega_g t \cos \omega_g t \frac{dB_0}{dx} \right\rangle$$

$$v_{\nabla y} = -\frac{v_\perp r_g}{B_0} \left\langle \sin^2 \omega_g t \frac{dB_0}{dx} \right\rangle \tag{2.34}$$

Taking the gyroperiod average, $v_{\nabla x}$ will vanish since it contains the product of sine and cosine terms. Averaging over the $\sin^2$ term yields a factor 1/2. Thus the drift will have only a y component

$$\mathbf{v}_\nabla = \pm \frac{v_\perp r_g}{2B_0} \frac{\partial B_0}{\partial x} \hat{\mathbf{e}}_y \tag{2.35}$$

where the direction of the motion depends on the sign of the charge. Since the direction of the magnetic field gradient was chosen arbitrarily, this can be written in general form

$$\boxed{\mathbf{v}_\nabla = \frac{m v_\perp^2}{2qB^3} (\mathbf{B} \times \nabla B)} \tag{2.36}$$

showing that a magnetic field gradient leads to a magnetic *gradient drift* perpendicular to both the magnetic field and its gradient, as sketched in Fig. 2.4.

Equation (2.36) shows that ions and electrons drift into opposite directions and that, furthermore, the gradient drift velocity is proportional to the perpendicular energy of the particle, $W_\perp = \frac{1}{2}mv_\perp^2$. Particles of higher perpendicular energy drift faster, since they have a larger gyroradius and experience more of the inhomogeneity of the field.

As in the case of the polarization drift, the opposite drift directions of electrons and ions lead to a transverse current. Introducing the *magnetic moment* μ of a particle

$$\mu = \frac{mv_\perp^2}{2B} = \frac{W_\perp}{B} \qquad (2.37)$$

as the ratio of perpendicular energy and field, the *gradient drift current* becomes

$$\mathbf{j}_\nabla = n_e e \left(\mathbf{v}_{\nabla i} - \mathbf{v}_{\nabla e}\right) = \frac{n_e(\mu_i + \mu_e)}{B^2} \mathbf{B} \times \nabla B \qquad (2.38)$$

Ion and electron moments add because their gradient drift velocities have opposite signs.

2.4.2 General Force Drift

By using Eq. (2.1) to replace $\mathbf{E}$ in Eq. (2.19) by $\mathbf{F}/q$, we obtain a completely general form of the guiding center drift of a charged particle, which is valid not only for the Coulomb force, but for any arbitrary force $\mathbf{F}$ acting on the particle in a magnetic field

$$\mathbf{v}_F = \frac{1}{\omega_g}\left(\frac{\mathbf{F}}{m} \times \frac{\mathbf{B}}{B}\right) \qquad (2.39)$$

All particle drifts can be described this way by using the appropriate force terms, whenever the drift velocity of the particle is much smaller than its gyrovelocity. For the gradient, the polarization, and the gravitational drift, these forces can be written as

$$\mathbf{F}_\nabla = -\mu \nabla B \qquad (2.40)$$

$$\mathbf{F}_P = -m \frac{d\mathbf{E}}{dt} \qquad (2.41)$$

$$\mathbf{F}_G = -m\mathbf{g} \qquad (2.42)$$

where $\mathbf{F}_\nabla$ denotes the gradient and $\mathbf{F}_P$ the polarization force. $\mathbf{F}_G$ is the gravitational force, which is typically much weaker than the other forces. Except for processes near the solar surface, it can usually be neglected.

Equation (2.39) shows that all drifts generated by a force other than the Coulomb force depend on the sign of the charge, since ω_g carries this sign. Hence, for these drifts ions and electrons run into opposite directions, creating a transverse current. Moreover, these drifts depend on the mass of the charge carriers and thus the drift velocities are typically quite different for ions and electrons. They also depend on the 'coupling constants' contained in the expressions for the forces.

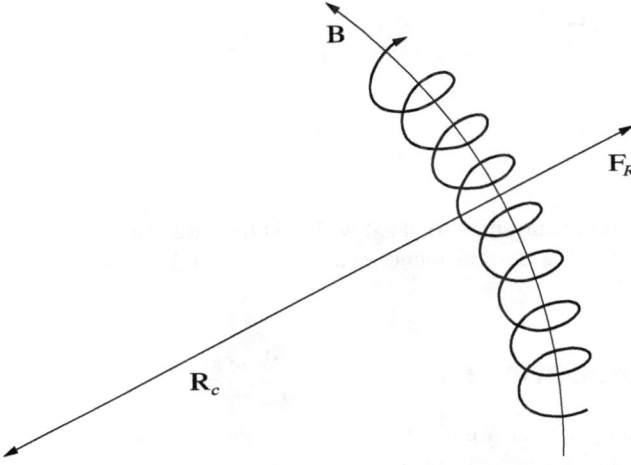

Figure 2.5: Centrifugal force felt by a particle moving along a curved field line.

2.4.3 Curvature Drift

The gradient drift is just one component of the particle drift in an inhomogeneous magnetic field. When the field lines are curved or bent, a magnetic *curvature drift* appears. Due to their parallel velocity, $v_{\parallel}$, the particles experience a centrifugal force

$$\mathbf{F}_R = mv_{\parallel}^2 \frac{\mathbf{R}_c}{R_c^2} \tag{2.43}$$

where $\mathbf{R}_c$ is the local radius of curvature. This situation is depicted in Fig. 2.5. Inserting Eq. (2.43) into (2.39) directly yields the wanted expression for the curvature drift

$$\boxed{\mathbf{v}_R = \frac{mv_{\parallel}^2}{q} \frac{\mathbf{R}_c \times \mathbf{B}}{R_c^2 B^2}} \tag{2.44}$$

The curvature drift is proportional to the parallel particle energy, $W_{\parallel} = \frac{1}{2}mv_{\parallel}^2$; it is perpendicular to both the magnetic field and its curvature radius. Again, it creates a transverse current since ion and electron drifts have opposite signs. The magnetic *curvature drift current* has the form

$$\mathbf{j}_R = n_e e \left(\mathbf{v}_{Ri} - \mathbf{v}_{Re}\right) = \frac{2n_e \left(W_{i\parallel} + W_{e\parallel}\right)}{R_c^2 B^2} \left(\mathbf{R}_c \times \mathbf{B}\right) \tag{2.45}$$

Like the associated drift direction, the curvature current flows perpendicular to both the curvature of the magnetic field and the magnetic field itself.

In a cylindrically symmetric field, it turns out that $-\nabla B = (B/R_c^2)\,\mathbf{R}_c$. Thus we may add the gradient to the curvature drift to obtain a simplified form of the total magnetic drift

$$\mathbf{v}_B = \mathbf{v}_R + \mathbf{v}_\nabla = (v_\parallel^2 + \tfrac{1}{2}v_\perp^2)\,\frac{\mathbf{B}\times\nabla B}{\omega_g B^2} \qquad (2.46)$$

It is the transverse current associated with this total magnetic drift which creates the magnetospheric ring current mentioned in Section 1.3 and further detailed in Sections 3.2–3.6.

2.5 Adiabatic Invariants

In the preceding section we encountered the magnetic moment $\mu = W_\perp/B$ of a particle. This quantity has been treated as a characteristic constant of the particle. Such quantities are called *adiabatic invariants*. Adiabatic invariants are not 'absolute' constants like total energy or total momentum, but may change slowly both in space and time. There essence is, however, that they change very slowly compared with some typical periodicities of the particle motion.

For particles in electromagnetic fields, adiabatic invariants are associated with each type of motion the particle can perform. In three spatial dimensions, the magnetic moment, μ, is associated with the particle gyration about the magnetic field, the longitudinal invariant, J, with the longitudinal motion along the magnetic field, and the third invariant, Φ, with the perpendicular drift on a closed orbit around the magnetic field.

Whenever these motions are periodic (i.e. the particle returns periodically to its approximate initial location with only small systematic displacements allowed) and changes in the system have angular frequencies much smaller than the oscillation frequency of the particle corresponding to one of the above motions, then the action integral

$$J_i = \oint p_i dq_i \approx \text{const} \qquad (2.47)$$

is a constant of the motion and describes an adiabatic invariant. The pair of variables (p_i, q_i) are the generalized momentum p_i and coordinate q_i of Hamiltonian mechanics, and the integration has to be performed over one full cycle of the period of q_i.

2.5.1 Magnetic Moment

To demonstrate that the magnetic moment of a particle does not change when the particle moves into stronger or weaker magnetic fields, we consider the energy conservation equation

$$W = W_\parallel + W_\perp \qquad (2.48)$$

Since, in the absence of electric fields, W is a strict constant, its time derivative vanishes

$$\frac{dW_{\parallel}}{dt} + \frac{dW_{\perp}}{dt} = 0 \tag{2.49}$$

For the transverse energy, we can use Eq. (2.37) and obtain

$$\frac{dW_{\perp}}{dt} = \mu \frac{dB}{dt} + B \frac{d\mu}{dt} \tag{2.50}$$

Here $dB/dt = v_{\parallel} dB/ds$ is the variation of the magnetic field as seen by the particle along its guiding center trajectory. The magnetic field itself is assumed to be constant. The parallel particle energy can be derived from the parallel component of the gradient force in Eq. (2.40) which gives the parallel equation of motion

$$\boxed{m \frac{dv_{\parallel}}{dt} = -\mu \nabla_{\parallel} B = -\mu \frac{dB}{ds}} \tag{2.51}$$

Multiplying the left-hand side of Eq. (2.51) with $v_{\parallel}$ and the right-hand side with its equivalent, ds/dt, one finds for the time derivative of the parallel energy

$$\frac{dW_{\parallel}}{dt} = -\mu \frac{dB}{dt} \tag{2.52}$$

Adding Eqs. (2.50) and (2.52) and observing Eq. (2.49), we obtain

$$\frac{dW_{\parallel}}{dt} + \frac{dW_{\perp}}{dt} = B \frac{d\mu}{dt} = 0 \tag{2.53}$$

which yields immediately that the magnetic moment is an invariant of the particle motion. The magnetic moment is not affected by small changes in the cyclotron frequency or the gyroradius which occur when the magnetic field changes along the particle path.

Up to now we have neglected electric fields, which can accelerate particles. Thus we have neglected temporal changes of the magnetic field, since $\partial \mathbf{B}/\partial t = -\nabla \times \mathbf{E}$. However, when the magnetic field fluctuations are slow enough, the magnetic moment is still conserved. This can be demonstrated by considering the change in perpendicular particle energy caused by an electric field. This change is calculated by taking the dot product of Eq. (2.7) with $\mathbf{v}_{\perp}$

$$\frac{dW_{\perp}}{dt} = q (\mathbf{E} \cdot \mathbf{v}_{\perp}) \tag{2.54}$$

The gain in energy over one gyration is obtained by integrating over the gyroperiod

$$\Delta W_{\perp} = q \int_{0}^{2\pi/\omega_g} (\mathbf{E} \cdot \mathbf{v}_{\perp}) \, dt \tag{2.55}$$

If the field changes slowly, the particle orbit is closed and we can replace the time integral by a line integral over the unperturbed orbit. Using Stokes' theorem (see Appendix B.4) and Maxwell's equations, we obtain

$$\Delta W_\perp = q \oint_C \mathbf{E} \cdot d\mathbf{s} = q \int_A (\nabla \times \mathbf{E}) \cdot d\mathbf{A} = -q \int_A \frac{\partial \mathbf{B}}{\partial t} \cdot d\mathbf{A} \qquad (2.56)$$

where $d\mathbf{s} = \mathbf{t}\,ds$ is the product of a line element, ds, of the closed gyratory orbit, C, with the line element's tangent vector, $\mathbf{t}$, and $d\mathbf{A} = \mathbf{n}\,dA$ is the product of a surface element, dA of the plane, $\mathbf{A}$, enclosed by the orbit and the surface element's normal vector, $\mathbf{n}$. For changes in the field much slower than the gyroperiod, $\partial \mathbf{B}/\partial t$ can be replaced by $\omega_g \Delta B / 2\pi$, with ΔB the average change during one gyroperiod

$$\Delta W_\perp = \tfrac{1}{2} q \omega_g r_g^2 \Delta B = \mu \Delta B \qquad (2.57)$$

Here we have inserted the expression for ω_g and r_g from Eqs. (2.12) and (2.14) and used the definition of μ in Eq. (2.37). On the other hand, in Eq. (2.50) the change in perpendicular energy is given by

$$\Delta W_\perp = \mu \Delta B + B \Delta \mu \qquad (2.58)$$

Comparing the last two equations, we find again that

$$\Delta \mu = 0 \qquad (2.59)$$

demonstrating that the magnetic moment is invariant in the approximation of slowly varying fields even then when the particles are accelerated in induction electric fields.

Similarly, slow temporal variations of the electric field do not violate the invariance of the magnetic moment, since they will produce only second order time variations of the magnetic field, $\partial^2 B/\partial t^2$, on the right-hand side of Eq. (2.56), which can be neglected. For all slow variations the magnetic moment is a constant of motion. Changes in the fields merely lead to the different types of particle drifts but conserve μ.

From the adiabatic invariance of the magnetic moment, it follows that also the magnetic flux Φ_μ through the surface encircled by the gyrating particle does not change. This flux is given by $\Phi_\mu = B \pi r_g^2$ (see Appendix B.5), or inserting the expression for r_g from Eq. (2.14) and using the definition of μ in Eq. (2.37)

$$\boxed{\Phi_\mu = \frac{2\pi m}{q^2} \mu = \text{const}} \qquad (2.60)$$

Hence, as a particle moves into a region of stronger magnetic field, the gyroradius of the particle will get increasingly smaller, so that the magnetic flux encircled by the orbit remains constant.

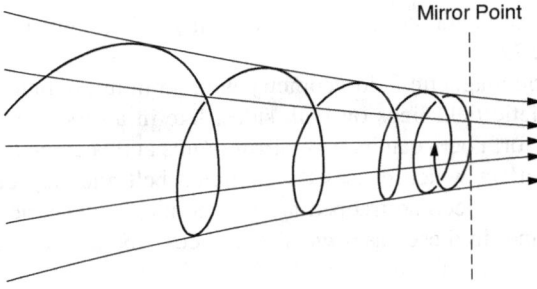

Figure 2.6: The orbit of an ion along a converging magnetic field and being reflected at its mirror point. The orbit is a spiral consisting of gyrocircles whose radii decrease in the increasing magnetic field while the parallel energy of the ion decreases until all particle energy is in the perpendicular direction and the parallel motion stops and reverses at the mirror point. This effect is due to the action of the mirror force $\mathbf{F}_{mirr}(s)$ which retards the ion and ultimately pushes it back into the weak magnetic field region.

2.5.2 Magnetic Mirror

Let us follow the guiding center of a particle moving along an inhomogeneous magnetic field by considering its magnetic moment

$$\mu = \frac{mv^2 \sin^2 \alpha}{2B} \tag{2.61}$$

where we have replaced $v_\perp$ by $v \sin \alpha$, using the pitch angle defined in Eq. (2.15). Since the magnetic moment is invariant and the total energy is a constant of motion, only the pitch angle can change when the magnetic field increases or decreases along the guiding center trajectory. The above equation also shows that the pitch angles of a particle at different locations are directly related to the magnetic field strengths at those locations according to

$$\frac{\sin^2 \alpha_2}{\sin^2 \alpha_1} = \frac{B_2}{B_1} \tag{2.62}$$

Thus knowing the pitch angle of a particle at one location, we can calculate this quantity at all other locations along the magnetic field line.

In a converging magnetic field geometry, a particle moving into regions of stronger fields will have its pitch angle increase and, therefore, have its transverse energy $W_\perp$ increase at the expense of its parallel energy $W_\parallel$. If B_m is a point along the field line where the pitch angle reaches $\alpha = 90°$, the particle is reflected from this *mirror point*. Here, all of the particle energy is in $W_\perp$ and the particle cannot penetrate any further, but is pushed back by the parallel component of the gradient force given in Eq. (2.40), the so-called *mirror force*,

$$\mathbf{F}_{mirr} \equiv \mathbf{F}_\nabla = -\mu \nabla_\parallel B(s) \tag{2.63}$$

where s is the coordinate along the magnetic field. This mirroring of a particle is visualized in Fig. 2.6.

In a symmetric magnetic field geometry with a minimum field in the middle and converging magnetic field lines on both sides, like in a dipole field, a particle may bounce back and forth between its two mirror points and become *trapped* in the field. It is this mirror effect which causes the radiation belt and ring current particles in the magnetosphere to become trapped around the magnetospheric field minimum in the equatorial plane. In these cases we also can describe the particle's pitch angle at a specific location along the magnetic field as a function of the local magnetic field strength

$$\sin \alpha(B) = \left(\frac{B}{B_m} \right)^{\frac{1}{2}}$$
(2.64)

i.e., the pitch angle of the particle becomes a function of the ratio of the respective local and mirror point field strengths showing its variation along the magnetic field.

2.5.3   Adiabatic Heating

The mirror effect described above is a consequence of the invariance of the magnetic moment when particles move along magnetic field lines. Conservation of the magnetic moment has also an important effect when particles drift across field lines. Consider a particle moving along its drift path from a region with magnetic field strength B_1 into a region of increasing field strength $B_2 > B_1$ being transported, for instance, convectively by E×B-drift across the magnetic field. Since the magnetic moment is conserved in this convective motion, we have for the ratio of the perpendicular energies at the final and initial locations of the particle

$$\frac{W_{\perp 2}}{W_{\perp 1}} = \frac{B_2}{B_1}$$
(2.65)

Hence on has $W_{\perp 2} > W_{\perp 1}$. Due to conservation of its magnetic moment the particle has gained energy. However, in contrast to the mirror case, the particle will not bounce back and the gain in perpendicular energy will remain.

The convective transport of particles into stronger magnetic fields therefore ends up with a gain of energy in the transverse direction. The work required for this process is of course taken from the drift motion which transports the particle into the stronger field. This type of particle energization is called *adiabatic heating* and is some kind of so-called *betatron acceleration*. It increases the perpendicular energy of the particle without affecting its parallel energy. This results in an anisotropy of the particle energy. However, since magnetic fields may vary also in the parallel direction, we will discuss the generation of such an anisotropy in a more general way after introducing the second adiabatic invariant.

2.5.4   Longitudinal Invariant

If the field has a mirror symmetry where the field lines converge on both sides as in a dipole field, there is the possibility for a second adiabatic invariant, J. A particle moving in such a converging field will be reflected from the region of strong magnetic field and can oscillate in the field at a certain *bounce frequency*, ω_b. The *longitudinal invariant* is defined by

$$J = \oint mv_{\parallel}ds \tag{2.66}$$

where $v_{\parallel}$ is the parallel particle velocity, ds is an element of the guiding center path and the integral is taken over a full oscillation between the mirror points.

For electromagnetic variations with frequencies $\omega \ll \omega_b$, the longitudinal invariant is a constant, irrespective of weak changes in the path of the particle and its mirror points due to slow changes in the fields.

The longitudinal invariant can also given a different representation when replacing the parallel velocity with the parallel energy and using the fact that the total energy W is conserved. Then the above integral becomes

$$J = \sqrt{2m} \oint \left[W - \mu B(s)\right]^{\frac{1}{2}} ds \tag{2.67}$$

where, as before, $\mu = $ const, and the magnetic field strength $B(s)$ must be known as function of the length s of the local magnetic field line along which the particle is bouncing back and forth. Another representation is obtained by exchanging the line element $ds = v_{\parallel}dt$ in the above integral against the parallel velocity. In this case the integral becomes an integral with respect to time

$$J = 2 \int_{-\tau_b/4}^{\tau_b/4} mv_{\parallel}^2 \, dt = W\tau_b - 2\mu \int_{-\tau_b/4}^{\tau_b/4} B[s(t)]dt \tag{2.68}$$

where we introduced the bounce time τ_b of the particle between the mirror points s_m along the field line and again replaced $W_{\parallel}$ through the perpendicular energy. Now $s(t)$, the particle location as function of time has to be known.

The bounce time, on the other hand, is obtained from the inverse parallel velocity of the particle when integrating between the two mirror points s_{m1}, s_{m2} which in the symmetric field case becomes

$$\tau_b = 2 \int_{-s_m}^{s_m} \frac{ds}{v_{\parallel}} = \sqrt{2m} \int_{-s_m}^{s_m} \frac{ds}{\sqrt{W - \mu B(s)}} \tag{2.69}$$

In the next chapter we will make use of these expressions when applying them to the geomagnetic dipole field.

2.5.5   Energy Anisotropy

Invariance of the magnetic moment in magnetic fields which change in the perpendicular direction may lead to an increase in the perpendicular energy of the particle. When the magnetic field also varies in the direction parallel to the field lines, conservation of the longitudinal invariant implies that the parallel energy of the particle will also change during the combined drift and bounce motion of the particle along and across the field.

Let us define the total length of the field line between the two mirror points of the particle as ℓ, and the average parallel velocity along the field line as $\langle v_\parallel \rangle$. In terms of these quantities the longitudinal invariant can be expressed as

$$J = \oint m v_\parallel \, ds = 2 m \ell \langle v_\parallel \rangle \tag{2.70}$$

During its drift motion from weaker into stronger fields, the particle necessarily moves from one field line of length ℓ_1 onto another field line of length ℓ_2. At the same time its average parallel velocity changes from $\langle v_\parallel \rangle_1$ to $\langle v_\parallel \rangle_2$. Conservation of the longitudinal invariant then implies that the averaged parallel energy, $\langle W_\parallel \rangle$, changes according to

$$\frac{\langle W_\parallel \rangle_2}{\langle W_\parallel \rangle_1} = \frac{\ell_1^2}{\ell_2^2} \tag{2.71}$$

If the length of the bounce path decreases, the parallel energy of the particle increases. This is the basic element of *Fermi acceleration*.

This result can be combined with the simultaneous increase in the perpendicular energy from Eq. (2.65) to determine the anisotropy in the energy attained by the particle. Let us define this anisotropy as

$$A_W = \frac{\langle W_\perp \rangle}{\langle W_\parallel \rangle} \tag{2.72}$$

Then the *energy anisotropy* of the particle changes according to

$$\boxed{\frac{A_{W2}}{A_{W1}} = \frac{B_2}{B_1} \frac{\ell_2^2}{\ell_1^2}} \tag{2.73}$$

which shows that the anisotropy increases if only the square of the field line length decreases less than the magnetic field increase. As for an example, a particle that starts with an isotropic energy distribution, i.e. with $A_{W1} = 1$, will develop an energy anisotropy

$$A_{W2} = (B_2/B_1)(\ell_2/\ell_1)^2 \tag{2.74}$$

when changing its magnetic flux tube. This anisotropy can be larger or smaller than one depending on whether the particle is moving into weaker stretched or stronger compressed fields. In all cases, however, the energy of the particle will become anisotropic.

2.5.6 Drift Invariant

The third invariant, Φ, is simply the conserved magnetic flux encircled by the periodic orbit of a particle trapped in an axisymmetric mirror magnetic field configuration when it performs closed *drift shell* orbits around the magnetic field axis. This *drift invariant* can be written as

$$\Phi = \oint v_d r \, d\psi \tag{2.75}$$

where v_d is the sum of all perpendicular drift velocities, ψ is the azimuthal angle, r the radius measured from the centre of symmetry, and the integration must be taken over a full closed about circular drift path of the particle.

Whenever the typical frequency of the electromagnetic fields is much smaller than the drift frequency, $\omega \ll \omega_d$, Φ is invariant and essentially equal to the magnetic flux enclosed by the orbit. This can be written like in Eq. (2.60) as

$$\boxed{\Phi = \frac{2\pi m}{q^2} M = \text{const}} \tag{2.76}$$

where M is the magnetic moment of the axisymmetric field.

2.5.7 Violation of Invariance

So far we have considered variations and motions which conserve the magnetic moment, the longitudinal invariant, and the drift invariant. In nature, however, all the fields may vary in such a way that the adiabatic invariance of one or the other invariant is violated.

Time variations faster than the gyrofrequency of the particle with frequencies $\omega > \omega_g$ violate the first adiabatic invariant μ, the magnetic moment of the particle. These are high-frequency variations in either the magnetic or electric field. In this case, the concept of a gyratory orbit about a guiding center becomes obsolete and the full particle motion must be considered.

On the other hand, for frequencies $\omega_g > \omega > \omega_b$ the magnetic moment μ is conserved and the guiding center approximation is useful for the drift motion. However, under these conditions the longitudinal invariant is not conserved but violated, and the particle motion cannot be described anymore as a simple oscillation along the magnetic field between mirror points.

Finally, for frequencies $(\omega_g, \omega_b) > \omega > \omega_d$ the first two invariants will be conserved while the drift invariant becomes violated. The particles gyrate and bounce but diffuse across drift shells under the influence of the variation in the magnetic field. All these cases are realized in nature. Some of them we will encounter later.

Adiabatic invariants are not only violated when the fields vary in time but also when they abruptly change over a length scale $L < r_a$ shorter than the characteristic radius r_a of the periodic motion related to the adiabatic invariant. This can be proven, for instance, for the gyration of a particle across a magnetic field gradient which is shorter than the particle gyroradius r_g. In such a case, $r_g/L > 1$. Using Eq. (2.14) to write the perpendicular velocity of the particle as

$$v_\perp = \omega_g r_g \tag{2.77}$$

and dividing both sides of this expression by L, the gradient length of the spatial change, one finds

$$\omega = \frac{v_\perp}{L} = \omega_g \frac{r_g}{L} > \omega_g \tag{2.78}$$

Hence, for the gyrating particle the effective frequency of change in the field, $\omega > \omega_g$, is higher than its gyrofrequency, and the magnetic moment of a particle gyrating in a magnetic field which varies strongly over a length of the order $L < r_g$ is not an adiabatic invariant. Similar arguments can be applied also to the two remaining invariants.

2.5.8 Summary of Guiding Center Drifts

For quick reference, we summarize the expressions of the guiding center drifts and the associated transverse currents.

E×B Drift:	$\mathbf{v}_E = \dfrac{\mathbf{E} \times \mathbf{B}}{B^2}$	
Polarization Drift:	$\mathbf{v}_P = \dfrac{1}{\omega_g B} \dfrac{d\mathbf{E}_\perp}{dt}$	$\mathbf{j}_P = \dfrac{n_e(m_i + m_e)}{B^2} \dfrac{d\mathbf{E}_\perp}{dt}$
Gradient Drift:	$\mathbf{v}_\nabla = \dfrac{mv_\perp^2}{2qB^3}(\mathbf{B} \times \nabla B)$	$\mathbf{j}_\nabla = \dfrac{n_e(\mu_i + \mu_e)}{B^2}(\mathbf{B} \times \nabla B)$
Curvature Drift:	$\mathbf{v}_R = \dfrac{mv_\parallel^2}{qR_c^2 B^2}(\mathbf{R}_c \times \mathbf{B})$	$\mathbf{j}_R = \dfrac{2n_e(W_{i\parallel} + W_{e\parallel})}{R_c^2 B^2}(\mathbf{R}_c \times \mathbf{B})$

2.6 Concluding Remarks

In concluding this chapter, a word of caution is mandatory. The guiding center theory assumes that the electromagnetic fields are prescribed. It can thus be used in geophysical plasmas where the external field is strong and will not be changed much by the motion of the particles themselves. In other words it is applicable only when the current which the different drifts generate produce magnetic field which are substantially weaker than the external ambient magnetic field. In the solar wind with its weak magnetic field, the drift approximation is not applicable. It does, however, apply fairly well to the very dilute energetic particle component trapped in the radiation belts in the magnetosphere and with a bit more caution also to some aspects of the ring current. It should not be used in weak field regions where the field will be substantially affected by the presence of charged particles and particle motion. There the concept of a guiding center looses its meaning.

Further Reading

For those readers who want to know more about the guiding center approach and to see the actual proof of the second and third invariant, we recommend reading one of the first two monographs listed below. Those readers who are interested in a highly mathematical rather than physical approach are pointed to the third reference in the list. There a complete though complicated derivation of the various drift motions in first approximation is given as well as mathematically rigorous proof of the constancy of the three adiabatic invariants. The fourth monograph is more of use for those among the readers who want to see application to observations in and observational data from the Earth's environment as well as learn about observational methods.

References

[1] H. Alfvén and C. G. Fälthammar, *Cosmical Electrodynamics, Fundamental Principles* (Clarendon Press, Oxford, 1963).

[2] F. F. Chen, *Introduction to Plama Physics and Controlled Fusion, Vol. 1: Plasma Physics* (2nd Edition, Plenum Press, New York and London, 1990).

[3] T. G. Northrop, *The Adiabatic Motion of Charged Particles* (Interscience Publishers, New York, 1963).

[4] G. K. Parks, *Physics of Space Plasmas: An Introduction* (Westview Press, Advanced Book Program, Boulder CO, 2004).

Problems

Problem 2.1 *What is the physical reason for the Lorentz force? Of which parts does it consist?*

Problem 2.2 *Compare the Coulomb force between two charges $\pm e$ with the gravitational force between two masses m_e and m_i. Which one is stronger and what is the reason for their differences?*

Problem 2.3 *Looking at the two electrodynamics field equations one finds a striking asymmetry between them. What is the reason for this asymmetry?*

Problem 2.4 *Show that a gyrating particle conserves its energy of motion.*

Problem 2.5 *Derive the cyclotron frequency of a particle in a constant magnetic field without explicitly solving for its equation of motion. Show that the cyclotron frequency is effectively an axial vector. Which direction does this vector assume?*

Problem 2.6 *Derive the gyro-radius of a particle without solving for the particle motion. What is its physical content? Why is the particle trajectory a spiral along the magnetic field? Hint: Think of the momentum of the particle.*

Problem 2.7 *Show that the electric field drift is not a plasma effect but a general consequence of the Lorentz transformation in crossed electric and magnetic fields. It is a streaming of plasma. Which practical use can be made of it?*

Problem 2.8 *What will happen if the transverse electric field $\mathbf{E}_\perp$ becomes extremely strong? What does this imply for an electric field in a magnetic field $\mathbf{B}$ of strength $B =$ 10 nT? Hint: Consider the gyrational orbit of the particle. Remember that velocities are bound from above.*

Problem 2.9 *Derive the expression for the finite Larmor radius correction to the electric field drift.*

Problem 2.10 *What are the 'coupling constants' in the gradient, polarisation, gravitation and curvature forces?*

Problem 2.11 *Show by direct calculation along the lines of deriving the gradient drift that the curvature drift includes the centrifugal force. What is the reason for the curvature drift in cylindrical symmetry to simplify and to assume the gradient drift form? Explain why it lacks the factor $\frac{1}{2}$.*

Problem 2.12 *Explain why for a constant of motion the action integral is constant. Where in the mathematical expression of the action integral is the approximation of quasi-periodicity hidden?*

Problem 2.13 *Derive the expression for the pitch angle during bounce. What is the condition for a particle being reflected? Which expression holds for the pitch angle at field minimum? Give an interpretation of the bounce time.*

Problem 2.14 *What does it mean that a particle gains energy and evolves an energy anisotropy when entering a field from the weak field side? What happens to the particle? What is the source of the energy gain and the gain in anisotropy? Hint: Think of the different components of its orbital motion.*

Problem 2.15 *In thinking of thermal equilibrium; do you have an idea of what would happen to the anisotropy in energy at later times?*

Problem 2.16 *What do you understand under violation of invariance of an adiabatic invariant? What will happen when the invariant will become violated?*

— 3 —

Trapped Particles

A *dipolar magnetic field* possesses a minimum of field strength in its equatorial plane caused by the divergence of the geomagnetic field lines with distance from central body, in the geomagnetic case the Earth, while the field lines converge in both hemispheres when approaching the Earth. As we have learned in Section 2.5, in such a configuration particles will be trapped and will bounce back and forth between their mirror points in the northern and southern hemispheres (see Fig. 3.1). In the case of the terrestrial magnetic field, which inside of about 6 R_E can, with sufficient accuracy, be approximated by a dipolar field, these trapped populations belong to the energetic particles in the Earth's *radiation belts* (see Fig. 1.4), known as Van Allan-belts, and to the magnetospheric *ring current*. Typical energies of the ions in these regions range between 3 and 300 keV, while the electrons have energies about an order of magnitude lower.

The radiation belt and also the ring current particles are trapped in the geomagnetic field. However they do not only gyrate and bounce, but undergo a slow azimuthal drift (see Fig. 3.1). This drift is caused by the gradient and curvature of the inhomogeneous dipole magnetic field as described in Eq. (2.46) and is oppositely directed for ions and electrons. The ions drift westward while the electrons move eastward around the Earth. It is the current associated with these different direction of the azimuthal drift that constitutes the *ring current*.

3.1 Dipole Field

The geomagnetic field at the Earth's surface is quite inhomogeneous. This is mainly due to the magnetic irregularities of the Earth's crust and mantle. Its inhomogeneity can be accounted for by an expansion of the geomagnetic field into multipoles. Their contributions to the magnetic field strength outside the body of the Earth decay according to odd powers $r^{-(2n+1)}$ of the inverse radial distance from Earth's centre. Here $n = 1, 2, ...$ Hence, at large though not too far distances from the Earth's surface, the geomagnetic field can be approximated by the lowest order multipole, i.e. by a dipole field. Introducing the *Earth's dipole moment*, $M_E = 8.05 \times 10^{22}$ Am^2, and choosing a spherical coordinate system with radius, r, and magnetic latitude, λ, we can write the expression of the dipolar field as

$$\mathbf{B} = \frac{\mu_0}{4\pi} \frac{M_E}{r^3} (-2\sin\lambda\,\hat{\mathbf{e}}_r + \cos\lambda\,\hat{\mathbf{e}}_\lambda) \tag{3.1}$$

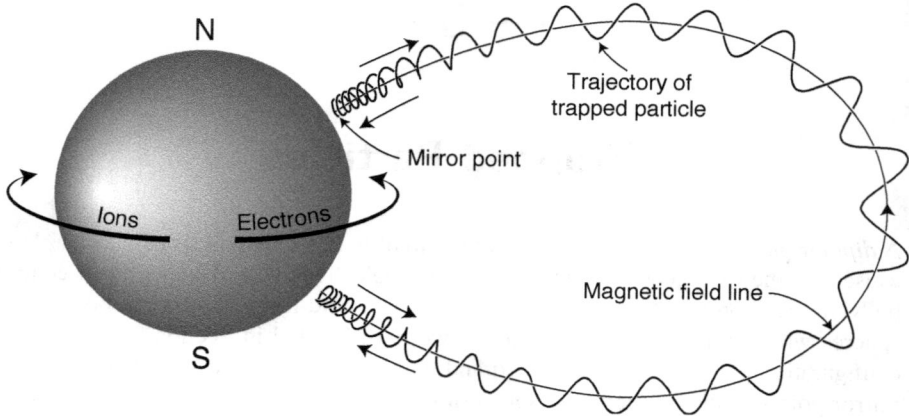

Figure 3.1: Trajectories of particles trapped on closed field lines, shown here for the Earth's dipole field. The particles are reflected from the mirror points after having spiralled along the magnetic field line. Due to the variation of the magnetic field strength which is minimum in the equatorial plane at largest distance from Earth the parallel velocity decreases from the equator towards the mirror points thereby gradually reducing the height of the spiral steps.

since a dipole field is symmetric about the azimuth. Here $\hat{\mathbf{e}}_r$ and $\hat{\mathbf{e}}_\lambda$ are unit vectors in the r and λ directions. The strength of the dipole field at a specific location can easily be obtained as

$$B = \frac{\mu_0}{4\pi} \frac{M_E}{r^3} (1 + 3\sin^2 \lambda)^{\frac{1}{2}} \qquad (3.2)$$

Typical dipolar field lines are shown in Fig. 3.2. In order to construct the field lines in this figure, and for many other applications, one needs to know the *field line equation*, $r = f(\lambda)$. If $d\mathbf{s}$ is an arc element, the lines of force are defined by the differential equation

$$d\mathbf{s} \times \mathbf{B} = 0 \qquad (3.3)$$

since the magnetic field vector is always tangent to the lines of force. For an axisymmetric field, this equation reduces to

$$\frac{dr}{B_r} = \frac{r d\lambda}{B_\lambda} \qquad (3.4)$$

Using the dipole field Eq. (3.1) we obtain

$$\frac{dr}{r} = -\frac{2\sin\lambda \, d\lambda}{\cos\lambda} = \frac{2d(\cos\lambda)}{\cos\lambda} \qquad (3.5)$$

Integration of Eq. (3.5) yields the dipole field line equation

$$r = r_{eq} \cos^2 \lambda \qquad (3.6)$$

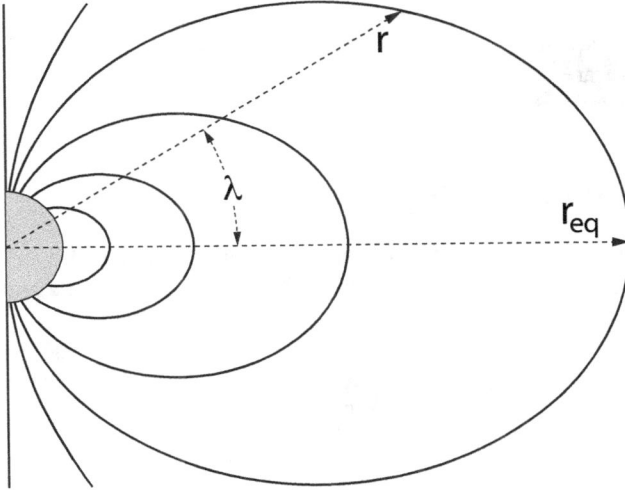

Figure 3.2: Dipolar magnetic field lines in the (r, λ) plane. Any point on the field line can be calculated from the field line equation (3.6). r_{eq} at $\lambda = 0$ is the field-line's largest distance.

where r_{eq} is an integration constant. Since $r = r_{eq}$ for $\lambda = 0$, r_{eq} becomes the radial distance to the particular field line in the equatorial plane and thus is equal to its greatest distance from the Earth's center.

The element of arc length along a field line is given by $(ds)^2 = (dr)^2 + r^2(d\lambda)^2$. Using Eqs. (3.4) and (3.6) to express $dr/d\lambda$ and r, respectively, one finds for the change of the arc element with magnetic latitude along the field line

$$\frac{ds}{d\lambda} = r_{eq} \cos \lambda (1 + 3 \sin^2 \lambda)^{\frac{1}{2}} \tag{3.7}$$

By integrating this equation between, say, the two footpoints of the field line on the surface of the Earth (or in the ionosphere) one can calculate the length of a field line with a given equatorial distance r_{eq}.

Often it is convenient to use the radius of the Earth, R_E, as the unit of distance and to introduce the *L-shell parameter* or *L-value*, $L = r_{eq}/R_E$ of a field line. The latter is the largest (equatorial) distance of the dipole field line as measured in Earth radii. With the equatorial magnetic field on the Earth's surface, $B_E = \mu_0 M_E/(4\pi R_E^3) = 3.11 \times 10^{-5}$ T, and by inserting the field line equation (3.6), we can rewrite Eq. (3.2) for the Earth's dipole field as

$$\boxed{B(\lambda, L) = \frac{B_E}{L^3} \frac{(1 + 3 \sin^2 \lambda)^{\frac{1}{2}}}{\cos^6 \lambda}} \tag{3.8}$$

Introducing L into Eq. (3.6) yields another useful relation

$$\boxed{\cos^2 \lambda_E = L^{-1}} \tag{3.9}$$

which defines the latitude, λ_E, where a dipolar field line of given L-value (or largest distance in the equatorial plane) intersects the Earth's surface. Since this is a quadratic equation there are two solutions expressing the symmetry of the dipole field with respect to the equator.

3.2 Bounce Motion

The most prominent motion of trapped particles in addition to their gyration is their bounce motion between the mirror points. The actual trajectory of a bouncing particle is characterised by its pitch angle as has been defined in Eq. (2.15).

In order to see that the motion of the particles along a magnetic field with mirror geometry like the dipole field of the Earth is indeed periodic we can use the parallel equation of motion (2.51) of the gyrating particle

$$\frac{dv_{\parallel}}{dt} = -\frac{\mu}{m}\frac{dB(s)}{ds} \tag{3.10}$$

which we derived in Chapter 2 and where the modulus of the magnetic field $B(s)$ is a function of the length s of the field line measured from some appropriate point on the field line. As such a point, playing the role of the origin $s = 0$ of particle motion, it is most convenient to chose the equatorial location of the field line at distance r_{eq}. At this point the field $B(r_{eq}, 0) = B_{min}$ is minimum. We expand $B(r_{eq}, s)$ around $s = 0$ and take into account that because of the minimum condition $dB(r_{eq}, s)/ds = 0$ at $s = 0$. Hence, to second order $B(r, s) \approx B_{min} + s^2(d^2B/ds^2)_{s=0}$. Inserting into the above parallel equation of motion and replacing the parallel velocity $v_{\parallel} = ds/dt$ yields

$$\frac{d^2s}{dt^2} = -\frac{\mu}{m}\left[\frac{d^2B(r,s)}{ds^2}\right]_{s=0} s^2 \tag{3.11}$$

Writing this equation in the canonical form of a harmonic oscillator equation

$$\ddot{s} + \omega_b^2 s^2 = 0 \qquad \text{with} \qquad \omega_b^2(r) \equiv \frac{\mu}{m}\left[\frac{d^2B(r)}{ds^2}\right]_{s=0} \tag{3.12}$$

we find that the particle indeed performs periodic oscillations along the magnetic field around its equatorial position on the field line, and the oscillation frequency is given by the *bounce frequency* $\omega_b(r)$ as function of the magnetic moment of the particle, the second derivative of the magnetic field $B''(0, r)$ along the field line taken at the equator, and on radius r. The above expressions hold for small-amplitude oscillations $s \ll R_c$ of the particle along the field line, where R_c is the curvature radius of the field, or as we will see below, for large equatorial pitch angles. When the particle oscillates along the magnetic field over much longer distances the curvature of the field must be taken into account. This makes the calculation more involved. The harmonic oscillator in this case becomes a physical oscillator as known from classical mechanics.

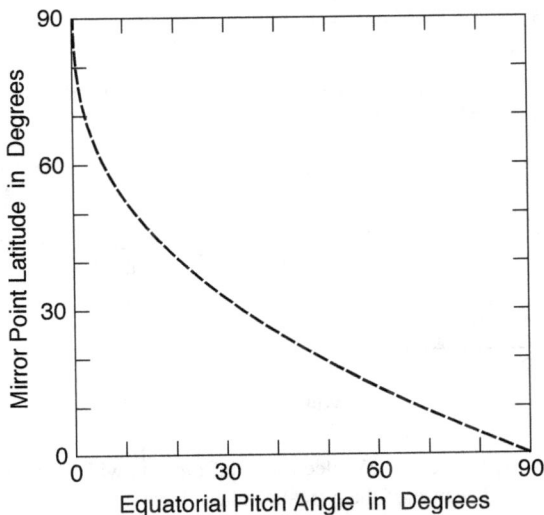

Figure 3.3: The latitudes λ_m of the mirror points of charged particles in a magnetic dipole field as function of the particle's equatorial pitch angle α_{eq}.

3.2.1 Equatorial Pitch Angle

From Eq. (2.64) we know that we can determine the pitch angle of a particle in a mirror field geometry anywhere along the field line from the ratio between the magnetic field at that location and the magnetic field at the particle's mirror point. One particularly interesting point along a dipolar field line is its intersection with the equatorial plane, where the field strength $B_{eq} = B_E/L^3$ along the field line is minimum. One is frequently interested in the value of the pitch angle α of all the trapped particles at this particularly exposed location.

Inserting Eq. (3.8) into Eq. (2.64), we obtain for this so called *equatorial pitch angle*, α_{eq} an expression relating it to the mirror latitude of the particle

$$\sin^2 \alpha_{eq} = \frac{B_{eq}}{B_m} = \frac{\cos^6 \lambda_m}{(1+3\sin^2 \lambda_m)^{\frac{1}{2}}} \tag{3.13}$$

where λ_m is the magnetic latitude of the particle's mirror point. Equation (3.13) shows that the equatorial pitch angle of a particle indeed depends only on the latitude of its mirror point and not on the equatorial distance of its field line or, equivalently, its L-value. Turning this argument around, one finds that the latitude of a particle's mirror point is completely fixed by its equatorial pitch angle and is independent of the L-value of the field line on which the particle is located.

Figure 3.3 shows the dependence of the magnetic latitude of a particle's mirror point on the equatorial pitch angle of that particle. Particles with small equatorial pitch angles have large parallel velocities and their mirror points are at high latitudes,

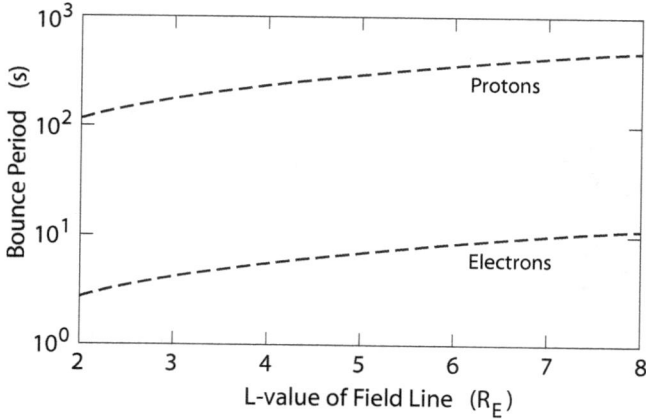

Figure 3.4: Bounce period for 1-keV electrons and protons with pitch angle $\alpha_{eq} = 30°$. At same energy, electrons, because of their smaller mass, have a much higher bounce frequency than ions.

close to the Earth. With increasing equatorial pitch angles, the mirror points move to more equatorial latitudes and the particles mirror close to the equatorial plane.

3.2.2 Bounce Period

The *bounce period*, τ_b, which we have already defined in Eq. (2.69) when discussing the longitudinal invariant J of a trapped particle, is the time it takes a particle to move from the equatorial plane to one mirror point, then to the other and back to the equatorial plane. It can be calculated by integrating $ds/v_\parallel$ over a full bounce path along the field line

$$\tau_b = 4 \int_0^{\lambda_m} \frac{ds}{v_\parallel} = 4 \int_0^{\lambda_m} \frac{ds}{d\lambda} \frac{d\lambda}{v_\parallel} \tag{3.14}$$

We can make use of Eqs. (2.62) and (3.13) to replace $v_\parallel = v [1 - (B/B_{eq}) \sin^2 \alpha_{eq}]^{1/2}$. Then Eqs. (3.7) and (3.8) can be used to substitute for $ds/d\lambda$ and the dipolar magnetic field ratio obtaining a rather involved expression which cannot be solved in closed form

$$\tau_b = 4 \frac{r_{eq}}{v} \int_0^{\lambda_m} \cos \lambda \, (1 + 3 \sin^2 \lambda)^{\frac{1}{2}} \left[1 - \sin^2 \alpha_{eq} \frac{(1 + 3 \sin^2 \lambda)^{\frac{1}{2}}}{\cos^6 \lambda} \right]^{-\frac{1}{2}} d\lambda \tag{3.15}$$

The integral must be solved numerically, but is usually approximated by the function

$$\Gamma_\alpha(\alpha_{eq}) \approx 1.30 - 0.56 \sin \alpha_{eq} \tag{3.16}$$

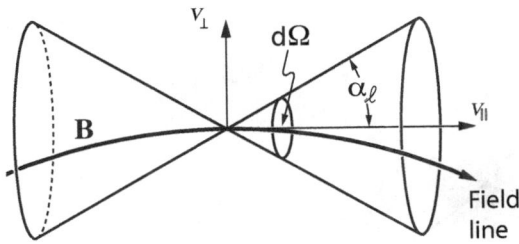

Figure 3.5: Definition of the loss cone α_ℓ of a gyrating particle moving with parallel velocity $v_\parallel$ along the geomagnetic field **B**.

Expressing equatorial distance r_{eq} and velocity v in terms of L and energy, W, one obtains

$$\tau_b \approx (3.7 - 1.6 \sin \alpha_{eq}) L R_E \left(\frac{m}{W} \right)^{\frac{1}{2}} \qquad (3.17)$$

The bounce period depends only weakly on the equatorial pitch angle, since particles with a smaller pitch angle have a larger velocity along the field line, yet have a longer way to their mirror points (see Fig. 3.3). Naturally, it is longer for longer field lines and shorter for particles with higher energies.

Figure 3.4 gives the bounce periods for 1-keV electrons and protons with $\alpha_{eq} = 30°$. The electrons, being much lighter, have typical bounce periods of some seconds, while the heavier protons take a few minutes to complete a full bounce cycle. As explained in Section 2.5, the bounce period must be much shorter than any fluctuation of the magnetic field for the integral $J = \oint m v_\parallel ds$ to be an adiabatic invariant. For keV electrons this is usually true, but for keV ions the second adiabatic invariant may be violated by, for example, geomagnetic pulsations which have periods of the order of the ion bounce period.

3.2.3 Loss Cone

Even when the longitudinal invariant is conserved, not all particles are actually trapped. If a particle's mirror point lies deep in the atmosphere, it will collide too often with neutral particles (see Sections 4.1 and 5.2) and, hence, will be absorbed by the atmosphere. The mirror point altitudes where particles are lost by collisions are those below about 100 km. For simplicity, we will use zero altitude since the magnetic field strength and mirror point latitude differ only by a few percent between the Earth's surface and the lower ionosphere. Under this assumption we can use Eq. (3.13) to define an *equatorial loss cone*

$$\sin^2 \alpha_\ell = \frac{B_{eq}}{B_E} = \frac{\cos^6 \lambda_E}{(1 + 3 \sin^2 \lambda_E)^{\frac{1}{2}}} \qquad (3.18)$$

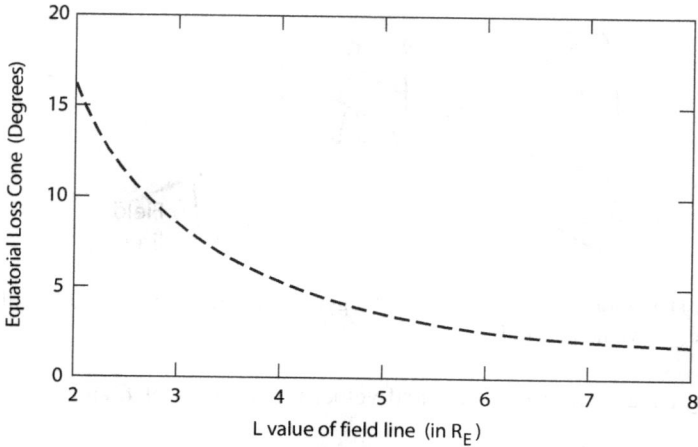

Figure 3.6: Equatorial loss cone as function of the L-value in the dipolar geomagnetic field.

Figure 3.5 shows the geometry of such an equatorial loss cone. All particles with equatorial pitch angles $\alpha < \alpha_\ell$ within the solid angle $d\Omega$ will be lost in the atmosphere. Since particles with $\alpha > 180° - \alpha_\ell$ will be lost in the other hemisphere, we get a double cone structure. Using Eq. (3.9) to express λ_E in terms of the L-value, we obtain

$$\sin^2 \alpha_\ell = \left(4L^6 - 3L^5\right)^{-\frac{1}{2}}$$

(3.19)

The width of the loss cone is independent from the charge, the mass, or the energy of the particles, but is purely a function of the field line radius. As can be seen in Fig. 3.6, the equatorial loss cone is typically rather small for equatorial distances of more than $3\,R_E$. At geostationary orbit ($6.6\,R_E$), the loss cone is less than $3°$ wide.

3.3   Drift Motion

A particle in a dipole field will gyrate, bounce, and drift at the same time. Thus one has to integrate over the former two motions if one is interested in the much slower drift motion. As long as we neglect electric fields, a particle will experience a purely azimuthal magnetic drift, v_d.

3.3.1 Magnetic Drift Velocity

The equatorial angular drift velocity is found by dividing the angular drift that occurs during one bounce cycle, $\Delta\psi$, by the bounce period. $\Delta\psi$ is computed by integrating $v_d/r\cos\lambda$ over one full bounce cycle. With $dt = ds/v_\parallel$ we obtain

$$\Delta\psi = 4\int_0^{\lambda_m} \frac{v_d}{r\cos\lambda}\frac{ds}{v_\parallel} \qquad (3.20)$$

We can replace the magnetic drift velocity by Eq. (2.46) and follow a similar procedure as used for the derivation of Eq. (3.15) to obtain the angular drift velocity, $2\pi\langle\Omega_d\rangle = \Delta\psi/\tau_b$, averaged over one full bounce cycle

$$\langle\Omega_d\rangle = \frac{3LW}{\pi q B_E R_E^2 \Gamma_\alpha}\int_0^{\lambda_m}\Psi(\alpha_{eq},\lambda)\,d\lambda \qquad (3.21)$$

where W is the particle energy and Γ_α is the integral introduced in Eqs. (3.15) and (3.16). The function Ψ is a rather complicated expression and the integral has to be solved numerically, but an approximate solution for the ratio of the integral over Ψ and Γ_α is

$$\Gamma_\alpha^{-1}\int_0^{\lambda_m}\Psi(\alpha_{eq},\lambda)\,d\lambda \approx 0.35 + 0.15\sin\alpha_{eq} \qquad (3.22)$$

Hence, we get an approximation for the average *drift period*, $\langle\tau_d\rangle = \langle\Omega_d\rangle^{-1}$

$$\langle\tau_d\rangle \approx \frac{\pi q B_E R_E^2}{3LW}(0.35 + 0.15\sin\alpha_{eq})^{-1} \qquad (3.23)$$

and for the average drift velocity, $\langle v_d\rangle = 2\pi L R_E/\tau_b$

$$\boxed{\langle v_d\rangle \approx \frac{6L^2 W}{q B_E R_E}(0.35 + 0.15\sin\alpha_{eq})} \qquad (3.24)$$

Both, drift period and drift velocity depend on particle charge, particle energy, and L-value, but not explicitly on the mass of the particle. Hence, electrons and protons with equal energies drift around the Earth with the same velocity, only in opposite directions. Since the drift velocity scales with L^2, the drift period is actually shorter on more distant L-shells. This is in contrast to the normal Keplerian motion, where more distant particles have lower azimuthal velocities. Figure 3.7, which gives a graphic representation of Eq. (3.23), clearly shows this effect. It also shows that the drift period (and the drift velocity) depend only weakly on the equatorial pitch angle, as was the case for the bounce period.

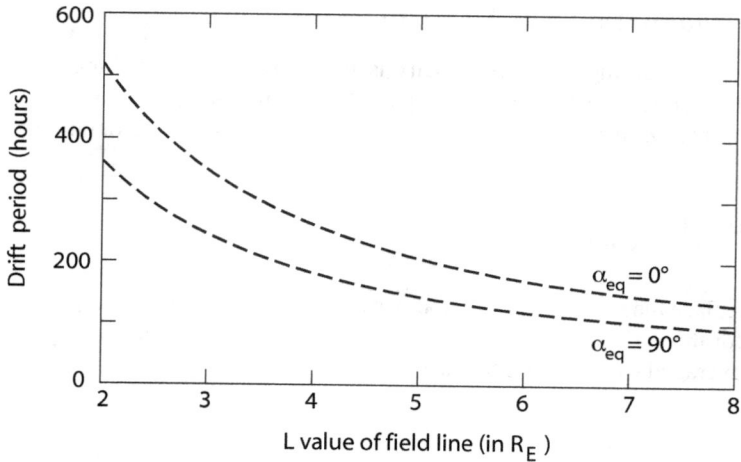

Figure 3.7: Magnetospheric drift period around Earth for geomagnetically trapped 1-keV ring current particles with different equatorial pitch angles under conservation of the drift invariant.

As can be seen in Fig. 3.7, the drift period for a 1-keV particle is of the order of several days. Since the magnetospheric field changes more frequently, it is very unlikely that the third adiabatic invariant, Φ, is conserved for typical ring current particles in the 10 keV energy range. Many ring current particles undergo *radial diffusion* across L-shells or do not complete a full revolution around the Earth. Only the most energetic ring current particles in the MeV range have a drift period short enough to perform closed orbits.

3.3.2 Electric Drift

However, even if the third invariant were not violated, particles in the outer ring current would not perform closed orbits. As we will see in the next section, the solar wind generates an electric field inside the magnetosphere, which is directed from dawn to dusk in the equatorial plane. Thus the particles will experience a sunward E×B drift. We can use Eqs. (2.19) and (3.8) to get an expression for the equatorial electric drift velocity, v_E, due to a uniform equatorial transverse electric field, E_{eq}, obtaining

$$
v_E = \frac{E_{eq}}{B_{eq}} = \frac{E_{eq}L^3}{B_E}
\tag{3.25}
$$

Electrons and ions with mirror points close to the equator will drift sunward with this velocity, in addition to their magnetic drift.

Because the magnetic drift of positive ions is directed westward, the two drifts are oppositely directed on the dawn side. For electrons, this holds on the dusk side. Since the magnetic drift velocity scales with L^2 while the E×B velocity scales with L^3, the

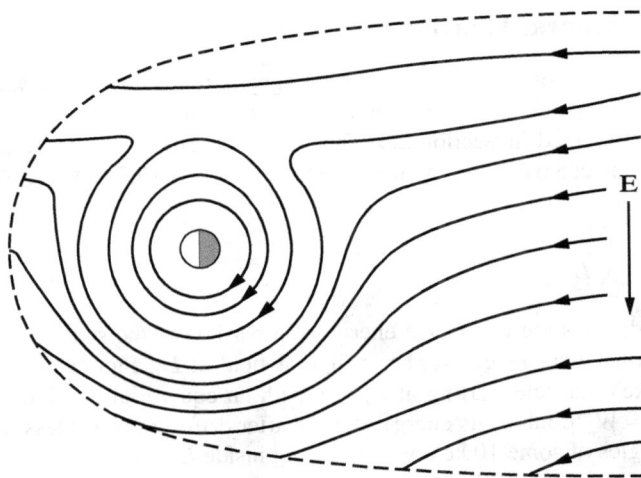

Figure 3.8: Energetic ion drift paths in the equatorial plane. Ion injected in the distant equatorial tail perform an E×B-drift toward the Earth. When approaching the Earth they start experiencing the gradient and curvature of the geomagnetic field thereby picking up a gradient-curvature drift term which close to the Earth becomes more important than the electric field drift forcing the ions into a nearly circular drift orbit around Earth. This drift becomes clockwise for ions which have started in the tail sufficiently far away from the morning boundary of the tail. The drift paths of these ions finally leave the magnetosphere on the dayside across the magnetopause back into the solar wind. The clockwise (westward) drift of the ions near Earth is what forms the ring current. Ions closer to Earth not having started in the tail perform a so-called corotating motion with Earth.

electric drift will typically overcome the magnetic drift outside some radial distance. Hence, for the combined electric and magnetic drift we have the situation sketched in Fig. 3.8. Close to the Earth, the magnetic drift forces prevail and we have a *symmetric ring current*. Far out, the particle trajectories are dominated by the E×B drift. In the intermediate region we get a *partial ring current*, caused by the deflection of sunward drifting particles around the Earth due to the gradient and curvature forces.

As we will see in the next section, the actual magnetospheric electric field is not quite uniform, but the energetic particle trajectories in a realistic electric field are similar to those shown in Fig. 3.8.

3.4   Sources and Sinks

Figure 3.8 also indicates that the source of the ring current is the tail plasma sheet (see Fig. 1.4). The particles are brought in from the tail by the electric drift. When they reach the stronger dipolar field in the ring current region, they start to experience the gradient and curvature forces.

3.4.1 Adiabatic Heating

Since the particles encounter stronger magnetic fields on their inward drift while keeping their magnetic moment, they are heated adiabatically by betatron acceleration as was described in Section 2.5. The transverse energy gain of a particle in the equatorial plane can easily be calculated by inserting Eq. (3.8) into Eq. (2.65)

$$\frac{W_\perp}{W_{\perp 0}} = \left(\frac{L_0}{L}\right)^3 \tag{3.26}$$

where $W_{\perp 0}$ denotes the transverse energy of a particle at the start when it begins to meet the strong inner magnetospheric dipolar field at L_0. Figure 3.9 visualizes this effect. A 1-keV particle starting at $L_0 = 8$ with an equatorial pitch angle $\alpha_{eq} = 90°$ and thus $W = W_\perp$ doubles its energy already after drifting inward less than 2 R_E and reaches energies of some 10 keV when coming inside $L = 3$.

The field lines on the radially inward path of the particle shorten, however, and thus the bounce paths along the crossed field lines get shorter the more the particle moves inward, electrons and ions with equatorial pitch angles $\alpha_{eq} < 90°$ also undergo Fermi acceleration as described in Eq. (2.71) to keep their longitudinal invariant constant. Calculating this energy gain is more involved, since it is somewhat difficult to calculate the length of the field line between the mirror points, but an approximate solution is

$$\frac{W_\parallel}{W_{\parallel 0}} = \left(\frac{L_0}{L}\right)^\kappa \tag{3.27}$$

where the exponent ranges between $\kappa = 2$ for particles with $\alpha_{eq} = 0°$ and $\kappa = 2.5$ for $\alpha_{eq} \to 90°$. Hence, as shown in Fig. 3.9, a 1-keV particle starting at $L_0 = 8$ with an equatorial pitch angle of $\alpha_{eq} \approx 0°$ and thus $W \approx W_\parallel$ doubles its energy after drifting inward by about 2.5 R_E and reaches 10 keV near $L = 2$.

This effect is the mechanism of ring current generation. It is the adiabatic heating during the convective E×B-drift inward transport of particles injected into the geomagnetic field which creates the 10–100 keV ring current ions from the 1–10 keV plasma sheet ions, and we immediately realise that for such an effective inward transport an electric field is required. This field must be applied to the magnetosphere from the outside as we will discuss later. But on their inward drift the particles do not only gain in energy but also their energy anisotropies as defined in Eq. (2.72) are increasing. Using an average $\kappa = 2.25$ we obtain

$$\frac{A_W}{A_{W0}} = \left(\frac{L_0}{L}\right)^{0.75} \tag{3.28}$$

Accordingly, the further one enters into the inner magnetosphere, the more energetic particles one will find with large pitch angles and the larger becomes the anisotropy of the energetic (ring current) particle component.

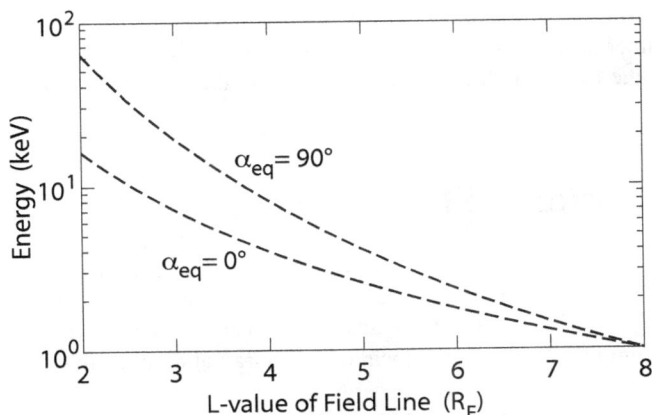

Figure 3.9: Adiabatic heating of 1-keV particles with different equatorial pitch angles.

3.4.2 Loss Processes

On average, the inward transport of particles from the tail must be balanced by loss processes in the inner magnetosphere. Ions are lost from the symmetric ring current mainly by a process called *charge exchange*. The outermost part of the neutral atmosphere has densities of some 100 cm^{-3} at ring current altitudes. This density is too low for direct collisions between the ring current ions and the neutral atoms (see Section 4.1). However, when a cold atom from the neutral atmosphere comes accidentally close to a hot ring current ion, it under certain conditions looses its electron to the ion by some resonance effect. In this charge exchange process, the hot ring current ion is suddenly turned into a neutral particle which either escapes from the magnetosphere, since its kinetic energy exceeds the gravitational energy, or heats the lower neutral atmosphere when entering it from above. The cold ion, on the other hand, does not contribute to the ring current anymore, but mixes into the plasmaspheric particle component. Typical life times of ring current ions before charge exchange span from a couple of hours to some days depending on energy and location.

Other loss processes are caused by electromagnetic variations with frequencies sufficiently far above the bounce and/or cyclotron frequencies. In the presence of such high-frequency waves, the respective second and first adiabatic invariants of a particle may become violated and particle's pitch angle may be altered in such a way that the particle falls into the 'loss cone'. This *pitch angle scattering* or *pitch angle diffusion* process, widens the otherwise narrow loss cone (see Fig. 3.5) and leads to an enhanced loss of particles in the lower neutral atmosphere. This process is discussed in more detail in our companion book, *Advanced Space Plasma Physics*. It works most effectively for the ring current electrons, since they are much lighter than the ions and thus more sensitive to violation of their adiabatic invariants. In the outer ring current the mean life time of ring current electrons before they are scattered into the loss cone is of the order of hours, while it takes several days before an ion is lost by precipitation. In the inner part of the ring current the loss cone is wider because of the greater proximity to the dense neutral atmosphere and the mean precipitation life times are shorter.

3.5 Radiation Belts

In the vicinity of Earth one encounters two regions which are populated by energetic particles the latter being sufficiently dilute to permit for a single particle treatment. These are the radiation (or after their discoverer also called Van Allan) belts and the ring current. Figure 3.10 shows the location of the radiation belts in the Earth's dipole field.

The radiation belts consist of two regions which closely follow the geometry of the magnetic dipole field. They consist of two parts, the outer and the inner radiation belt, respectively. If speaking about the radiation belt one has mainly electrons in mind. Of course, the magnetic dipole flux tubes are also populated by ions which are as well trapped and also needed for the purpose of overall charge neutralisation, but the ions, as will be described in the next section, carry also currents and are thus part of the magnetospheric ring current system.

The radial structure of the two radiation belts is determined by the particle energy. The outer belt contains electrons of lower energy $W_e > 40$ keV, still far above the background plasmaspheric electron energies which are below 1 keV, but substantially less than the inner belt electrons which have energies of few 100 keV and up to 1 MeV or, during particular storm events, even higher. One immediately realises that this energy structuring of the radiation belts is to some degree due to adiabatic heating during injection according to the processes described in the previous sections. However, the higher energies of the radiation belt electrons cannot be explained in this way, and other processes must be evoked which are related to massive violation of the adiabatic invariants when electrons interact with fluctuations in the electric and magnetic fields. These processes are of kinetic nature and require knowledge of wave particle interaction.

The inner and outer radiation belts are spatially separated by a region which is called the *radiation belt slot* and which is just populated by the background plasmaspheric electrons. This regions has for quite a while surprised the space physicists until they realised that it is produced by kinetic processes when the trapped electrons themselves generate fluctuations in the fields which have high enough frequencies to violate the magnetic moment and second invariant and as a consequence scatter them out of their orbits thereby causing the loss of radiation belt electrons from the slot region. This is a very localised effect. In addition, the inner belt also contains a high energy ion component also shown in the figure. These ions are trapped cosmic ray particles which bounce back and forth in the Earth's dipole field at the location where they become trapped. Moreover, a very energetic ion component exists at much closer distance to the Earth above the atmosphere. These ions are so-called secondaries which

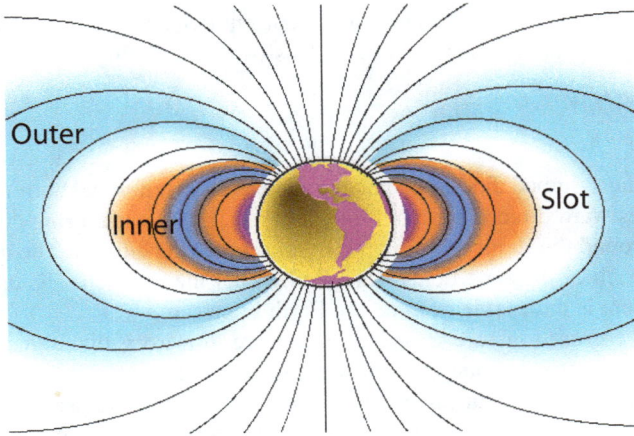

Figure 3.10: The location of the radiation belts in Earth's dipolar magnetic field geometry. One distinguishes the outer (light blue) and inner (orange) radiation belts with the outer containing lower energy particles than the inner. Between the two radiation belts a so-called slot region is found which is empty of high energy particles. The inner radiation belt also hosts a component of energetic trapped particles (dark blue) of cosmic ray origin. Very close to Earth's surface exists another (pink colour) region which contains energetic secondary ions which are produced by the bombardment of the upper atmosphere by cosmic rays and afterwards trapped in the geomagnetic dipole field.

are produced by the impact of cosmic ray particles onto the upper atmosphere. In this process atmospheric neutral gas particles (mostly oxygen O and nitrogen N, but also hydrogen H) are ionised and also gain energy. Those ions which are scatter back into space are subsequently trapped by the geomagnetic field close to Earth and bounce back and forth between their mirror points for a while until having lost their energies and merge into the upper ionosphere. However, this inner secondary ion belt is practically permanently present and in equilibrium because it is continuously refilled by the stationary stream of cosmic rays hitting the upper atmosphere such that its losses are compensated by newly produced secondaries.

3.6   Ring Current

We have already said that the ring current consists of the ion component of the outer radiation belt which in the inhomogeneous magnetic field after having been injected into the inner magnetosphere by the $E \times B$- drift start feeling the magnetic gradient and curvature and start performing an azimuthal drift around Earth. During this motion they constitute a current j_d. Using the equatorial drift velocity given in Eq. (3.24) and assuming $\alpha_{eq} = 90°$, we obtain for the current density j_d caused by ring current particles with a particular energy, W, and density, n, circulating on a given L-shell

$$j_d = \frac{3L^2 nW}{B_E R_E} \tag{3.29}$$

where j_d in the assumed dipolar geomagnetic field is an azimuthal current that is flowing in the westward direction around Earth. In fact, already each ring current particle on its drift around the Earth constitutes a microscopic ring current. The magnetic field induced by each of these particles is negligible, but the magnetic disturbance due to the total current is noticeable even on the Earth's surface.

The ring current ions are invisible. There are, however, two signatures of their presence. The first is the magnetic disturbance caused by the drifting ion ring current at the Earth's surface during geomagnetic storms which we are going to describe below. The second is optical. It is caused by the charge exchange between hot ring current ions and atmospheric neutral gas which we have noted in the section on the losses. Clearly, after charge exchange the newly created cold ions remain invisible. However, the hot ring current ions which have suddenly become hot neutrals are in an excited non-ionised state and are capable of emitting radiation in the ultraviolet light. This light has been monitored during magnetic storms by the IMAGE spacecraft, when fresh ring current ions have been injected into the magnetosphere. By this observation it was possible to make the ring current or better its by product, the hot neutrals, optically visible. Figure 3.11 shows an example of those observations with the secondary ring current neutrals surrounding the Earth as a nearly complete and broad ring and not yet having had time to disappear in space.

3.6.1 Magnetic Disturbance

The total current, I_L, caused by all ring current particles on a particular L-shell is related to the current density by $I_L dl = j_d dV$. When integrating over the circumference element dl and the volume element dV, we may note that the total energy of all ring current electrons and ions at that particular radial distance is given by $U_L = \int nW \, dV$, and that the total circumference is simply $\int dl = 2\pi L R_E$

$$I_L = \frac{3U_L L}{2\pi B_E R_E^2} \tag{3.30}$$

From Biot-Savart's law we can evaluate the magnetic field disturbance caused by such a circular current loop at the Earth's center (see Appendix B.5)

$$\delta B_d = -\frac{\mu_0 I_L}{2L R_E} = -\frac{\mu_0}{4\pi} \frac{3U_L}{B_E R_E^3} \tag{3.31}$$

where we have introduced the minus sign to account for the fact that the disturbance field δB_d of the westward ring current is directed opposite to the terrestrial dipole magnetic field. Since the magnetic disturbance does not depend on radial distance, we

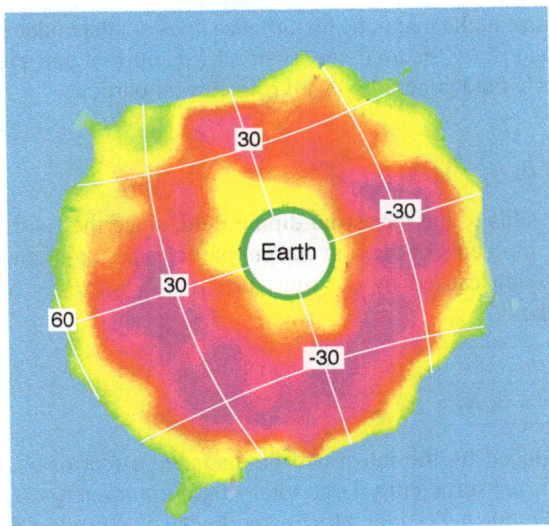

Figure 3.11: The ring current made visible in hot secondary neutrals by the NASA IMAGE satellite during a magnetic storm. The hot neutrals are created from ring current ions by absorbing an electron from cold atmospheric neutrals. One notices the nearly complete though asymmetric ring the neutrals form around Earth, not yet having had time to escape into space. The Earth is the circle in the centre.

can replace U_L by the total energy, U_R, and obtain for the disturbance field generated by the drift of all ring current particles

$$\Delta B_d = -\frac{\mu_0}{4\pi}\frac{3U_R}{B_E R_E^3} \tag{3.32}$$

The total magnetic field perturbation caused by the ring current must also include the diamagnetic contribution due to the cyclotron motion of the ring current particles. Again assuming $\alpha_{eq} = 90°$, we obtain the diamagnetic field δB_μ at the Earth's center caused by a charged particle orbiting on a particular L-shell around Earth by replacing the magnetic moment of Earth M_E in Eq. (3.1) by the particle magnetic moment

$$\delta B_\mu = \frac{\mu_0}{4\pi}\frac{\mu}{L^3 R_E^3} \tag{3.33}$$

Using the definition of the magnetic moment given in Eq. (2.37) and replacing the field strength by the dipole field value given in Eq. (3.8) we find

$$\delta B_\mu = \frac{\mu_0}{4\pi}\frac{W}{B_E R_E^3} \tag{3.34}$$

As for the drift current, the magnetic disturbance does not dependent on radial distance and we can replace W by the total energy of all ring current particles, U_R, and obtain for the diamagnetic field generated by all ring current particles

$$\Delta B_\mu = \frac{\mu_0}{4\pi} \frac{U_R}{B_E R_E^3} \tag{3.35}$$

This disturbance adds to the terrestrial dipole field, since the Earth's dipole moment and the magnetic moments of the ring current particles are co-aligned. The total magnetic field depression caused by the ring current $\Delta B_R = \Delta B_d + \Delta B_\mu$ at the Earth's center thus becomes

$$\boxed{\Delta B_R = -\frac{\mu_0}{2\pi} \frac{U_R}{B_E R_E^3}} \tag{3.36}$$

being slightly reduced by the integrated diamagnetic effect of all the gyrating ring current ions. Still, for sufficiently large total ring current energy U_R it contributes to a depression of the magnetic field which is particularly strongly felt during magnetic storms.

3.6.2 Magnetic Storms

At certain times more particles than usual are injected from the tail into the ring current, mainly by an enhanced duskward electric field $\mathbf{E}_T$ applied to the geomagnetic tail. We have described that this way the total energy of the ring current is increased when fresh accelerated particles are added from the magnetosphere causing an additional depression of the surface magnetic field at Earth. This additional depression can clearly be seen in near-equatorial magnetograms like the one shown in Fig. 3.12. For about one day, the equatorial terrestrial field was depressed by more than 300 nT, i.e., by more than 1% of its total value. Such strong depressions of the terrestrial field over several days have been noticed in magnetograms long before one knew about the ring current and have been called *magnetic storms*.

A magnetic storm has two distinct phases. For some hours or days, an enhanced electric field injects more and more particles into the inner magnetosphere, building up the strong storm-time ring current and the associated magnetic disturbance field. After a day or two, the electric field amplitude and the rate of injection get back to the normal level. Now the disturbance field starts to recover, since the ring current looses more and more storm-time particles due to charge exchange and pitch angle scattering. As can be seen in Fig. 3.12, this recovery phase typically lasts several days.

The depression of the terrestrial dipole field given in Eq. (3.36) can be used to estimate the amount of additional energy deposited in the ring current during a magnetic storm, if one approximates ΔB_R by the average magnetic field depression on the Earth's surface near the equator, usually taken from the *Dst index*.

The *Dst* index, which is always taken positive, represents the average (geomagnetic) storm time surface disturbance field $\langle |\Delta B_R| \rangle_{storm}$ taken at the Earth's equator.

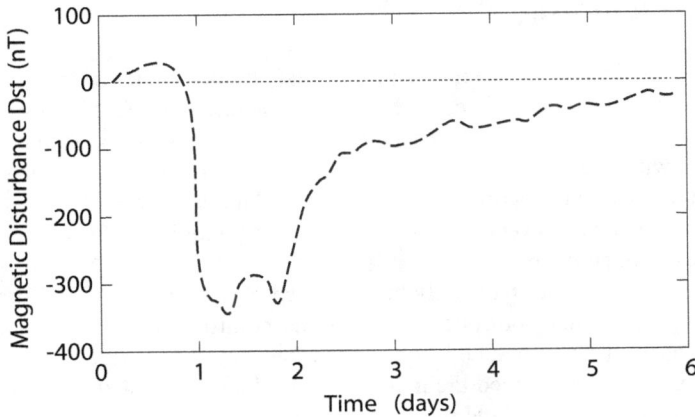

Figure 3.12: Magnetic field depression during a major magnetic storm as measured by *Dst*. The storm starts usually with a small positive pulse in the equatorial horizontal component of the geomagnetic field indicating a brief compression of the magnetosphere. It is followed by the build up of the ring current whose diamagnetic effect adds negatively to the surface field causing the storm time main-phase depression of the geomagnetic field which is also accompanied by weak inflation of the magnetosphere. This depression decays in the storm-recovery phase over days with the gradual decay of the ring current.

It is calculated on the basis of magnetic recordings from four low-latitude magnetic observatories (see Appendix C.3).

Since *Dst* is also influenced by sources other than the ring current (i.e. magnetic induction effects in the Earth's body, changes in the magnetopause topology, etc.), which may contribute more than 30% of its value, one may use roughly half the *Dst* value to derive from Eq. (3.36) a lower limit for the storm time increase in the total energy content of the ring current particle component

$$\Delta U_R \geq \frac{\pi}{\mu_0} B_E R_E^3 Dst \qquad (3.37)$$

Putting numbers into Eq. (3.37) one finds that a depression of 1 nT is caused by an increase in ring current energy of more than 2×10^{13} J. Assuming all current to be concentrated at $L = 5$ and using Eq. (3.30) we find that a ring current of about 4×10^4 A produces a 1 nT depression. During the big magnetic storm shown in Fig. 3.12 the ring current energy increased by more than 5×10^{15} J and the total current reached more than 10^7 A. Alternatively, we may note that 1 eV corresponds to 1.6×10^{-19} J and that more than 3×10^{30} particles of ~ 10 keV energy each have been injected into the ring current during that particular magnetic storm.

3.7   Concluding Remarks

In the present chapter we have applied the simplest plasma physics approach to trapped particles and the ring current. The single particle picture is useful when studying the basic behavior of these trapped particles, but readily reaches its limits when the particles interact with each other and cannot be considered anymore as independent. One of the sole applications was the magnetic storm. Another one is the physics of the radiation belts which, however, for its understanding requires knowledge of kinetic processes and collective processes and is thus deferred to a later place in this book. These collective processes have only be described here in very general terms as currents flowing in the magnetospheric system and contributing to the magnetic field while we not even have included these fields into any description of their motion. We have, so far, not mentioned them at all in this chapter. We will return to some of them after having developed the more sophisticated approaches to the description of plasmas.

Further Reading

Readers interested in the proper particle dynamics of the radiation belts and to more quantitative accounts of the various ring current effects on the structure of the magnetosphere during storms — and also during quiet times — are referred to the monographs and reviews listed below. Even though some of them are quite historical by now, there has not been very much added in the meantime to the non-collective knowledge in radiation belt and ring current physics such that these books and reviews still represent the state of the art what concerns the description of single particle dynamic applications to the close plasma environment of Earth.

References

[1] I. A. Daglis, *Ring current dynamics*, Space Science Rev. **124**, 183–202 (2006).
[2] L. R. Lyons and D. J. Williams, *Quantitative Aspects of Magnetospheric Physics* (D. Reidel Publ. Co., Dordrecht, 1984).
[3] A. Nishida, *Geomagnetic Diagnosis of the Magnetosphere* (Springer Verlag, Heidelberg, 1978).
[4] J. G. Roederer, *Dynamics of Geomagnetically Trapped Radiation* (Springer Verlag, Heidelberg, 1970).
[5] M. Schulz and L. J. Lanzerotti, *Particle Diffusion in Radiation Belts* (Springer Verlag, Heidelberg, 1974).
[6] M. Walt, *Introduction to Geomagnetically Trapped Radiation* (Springer Verlag, Heidelberg, 1994).
[7] R. A. Wolf, in *Solar-Terrestrial Physics*, eds. R. L. Carovillano and J. M. Forbes (D. Reidel Publ. Co., Dordrecht, 1983), p. 303.

Problems

Problem 3.1 *Derive the magnetic field strength of the dipolar magnetic field in spherical coordinates.*

Problem 3.2 *Why can the geometry of a field line in a magnetic field* **B** *be calculated from Eq. (3.3)? Show explicitly that for a dipole field this equation reduces to Eq. (3.5) and that its solution is Eq. (3.6).*
Hint: *Think of the definition of the Lorentz force and its physical content.*

Problem 3.3 *Show that the length element along the field line is given by* $(ds)^2 = (dr)^2 + r^2(d\lambda)^2$ *with* λ *the latitude measured from the equator. Derive the equation for* $s(\lambda)$ *for the geomagnetic dipole field.*

Problem 3.4 *Calculate the bounce frequency for a small amplitude oscillation of an electron and a proton along a field line of value* $L = 5$ *in the geomagnetic dipole field as function of particle energy. Compare these values with those calculated from the expressions for* τ_b *assuming several equatorial pitch angles.*

Problem 3.5 *In the expression for the loss cone the latitude* λ_E *is used. What is the reason for this? Would you believe it is exact enough or should one rather use another value, for instance the values which you calculated in Problem 4?*

Problem 3.6 *What happens to particles which have been injected in the tail, surround Earth as ring current particles and leave through the magnetopause? Do you believe this happens really?*

Problem 3.7 *Explain in words what happens during adiabatic heating when the particles enter from the tail into the ring current. You may have though about this already in Chapter 2. Where is the energy taken from?*

Problem 3.8 *Can you give a qualitative description of what happens in pitch angle scattering? Why is this considered to contribute to losses of trapped particles?*

Problem 3.9 *What is the reason for the ring current magnetic field at Earth's surface to be composed of two terms? Isn't the magnetic field generated by the particle drift current sufficient for the depression of the Earth's surface field during storm time ring currents?*

Problem 3.10 *As a corollary to the last problem: What is the diamagnetic effect of all the particles in the case of a homogeneous plasma embedded into a homogeneous magnetic field? Is this effect, if it exists, possible to measure?*

Problem 3.11 *Why can the ring current be mapped into UV after charge exchange and thus can be made visible?*

— 4 —

Collisions and Conductivity

So far we have considered only the motion of single particles in external and slowly variable electromagnetic fields, but have neglected any interaction between the particles. Interaction in plasmas is, however, unavoidable because the particles carry charges which fill the near-particle space with an electric potential field, and the particles move through space. This motion implies that each particle represents a microscopic current which carries it own magnetic field. As a consequence, each of the many particles in a plasma moves across the electric nd magnetic fields of all the other particles which necessarily affects its motion by the Lorentz force it experiences in these fields.

As we have argued in Chapter 1 the particles shield the individual fields within a distance of the Debye radius around each particle; this is necessary for keeping the plasma quasi-neutral to the external space. However it does not eliminate the internal microscopic fields inside the Debye sphere which each particle experiences when it moves across the plasma, passes many other particles and contributes to the shielding. During this motion it is permanently exposed to the collective fields produces by all the other particles in its vicinity. Hence, even if particles do not collide head on they interact through these mixtures of microscopic electromagnetic fields. We will, in later chapters, see that these collective interparticle-actions produce the most interesting effects in plasma.

In the present chapter we will, however, treat just the simplest kind of this interaction between single particles: the direct collision of two particles which each carry an electric charge. We do not consider in detail however the interaction between a charge plasma particle and a neutral atom which is one of the processes which lead to the already mentioned ionisation, charge exchange, and recombination which are important in not fully but partially ionized plasmas like the lower ionosphere. This would require a non-classical quantum mechanical treatment belonging into the domain of atomic physics.

The simple 'collision' between two charges considered in this chapter excluding all the processes mentioned with the exception of the distortion of the particle path and momentum exchange, on the other hand, can be described classically within the domain of electrodynamics. This kind of collisions, the so-called Coulomb collisions, is not yet a genuine collective plasma effect; it involves interactions between individual particles only and does not account for the interaction between large groups of particles in the plasma or the plasma as a whole, but under the presence of collisions the particles already behave quite differently from what would be expected when using the single particle picture.

4.1 Collisions

One distinguishes two types of plasmas: *collisional* and *collisionless*. In the latter direct collisions between two particles, i.e. so-called binary collisions, are so infrequent compared with any relevant variation in the fields or the time scales of particle dynamics that they can be safely neglected. In such plasmas the mean free path of a particle is very large, much larger than the typical scales of interest. Most plasmas in space are sufficiently dilute to belong to this type. At the contrary, in the former plasmas binary collisions are sufficiently frequent to influence or even dominate the behaviour of the plasma. Such plasmas are dense and are found only in particular regions in space like in the lower ionosphere.

Collisional plasmas can again be divided into two subclasses: *partially ionized* plasmas and *fully ionized* plasmas. Partially ionized plasmas like the ionosphere contain a large amount of residual neutral atoms or molecules, while fully ionized plasmas consist of electrons and ions only. It is clear that the types of collisions in both cases must be very different because neutrals do not respond to Coulomb fields, and thus a collision between a charged particle and a neutral is to good approximation independent of the charging of one of the participants in the collision. In partially ionized gases direct collisions between the charge carriers and neutrals dominate, while in fully ionized plasmas direct collisions are replaced by Coulomb collisions.

4.1.1 Partially Ionized Plasmas

In a partially ionized plasma most collisions occur between charged and neutral particles. Neutral particles affect the motion of charged particles by their mere presence as heavy compact obstacles. Hence, neglecting all quantum mechanical effects like excitation, charge exchange, ionisation and recombination, collisions between a charge particle and a neutral atom or molecule can to first approximation be treated as a head-on collision without accounting for the fact that one of the particles is charged (the ion or electron) and the other consists of a positively charged nucleus which is embedded into a negatively charge cloud of electrons. Such a head-on collision occurs only when a charged particle directly hits a neutral atom or molecule along its orbit. The *neutral collision frequency*, ν_n, that is the number of collisions per second, is therefore proportional to the number of neutral particles in a column of the cross-section of an atom or molecule, $n_n \sigma_n$, where n_n is the neutral particle density and $\sigma_n = \pi d_0^2$ the molecular cross-section, and to the average velocity, $\langle v \rangle$, of the charged particles

$$\nu_n = n_n \sigma_n \langle v \rangle \qquad (4.1)$$

The molecular cross-section can be approximated by $\sigma_n \approx 10^{-19}\,\text{m}^2$. Similarly, it is possible to define the *mean free path length* a charged particle can propagate between two collisions with neutrals as

$$\lambda_n = \frac{\langle v \rangle}{\nu_n} = (n_n \sigma_n)^{-1} \qquad (4.2)$$

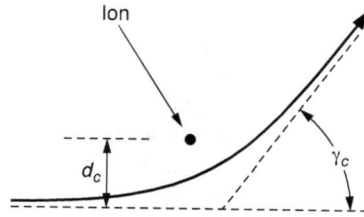

Figure 4.1: Electron orbit during a Coulomb collision with an ion at rest. The electron is deflected from its original straight orbit into a hyperbolic orbit of scattering angle γ_c passing the ion at distance d_c, the collision parameter.

The reason for using the average velocity instead of the actual velocities of the particles themselves in these definitions is that collisions are unpredictable and that it is meaningful only to define average collision frequencies and mean free path lengths.

As noted already, collisions between charged and neutral particles become important in the denser and partially ionized regions of space plasmas as the terrestrial ionosphere, where they contribute to ionisation, excitation, recombination, electrical resistivity, and current flow, and cause charged particle diffusion across the magnetic field.

4.1.2   Fully Ionized Plasmas

In fully ionized plasmas the charged particles interact via their electric Coulomb fields. The existence of these fields implies that the particles are deflected at interparticle distances much larger than the atomic radius. The Coulomb potential therefore enhances the cross-section of the colliding particles, but also leads to a preference for small angle deflections. Both these facts considerably complicate the calculation of a collision frequency in a fully ionized plasma.

A further complication arises from the fact that in a plasma with many particles in a Debye sphere, the Coulomb potential is screened and the electric field is approximately confined to the Debye sphere (see Fig. 1.1). One could think that the effective radius of the cross-section would become equal to the Debye length. But this is not the case because the Debye sphere is transparent for particles of sufficiently high energy. Since the potential increases steeply when approaching the center of the sphere, deflections will occur predominantly inside the Debye radius, but large angle deflections will still be rare. Formally, the *Coulomb collision frequency* in a fully ionized plasma has the same functional dependence as Eq. (4.1). The problem lies in determining the Coulomb collisional cross-section, σ_c. In the following we present a simplified derivation of the Coulomb collision frequency between electrons and ions

$$\nu_{ei} = n_e \sigma_c \langle v_e \rangle \tag{4.3}$$

in a fully ionized plasma and later include the modification introduced by the overwhelming predominance of the much more frequent small angle deflections.

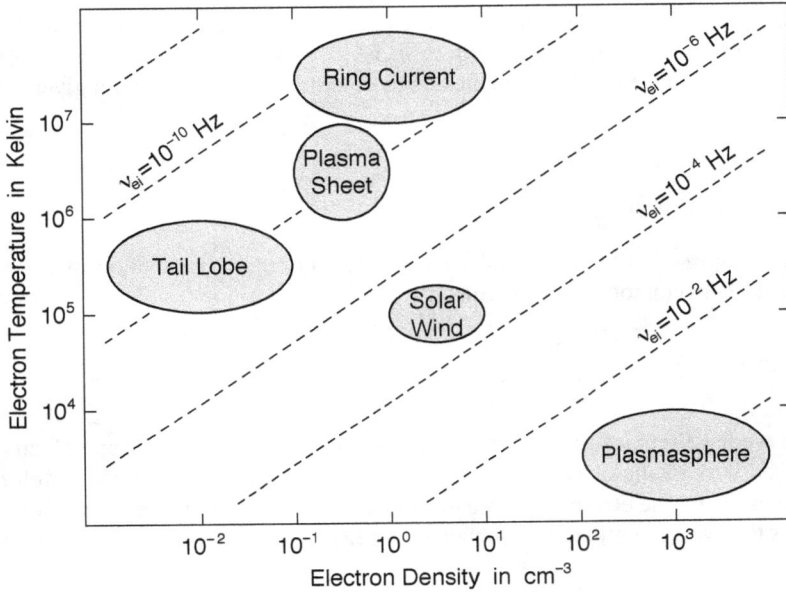

Figure 4.2: Typical Coulomb collision frequencies for geophysical plasmas.

Consider the collision between a single heavy ion and an electron. Because of the much larger mass, the ion can be considered at rest. When the electron approaches the ion it will be deflected in the Coulomb field of the ion as shown in Fig. 4.1 due to its attraction toward the ion. In a fully ionized plasma the temperature and consequently the energy of the electron is so high that the ion cannot trap the electron and recombine to a neutral atom. The electron will turn around the ion on a nearly Keplerian orbit in the central Coulomb field of the ion and escape into the surrounding plasma. Its orbit of escape is a hyperbola which at large distances from the ion can be approximated by straight lines and close to the ion by a section of a circle of radius d_c. It is this radius which we are going to determine.

The distance d_c is called *collision parameter* or *impact parameter*. The simplest method to determine this quantity is to consider the Coulomb force an ion is exerting on an electron of mass m_e, charge $q = -e$, and velocity v_e

$$F_C = -\frac{e^2}{4\pi\varepsilon_0 d_c^2} \tag{4.4}$$

This force is felt by the electron only during an approximate average time $\tau \approx d_c/v_e$ when it passes close to the ion. The change in momentum $|\Delta(m_e v_e)|$ it experiences during this time is approximately given by the product $\tau|F_C|$ or

$$|\Delta(m_e v_e)| \approx \frac{e^2}{4\pi\varepsilon_0 v_e d_c} \tag{4.5}$$

For large deflection angles, $\gamma_c \approx 90°$, the change in the particle momentum is of the same order as the momentum itself, $\Delta(m_e v_e) \approx m_e v_e$. Inserting this crude approximation in the above equation enables us to determine d_c as function a given electron velocity

$$d_c \approx \frac{e^2}{4\pi\varepsilon_0 m_e v_e^2} \tag{4.6}$$

yielding the maximum collisional cross-section, i.e. the cross section of large angle collisions between ions and electrons, as

$$\sigma_c = \pi d_c^2 \approx \frac{e^4}{16\pi\varepsilon_0^2 m_e^2 \langle v_e \rangle^4} \tag{4.7}$$

where we have replaced v_e in the denominator by the average electron velocity, $\langle v_e \rangle$, since the bulk of the electrons move with the average thermal velocity. Multiplying this equation by the electron plasma density, n_e, and the average electron velocity, one obtains the average collision frequency between electrons and ions defined in Eq. (4.3)

$$\nu_{ei} = n_e \sigma_c \langle v_e \rangle \approx \frac{n_e e^4}{16\pi\varepsilon_0^2 m_e^2 \langle v_e \rangle^3} \tag{4.8}$$

Moreover, one can use the average thermal electron energy given by $k_B T_e = \frac{1}{2} m_e \langle v_e \rangle^2$ and apply the formula for the plasma frequency given in Eq. (1.6) to obtain the simplified expression

$$\nu_{ei} \approx \frac{\sqrt{2}\,\omega_{pe}^4}{64\pi n_e} \left(\frac{k_B T_e}{m_e} \right)^{-\frac{3}{2}} \tag{4.9}$$

The collision frequency turns out to be proportional to the plasma density and inversely proportional to the 3/2 power of the electron temperature. It increases with increasing density and decreases with increasing electron temperature.

This formula is not exact, insofar as we have to include a correction factor $\ln\Lambda$ to correct for the predominance of weak deflection angles as well as for the different velocities electrons assume in thermal equilibrium in the plasma (see Appendix C.1). Λ is within a factor 4π equal to the plasma parameter introduced in Eq. (1.5) which is proportional to the number of particles in a Debye sphere. Multiplying Eq. (4.9) with $\ln\Lambda$ and using the definition of Λ in Eq. (1.5) to simplify the expression, we obtain

$$\boxed{\nu_{ei} \approx \frac{\omega_{pe}}{16\pi} \frac{\ln\Lambda}{\Lambda}} \tag{4.10}$$

which gives a reasonably good estimate of the electron-ion collision frequency. $\ln\Lambda$ is called the *Coulomb logarithm*. Because Λ is usually a very large number, $\ln\Lambda$ is of

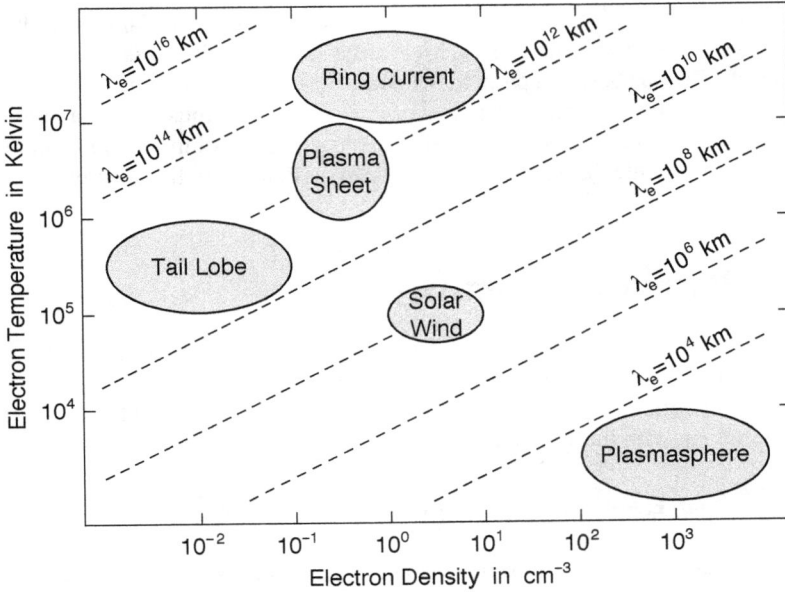

Figure 4.3: Typical Coulomb mean free path lengths for geophysical plasmas.

the order of 10–30 (see Appendix C.2). As in the case of neutral particle collisions, a mean free path length can be defined as

$$\lambda_e = \frac{\langle v_e \rangle}{v_{ei}} \approx \frac{64\pi n_e}{\omega_{pe}^4 \ln \Lambda} \left(\frac{k_B T_e}{m_e} \right)^2 \qquad (4.11)$$

Since this length is proportional to the square of the electron temperature, the mean free path of an electron will be relatively short in a cold, but very long in a hot plasma. Again, simplifying the expression by introducing the plasma parameter $\Lambda \gg 1$ yields

$$\boxed{\frac{\lambda_e}{\lambda_D} \approx \frac{16\pi \Lambda}{\ln \Lambda} \gg 1} \qquad (4.12)$$

which shows that the ratio of the mean free path to the Debye length in a plasma is indeed a very large number. An electron thus between two collisions passes a large number of Debye spheres and thus helps screening the Coulomb electric fields of may ions on its path whose vicinity it visits only for a short time of passage.

Figures 4.2 and 4.3 show typical ranges for the collision frequency and the mean free path length for some geophysical plasmas. One can immediately see that the mean free path length is much larger than the dimensions of the plasma regions themselves. Moreover, the collision frequencies are much smaller than the plasma frequencies (see Fig. 1.2) in all regions. Hence, most geophysical plasmas can be considered as collisionless.

Collisions occur also between electrons and electrons or ions and ions. These particles have equal masses and in each case the collision affects the motion of both particles. Collisions between particles of equal masses quickly equilibrize their velocities via momentum exchange. Particles of one species will thus readily assume a well defined common average energy allowing to assign them common temperatures T_e or T_i, respectively, while the temperatures between particles of unlike masses may remain quite different, e.g. $T_e \neq T_i$.

4.2   Plasma Conductivity

In the presence of collisions we have to add a collisional term to the equation of motion Eq. (2.7) for a charged particle under the action of the Coulomb and Lorentz forces. Assuming all collision partners to move with the velocity $\mathbf{u}$, we obtain for a charged particle moving with the velocity $\mathbf{v}$

$$m\frac{d\mathbf{v}}{dt} = q\left(\mathbf{E} + \mathbf{v} \times \mathbf{B}\right) - m v_c (\mathbf{v} - \mathbf{u}) \tag{4.13}$$

The collisional term on the right-hand side describes the momentum lost through collisions occurring at a frequency v_c. This addition term is called frictional term since it impedes motion in the same way as a friction the particle experiences in a viscous fluid. Equation (4.13) holds both for Coulomb and also for neutral collisions. In the absence of any electromagnetic fields the Lorentz force would vanish, and the friction term would be the only remaining force term. This is the case in very dense and sufficiently cold plasmas in the absence of any external electromagnetic fields and no external magnetic field when v_c becomes large. The long term behaviour of such plasmas is that of simple fluids. This, however changes in the presence of an electric field.

4.2.1   Unmagnetized Plasma

Let us assume a steady state in an unmagnetized plasma with $\mathbf{B} = 0$, where all electrons move with the velocity $\mathbf{v}_e$ and all collision partners (ions in the case of a fully ionized or neutrals in a partially ionized plasma) are at rest. Then we get

$$\mathbf{E} = -\frac{m_e v_c}{e}\mathbf{v}_e \tag{4.14}$$

Since the electrons move with respect to the ions, they carry a current

$$\mathbf{j} = -e n_e \mathbf{v}_e \tag{4.15}$$

Combining these two equations yields for the electric field

$$\mathbf{E} = \frac{m_e v_c}{n_e e^2}\mathbf{j} \tag{4.16}$$

which is the familiar Ohm's law with

$$\eta = \frac{m_e v_c}{n_e e^2} \tag{4.17}$$

where η is the *plasma resistivity*. It has the same form for fully and partially ionized plasmas and differs only in the collision frequency used.

For a fully ionized plasma, we may introduce the Coulomb collision frequency from Eq. (4.10) into Eq. (4.17) to get for the celebrated *Spitzer resistivity* in fully ionized plasma

$$\eta_s = \frac{1}{16\pi\varepsilon_0\omega_{pe}}\frac{\ln\Lambda}{\Lambda} \tag{4.18}$$

The Spitzer resistivity is actually independent from the plasma density, since ω_{pe} is directly and Λ inversely proportional to the square root of the electron density. This has its roots in the fact that if one tries to increase the current by adding more charge carriers one also increases the collision frequency and the frictional drag and, by this, decreases the velocity of the charge carriers and thus the current.

4.2.2 Magnetized Plasma

In the magnetized case, the plasma may move with velocity $\mathbf{v}$ across a magnetic field and we must add the $-\mathbf{v} \times \mathbf{B}$ electric field resulting from the Lorentz transformation to Eq. (4.16), yielding

$$\mathbf{j} = \sigma_0(\mathbf{E} + \mathbf{v} \times \mathbf{B}) \tag{4.19}$$

where we replaced the resistivity by its inverse, the *plasma conductivity*

$$\sigma_0 = \frac{n_e e^2}{m_e v_c} \tag{4.20}$$

Equation (4.19) is a simple form of the *generalized Ohm's law*, which is valid in all fully ionized geophysical plasmas where the typical collision frequencies are extremely low (see Fig. 4.2) and the plasma conductivity can be taken as near-infinite.

While treating the plasma conductivity as a scalar is warranted in the dilute, fully ionized magnetospheric and solar wind plasmas with their near-infinite conductivity, there is one place where we have to take the anisotropy introduced by the presence of the magnetic field into account. This is the lower part of the partially ionized terrestrial ionosphere where abundant collisions between the ionized and the neutral part of the upper atmosphere in the presence of a strong magnetic field lead to a finite anisotropic conductivity tensor.

Starting again from Eq. (4.13) and assuming a steady state, where all electrons move with the velocity $\mathbf{v}_e$ and all collision partners are at rest, but now in a magnetized plasma, we obtain

$$\mathbf{E} + \mathbf{v}_e \times \mathbf{B} = -\frac{m_e v_c}{e} \mathbf{v}_e \tag{4.21}$$

Using the definition of σ_0 in Eq. (4.20) and Eq. (4.15) to express $\mathbf{v}_e$ by the current yields another form of Ohm's law

$$\mathbf{j} = \sigma_0 \mathbf{E} - \frac{\sigma_0}{n_e e} \mathbf{j} \times \mathbf{B} \tag{4.22}$$

Let us now assume that the magnetic field is aligned with the z axis, $\mathbf{B} = B\hat{\mathbf{e}}_z$. Taking into account the definition of the electron cyclotron frequency given in Eq. (2.12) and remembering that the cyclotron frequency carries the sign of the charge, we obtain

$$
\begin{aligned}
j_x &= \sigma_0 E_x + \frac{\omega_{ge}}{v_c} j_y \\
j_y &= \sigma_0 E_y - \frac{\omega_{ge}}{v_c} j_x \\
j_z &= \sigma_0 E_z
\end{aligned}
\tag{4.23}
$$

Combining the first two equations to eliminate j_y from the first and j_x from the second equation yields

$$
\begin{aligned}
j_x &= \frac{v_c^2}{v_c^2 + \omega_{ge}^2} \sigma_0 E_x + \frac{\omega_{ge} v_c}{v_c^2 + \omega_{ge}^2} \sigma_0 E_y \\
j_y &= \frac{v_c^2}{v_c^2 + \omega_{ge}^2} \sigma_0 E_y - \frac{\omega_{ge} v_c}{v_c^2 + \omega_{ge}^2} \sigma_0 E_x \\
j_z &= \sigma_0 E_z
\end{aligned}
\tag{4.24}
$$

This set of component equations can be written in dyadic notation (see Appendix B.4)

$$\mathbf{j} = \sigma \cdot \mathbf{E} \tag{4.25}$$

For a magnetic field aligned with the z direction, the conductivity tensor reads

$$
\sigma = \begin{pmatrix} \sigma_P & -\sigma_H & 0 \\ \sigma_H & \sigma_P & 0 \\ 0 & 0 & \sigma_\parallel \end{pmatrix} \tag{4.26}
$$

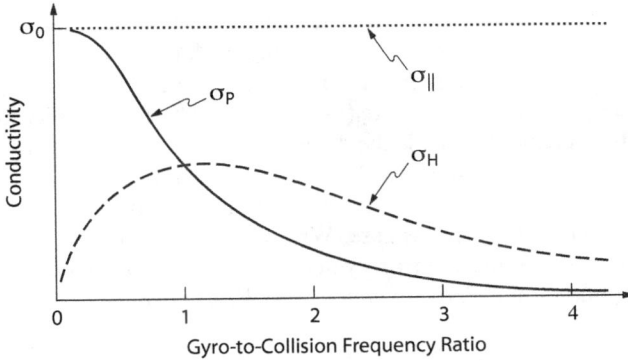

Figure 4.4: Dependence of the conductivities on the frequency ratio ω_g/ν_c. For very large collision frequencies $\nu_c \gg \omega_{ge}$ the plasma becomes isotropically conducting with negligible Hall conductivity. For very large electron-gyro frequency $\omega_{ge} \gg \nu_c$ the Pedersen conductivity vanishes while the Hall conductivity approaches the constant value $\sigma_H = en_e/B$ which is small compared with the parallel conductivity. For $\omega_{ge} = \nu_c$ the values of Pedersen and Hall conductivities are just half the parallel conductivity.

and the tensor elements are given by

$$
\begin{aligned}
\sigma_P &= \sigma_0 \, \frac{\nu_c^2}{\nu_c^2 + \omega_{ge}^2} \\[2mm]
\sigma_H &= \sigma_0 \, \frac{\omega_{ge}\nu_c}{\nu_c^2 + \omega_{ge}^2} \\[2mm]
\sigma_\| &= \sigma_0 = \frac{n_e e^2}{m_e \nu_c}
\end{aligned}
\tag{4.27}
$$

The tensor element σ_P is called *Pedersen conductivity* and governs the *Pedersen current* in the direction of that part of the electric field, $\mathbf{E}_\perp$, which is transverse to the magnetic field. The *Hall conductivity*, σ_H, determines the *Hall current* in the direction perpendicular to both the electric and magnetic fields, i.e. in the $-\mathbf{E} \times \mathbf{B}$ direction (remember that ω_{ge} is a negative number). The element $\sigma_\|$ is called *parallel conductivity* since it governs the magnetic *field-aligned current* driven by the parallel electric field component, $E_\|$. The parallel conductivity is equal to the plasma conductivity in the unmagnetized case.

When the magnetic field has an arbitrary angle to the axes of the chosen coordinate system, one can rewrite Eq. (4.25) into the form

$$
\mathbf{j} = \sigma_\| \mathbf{E}_\| + \sigma_P \mathbf{E}_\perp - \sigma_H (\mathbf{E}_\perp \times \mathbf{B})/B
\tag{4.28}
$$

This expression can be derived directly from Eq. (4.22) by taking the cross-product of Eq. (4.22) with $\mathbf{B}$ and using the result to eliminate the $\mathbf{j} \times \mathbf{B}$ term from Eq. (4.22).

The dependence of the conductivity tensor elements on the ratio of the cyclotron frequency to the collision frequency is shown in Fig. 4.4. In a highly collisional plasma containing a weak magnetic field we have $|\omega_{ge}| \ll v_c$. The set of Eqs. (4.27) then shows that $\sigma_P = \sigma_{\|} = \sigma_0$ and $\sigma_H = \sigma_0(\omega_{ge}/v_c) \approx 0$. The conductivity tensor becomes isotropic with no elements outside the diagonal and reduces to a scalar.

For a dilute, nearly collisionless plasma embedded into a strong magnetic field, on the other hand, we are in the opposite regime $|\omega_{ge}| \gg v_c$. It is instructive to first consider the Hall conductivity in this case. Writing it as $\sigma_H = \sigma_0(v_c/\omega_{ge})$ and inserting for σ_0 and the value of the electron-cyclotron frequency $\omega_{ge} = eB/m_e$ one finds the important formula for the *Hall conductivity in ideally conducting* plasma

$$|\sigma_H| = \frac{en_e}{B} \tag{4.29}$$

The Hall conductivity depends only on the ratio of density to magnetic field strength in this case, and the perpendicular motion of the current carrying electrons is the $E \times B$ drift. (For the interested reader we may note in passing that it is this insight which led to the discovery of the quantum Hall effect in two-dimensional non-plasma systems.) It is thus convenient to express the other components of the conductivity tensor through the Hall conductivity. We then have

$$\sigma_{\|} \equiv \sigma_0 = \sigma_H \frac{\omega_{ge}}{v_c} \gg \sigma_H, \qquad \sigma_P = \sigma_H \frac{v_c}{\omega_{ge}} \ll \sigma_H \tag{4.30}$$

Hence, in such a plasma the current flows essentially along the field lines and the perpendicular current flow is restricted mainly in the direction of the $E \times B$-drift, which is the direction of the Hall current.

An *ideally conducting* plasma thus carries field aligned and Hall currents with the Hall current being weaker than the field aligned current. In this case, the electrons will, on average, move at an angle with both the direction of the transverse electric field and the $E \times B$ direction. In the physical reality, however, there will always be some weak scattering of the electrons such that a small perpendicular conductivity remains. From very fundamental quantum mechanical considerations it can even be shown that the Pedersen conductivity cannot become larger than its upper limiting value

$$\sigma_{P,max} = \frac{m_e}{\mu_0 h} \approx 10^{-9} \ \frac{S}{m} \tag{4.31}$$

where h is Planck's quantum of action.

The conductivity is highly anisotropic for plasmas with $|\omega_{ge}| \approx v_c$ in which case one has $\sigma_0 = \varepsilon_0 \omega_{ge}$ and $\sigma_P = \sigma_H = \frac{1}{2}\sigma_0$ as seen from Fig. 4.4. The electrons are scattered about once per gyration. Hence, both $E \times B$ drift and motion along the transverse electric field are equally important and the Pedersen and Hall conductivities are of the same order. For $|\omega_{ge}| < v_c$ the Pedersen conductivity dominates, since in such a domain the electrons are scattered in the direction of the electric field before they can start to gyrate about the magnetic field. For $|\omega_{ge}| > v_c$ the electrons experience the

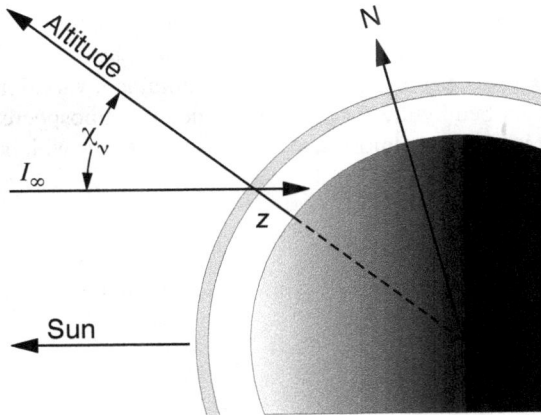

Figure 4.5: The angular dependence of the solar UV absorption in the ionosphere which here is shown as the thin spherical shell at distance z above Earth's surface. Absorption is restricted to the illuminated half-sphere and is maximum at the equator. This picture is drawn for north polar summer with the polar cap being partially illuminated.

E×B drift for many gyrocycles, before a collision occurs, and the Hall conductivity starts dominating unless it becomes the sole perpendicular conductivity as noted above for negligible collision frequencies.

4.3 Ionosphere Formation: The Chapman Layer

The *ionosphere* forms the base of the magnetospheric plasma environment of the Earth. It is the transition region from the fully ionized magnetospheric plasma to the neutral atmosphere. This implies that it consists of a mixture of plasma and neutral particles and will therefore have an electrical conductivity to which Coulomb and especially neutral collisions may contribute.

Before doing so, we need to know how the plasma density and the collision frequency varies in dependence of height, latitude, and time of day, and possibly also during times of magnetospheric disturbance. What interests us first is, how the plasma density depends on height and how the ionization of the ionosphere is created. Two main sources of ionization can be identified: the solar ultraviolet radiation and energetic particle precipitation from the magnetosphere into the atmosphere.

4.3.1 Solar Ultraviolet Ionization

In order to produce ionization, the solar photons must have energies higher than the ionization energy of the atmospheric atoms. Thus the photons should come from the ultraviolet spectral range or higher, but at higher frequencies (or photon energies) the solar radiation intensity becomes very weak and sporadic and is therefore unimportant when considering the average state of the ionosphere.

The ionosphere is horizontally structured. Its dominant variation occurs with altitude, z, and is prescribed by the variation of the neutral atmosphere density, $n_n(z)$. In a one-component isothermal atmosphere the density changes with height according to the *barometric law*

$$n_n(z) = n_0 \exp\left(-\frac{z}{H}\right) \tag{4.32}$$

H is the *scale height* for an isothermal atmosphere with atoms of mass m_n and temperature T_n. It is defined as

$$\boxed{H = \frac{k_B T_n}{m_n g}} \tag{4.33}$$

where g is the gravitational acceleration and n_0 is the atmospheric density at $z = 0$.

Solar ultraviolet radiation impinging onto the atmosphere at height z under an angle χ_v (see Fig. 4.5) hits atmospheric atoms of density n_n and looses its energy due to ionization. At this height the radiation is partially absorbed. The interaction of radiation with the atmospheric atoms takes place along the oblique ray path, i.e., along $z/\cos\chi_v$. The diminution of radiation intensity, I, with altitude z along the ray path element $dz/\cos\chi_v$ is given by

$$dI = \sigma_v n_n \frac{dz}{\cos\chi_v} I \tag{4.34}$$

where σ_v is the radiation absorption cross-section. The equation shows that the differential decrease in radiation intensity is proportional to the incident intensity, to the number density of absorbing neutral gas particles, to the absorption cross-section, and to the path length of the radiation in the atmosphere. Using the barometric law (4.32), one may integrate Eq. (4.34) to find the height variation of the intensity

$$\int_{I_\infty}^{I(z)} \frac{dI}{I} = \int_\infty^z \exp\left(-\frac{z}{H}\right) \frac{\sigma_v n_0}{\cos\chi_v} dz \tag{4.35}$$

where I_∞ is the solar flux at the top of the atmosphere. Solving for $I(z)$ yields

$$I(z) = I_\infty \exp\left[-\frac{\sigma_v n_0 H}{\cos\chi_v} \exp\left(-\frac{z}{H}\right)\right] \tag{4.36}$$

which shows the exponential decrease of the solar radiation intensity with decreasing altitude due to atmospheric absorption that is schematically plotted in Fig. 4.6.

The number of electron-ion pairs locally produced by the solar ultraviolet radiation, the *photoionization rate* per unit volume at a particular height, $q_v(z)$, is proportional to the absorbed fraction of radiation in the altitude interval dz and to the

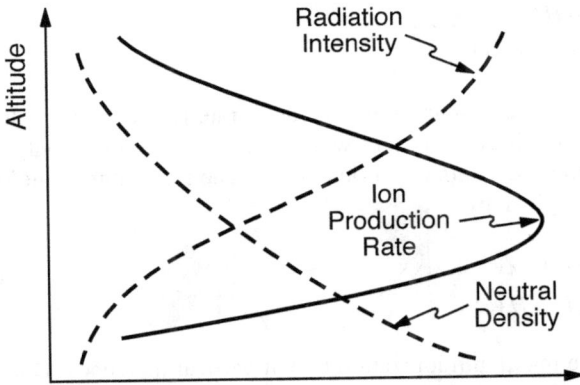

Figure 4.6: Formation of a Chapman ionization layer in Earth's upper atmosphere by the absorption of solar UV radiation in the atmosphere. Progressive absorption of the solar radiation in the exponentially increasing atmospheric density causes the exponential decay of intensity. The interplay of increasing absorption and decreasing radiation intensity with altitude causes a maximum of the ionisation at height z_m. The location of this maximum depends on local time, season as well as atmospheric conditions and the solar cycle.

photoionization efficiency, κ_v, the fraction of the absorbed radiation that goes into ionization

$$q_v(z) = \kappa_v \cos \chi_v dI/dz \tag{4.37}$$

Using Eq. (4.34) to replace the solar intensity gradient dI/dz through the intensity itself, one obtains

$$q_v(z) = \kappa_v \sigma_v n_n I \tag{4.38}$$

Equations (4.32) and (4.36) can be used to introduce the explicit height dependence of the neutral density and the ray intensity

$$q_v(z) = \kappa_v \sigma_v n_0 I_\infty \exp\left[-\frac{z}{H} - \frac{\sigma_v n_0 H}{\cos \chi_v} \exp\left(-\frac{z}{H}\right) \right] \tag{4.39}$$

This *Chapman production function* of ions can be given a simpler form

$$q_v(\zeta) = q_{v0} \exp\left[1 - \zeta - \frac{\exp(-\zeta)}{\cos \chi_v} \right] \tag{4.40}$$

To obtain this expression one considers the variation of the ionization with altitude. As shown in Fig. 4.6, the density decreases with height while the solar intensity increases. Thus it is clear that the ionization will have a pronounced maximum at a particular altitude z_m. The value of z_m is obtained setting the derivative of Eq. (4.39) to zero

$$z_m = z_0 + H \ln (\cos \chi_v)^{-1}$$
$$z_0 = H \ln (\sigma_v n_0 H) \tag{4.41}$$

Here z_0 is the altitude of maximum ionization rate for vertical incidence of the solar radiation ($\chi_v = 0$). It is a constant which depends only on gravity, ion mass, scale height, and ground level atmospheric density. The maximum value of the ionization rate, q_{vm}, is then given by

$$q_{vm} = q_{v0} \cos \chi_v$$
$$q_{v0} = \kappa_v I_\infty (eH)^{-1} \tag{4.42}$$

where q_{v0} is the maximum ionization rate at vertical incidence. One now introduces the new variable $\zeta = (z - z_0)H^{-1}$, inserts it into Eq. (4.39) and obtains Eq. (4.40).

Equation (4.40) must be evaluated numerically. The results are shown schematically in Fig. 4.6. One finds that the height of maximum ionization, z_m, is restricted to a narrow range of altitudes in the so-called Chapman layer (after Sydney Chapman who in 1931 in a Bakerian lecture to the Royal Society developed the above theory). Its position depends on χ_v in such a way that for smaller χ_v the maximum is found at lower altitudes. Moreover, the maximum weakens with increasing χ_v. Since χ_v is a function of geographic latitude and longitude, the photoionization layer in the ionosphere exhibits a strong dependence on geographic latitude, time of day, and season.

4.3.2   Ionization by Energetic Particles

In addition to photoionization, ionospheric ionization is produced in the regions where sufficiently energetic particles, in the first place electrons, impinge onto the atmosphere. Because such particles must follow the magnetic field lines, one naturally expects that this type of ionization will dominate at high magnetic latitudes in the auroral zone (see Section 1.2), where photoionization becomes less important. Also during nighttime, when photoionization ceases, ionization due to particle impact can maintain the ionosphere.

Ionization by electrons precipitating into the atmosphere along magnetic field lines from the magnetosphere is collisional. It requires electron energies $W_e > W_{ion}$, where W_{ion} is the ionization energy needed to extract an electron out of an atom or molecule. For oxygen atoms the ionization energy is about 35 eV. The *collisional ionization rate* per unit volume at a particular height, $q_e(z)$, is proportional to the energy loss, $dW_e(z)$, a precipitating electron will experience at this altitude. Hence, it is proportional to the product of ionization energy, W_{ion}, collisional ionization efficiency, κ_e, and the number of collisions per unit height, v_c/v_z, at this altitude. How the two latter are distributed over altitude requires precise knowledge of the altitude profile of atmospheric density and composition. Moreover, the number of collisions per unit height depends on the pitch angle of the precipitating electron, which determines how much time a particle spends at a given altitude.

However, one can get a fairly good idea on the altitude variation of the ionization rate changes if one considers a field-aligned electron beam impinging vertically onto a neutral atmosphere governed by a simple barometric law and assumes the ionization efficiency to be constant. For an electron moving purely vertical, the number of collisions per unit height is given by the inverse mean free path length defined in Eq. (4.2), yielding for the energy loss

$$dW_e(z) = \kappa_e W_{ion} \sigma_n n_n dz \tag{4.43}$$

The variation of the energy loss with altitude does not depend on the original energy of the precipitating particle. The collisional ionization rate in the height interval dz produced by a flux of precipitating electrons, F_e, is

$$q_e(z) = F_e \frac{dW_e}{dz} \tag{4.44}$$

Replacing the energy loss by Eq. (4.43) and the neutral density profile by the barometric law (4.32), we obtain

$$q_e(z) = \kappa_e F_e W_{ion} \sigma_n n_0 \exp\left(-\frac{z}{H}\right) \tag{4.45}$$

Accordingly, the height profile of the collisional ionization rate is simply determined by the altitude variation of the neutral density. Independent of particle energy, the ion production rate increases exponentially with decreasing altitude.

However, the energy of the precipitating electron enters when considering the lowest altitude reached. An electron with energy W_e can penetrate only down to a stopping height, z_s, where it will have lost all its energy by collisions. Naturally, more energetic electrons penetrate deeper into the atmosphere and produce more electron-ion pairs by collisions because more energy can be distributed. The stopping height can be calculated by integrating the energy loss

$$W_e = \int_0^{W_e} dW_e = \kappa_e W_{ion} \sigma_n n_0 \int_\infty^{z_s} \exp\left(-\frac{z}{H}\right) dz \tag{4.46}$$

Solving for z_s, one finds z_s logarithmically decreasing with electron energy W_e

$$z_s = z_{s0} - H \ln\left(\frac{W_e}{W_{ion}}\right) \quad \text{with} \quad z_{s0} \equiv H \ln(\kappa_e \sigma_n n_0 H) \tag{4.47}$$

showing that particles of higher energy reach lower altitudes. z_{s0} is the altitude of an electron of ionisation energy $W_e = W_{ion}$. In order to reach down to, say, a stopping altitude of z_{min} the energy of the electron should amount to

$$\frac{W_e(z_{min})}{W_{ion}} = \exp\left(\frac{z_{s0} - z_{min}}{H}\right) \tag{4.48}$$

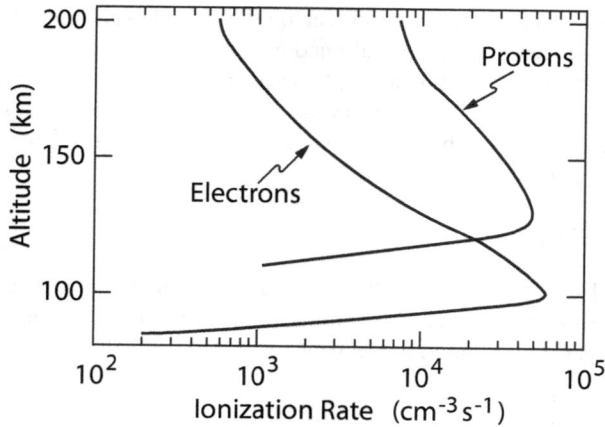

Figure 4.7: Total ion production due to precipitating electrons and protons. At altitudes above 120 km, the ionization is mainly caused by protons. Below that altitude, electrons contribute more to the ionization than protons. These processes are most important in the night time auroral ionosphere at the latitudes of the auroral oval where precipitation of energetic particles is strongest. Since precipitation is not a stationary process, the auroral ionospheric content is highly variable both in time and location.

The energy to reach a certain lowest altitude z_{min} thus increases exponentially with decreasing stopping height. The energy required to reach the lower atmosphere is of the order of $W_e(0) \approx W_{ion} \exp(z_{s0}/H)$ and is of the order of cosmic ray energies in the far GeV domain.

At the stopping altitude the electrons encounter the largest number of neutral particles and also deposit the largest fraction of their energy thus producing the largest fraction of electron-ion pairs. In a realistic atmosphere electrons of 300 keV energy penetrate down to about 70 km, while electrons of 1 keV energy are stopped at an altitude of about 150 km. Due to the exponential ionization rate profile, the ionization maximum is more pronounced for more energetic electrons.

Our rather crude model predicts the shape of real ionization rate profiles (see Fig. 4.7) quite well. Here, the ionization rate profile labelled 'electrons' was computed from rocket observations of precipitating electrons with energies of about 10 keV, using realistic profiles of ionization efficiency, atmospheric density, and composition. As also indicated in Fig. 4.7, ions are stopped at greater heights than electrons of the same energy, because their ionization efficiency is lower. The electrons are responsible for the ionization measured near 100 km height, while the precipitating ions contribute most to the collisional ionization at heights above 130 km.

4.3.3 Recombination and Attachment

The production of ionization in the ionosphere either by solar ultraviolet radiation or by energetic particles would, if it continued endlessly, lead to full ionization of the upper atmosphere. However, in reality two processes counteract the ionization and in equilibrium limit it to its observed values. These processes are the *recombination* of ions and electrons to reform neutral atoms and the *attachment* of electrons at neutral atoms or molecules to form negative ions. Formally these two processes can be described by two coefficients, α_r and β_r, the recombination and attachment coefficients, respectively. These coefficients determine how many electrons and ions per second recombine and how many electrons per second attach to neutral particles.

Because recombination and attachment are effective losses of ionization, they contribute negatively to ionization of the atmosphere. Recombination is proportional to both the number density of electrons and of ions which can recombine, while attachment is proportional only to the number of electrons available to attach to a neutral particle. Observing that in equilibrium the ionospheric plasma should be quasineutral, $n_i \approx n_e$, the continuity equation for the electron density becomes

$$\frac{dn_e}{dt} = q_{v,e} - \alpha_r n_e^2 - \beta_r n_e \tag{4.49}$$

The first term on the right-hand side is the ionization due to solar ultraviolet radiation and collisions with precipitating particles. It acts as the source of the electron density, while the two other terms are sinks of ionization. Recombination is proportional to the square of the electron density, while attachment is a linear effect. The coefficients α_r and β_r contain a number of complicated photochemical processes which are responsible for the differences of the ionospheric composition at different heights.

In equilibrium $dn_e/dt = 0$. Hence, letting the left-hand side of Eq. (4.49) vanish, one finds the equilibrium electron density of the ionosphere. At low altitudes, recombination dominates attachment, and with $\beta_r = 0$ we obtain

$$n_e = \left(\frac{q_{v,e}}{\alpha_r}\right)^{1/2} \tag{4.50}$$

The electron density in the lower ionosphere at equilibrium varies as the square root of the ratio between the ion production rate and the recombination coefficient. At higher altitudes attachment is the dominant loss process. Here we neglect the recombination term, finding that the upper ionospheric electron density is proportional to the ion production rate divided by the coefficient of attachment

$$n_e = \frac{q_{v,e}}{\beta_r} \tag{4.51}$$

4.3.4 Ionospheric Layers

The real structure of the Earth's ionosphere is not as simple as given by Eq. (4.40) and sketched in Fig. 4.6. The actual electron density profile is determined by the

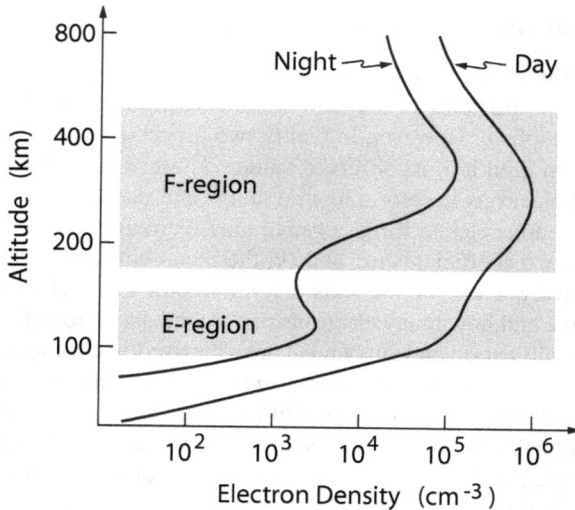

Figure 4.8: Vertical profiles of mid-latitude electron density day time and night time. The iono-sphere is roughly structured into two layers, the E layer with local maximum around 110 km altitude and the F layer with broad local maximum around 300 km altitude. This structure is quite clear during night time while it is smoothed in altitude during day time when the ion content of the ionosphere is increased by up to two orders of magnitude.

specific absorption properties of the gaseous constituents of the atmosphere, which due to the different barometric laws for different molecular components vary dras-tically. In addition, the altitude variation of the recombination and attachment coef-ficients has to be taken into account. A number of chemical reactions must also be considered when investigating the formation of the ionosphere. Those reactions are the subject of atmospheric physics and depend heavily on the presence of various minor-ity gases and metallic ions as well as on the dynamics of the atmosphere. Neither of them will be considered here. As a result of these processes, the electron density in the Earth's ionosphere exhibits roughly spoken three different layers.

The lower ionosphere below an altitude of about 90 km is called the *D-region*. It is very weakly ionized and due to high collision frequencies mostly dominated by the neutral gas dynamics and chemistry and cannot be considered a proper plasma even though it is very important for the coupling processes with the atmosphere, iono-spheric losses and transfer of momentum between the atmosphere and the ionosphere (see Section 1.1).

The upper ionosphere is the region above 90 km. It is highly but still partially ion-ized, containing a substantial contribution of neutral gas up to about 500 km altitude. The upper ionosphere consists of two well separated layers of ionized matter, the *E-region*, which has its ionization peak at about 110 km, and the *F-region* which peaks around 300 km altitude. Figure 4.8 shows height profiles of the ionospheric electron density during day and night hours in the mid-latitudes. The distinction between the two layers is obvious from the nightside profiles. During daytime the gap between the E- and F-region is partially filled due to the additional ionization produced during the day and also due to the stronger dynamics of the atmosphere which induces motions and mixing also in the ionospheric plasma.

The E-region is formed in a variant of the Chapman process by the absorption of longer wavelength ultraviolet radiation (approximately 90 nm) which passes the higher altitudes until the density of molecular oxygen becomes high below about 150 km height. Thus oxygen ions dominate the E-region. At higher latitudes ionization due to precipitating energetic electrons and protons contributes significantly to the formation of the E-region (see Fig. 4.9). The E-region is also the site where during meteorite impact onto the atmosphere metallic ions are deposited which lead to additional ionization and a particular irregular and highly variable structuring of the E-region called *sporadic* E which we will not describe in detail. Due to the presence of sometimes large amounts of metallic ions it is subject to chemical processes and thermodynamic effects.

The F-region splits into two layers, the *F1-region* at around 200 km, actually the layer that was proposed by Chapman in 1931, and the *F2-region* around 300 km height. The former is a dayside feature created in the same way as the E-region, but the absorbed ultraviolet wavelengths are shorter (20–80 nm) because of different absorbing molecules. The more important layer is the F2-region. Its formation is basically determined by the height variation of the neutral densities and the recombination and attachment rates for the different atmospheric constituents. In the lower F2-region ionization of atomic oxygen and recombination play the key roles. At greater altitudes the decreasing neutral density and attachment limit the increase in electron density. In this way the competition between ionization and attachment leads to the F2-region peak at roughly 300 km seen in Fig. 4.8. The F2-region peak contains the densest plasma in the Earth's environment, with electron densities up to 10^6 cm^{-3}.

4.4 Ionospheric Conductivity

In Section 4.2 we derived the conductivity tensor due to collisions between moving electrons and unspecified scatter centers at rest. For a partially ionized ionosphere, the collision partners are the neutral atmosphere particles and the general collision frequency ν_c in Eq. (4.27) is replaced by the *electron-neutral collision frequency*, ν_{en}. A fully general theory must include both types of collisions, those between ions and electrons and those of charged particles with neutrals. In the following we consider only the case when neutrals dominate but include collisions with both kinds of charged particles.

4.4.1 Conductivity Tensor

In the terrestrial ionosphere, not only the electrons are scattered by the neutrals but also the ions. Since the current caused by the finite *ion-neutral collision frequency*, ν_{in},

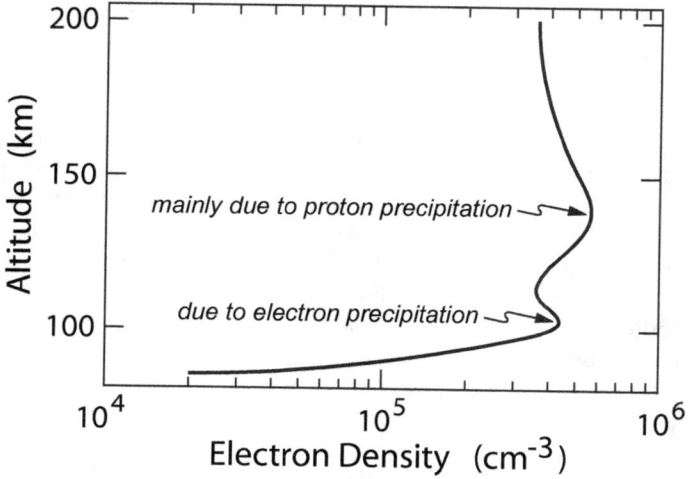

Figure 4.9: Vertical profile of electron density over an about stationary diffuse aurora exhibiting the double structure of the auroral ionosphere generated by the precipitation of the two kinds of charged particles.

is governed by the same equation as the current carried by the electrons, we can retain the generalized Ohm's law first given in Eq. (4.28)

$$\mathbf{j} = \sigma_\parallel \mathbf{E}_\parallel + \sigma_P \mathbf{E}_\perp - \sigma_H (\mathbf{E} \times \mathbf{B})/B \tag{4.52}$$

if we add the ion contribution to the electron conductivity tensor elements given in Eq. (4.27). The ion conductivities are simply found by replacing ω_{ge} and ν_{en} by ω_{gi} and ν_{en}. Since we have defined the cyclotron frequency as carrying the sign of the charge, the latter is automatically taken care of and we get

$$
\begin{aligned}
\sigma_P &= \left(\frac{\nu_{en}}{\nu_{en}^2 + \omega_{ge}^2} + \frac{m_e}{m_i} \frac{\nu_{in}}{\nu_{in}^2 + \omega_{gi}^2} \right) \frac{n_e e^2}{m_e} \\
\sigma_H &= \left(\frac{\omega_{ge}}{\nu_{en}^2 + \omega_{ge}^2} + \frac{m_e}{m_i} \frac{\omega_{gi}}{\nu_{in}^2 + \omega_{gi}^2} \right) \frac{n_e e^2}{m_e} \\
\sigma_\parallel &= \left(\frac{1}{\nu_{en}} + \frac{m_e}{m_i} \frac{1}{\nu_{in}} \right) \frac{n_e e^2}{m_e}
\end{aligned}
\tag{4.53}
$$

where we have used the simplified assumption that there is only one type of ions in the terrestrial ionosphere.

4.4.2 Conductivity Profile

Figure 4.10 shows typical altitude profiles of ion and electron cyclotron and collision frequencies in the E-region ionosphere at mid latitudes. In the narrow altitude

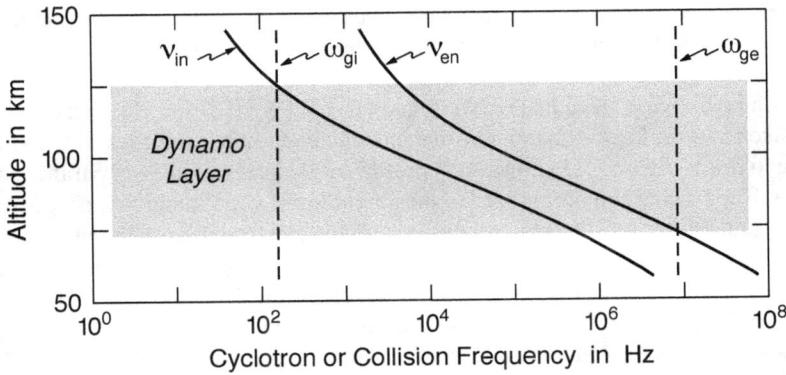

Figure 4.10: Cyclotron (vertical dashed lines) and collision frequencies in the mid-latitude ionosphere. The collision frequency between ions and neutrals v_{in} exceeds the ion cyclotron frequency ω_{gi} up to an altitude of $\sim$125 km. The ionospheric electrons become independent of the collisions with neutral gas already above $\sim$75 km where v_{en} drops below ω_{ge}. The shaded domain between these two altitudes (the dynamo layer) is the region where the drag of the neutral atmospheric winds on the ionized gas component in the ionosphere is large enough to drive so-called dynamo currents in the magnetic field of which generates a susceptible variation of the geomagnetic field. The electrons in this region are practically free accounting for charge neutrality but being frozen to the magnetic field which allows them to move only along the magnetic field lines.

range shown the value of the geomagnetic dipole field is about constant, and correspondingly the cyclotron frequencies are constant as indicated by the dashed vertical lines in Fig. 4.10 (note that the ion cyclotron frequency is governed by the heavy oxygen ions which dominate the E-region). The collision frequencies decrease across the E-region. Below altitudes of about 75 km the electron collision frequency exceeds the electron cyclotron frequency. Inside the shaded region in Fig. 4.10, the electron collision frequency is lower than the electron cyclotron frequency, but the ion collision frequency is still larger than the ion cyclotron frequency. Ions are therefore coupled to the neutral gas while electrons are partially decoupled.

Figure 4.10 explains how the Hall and Pedersen currents are generated and which are the primary charge carriers for these currents. Except for the lower bottom of the shaded region, where the density of the ionized component is too low to allow any appreciable current, the lower two thirds of the *dynamo layer* are governed by $v_{in} \gg \omega_{gi}$ and $v_{en} \ll \omega_{ge}$. Around 100 km the electrons can already do about a hundred gyrocycles before they collide with neutrals and experience a somewhat impeded $E \times B$ drift. The ions, on the other hand, still collide with neutrals about a hundred times per gyrocycle and thus move with the neutrals. Hence, a Hall current is carried by the electron motion transverse to the electric and magnetic fields. At the top of the dynamo layer, around 125 km altitude, the ion cyclotron and collision frequencies become comparable and the ions will not be perfectly coupled to the neutrals anymore. Instead they will move in the direction of the electric field and carry a Pedersen current, while the electrons still move at right angles to the fields and carry a Hall current. Above the E-region, both electrons and ions will undergo the same $E \times B$ drift and no current flows.

Using the cyclotron and collision frequencies in Fig. 4.10 and electron density profiles like those in Figs. 4.8 and 4.9, one can calculate height profiles of the E-region conductivities. Figure 4.11 shows such profiles of Pedersen, Hall, and parallel conductivities. They have been normalized to the maximum Hall conductivity, σ_{peak}, since the actual value of σ_{peak} scales about linearly with the electron density at the center of the E-region, while the height dependence is similar for the day and nightside ionosphere at all latitudes. The Pedersen conductivity peaks around 130 km and the Hall conductivity has its peak around 100 km altitude, in accordance with our above reasoning. The peak Hall conductivity is always larger than the peak Pedersen conductivity.

The actual peak value of the conductivity differs widely between the dayside low-latitude ionosphere, where $n_e \approx 2 \times 10^5$ cm^{-3} and $\sigma_{peak} \approx 10^{-3}$ S/m, and the nightside ionosphere, where $n_e \approx 2 \times 10^3$ cm^{-3} and $\sigma_{peak} \approx 10^{-5}$ S/m.

In the auroral zone, the electron density caused by energetic electrons precipitating from the magnetosphere may be even higher, up to $10^6 - 10^7$ cm^{-3}, and the peak Hall conductivity can reach values of more than 10^{-2} S/m. The peak Pedersen conductivity is typically half the peak Hall conductivity. The parallel conductivity is much higher than the other two. In the E-region it may reach 10^2 S/m and in the F-region and above it approaches the conductivity of a fully ionized plasma and can usually be taken as infinite. However, because electron precipitation is not a stationary effect but depends strongly on the conditions of disturbance in the magnetosphere, the auroral conductivities are highly variable both in time and location.

4.5   Ionospheric Currents

We have already mentioned in the previous section that the ions and, to a lesser degree, also the electrons in the E-region are coupled to the neutral components of the atmosphere and follow their dynamics. Atmospheric winds and tidal oscillations of the atmosphere force the E-region ion component to move across the magnetic field lines, while the electrons move much slower at right angles to both the field and the neutral wind. The relative movement constitutes an electric current and the separation of charge produces an electric field, which in turn affects the current. Because of this the E-region bears the name dynamo layer, the generator of which is the atmospheric wind motion.

To see the relation between current, conductivity, electric field, and neutral wind velocity, we have to write down Ohm's law. As in Section 4.2, we have to add a $\mathbf{v}_n \times \mathbf{B}$ term to the Ohm's law given in Eq. (4.25)

$$\mathbf{j} = \sigma \cdot (\mathbf{E} + \mathbf{v}_n \times \mathbf{B}) \tag{4.54}$$

This relation is valid throughout the Earth's ionosphere. For mid- and low-latitude dynamo currents, the dominant driving force for the current is the $v \times B$ field induced

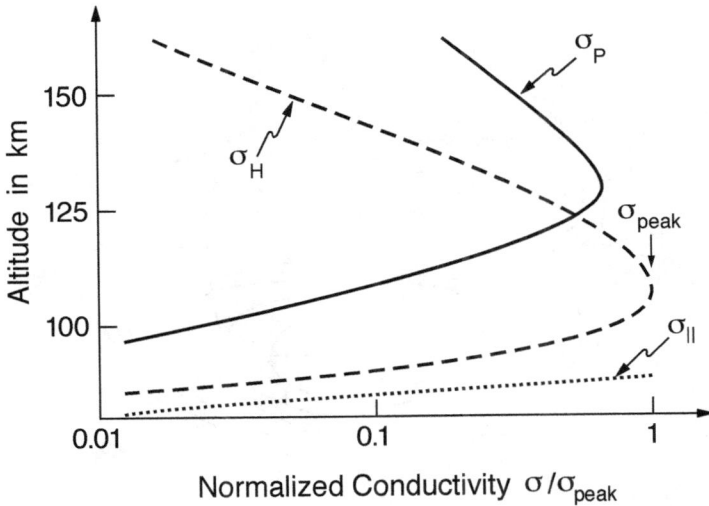

Figure 4.11: Height profiles of normalized conductivities. The three conductivities $\sigma_P, \sigma_H, \sigma_\|$ in this figure have been normalized to the peak of the Hall conductivity (the dashed curve). Note the strong altitude variation of the conductivities. In the E-region $\sigma_\|$ dominates all other conductivities, but of the perpendicular conductivities σ_H is largest at low altitudes around 100–110 km while above 125 km σ_P takes over.

by the ion motion across the magnetic field. For auroral oval current systems at higher latitudes, on the other hand, the neutral wind term is usually much smaller than the electric field term and can safely be neglected. In the present section we will study two current systems confined to the mid and low-latitude ionosphere. The auroral oval current systems will be described later in Sections 5.4 through 5.7, after we have discussed their external driving force, the convection electric field.

4.5.1 Sq Current

The most important dynamo effect at mid-latitudes is the daily variation of the atmospheric motion caused by the tides of the atmosphere. The tides with the lowest and the largest amplitudes are the diurnal and semi-diurnal oscillations which are excited by the solar radiation heating of the atmosphere. The current system created by this tidal motion of the atmosphere is called solar quiet or *Sq current*.

The *Sq* currents create a magnetic field disturbance, which can be measured on the ground by magnetometers and permits to determine the extent of the currents. Records of these daily magnetic variations obtained at many different stations distributed across the globe can be used to construct the *Sq* current system, using Biot-Savart's law and methods of potential theory. More sophisticated methods use measured wind patterns, conductivities, and disturbance magnetic fields and calculate electric fields and currents based on Eq. (4.54) and Biot-Savart's law (see Appendix B.5).

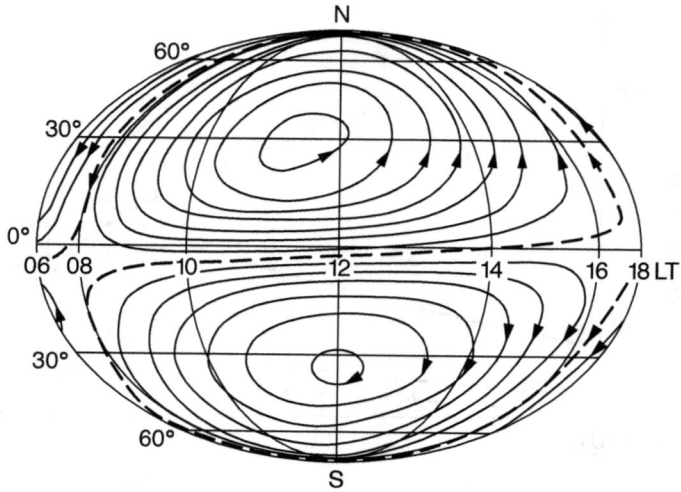

Figure 4.12: Global view of the average Sq current system which is driven by the ionospheric dynamo, the driving force of which is neutral atmospheric, mainly tidal wind system in the ionospheric dynamo layer. The Sq currents form two vortices in the hemispheres with their centres at mid latitudes. The currents of these vortices touch at the equator where they both flow in east-west direction from morning to evening forming a concentrated equatorial jet current. This equatorial electrojet is amplified by the enhanced equatorial conductivity, the Cowling conductivity.

Figure 4.12 presents a global view of the average Sq current system as seen from above the terrestrial ionosphere. In this figure, the lines give the direction of the current while the distance between the lines is inversely proportional to the height-integrated current density. The Sq currents form two vortices, one in the northern and the other in the southern hemisphere, which touch each other at the geomagnetic equator. In accordance with the day-night contrast in the low and mid-latitude E-region electron densities (see Fig. 4.8), the Sq currents are concentrated in the dayside region.

4.5.2   Equatorial Electrojet

At the geomagnetic equator, the Sq current systems of the southern and northern hemispheres touch each other and form an extended nearly jet-like current in the ionosphere, the *equatorial electrojet*. However, the electrojet would not be so strong as it is if it were formed only by the concentration of the Sq current. The special geometry of the magnetic field at the equator together with the nearly perpendicular incidence of solar radiation cause an equatorial enhancement in the effective conductivity which leads to an amplification of the jet current.

To see this combined action consider a situation where the magnetic field is about horizontal to the Earth's surface as is the case at the equator. The direction of the magnetic field is from south to north, along the x axis. The primary Sq Pedersen current flows eastward in the y direction, parallel to the primary ionospheric electric field, E_{py}. As sketched in Fig. 4.13, this primary electric field drives a Hall current which flows vertically downward in the z direction, causing a charge separation in the equatorial ionosphere with negative charges accumulating on the top boundary and positive charges accumulating at the bottom of the highly conducting layer. This space charge distribution creates a secondary *polarization electric field*, E_{sz}, vertically directed from the bottom to the top of the conducting ionosphere. The polarization electric field drives a vertical Pedersen current opposing the Hall current until it compensates it. The resulting equilibrium condition in which no vertical current flows is

$$j_z = \sigma_H E_{py} + \sigma_P E_{sz} = 0 \tag{4.55}$$

which yields for the secondary vertical electric field

$$E_{sz} = -\frac{\sigma_H}{\sigma_P} E_{py} \tag{4.56}$$

In addition, the secondary polarization electric field component generates a secondary Hall current component flowing into the y direction

$$j_{sy} = -\sigma_H E_{sz} = \frac{\sigma_H^2}{\sigma_P} E_{py} \tag{4.57}$$

The total current into the eastward direction consists of the sum of the primary Pedersen and the secondary Hall current

$$j_y = j_{py} + j_{sy} = \left(\sigma_P + \frac{\sigma_H^2}{\sigma_P} \right) E_{py} \tag{4.58}$$

The conductivity term appearing on the right-hand side is called *Cowling conductivity*

$$\boxed{\sigma_C = \sigma_P + \frac{\sigma_H^2}{\sigma_P}} \tag{4.59}$$

For typical Hall-to-Pedersen conductivity ratios of 3–4 the Cowling conductivity is an order of magnitude higher than the Pedersen conductivity, explaining the amplification and concentration of the equatorial electrojet current above the equator. The strong horizontal jet current causes a magnetic field disturbance which weakens the horizontal terrestrial magnetic field at the Earth's surface over a distance of about 600 km across the equator (similar to the effect of the ring current field; see Section 3.6). Typical disturbance fields near the noon magnetic equator are of the order of 50–100 nT.

4.6 Auroral Emissions

The Sq current vortices in mid-latitudes and the equatorial electrojet are only part of the ionospheric current system. There are similar ionospheric current systems in the

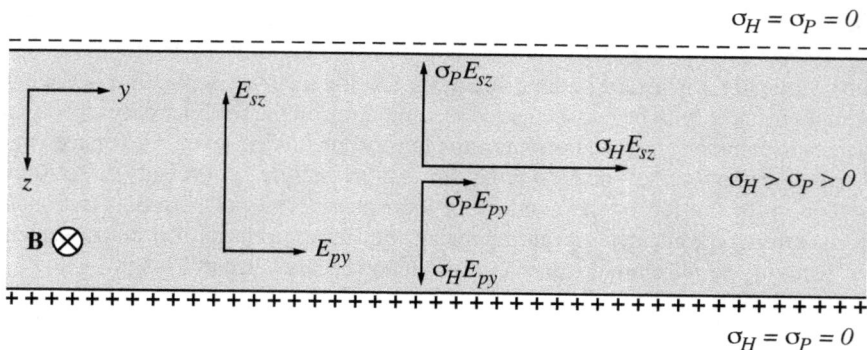

Figure 4.13: Enhancement of the effective conductivity at the magnetic equator due to the alignment of the primary Pedersen and secondary Hall currents which both add to the equatorial jet current. The Sq jet current is thereby amplified to become the equatorial jet.

polar cap and in the auroral region which can flow in the ionosphere because of the finite ionospheric conductivities in these regions.

As we have shown, these high latitudinal conductivities are caused only to a limited extent by the absorption of solar light. During polar night the polar cap is for long time not illuminated at all, and there is practically no optical ionisation during this long period of polar winter darkness.

Also in the auroral zone illumination is of secondary importance for the generation of finite conductivities. The main reason for a high ionospheric conductivity in the auroral region in particular is the precipitation of energetic particles from the magnetosphere which is depositing charges in the ionosphere as shown in Fig. 4.9 during a diffuse aurora, the least variable form of an auroral event.

Usually, this precipitation is, however, rather variable and locally limited causing the conductivity to fluctuate very strongly both in time and in space. As a consequence the auroral current system is as well highly variable and spatially irregular. We will discuss the related effects in the next chapter in connection with the presentation of the auroral electrodynamics. In the polar cap, on the other hand currents are mainly cause by convective motion in the lobes of the magnetosphere. They are, to say it metaphorically, the signatures of the moving foot points of the magnetospheric lobe field lines which are advected with the general magnetospheric convective motion.

Precipitating electrons do not only produce ionization (see Section 4.3). At higher latitudes, where the geomagnetic field lines map to the plasma sheet, collisions between precipitating energetic electrons and neutral atmosphere ions also produce radiation in the visible wavelength range. This light has been named *aurora* and can be seen from the ground with the naked eye. Intense aurora has an *emission rate* of several million Rayleigh (1 R = 10^6 photons/cm^{2}s). The aurora is visible at latitudes of about 70° all around the globe inside a belt called *auroral oval* (see Section 1.2 and Fig. 1.5). However, the aurora is not a kind of continuous glow inside this belt, but organized into distinctive structures or auroral forms. Most often it appears as shown in the 180°

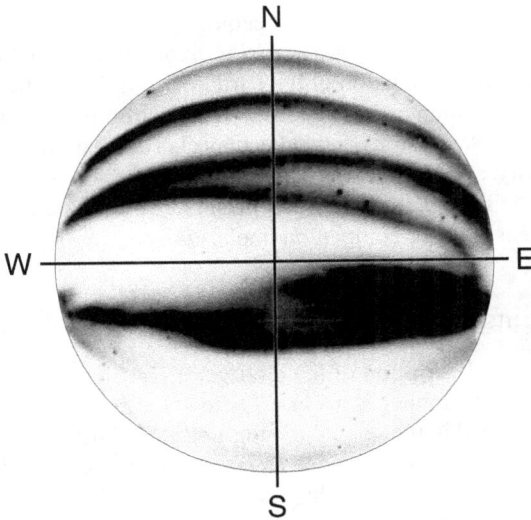

Figure 4.14: Auroral arcs over Kiruna, Sweden.

all-sky camera picture displayed in Fig. 4.14, namely as a couple of thin band-like structures which are aligned with the east-west direction and called *auroral arcs*.

The auroral emissions are the result of processes where neutral atoms or molecules are excited by collisions with precipitating electrons and fall back into their ground state by emitting photons. Often the excitation is accompanied by ionization. The commonly observed green color in the aurora is due to the *auroral green line* of atomic oxygen at 557.7 nm, typically observed at altitudes between 100 and 200 km. At higher altitudes one may observe the *auroral red line* of atomic oxygen at 630.0 nm. Both lines represent so-called forbidden lines. The excited energy states are relatively long-lived (metastable) and under normal atmospheric pressure the excited atom would loose its energy by collisions with other particles rather than by light emission.

Molecular nitrogen produces some weaker violet or blue auroral emission lines at 391.4, 427.0 and 470.0 nm. In the ultraviolet range one finds molecular nitrogen emission lines (around 150 nm) and an atomic oxygen line at 130.4 nm. Auroras also have infrared emissions, for example the molecular oxygen lines at 1270 and 1580 nm.

4.7 Concluding Remarks

Except for the partially ionized ionosphere where collisions with neutrals play an important role, most geophysical plasmas are fully ionized and collisionless, in the sense that the Coulomb collision frequency is much lower than the plasma frequency. However, in these plasmas collective interactions, in which the self-generated fields of the particles take over a correlative role in scattering the electrons, may lead to *anomalous collisions*. The plasma may become collisional again, with the collision frequency replaced by an *anomalous collision frequency*, ν_{an}, which must be calculated on the basis of the interaction of the particles with the electric field fluctuations. A good example for anomalous collisions is the electron pitch angle scattering mentioned in Section 3.4. Anomalous collisions may also decrease the near-infinite normal plasma conductivity and thus lead to *anomalous resistivity*, which can have a critical influence on the interrelation between plasma and fields in certain regions of the Earth's environment (see our companion book, *Advanced Space Plasma Physics*).

Further Reading

A more complete description of the physics of collisions is found in the first and especially the last of the monographs listed below. Further material on the physics of ionospheres is contained in the second monograph. The third monograph gives a comprehensive treatment of the phenomena associated with the optical aurora, including collisional ionization. The general physics of the ionosphere is treated in the fourth and fifths monograph with different emphases on low and middle and high latitudes, respectively.

References

[1] J. A. Bittencourt, *Fundamentals of Plasma Physics* (Pergamon Press, Oxford, 1986).

[2] J. K. Hargreaves, *The Solar-Terrestrial Environment* (Cambridge University Press, Cambridge, 1992).

[3] A. V. Jones, *Aurora* (D. Reidel Publ. Co., Dordrecht, 1974).

[4] M. C. Kelley and R. A. Heelis, *The Earth's Ionosphere: Plasma Physics and Electrodynamics* (Academic Press, New York, 1989).

[5] J. A. Rishbeth and O. K. Garriot, *Introduction to Ionospheric Physics* (Academic Press, New York, 1969).

[6] L. Spitzer, *Physics of Fully Ionized Gases* (Interscience Publishers, New York, 1962).

Problems

Problem 4.1 *Give a physical explanation of the definition of the collision frequency* ν_n, *the mean free path* λ_n *and the cross section* σ_n. *Make some drawings which show the meaning of each of these quantities.*

Problem 4.2 *Solve the electron equation of motion when the electron is coming with velocity* v_x *from* $x = -\infty$ *and collides with an ion of mass* $m_i \gg m_e$ *at* $x = 0$. *Does the electron hit the ion?*

Problem 4.3 *Under which condition would the ion trap the electron and what orbit wold the electron classically perform after trapping?*

Problem 4.4 *Derive the average collisional cross section for the case of the former scattering problem. Keep in mind that most electrons are scattered at small angles. Hint: Find the angular distribution of the electron scattering and average over the angle.*

Problem 4.5 *Write the expression for the cross section and collision frequency down when the electron has its full velocity v_e and not its thermal velocity $\langle v_e \rangle$. What do you conclude from the expression you obtain?*

Problem 4.6 *Find the expressions for the cross sections of electron-electron and ion-ion collisions and compare them with the cross section for electron-ion collisions given in the text.*

Problem 4.7 *Which assumptions have to be made in order to be able to derive an expression for the resistivity in Ohm's law of an unmagnetised plasma?*

Problem 4.8 *Show explicitly that the Hall current flows perpendicular to the magnetic field and the electric field. How does this current close? What does the negative sign mean?*

Problem 4.9 *Derive the dyadic representation of Ohm's law in a magnetised plasma.*

Problem 4.10 *Prove that in a strong magnetic field the Hall conductivity is finite while the Pedersen conductivity vanishes.*

Problem 4.11 *Derive the scale height H for an atmospheric gas in the Earth's gravitational field for oxygen and nitrogen. What is the difference?*

Problem 4.12 *Prove the expression for the differential absorption dI of the UV light for the spherical Earth as function of height dz.*

Problem 4.13 *Show that no stationary ionosphere would exist without recombination and attachment.*

Problem 4.14 *Give a derivation of the Cowling conductivity. Take into account that the geomagnetic field is strictly tangential only above the equator. At what latitude away from the equator is the expression for the Cowling conductivity becoming obsolete?*

— 5 —

Convection and Substorms

In Chapter 3 we found that energetic particles move across magnetic field lines under the influence of magnetic gradient and curvature forces. On the other hand, these forces are proportional to the particle energy and cold plasma particles with near-zero energy do not feel them because their energy is too low. They perform only an $E \times B$ drift in the crossed electric and magnetic fields. In the absence of external electric fields, cold particles do not drift at all but stay close to the field line they gyrate about. But that does not mean that the cold plasma in the Earth's magnetosphere is stagnant.

Actually, the cold magnetospheric plasma and the magnetic field lines move together and circulate under the influence of two external forces which are applied from the outside to the magnetosphere. The first of these forces is provided by the solar wind whose kinetic energy is the major energy source for the circulation of plasma and field lines in the outer magnetosphere. This force drives the general *magnetospheric convection*. The second force exerted onto the magnetospheric plasma is provided by the daily rotation of the Earth which, in the inner magnetosphere, forces the plasma into *corotation* on circular orbits around the Earth.

5.1 Diffusion and Frozen Flux

Cold plasma particles in the collisionless magnetospheric plasma are bound to a specific field line — more precisely, a magnetic flux tube with radius of the cold plasma gyro radius — which they cannot leave unless collective effects like the anomalous collisions mentioned on p. 85 violate the adiabatic invariants, i.e. the magnetic moment of the particles in the case of a cold plasma. Even the more energetic particles will stay with the magnetic field line they gyrate about in regions without strong magnetic field gradients or curvature when the gradient and curvature drifts are absent, as long as the applied electric field does not vary in time or space and no polarisation drifts are created. This has an important consequence for the plasma. Whenever a field line moves due to the action of external forces, the cold plasma tied to that field line is also set into motion together with the field line. The same holds for a moving plasma. Since it cannot leave the magnetic flux tubes it is sitting on, any moving cold plasma transports its internal magnetic field along with it. Hence, the motions of the plasma and of the associated flux tubes are intimately related and cannot be separated. Speaking of cold plasma flow implies that the embedded magnetic field is flowing together with the plasma.

In order to study the transport of field lines and plasma more quantitatively, we may use the generalized Ohm's law (4.19) to eliminate the electric field in Faraday's law (2.4), obtaining

$$\frac{\partial \mathbf{B}}{\partial t} = \nabla \times \left(\mathbf{v} \times \mathbf{B} - \frac{\mathbf{j}}{\sigma_0} \right) \tag{5.1}$$

Using the quasi-stationary Ampère's law (2.3) dropping the $\partial \mathbf{E}/\partial t$ term and noting that $\nabla \cdot \mathbf{B} = 0$, we get a general induction equation for the magnetic field (for the differential vector algebra see Appendix B.4)

$$\frac{\partial \mathbf{B}}{\partial t} = \nabla \times (\mathbf{v} \times \mathbf{B}) + \frac{1}{\mu_0 \sigma_0} \nabla^2 \mathbf{B} \tag{5.2}$$

where σ_0 is the (isotropic) plasma conductivity due to Coulomb or neutral collisions defined in Eq. (4.20). The physical content of this general magnetic *induction equation* is that the magnetic field at a point in the plasma can be changed by either the motion of the plasma with velocity $\mathbf{v}$, which is described in the first term on the right-hand side, or it can also be changed by diffusion due to the second term on the right-hand side.

5.1.1 Magnetic Diffusion

Assuming the plasma to be at rest and dropping the first term on the right-hand side, Eq. (5.2) becomes a diffusion equation now not for the particles or the fluid but instead for the magnetic field

$$\frac{\partial \mathbf{B}}{\partial t} = D_m \nabla^2 \mathbf{B} \tag{5.3}$$

The constant D_m on the right-hand side of this equation has dimension Length^2/Time and is identified as the *magnetic diffusion coefficient*. It is given by

$$\boxed{D_m = \frac{\eta_0}{\mu_0}, \qquad [D_m] = \frac{\mathrm{m}^2}{\mathrm{s}}} \tag{5.4}$$

Under the influence of a finite resistance η_0 in the plasma the magnetic field tends to diffuse across the plasma and to smooth out any local magnetic inhomogeneities or to release magnetic stresses.

Consider the sketch in Fig. 5.1, where we start out with magnetic field lines confined to small regions of space at time t_1. With time the field lines expand away in a diffusive manner, moving through the plasma. When field lines from different regions come to overlap, at a later time t_2, they must be added together vectorially, thereby possibly changing the topological structure of the magnetic field.

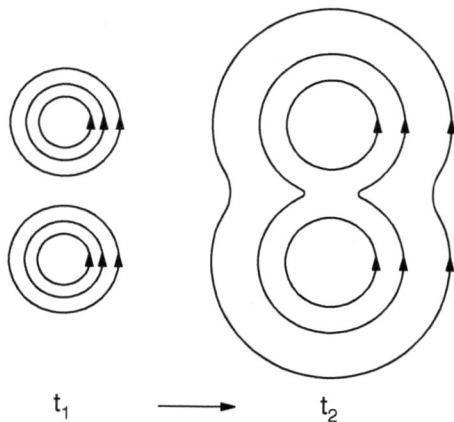

Figure 5.1: Diffusion of magnetic field lines. Two magnetic flux tubes which originally at time t_1 had been separated diffuse across the plasma until at time t_2 merging into one single flux tube.

The characteristic time of the magnetic field diffusion is found by replacing the vector derivative by the inverse of the characteristic gradient of the magnetic field, L_B. Then the local solution of the diffusion equation is

$$B = B_0 \exp(\pm t/\tau_d) \tag{5.5}$$

where the magnetic *diffusion time* τ_d is given by

$$\boxed{\tau_d = \mu_0 \sigma_0 L_B^2} \tag{5.6}$$

Whenever either $\sigma_0 \to \infty$ or when the characteristic length, L_B, is very large, the decay or diffusion time can become extremely long, in which case the magnetic field will not able to diffuse efficiently across the plasma over the length L_B.

Such situations may be realized in many geophysical plasmas where the conductivities are high while at the same time the characteristic length scales are huge. Sometimes the combination of both yields diffusion times longer than the age of the object or even longer than the age of the universe. It is then clear that the magnetic field does not efficiently diffuse across the plasma.

For an example we consider the solar wind. Its density is of the order of 5 cm^{-3} and its electron temperature is about 50 eV. Since there are no neutral particles, the sole collision frequencies entering the problem are Coulomb collisions between electrons and protons. Using Eqs. (4.18) and (5.6) permits us to estimate the magnetic diffusion time as $\tau_d \approx 0.3 L_B^2$ (if τ_d is given in seconds and L_B is given in meters). The time the solar wind needs to flow with its typical velocity of 500 km/s across the Sun-Earth distance of 1.5×10^{11} m is $\tau_{sw} \approx 3 \times 10^5$ s or 3.5 days. Setting this transit time equal to τ_d, we find that the magnetic field is permitted to diffuse across the solar wind over

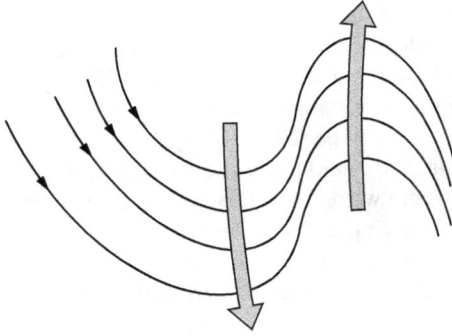

Figure 5.2: Magnetic field lines moving with the plasma.

the very short distance of only $L_B \approx 1.9 \sqrt{\tau_{sw}} \approx 10^3$ m during the 3.5 days it needs to travel across the Sun-Earth distance. Hence, the magnetic field is practically frozen in the solar wind and carried along with the particle stream (see Section 8.1.3).

The situation changes in the lower E-region where the collision frequency of charged particles and neutrals is high due to the presence of a dense neutral atmosphere (see Fig. 4.10) and the conductivities are of the order of 10^{-3} S/m. Here $\tau_d \approx 10^{-9} L_B^2$, and structures with widths of the order of 10 km become diffusive in times of the order of 1 s. All smaller structures will readily become diffusive, implying that on these temporal and spatial scales the magnetic field can slip through the plasma and vice versa. Narrow structures persisting for longer than a second are not magnetized any more and may move independent of the magnetic field.

5.1.2 Hydromagnetic Theorem

In cases where the magnetic diffusion is negligible one speaks about *frozen-in magnetic flux*. We will discuss the meaning of this term by considering the other limiting case where the plasma is in motion but has negligible electrical resistance. In collisionless plasmas with infinite conductivity Eq. (5.2) reduces to

$$\frac{\partial \mathbf{B}}{\partial t} = \nabla \times (\mathbf{v} \times \mathbf{B}) \tag{5.7}$$

When $\mathbf{B}$ is replaced with fluid vorticity ϖ, this equation for the magnetic field is recognised as being identical with the equation for the vorticity in the theory of non-viscous fluids. In that theory it is interpreted as implying that any vortex lines in the fluid will move with the fluid velocity $\mathbf{v}$ downstream. Equivalently, Eq. (5.7) implies that any field changes are such as if the magnetic field lines are constrained to move with the plasma. For example, if patches of plasma populating different sections of a bundle of field lines move into different directions, the field lines will be deformed in the manner shown in Fig. 5.2.

In fact, this intuitive and obvious fact can be proved by a lengthy mathematical calculation which shows that Eq. (5.7) implies that, in the absence of any diffusive process and conservation of the first adiabatic invariant (i.e. conservation of the magnetic moment μ) in an ideally conducting plasma the total magnetic induction field **B** encircled by a closed loop remains unchanged even if each point on this closed loop moves with a different local plasma velocity **v**. The field lines are called to be *frozen* to the plasma and can actually be identified by the plasma volume that is glued to it because plasma and magnetic field cannot be separated. We will call these identifiable field lines *flux tubes*, where we define a flux tube to be a volume bounded by a surface which is generated by moving a closed loop sliding it parallel to the magnetic field lines **B** it embraces at a given time. In this sense, a flux tube is a macroscopic entity filled with a large number of particles of gyro-radii much smaller than the flux tube radius.

Thus a macroscopic flux tube is kind of a generalised cylinder of cross section A containing a constant amount of magnetic flux which is a scalar defined as the integral

$$\Phi = \int_A \mathbf{B} \cdot d\mathbf{A} \tag{5.8}$$

taken over the cross section A perpendicular to the magnetic field **B**, i.e. adding up all magnetic field lines in the cross section (see also Appendix B.5). Clearly, since the magnetic field does not change with time one also has $d\Phi/dt = 0$, and the flux across A and thus in the flux tube is conserved even if the volume is transported across space by the plasma at E×B velocity. A much more fundamental definition of a flux tube requires a quantum-mechanical treatment which shows that the magnetic flux is quantised with quantum $\Phi_B = 2\pi\hbar/e$. Such flux quanta define magnetic field lines of diameter $d_B = \sqrt{8\hbar/eB}$ inversely proportional to the magnetic field strength. The stronger the field, the narrower a field line. Even a cold-particle gyro-circle contains a huge number of such field-line flux-tubes.

The frozen-in concept implies that all particles and all magnetic flux contained in a certain flux tube at a certain instant will stay inside the flux tube at all later instants, independent from any motion of the flux tube or any change in the form of its bounding surface as a simple consequence of the fact that the number of field lines inside the flux tube does not change as long as the magnetic moments of the particles on these field lines are conserved.

Due to the analogy with hydrodynamics, Eq. (5.7) is usually called the *hydromagnetic theorem*. One also finds the name *frozen-in flux theorem* and often Eq. (5.7) is represented by its equivalent

$$\mathbf{E} + \mathbf{v} \times \mathbf{B} = 0 \tag{5.9}$$

where we used Faraday's law (2.4) to replace $\partial\mathbf{B}/\partial t$. Equation (5.9) shows that in an infinitely conducting plasma there are no perpendicular electric fields in the frame moving with the plasma. Perpendicular electric fields can only result from a Lorentz transformation to a system moving at relative velocity to the plasma (see Section 2.3).

Moreover, Eq. (5.9) contains another important point. Since the cross-product between any velocity component parallel to the magnetic field and the field itself is zero, we can immediately see that any component of the electric field parallel to the magnetic field must vanish in a plasma with isotropic infinite conductivity. Parallel electric fields are shortened out in this case, and all magnetic field lines become lines of constant electric potential. Such parallel electric fields can be maintained in a collisionless plasma only when other forces act on the plasma which either induce charge separation or cause inductive magnetic fields. All such processes are highly time variable and can exist only locally.

5.1.3 Magnetic Reynolds Number

The term frozen-in can be given a more precise definition. The induction Eq. (5.2) can be rewritten in simple dimensional form as

$$\frac{B}{\tau} = \frac{VB}{L_B} + \frac{B}{\tau_d} \tag{5.10}$$

In this equation B is the average magnetic field strength and V represents the average plasma velocity perpendicular to the field, while τ denotes the characteristic time of magnetic field variations, and L_B is again the characteristic length over which the field varies. The second term on the right-hand side describes the diffusion of the magnetic field through the medium, while the first term has the form of a convective derivative, which describes the convective motion of the field with the plasma. The ratio of the first and second term yields the so-called *magnetic Reynolds number*

$$\boxed{R_m = \mu_0 \sigma_0 L_B V} \tag{5.11}$$

This number is very useful in deciding if a medium is dominated by diffusion or flow. In particular, when $R_m \gg 1$ the diffusion term in the induction equation can be entirely neglected. In this case the flow dominates, and the magnetic field simply moves together with the flow: it is frozen-in into the flow as we have described above. For example, the solar wind magnetic Reynolds number is about $R_m \approx 7 \times 10^{16}$, indeed very much larger than one and justifying our previous estimate about the negligible diffusion of the magnetic field in the solar wind. Of course, only the perpendicular velocity enters the frozen-in convective term. Any plasma flow parallel to the magnetic field has no consequences. On the other hand, when $R_m \approx 1$, diffusion becomes important and may dominate. The magnetic field is not any more frozen into the plasma and may slip across the plasma. In a completely diffusion dominated region in particular, the plasma can freely stream across the magnetic field without any remarkable effect on the latter, a situation realised in the D-region of the lower ionosphere.

The magnetic Reynolds number R_m is proportional to the conductivity, the length scale, and the velocity. Increase in any of these quantities therefore necessarily leads to the dominance of frozen-in conditions. In most large-scale and dilute plasmas of natural origin one therefore finds that the magnetic fields are frozen-in for most of the time scales of interest. Reasonable plasma flow velocities are normally restricted to low values. So the dominant quantities determining the frozen-in behaviour are the

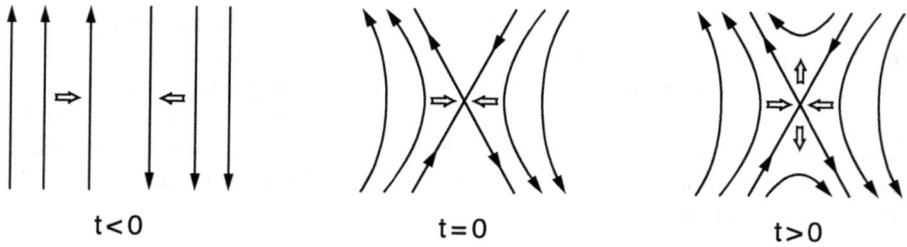

Figure 5.3: Evolution of the magnetic field topology in the process of merging of antiparallel magnetic field lines embedded into two approaching plasma streams (oppositely directed open horizontal arrows) which at time $t < 0$ have been separated, encounter each other at time $t = 0$ and have evolved a final state at later times $t > 0$ into an X-fan configuration in which the inflow from both sides is balanced by a perpendicularly directed outflow (open vertical arrows).

plasma conductivities and length scales of interest, the latter effectively being the only 'free parameter' which is determined by the scales of the problem.

5.1.4   Magnetic Merging

There is one particular situation where both the frozen-in flux concept and its breakdown are equally important. This is the process of *magnetic merging*, where field lines are cut and reconnected to other field lines, thus changing the magnetic topology. This process is theoretically quite complicated and will be detailed in a later chapter. Here, we will deal only with the basic structure of this process as illustrated in Fig. 5.3.

Consider a magnetic topology with antiparallel field lines frozen into the plasma, like sketched in the left-hand diagram of Fig. 5.3. Such a topology exists around thin current sheets like at the magnetopause and in the tail neutral sheet (see Fig. 1.6). If the field lines on both sides of the current sheet are stagnant and do not move, such a topology may be quite stable over long times.

However, when plasma and field lines on both sides move toward the current sheet, the situation may change. Whenever the magnetic Reynolds number becomes equal or greater than one in even a small volume of space due to anomalous collisions (see p. 85), the magnetic field may vanish due to diffusion at a particular point. This results in the X-type configuration shown in the middle panel of Fig. 5.3, with the magnetic field being zero at the center of the X, the magnetic *neutral point*. The two kinked field lines forming the X and passing through the neutral point are each called *separatrix*.

As a result we will get the situation sketched on the right-hand side of Fig. 5.3. Plasma and field lines are being transported toward the neutral point from either side. At the neutral point the antiparallel field lines are cut into halves and the field-line halves from one side are reconnected with those from the other side. The merged field lines are then expelled from the neutral point. The merged field lines will be populated by a mixture of plasma from both sides of the current sheet carrying this plasma out and away from the neutral point into the direction perpendicular to the original flow

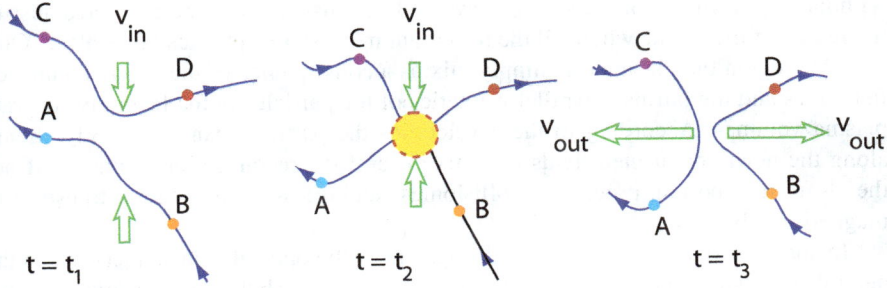

Figure 5.4: Exchange of identities of frozen-in plasma volume elements during merging of two oppositely directed field lines. Volume elements A-B and C-D are found after reconnection on separate field lines even though belonging before merging to one and the same field line respectively. This important property of reconnection demonstrates that during field line merging and reconnection the plasmas become mixed even without any necessity of diffusion. The diffusion process is restricted to the X-line only.

thereby forming two out-flowing fans or — about symmetric — jets which are called *reconnection jets*. As long as oppositely directed flux tubes are being pushed toward each other from both sides and as long as anomalous resistivity lets the magnetic field vanish inside a small volume of space around the X-line, the process of field line merging continues.

For the plasma, merging and reconnection have another very important implication which at first glance is not obvious from the simple changes of topology in the magnetic geometry depicted in Fig. 5.5. In order to explain them we must refer to the frozen-in state of the two encountering plasmas. This state is maintained until the two plasmas collide at the X-line and merge or reconnect.

However, sufficiently far outside the neutral line the plasma is definitely frozen into the magnetic field (or vice versa the magnetic field is frozen into the plasma streams) as is shown schematically in Fig. 5.4. This figure shows essentially the same three cases as in Fig. 5.5 just with the direction of axes exchanged. This time the two colliding flows come along the vertical. The letters A-D indicate plasma elements which are frozen to the magnetic field lines around which their particles gyrate. At times t_1 and t_2 the two elements A-B belong to the lower, the two elements C-D to the upper field lines, respectively. After merging at time t_3 the situation has, however, changed. Now, the particles in plasma volumes A-C belong to the newly formed reconnected left field line while the particles of plasma volumes B-D belong to the newly formed reconnected right field line which both move into opposite horizontal directions away from the X-line. Obviously, during merging the plasma volumes have exchanged their identities and plasma have become mixed even though the plasma contained in these elements has not passed across the X-line. The implication for a plasma is very important. The exchange of plasma elements during reconnection shows that reconnection allows plasmas to mix along large parts of the field line without any diffusive process to be involved. Diffusion might be restricted just to the region of the X-line where all the important microscopic physics takes place. Outside this region the plasma can simply mix as a consequence of the newly connected field lines and the intrinsic parallel velocities of the particles in the formerly separate plasma volumes. According to these velocities the particles can now freely stream along the newly connected fields and mix. Therefore, reconnection is one — if not the — most important process of collisionless and diffusionless plasma transport in magnetised plasmas.

In addition to this, reconnection also generates the opposite plasma jets which the neutral line expels, and it causes plasma heating and turbulence, two processes we only mention at this place without going into detail. One can, however, guess that the mixing of plasmas described above necessarily causes interaction and turbulence, while the annihilation of the field lines along the X-line destroys magnetically stored energy in the flow which for the required energy conservation must go into plasma heating.

5.2   The Convection Electric Field

The concurrent drift of plasma an the field lines as a whole is called plasma *convection*. The infinite conductivity lets the electric field vanish in the frame of reference moving at velocity $\mathbf{v}_c$ with the plasma and flux tubes. However, an observer in the Earth's fixed frame of reference will, according to the Lorentz transformation (2.21), measure an electric field

$$\mathbf{E}_c = -\mathbf{v}_c \times \mathbf{B} \tag{5.12}$$

It is this electric field which is conventionally referred to by the name *convection electric field*.

5.2.1   Merging and Reconnection

The ultimate source of the *magnetospheric convection* is the momentum of the solar wind flow. According to Eq. (5.12) the flow of the magnetized solar wind represents an electric field in the Earth's frame of reference. But since the solar wind cannot penetrate the magnetopause (see Section 1.2), this electric field cannot directly penetrate into the magnetosphere either. However, when the interplanetary magnetic field has a southward component, the northward directed terrestrial field lines at the dayside magnetopause are allowed to merge with the interplanetary magnetic field (see Section 5.1).

When a southward directed interplanetary field line (denoted by 1 in Fig. 5.5) encounters the magnetopause, it will merge with the closed terrestrial field line 1, which has both footpoints on the Earth and is transported to the magnetopause from the inside. The merged field lines will split into the two open field lines marked by 2, each of which has one end connected to the Earth and the other stretching out into the

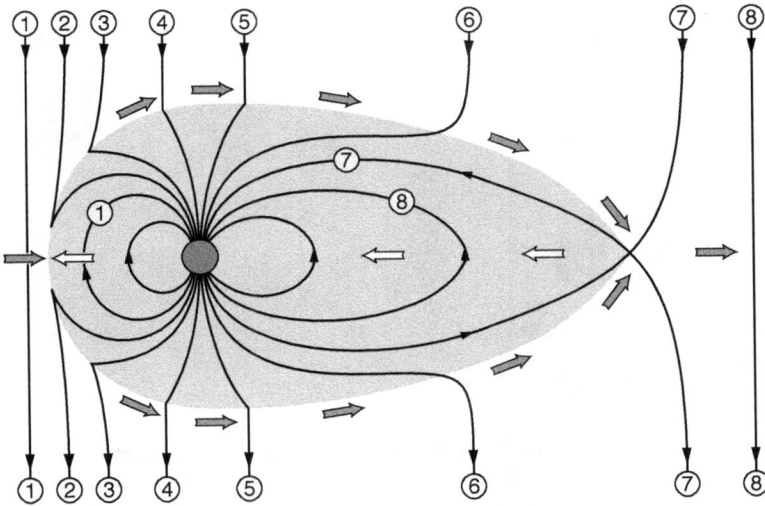

Figure 5.5: Field line merging and reconnection at the magnetopause.

solar wind. Subsequently, the solar wind will transport this field line down-tail across the polar cap (field lines marked 3–6) and due to the stiffness of the field line, the magnetic tension (see Section 7.3), the magnetospheric part of the field line (inside the shaded region) will also be transported down-tail.

At the nightside end of the magnetosphere, around 100–200 R_E downtail, the two open field line halves will meet again and reconnect, leaving a closed but stretched terrestrial field line in the magnetotail and an open solar wind field line down-tail of the magnetosphere (denoted by 7 and 8 in Fig. 5.5). Due to magnetic tension, the stretched tail field line marked by 8 will relax and shorten in the Earthward direction. During this relaxation it transports the plasma to which it is frozen toward the Earth. This is the reason for the Earthward convective flow of plasma in the magnetotail.

Under equilibrium conditions, the field line will eventually be brought back to the frontside magnetosphere and replace the terrestrial field line denoted by 1 in Fig. 5.5, since otherwise the dayside magnetosphere would soon be devoid of magnetic flux and collapse. Provided the interplanetary magnetic field still has a southward component, which is almost always the case simply because the magnetic field in the magnetosheath transition region between the bow shock and the magnetopause is in a turbulent state with fluctuating magnetic fields, the same cycle can be repeated. Thus in the average the general magnetospheric convective circulation acts to maintain the stationary state of the presence of the magnetosphere as a magnetic cavity and shield of Earth.

Frontside merging at the magnetopause and tail *reconnection* do not occur at singular points but rather along extended lines. Such a line is called an *X-line*, since along this line the magnetic field lines have a topology resembling this letter (see Fig. 5.3). The X-line is aligned more or less perpendicular to the plane shown in Fig. 5.5. It also

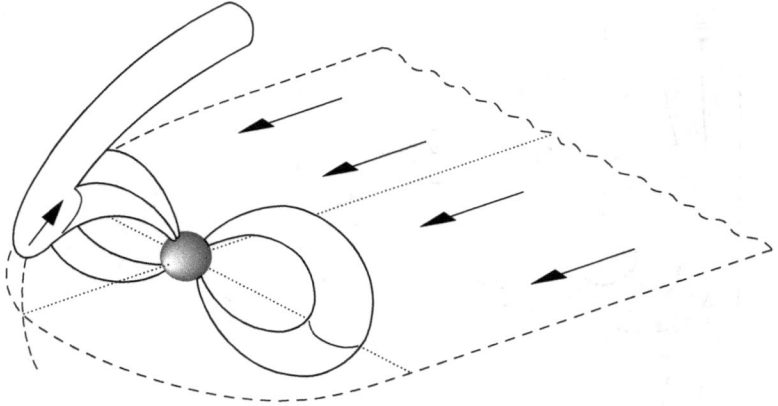

Figure 5.6: Flux tube and plasma convection caused by magnetic merging at the magnetopause. In this process the physics of which will be discussed later in this book, the merging flux tube becomes heavily kinked and pulled by convection over the pole to the magnetospheric tail.

bears the (somewhat unfortunate) name *neutral line* since the magnetic field strength vanishes at the points forming this line. The latter name is typically used for the nightside X-line, the term preferred in the present treatise.

5.2.2   Convection Electric Field

As explained above, the circulation from the frontside over the polar cap to the magnetotail and back to the dayside near the equatorial plane is not only experienced by the terrestrial field lines but also by all the particles gyrating about them. Hence, as visualized in Fig. 5.6, magnetic merging will drive a tailward plasma flow on open field lines across the polar caps and the magnetospheric lobes and complete the cycle by sunward convection in the inner magnetosphere.

Magnetic merging works most efficiently in the presence of a southward orientation of the interplanetary magnetic field. In fact, there is a wealth of experimental evidence that magnetospheric convection is strongly enhanced during periods of southward interplanetary magnetic field. For example, the magnetic storms discussed in Section 3.6, which are caused by enhanced magnetospheric convection, occur mainly during periods of prolonged southward interplanetary magnetic field. The reader can now easily understand the reason for the southward interplanetary magnetic field drives merging at the frontside magnetosphere which couples the magnetospheric field to the interplanetary field. Since the interplanetary field cannot stopped when convecting out with the solar wind into the distant heliosphere, it ties the magnetospheric field with it, and the motion of the magnetospheric field convects the magnetospheric plasma that is frozen into it downtail until it reconnects in the nightside X-line.

The sunward transport of plasma in the inner magnetosphere caused by magnetic merging or reconnection at the Earth's magnetopause is, for an observer on the Earth, equivalent to an electric convection field that is applied to the magnetosphere by the solar wind flow. In fact, this convection electric field is some kind of a measure for the efficiency of the merging process at the magnetopause. Since the electric field is the gradient of a potential, one can describe the magnetospheric convection by an electric potential difference across the magnetosphere. The total potential difference between the dawn and dusk magnetopause (or, equivalently, across the polar cap) corresponds to about 50–100 kV. Taking an average cross-section of the magnetosphere of about 30 R_E, this amounts to a dawn-to-dusk directed field of some 0.2–0.5 mV/m which is about 10% of the solar wind electric field thus suggesting that reconnection at the magnetopause is in the average a very efficient process. It consumes roughly one tenth of the magnetic energy stored in the solar wind flow. This implies also that in the average the southward magnetic field component in the magnetosheath amount to about one third of the total magnetosheath magnetic field.

The dawn-to-dusk directed convection electric field can be assumed as homogeneous in a first approximation. The same holds for the associated convection electric potential defined by

$$\mathbf{E}_c = -\nabla \phi_c \tag{5.13}$$

In polar coordinates, the convection potential in the equatorial plane is expressed as

$$\boxed{\phi_c = -E_c L R_E \sin \psi} \tag{5.14}$$

where E_c is the uniform convection electric field strength in the equatorial plane, $L R_E$ is the radial distance, and ψ denotes the azimuthal angle (or geomagnetic local time).

5.2.3 Shielding

In the inner magnetosphere the convection electric potential is somewhat weaker than described by Eq. (5.14). This *shielding effect* is caused by the different magnetic drift paths of energetic electrons and protons. As detailed in Section 3.3, energetic ions being brought in from the tail tend to drift toward the dusk side of the inner magnetosphere due to the magnetic gradient and curvature force (see Fig. 3.8), while energetic electrons tend to be found on the dawn side. The different drift paths lead to a weak charge separation with a surplus of positive charges on the dusk side and of negative charges on the dawn side. The polarization electric field associated with this charge separation is directed from dusk to dawn and will partially shield the inner magnetosphere from the full dawn-to-dusk directed cross-tail convection electric field. Taking the shielding effect into account leads to a more realistic form of the convection potential

$$\boxed{\phi_{cs} = -A_\gamma (L R_E)^\gamma \sin \psi} \tag{5.15}$$

where γ is the shielding exponent and A_γ is a constant shielding factor given by

$$A_\gamma = 0.5 \Delta\phi \, \Delta y^{-\gamma} \tag{5.16}$$

with $\Delta\phi$ denoting the cross-tail potential difference and Δy half the distance between the dawn and dusk magnetopause along the $\psi = \pm 90°$ axis. Under typical conditions, the shielding exponent γ ranges between 2 and 3. Calculating the electric field amplitude from Eq. (5.15) one obtains

$$E_{cs} = A_{\gamma}(LR_E)^{\gamma-1} \left[(\gamma^2 - 1) \sin^2 \psi + 1 \right]^{1/2} \tag{5.17}$$

For $\gamma = 1$ we recover a uniform electric field, while for a realistic shielding factor $\gamma \approx 2 - 3$ the electric field amplitude decreases toward the inner magnetosphere and varies with local time.

5.3   Corotation and Plasmasphere

To obtain the full magnetospheric plasma motion, one must add yet another electric field to the convection field. This field is again caused by a movement of the field lines and the plasma tied to it. But in this case the plasma motion is not caused by the solar wind but by the Earth's rotation. Both the Earth's atmosphere and magnetic field rotate together with the Earth around its axis. Hence, when the ionospheric and magnetospheric plasmas couple to either of them they are forced into corotation together with the Earth.

The ionospheric plasma is only partially ionized and the neutral collision frequency defined in Section 4.1 is rather high in the lower ionospheric layers. Here, the neutral atmospheric particles which corotate with the Earth will force the ionospheric plasma into *corotation* via collision between neutrals and ions and also neutrals and electrons. We have seen in Section 5.1 that the ionospheric conductivity in these layers is finite due to the high plasma-neutral collision frequency. As a consequence the magnetic field is not completely frozen-in into the ionospheric plasma. This allows the plasma to slip with respect to the rigidly corotating magnetic field as long as no other external forces act on the magnetic field and break its rotation. Clearly any relative motion of the ionospheric plasma against the magnetic field that is caused by the neutral drag will produce a corotation field and also cause currents to flow in the ionosphere. Such condition typically prevail in the low and mid-latitude ionosphere. In the high-latitude ionosphere, the solar wind-induced convection dominates over rotation which in the polar cap comes to rest anyway. This convection exerts a controlling influence on the field lines and causes them to slip through the collision-dominated and thus not infinitely conducting ionospheric plasma (see p. 91 and Section 5.4) thereby generating a typical electric field configuration in the auroral region and the polar cap. At lower than auroral latitudes and larger than ionospheric altitudes, however, the rigidly rotating magnetic field carries the cold and collisionless frozen-in plasma with it on its circular motion around Earth. This motion is accompanied by an induced corotation electric field.

5.3.1 The Corotation Electric Field

If we neglect the aforementioned slippage of the magnetic field lines at auroral and polar latitudes, the corotation of plasma and flux tubes is, for a non-corotating observer (spacecraft, earth-centred local-time reference frame etc.), equivalent to the presence of another electric field component which is solely produced by the rigid rotation of the Earth and reads

$$\mathbf{E}_{cr} = -(\mathbf{\Omega}_E \times \mathbf{r}) \times \mathbf{B} \tag{5.18}$$

with $\Omega_E = 7.27 \times 10^{-5}$ rad s^{-1} the angular velocity of Earth's rotation. Using Eq. (3.8) to express the magnetic field in the equatorial plane, we obtain for the equatorial corotation electric field strength in the magnetosphere as function of L

$$E_{cr}(L) = \frac{\Omega_E B_E R_E}{L^2} \tag{5.19}$$

The equatorial corotation field in the magnetosphere is directed radially inward and decreases with the square of the L-value. Multiplying with LR_E one obtains the electric corotation potential in the equatorial plane

$$\boxed{\phi_{cr}(L) = -\frac{\Omega_E B_E R_E^2}{L}} \tag{5.20}$$

where the term $\Omega_E B_E R_E^2$ amounts to 92 kV. In the equatorial plane the corotation potential distribution is radially symmetric and consists of concentric circles with the inter-circle distances decreasing as L^{-1}. These circles are the corotating plasma orbits. In the absence of magnetospheric convection the plasma would corotate on such circles out to large distances where the centrifugal force would break the frozen-in state of the plasma. At such distances the plasma would be expelled from the region of corotation. Under real condition this does not happen in the magnetosphere as we will show below because of the presence of the magnetospheric convection which inhibits corotation to extend farther than just the inner magnetosphere and, in addition, deforms the circular corotation orbits.

5.3.2 The Plasmasphere

The radial decrease of the corotation potential leads to a weakening of the influence of corotation with distance. At larger distances the convection potential takes over and dominates the drift of the (cold) plasma. Neglecting the shielding effect, the total electric potential in the equatorial plane is given by the sum of the two potentials $\phi_{tot} = \phi_c + \phi_{cr}$ and is depicted in Fig. 5.7.

Figure 5.7 shows two topologically different regions. Close to the Earth a region of closed equipotential contours exists, marked by the shading. Here, within the *plasmasphere*, the electric field is directed inward and the corotation dominates. The plasma

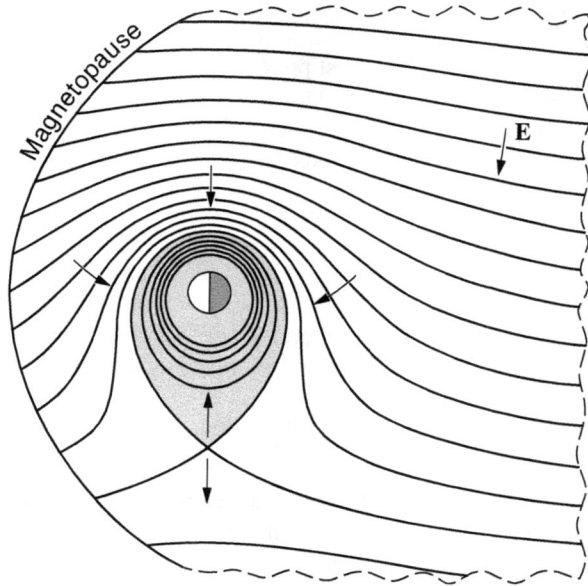

Figure 5.7: Electric equipotential contours in the equatorial plane. These contours are the electric drift paths of the cold (thermal) magnetospheric plasma. Note the distinction between the general convection in the magnetosphere and the inner corotation region.

in this region is to a large extent independent of the convection in the outer magnetosphere. The effect of the magnetospheric convection is seen in the confinement of corotation to the inner magnetosphere only and in the cause of asymmetry of the corotation orbits which are the closed equi-potentials of the electric field. These orbits are far from being circular the farther out from the ionosphere.

The plasma content of a plasmaspheric flux tube is nearly constant and, accordingly, the plasma density is rather high, several 10^3 cm^{-3}. The plasmaspheric particles originate from the ionosphere. They are 'cold', with energies in the range of 1 eV capable of evaporating from the ionosphere to higher altitudes. Outside the plasmasphere, the potential contours are open corresponding to the convective flow of the cold plasma in the magnetosphere. The magnetospheric electric field has a strong duskward component. A magnetospheric flux tube will, at some time, encounter the dayside magnetopause and loose its plasma to the magnetosheath. This is the reason for the sharp density gradient between the plasmasphere and the outer region, where less than one particles is found per cubic-centimeter.

When referring to the plasma flow in the magnetosphere, one should, however, keep in mind that the magnetospheric electric equi-potentials, which are also the flow lines of plasma, hold only for the cold background component of the plasma in the magnetosphere. These flow lines are nearly straight on the dawn side of the magnetosphere. We have seen earlier in Section 3.3 that the more energetic component of the

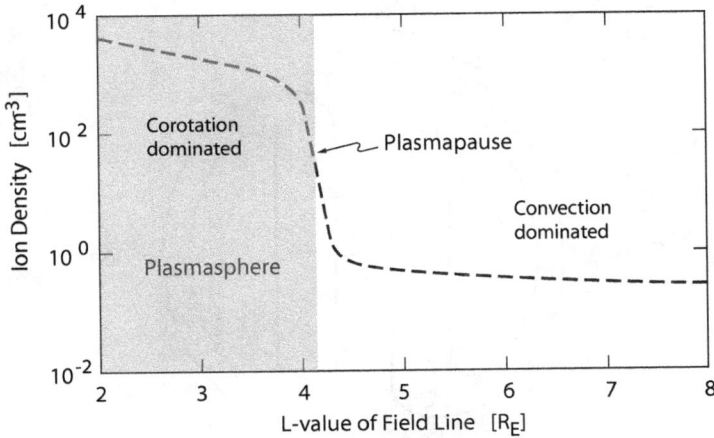

Figure 5.8: Average thermal ion density profile in the night-side magnetosphere. The drop in the ion density at $L \sim 4\,R_E$ indicates the location of the plasmapause as the boundary between the corotation and convection dominated regions of the equatorial magnetosphere.

magnetospheric plasma does not follow these flow lines anymore when entering into the inner magnetosphere where they experience the gradient and curvature drifts and their orbits deviate from the general convection flow in Fig. 5.7 and accumulate in the duskside magnetosphere to create the ring current as was shown in Fig. 3.8. The complete plasma flow lines are in fact the superposition of drift orbits and equi-potentials.

Figure 5.8 shows an average ion density profile in the night-side magnetosphere. The sharp density gradient at the outer boundary of the plasmasphere is called *plasmapause* and not only observed on the night side, but everywhere along the border of the shaded region in Fig. 5.7.

An interesting region is the *stagnation point* in Fig. 5.7. This is the point on the evening side, where the plasmasphere has a *bulge*, the farthest distance of the plasmapause from Earth. Here the outermost corotation equipotential contour 'crosses itself' because the eastward directed corotation and the westward directed convection have the same velocity and the plasma becomes stagnant. Since $\phi_c = \phi_{cr}$ and $\sin \psi = 1$ at this point, the distance of the stagnation point from the Earth's surface can easily be calculated as

$$L_{sp} = \left(\frac{\Omega_E B_E R_E}{E_c} \right)^{\frac{1}{2}} \tag{5.21}$$

where $\Omega_E B_E R_E = 1.45 \times 10^{-2}$ Vm. Thus, for a convection electric field of 1 mV/m the stagnation point is located at a distance of 3.81 R_E. The radial distance of the stagnation point depends on the strength of the convection electric field: whenever the convection electric field increases, the radius of the plasmasphere decreases and vice-versa.

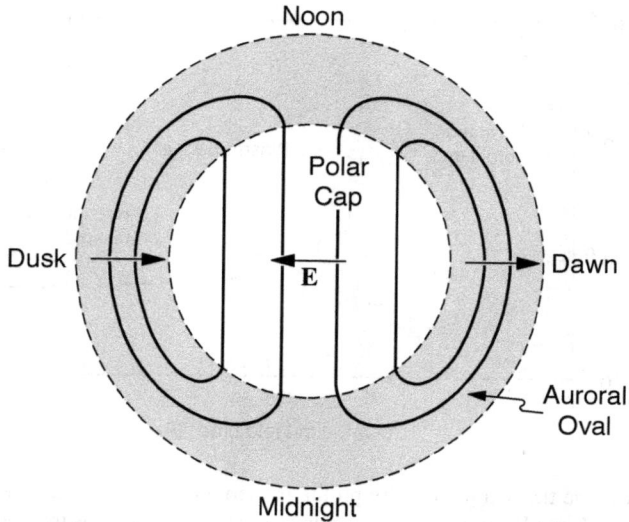

Figure 5.9: Equipotential contours of the high-latitude electric field. These contours are the flow lines of the foot points of the polar geomagnetic field lines which after merging with the solar wind field are carried by the solar wind flow outside the magnetosphere from noon across the polar cap to midnight and return to noon with the general magnetospheric convection after reconnection in the distant tail in two loops either through dawn or through dusk. The return paths correspond to the auroral oval.

5.4 High-Latitude Electrodynamics

Until now we have considered the motion of plasma and flux tubes in the equatorial plane of the magnetosphere. But the circulation of the flux tubes is not restricted to the equatorial plane, it also affects the lower ends of the flux tubes, the foot points of the field lines in the high-latitude ionosphere. Since the ionosphere is conducting, the electric field associated with the ionospheric convection will drive ionospheric currents.

5.4.1 Ionospheric Convection

The motion of the flux tubes across the polar cap due to magnetic merging depicted in Fig. 5.5 also moves the ionospheric footpoint of the flux tube and the plasma tied to it across the polar cap to the nightside. Similarly, the sunward convection of magnetospheric flux tubes shown in Fig. 5.6 leads to a sunward convection of the foot points of these flux tubes in the dawn and duskside high-latitude ionosphere, inside the auroral oval shown in Fig. 1.5. This leads to a polar ionospheric two-cell convection pattern.

The convection pattern is equivalent to an electric potential pattern. This can again be recognized from the definition of the E×B drift in Eq. (2.19) when using Eq. (5.13) to substitute the electric field by the gradient of the electric potential

$$\mathbf{v}_E = -(\nabla\phi \times \mathbf{B})/B^2 \tag{5.22}$$

Because the gradient of the electric potential, $\nabla\phi$, is perpendicular to the contours of constant potential, ϕ, and the E×B drift is perpendicular to both $\nabla\phi$ and $\mathbf{B}$, the convection streamlines are also equipotential contours. Cold particles will drift along these contours. Drawing equipotential contours and drawing E×B drift trajectories of the plasma is therefore equivalent. We already made use of this fact in earlier discussions.

Hence, we can translate the two-cell convection pattern into a two-cell pattern of equipotential contours shown in Fig. 5.9. This equipotential pattern is equivalent to an ionospheric electric field that is directed toward dusk in the northern polar cap. Inside the northern hemisphere auroral oval the electric field is directed toward the pole on the duskside, while it has a southward direction in the morning hours.

5.4.2   Height-Integrated Ohm's Law

Since the ionospheric conductivity is a tensor with three different components (see Section 4.2), three types of currents will be generated by the convection electric field. The first type are the field-aligned currents flowing parallel to the magnetic field into and out of the ionosphere. Secondly, there are the Pedersen currents which flow perpendicular to the magnetic field lines and parallel to the ionospheric convection field. Finally, Hall currents will flow perpendicular to both the magnetic and the electric field.

As discussed in Section 4.2, the ionospheric conductivity along the geomagnetic field lines, $\sigma_\parallel$, is always much higher than the transverse conductivities, σ_P and σ_H. For reasons of current continuity, the electric field component along the magnetic field lines, $E_\parallel$, will, therefore, generally be much smaller than the perpendicular field, $E_\perp$. Often the magnetic field lines may be considered as perfectly conducting and the parallel electric field vanishes (see p. 93). At high latitudes, where the field lines are nearly vertical, the horizontal electric field becomes almost height-independent and one may introduce the height-integrated quantities

$$\begin{aligned}
\Sigma_P &= \int \sigma_P dz \\
\Sigma_H &= \int \sigma_H dz \\
\mathbf{J}_\perp &= \int \mathbf{j}_\perp dz
\end{aligned} \tag{5.23}$$

The height-integrated conductivities or *conductances* and the height-integrated horizontal current density are related through the height-integrated version of Ohm's law. The latter follows by integrating the transverse component of Eq. (4.28), yielding

$$\mathbf{J}_\perp = \Sigma_P \mathbf{E}_\perp - \Sigma_H (\mathbf{E}_\perp \times \mathbf{B})/B \tag{5.24}$$

Here, we have neglected the contribution of the neutral wind, which is important at low- and mid-latitudes (see Section 4.5), but at high latitudes is much weaker than the motion due to the convection electric field. Also, the altitude dependence of the geomagnetic field was ignored which is valid for a thin ionosphere.

The field-aligned current is not specified by Eq. (5.24). But since the total current flow must be continuous, it can be calculated from the divergence of the height-integrated horizontal current. For vertical field lines one obtains

$$j_\| = \nabla_\perp \cdot \mathbf{J}_\perp \qquad (5.25)$$

where $\nabla_\perp$ denotes the vector derivative in the horizontal plane. Inserting Eq. (5.24) into (5.25) we obtain after some vector algebra (see Appendix B.4)

$$j_\| = (\nabla_\perp \Sigma_P) \cdot \mathbf{E}_\perp - (\nabla_\perp \Sigma_H) \cdot (\mathbf{E}_\perp \times \mathbf{B})/B + \Sigma_P (\nabla_\perp \cdot \mathbf{E}_\perp) \qquad (5.26)$$

where we haven't taken into account that the ionospheric magnetic field is both stationary and uniform on the typical scales involved and thus $\nabla_\perp \cdot (\mathbf{E}_\perp \times \mathbf{B}) = 0$. Accordingly, field-aligned currents are generated at gradients of the Pedersen conductance along the electric field direction, at gradients of the Hall conductance which are aligned perpendicular to the horizontal electric field, and in regions where the divergence of the electric field is non-zero.

5.4.3   Polarization Electric Fields

The primary electric field in Eq. (5.24) is the magnetospheric convection electric field mapped down to the ionosphere as sketched in Fig. 5.9. Current continuity at conductivity gradients is usually served for by field-aligned currents. However, despite the fact that the parallel conductivity is near-infinite, the field-aligned current density cannot grow above certain levels. As we will see in more detail later, the field-aligned current density is restricted by the number of charge carriers that can be moved along the field line with a high-enough velocity. Moreover, the field-aligned currents have to be closed somewhere in the magnetosphere and the current transverse to the magnetic field may also be limited.

At strong conductance gradients the field-aligned current density due to the first two terms on the right-hand side of Eq. (5.26) may become larger than the limiting value. It is hardly possible to alter the conductivity pattern since it is imposed by the structure of the particle precipitation. But field-aligned current resulting from the $\nabla\Sigma$ terms may be balanced by altering the electric field distribution in the last term on the right-hand side of Eq. (5.26).

Maxwell's equation (2.6) tells us that changing the divergence of the electric field is equivalent to adding polarization charges. Hence, the secondary electric field that has to be added to the convection electric field in order to limit the field-aligned current is a polarization electric field introduced in Section 4.5.

5.4.4 Joule Heating

Ionospheric currents may heat the atmosphere by Ohmic dissipation or Joule heating. The *Joule heating rate* is proportional to the current flowing parallel to the electric field. The height-integrated Joule heating rate can be written as

$$Q_J = \mathbf{J}_\perp \cdot \mathbf{E}_\perp \tag{5.27}$$

Since the Hall currents flow perpendicular to the electric field, they do not contribute to Ohmic dissipation. Thus Eq. (5.27) can be rewritten as

$$Q_J = \Sigma_P E_\perp^2 \tag{5.28}$$

This expression shows that the amount of Ohmic heat produced in the auroral ionosphere depends more on the level of convection than on the number of precipitating particles.

5.5 Auroral Electrojets

Since particles precipitating into the auroral oval cause significant ionization (see Section 4.3), its conductivity is much higher than that of the polar cap. As a result, the high-latitude current flow is concentrated inside the auroral oval, where it forms the *auroral electrojets*. The auroral electrojets are the most prominent currents at auroral latitudes. They carry a total current of some 10^6 A. This is the same order of magnitude as the total current carried by the ring current discussed in Section 3.6, but since the auroral electrojets flow only 100 km above the Earth's surface, they create the largest ground magnetic disturbance of all current systems in the Earth's environment. The disturbance fields have typical magnitudes of 100–1000 nT, but may reach 3000 nT during the largest magnetic storms. The latter is a sizable fraction (about 5%) of the terrestrial dipole field at high latitudes. The present knowledge about the conductivity structure, the electric field distribution and the current flow associated with the auroral electrojet system is summarized in Fig. 5.10.

5.5.1 Conductances and Electric Fields

Inside the auroral oval, which is approximately an off-center ring shifted by an average 5° from the magnetic pole toward magnetic midnight, the ionospheric conductivity is enhanced above the solar ultraviolet-induced level due to the ionization of neutral atoms and molecules by precipitating electrons and ions (see Sections 4.3 and 4.4). The energetic particles drift toward and around the Earth (see Fig. 3.8) and precipitate, depending on their energy and pitch angle, in different local time sectors. The precipitation pattern is reflected in the conductivity structure. The weakest conductivities are found near the noon sector while the conductivity maximum lies in the midnight sector where typical values of 7–10 S and 10–20 S for Pedersen and Hall conductances, respectively, are found.

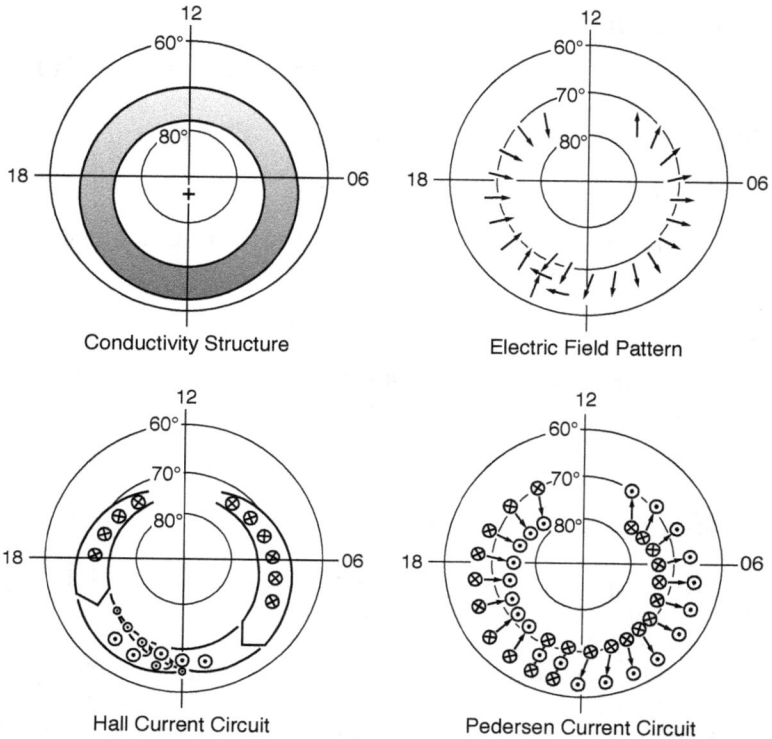

Figure 5.10: Synopsis of the ingredients of the auroral electrojet system which consists of the ionospheric conductivity structure at high latitudes, the convection electric field pattern, the Hall current circuit which is closed in itself, and the perpendicular Pedersen currents which are the closure currents of the auroral field-aligned current system.

The electric field pattern in the auroral oval reflects the large-scale pattern of magnetospheric plasma convection. The transport of open and closed flux tubes results in a convection pattern with two cells sketched in Fig. 5.9. In reality, the convection pattern is slightly more complicated but still resembles a two-cell pattern. The auroral zone electric field associated with the two-cell system of plasma transport has typical values between 20 and 50 mV/m. It is poleward directed in the afternoon and early evening sector, points equatorward in the postmidnight and morning sector and rotates from north over west to south in the premidnight sector. This region of field rotation is called the *Harang discontinuity* region. The clockwise rotation of the pattern, as compared to the idealized picture in Fig. 5.9, has its origin in a weak polarization electric field (see p. 106), which is directed from midnight to noon and stems from electric charges built up at the boundary between the highly conducting auroral zone and the polar cap.

5.5.2 Ionospheric and Field-Aligned Currents

As summarized in the lower left panel of Fig. 5.10, the eastward and westward electrojets are primarily Hall currents which originate around noon and are fed by downward field-aligned currents. Typical sheet current densities range between 0.5 and 1 A/m and increase toward midnight due to the increasing Hall conductance. The eastward electrojet flows in the afternoon sector and terminates in the region of the Harang discontinuity where it partially flows up magnetic field lines and partially rotates northward, joining the westward electrojet. The westward electrojet flows through the morning and midnight sector and typically extends into the evening sector along the poleward border of the auroral oval where it also diverges as upward field-aligned currents.

The Pedersen currents shown in the lower right panel of Fig. 5.10 have typical densities of 0.3–0.5 A/m. They flow northward in the eastward electrojet region and are connected to sheets of downward and upward field-aligned currents in the southern and northern half of the northern hemispheric afternoon-evening auroral oval, respectively. In the midnight-to-noon sector the Pedersen current flows equatorward and field-aligned currents provide continuity by flowing upward in the southern and downward in the poleward half of the auroral oval. The sheets of field-aligned currents in the poleward half of the auroral oval were named *Region-1 currents* and those in the equatorward half are called *Region-2 currents*. Inside the Harang discontinuity region, the evening and morning side Pedersen current circuits overlap, leading to three sheets of field-aligned currents.

As sketched in Fig. 1.6, the Region-1 and -2 field-aligned current belts extend into the magnetosphere. The Region-2 currents in the equatorward belt are closed by the westward ring current in the near-Earth equatorial plane. On the other hand, the Region-1 currents in the poleward half flow along the high-latitude boundary of the plasma sheet. Deep in the magnetotail, they merge with the neutral sheet current. The upward and downward directed field-aligned currents continuing the Hall currents in the midnight and noon sector are closed in the magnetosphere by the evening-side partial ring current mentioned in Section 3.3.

5.5.3 Joule Heating and AE Index

Inserting the average values for the Pedersen conductivity and the electric field given above into Eq. (5.28), one finds that typical Joule heating rates due to the auroral electrojets are in the range 5–50 mW/m^2. While this number does not seem large, the total heat input into the auroral zone is quite significant. The total Joule heat input into one hemisphere can be estimated from the *auroral electrojet index*, which was introduced as a measure of global auroral electrojet activity (see Appendix C.3).

The *AE* index is based on readings of the northward magnetic disturbance component from twelve auroral zone observatories located in different local time zones and calculated for any given instant of universal time as the difference between the maximum northward and southward disturbance field (which are thought to represent the maximum current in the eastward and westward electrojet). Empirically one has found that a total hemispheric Joule heat input of 0.3 GW is equivalent to 1 nT in AE.

Taking into account both hemispheres and adding the heating of the upper atmosphere due to particle precipitation, which empirically is known to be about half as effective as the Joule heating for the same level of AE, we find that the total heating of the upper atmosphere amounts to about 1 GW per nT in AE. During major magnetic storms, when AE can easily reach 1000 nT and above, the total heat input at high latitudes amounts to more than 10^{12} W.

5.6   Magnetospheric Substorms

Until now we have treated convection as a stationary process. First, we tacitly assumed that magnetic merging between interplanetary and terrestrial field lines at the frontside magnetopause occurs always at the same rate. Second, it was implicit in our scenario shown in Fig. 5.5 that dayside merging and reconnection in the distant magnetotail are in equilibrium. Both assumptions hold only approximately and have to be challenged when being confronted with the magnetospheric reality.

Let us now look closer at what really happens on the dayside. The amount of dayside magnetic flux merged per unit time, the dayside *reconnection rate*, depends on the number of southward oriented interplanetary field lines that get into contact with the Earth's magnetopause during a given time interval. Thus it depends on the solar wind velocity and on the magnitude of the southward interplanetary magnetic field component. In particular the latter is rather variable. Moreover, for considerable periods of time the interplanetary field is directed northward, rendering any dayside merging impossible. Hence, we have intervals when the magnetosphere is very quiet and convection will cease as well as other periods where a lot of flux is merged on the dayside and the magnetospheric plasma will be active.

Eventually, all magnetic flux transported to the tail has to be reconnected and convected back to the frontside magnetosphere. But there is no need for identical instantaneous dayside and nightside reconnection rates, only the average rates must be equal. Actually, it has been found that only part of the flux transported into the tail is reconnected instantaneously and convected back to the dayside. The remaining field lines become added to the tail lobes, where they increase the magnetic flux density. After about one hour these intermediately stored field lines are suddenly reconnected in the tail and their magnetic energy is explosively released.

The sudden reconnection of previously stored flux tubes has rather dramatic effects on the magnetospheric plasma and associated phenomena like aurora and magnetospheric and ionospheric currents. These effects, which last for about one or two hours and will be detailed below, have been summarized under the term *magnetospheric substorm*, since they form the basic element of the magnetic storm described in Section 3.6. Whenever a couple of major substorms occur in a row, the ring current increases significantly and the low-latitude magnetograms exhibit a magnetic storm.

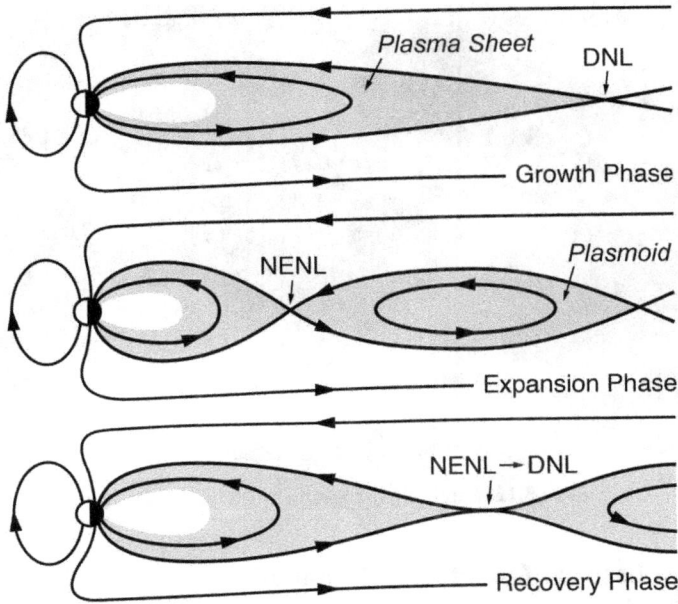

Figure 5.11: Reconfiguration of the plasma sheet during a substorm.

5.6.1 Substorm Growth

The substorm starts when the dayside merging rate is distinctively enhanced, typically due to a southward turning of the interplanetary magnetic field. The flux eroded on the dayside magnetopause is transported into the tail. Part of the flux is reconnected and convected back to the frontside of the magnetosphere. The enhanced convection due to the driven substorm component causes enhanced current flow in the auroral electrojets and an associated growth of the *AE* index.

At the same time the magnetic flux eroded from the dayside magnetopause which is not reconnected is added to the tail lobes. Since the magnetic field in the tail lobes and the neutral sheet current are related by Biot-Savart's law (see Appendix B.5), the growth of the tail lobe magnetic field must be accompanied by a growing neutral sheet current. The growth of the latter will also stretch the field lines threading the plasma sheet into a more tail-like configuration. This is sketched in the upper panel of Fig. 5.11.

The period of enhanced convection and loading of the tail with magnetic flux is called *substorm growth phase*. It typically lasts for about one hour. After that time period too much magnetic flux and thus magnetic energy has been accumulated in the tail. The tail becomes unstable and tries to get rid of the surplus energy. This is the time of *substorm onset* and the beginning of the *substorm expansion phase*.

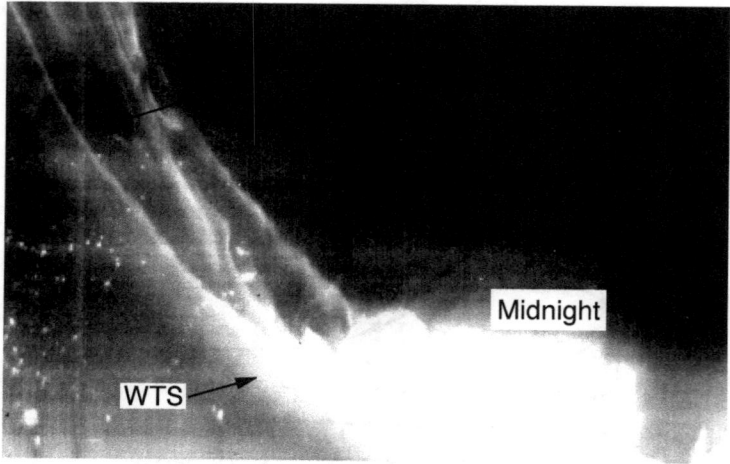

Figure 5.12: Satellite photograph of evening-side aurora.

5.6.2 Substorm Onset and Expansion

During the substorm expansion phase, which typically lasts about 30–60 min, rather dramatic changes are seen in the magnetosphere and auroral zone ionosphere. One of those can even be observed with the naked eye, or with a satellite camera as in Fig. 5.12, which shows a negative image of the evening auroral oval over Scandinavia and western Russia taken by a satellite flying high above the auroral oval. During the growth phase the aurora appears in the form of *auroral arcs*. These are structures which are widely stretched azimuthally along the auroral oval, but very thin in their north-south extent (see Figs. 4.14 and 5.12). At substorm onset one of these arcs suddenly brightens and fills the whole sky. The *auroral break-up* first occurs in the midnight sector, but then expands rapidly northward and, in particular, westward. As can be seen from Fig. 5.12, the westward motion of the break-up aurora looks like that of a surge. Accordingly, this type of aurora has been named *westward traveling surge* or WTS.

Not only the aurora changes dramatically at substorm onset. The sharp increase in the *AE* index on the left-hand side of Fig. 5.13 to values of about 500 nT indicates that the ionospheric current flow is strongly enhanced. Moreover, the stretched magnetic field in the plasma sheet suddenly becomes more dipolar again, as is evident from the right-hand panel of Fig. 5.13 where the average magnetic field elevation in the plasma sheet, i.e., the angle between the magnetic field direction and the equatorial plane, rises from less than $10°$ to more than $30°$ during the first 20 min after substorm onset.

The dipolarization of the magnetotail field is the signature of a dramatic reconfiguration of the plasma sheet sketched in the middle panel of Fig. 5.11. Around 30 R_E downtail, a new neutral line is formed. To distinguish it from the distant neutral line at around 100–200 R_E, the newly formed X-line is usually called *near-Earth neutral*

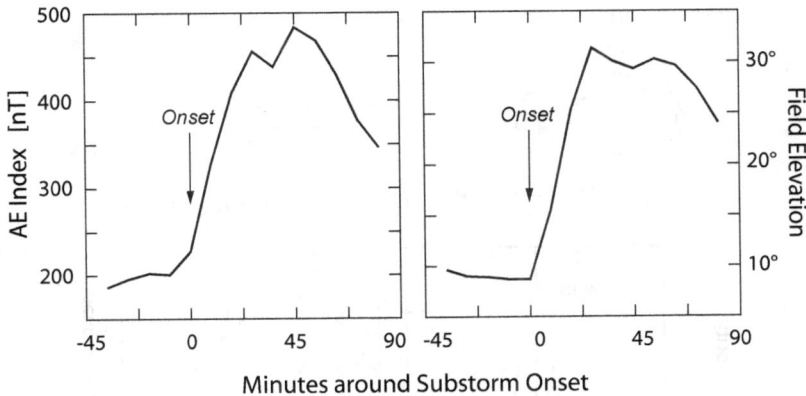

Figure 5.13: Variation of AE index and plasma sheet magnetic field elevation during substorms.

line. The excess magnetic flux deposited in the tail during the growth phase is reconnected along this new X-line during the expansion phase. After the stretched field lines are reconnected, they move back to their normal more dipolar shape.

The large region of the tail between the two neutral lines forms a *plasmoid.* The magnetic field inside the plasmoid has a peculiar structure. Its field lines are neither connected to the terrestrial nor to the interplanetary magnetic field, but form closed loops.

We should add a word of caution. The exact configuration of the magnetic field in the vicinity of the near-Earth neutral line is not fully known yet. It may be a single large-scale X-line like the one shown in Fig. 5.11. But there is also the possibility that reconnection proceeds along multiple neutral lines. Similarly, if the magnetic field in the plasmoid has a dawn-dusk component, the field lines may form a three-dimensional helix rather than a two-dimensional closed loop.

5.6.3 Substorm Recovery

Figure 5.13 shows that about 45 min after substorm onset the ionospheric current flow and the strong dipolar field orientation in the tail plasma sheet start to decrease again. At this instant also the aurora starts to fade and retreats to higher latitudes. In general, reconnection at the near-Earth neutral line ceases and substorm activity settles. This is the begin of the *substorm recovery phase.* This phase lasts for about 1–2 hours and ends when the magnetosphere has returned to a quiet state. However, during intervals when the interplanetary magnetic field has a stable southward orientation, the recovery phase of one substorm may coincide with the growth phase of the next substorm.

Besides the general decrease in activity, satellite observations have identified an important process that occurs during the recovery phase. As sketched in the lower panel of Fig. 5.11, the near-Earth neutral line starts to retreat tailward at the time of

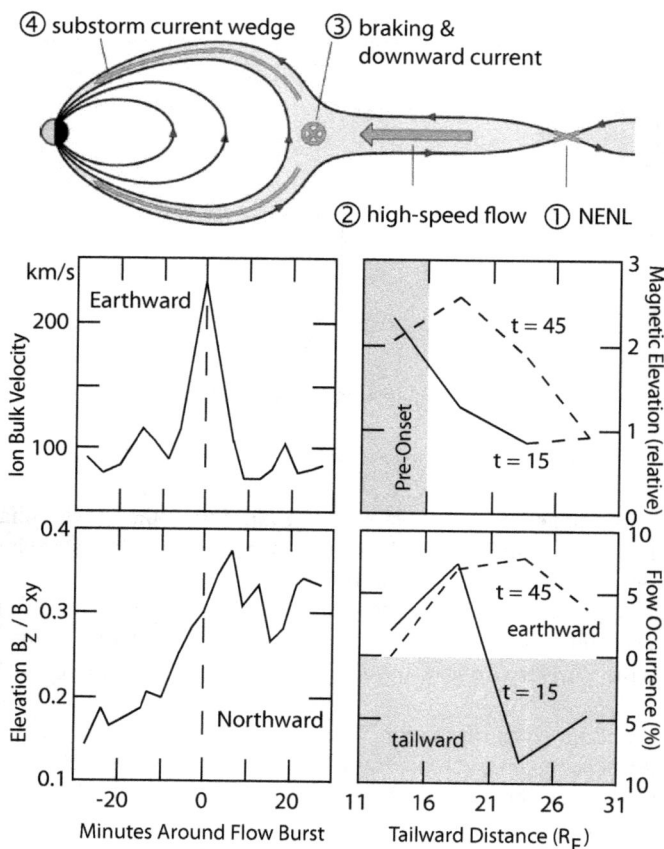

Figure 5.14: Bursty bulk flow, flux pile-up and braking of reconnection — the onset of the recovery phase. *Top*: The braking of the high-speed reconnection flows generated in the near-Earth neutral line (NENL) by pile-up of flux tubes at the inner tailward plasmapause. The generated downward braking current is indicated as well as the substorm current wedge. *Bottom Left*: The typical variation of earthward flow velocity and northward magnetic field dipolarization during bursty bulk flow events on the time-scale around the flow burst. Typical duration times are 20 minutes. *Bottom Right*: Radial profiles of the normalised magnetic field elevation and fast flow occurrence rates during expansion and recovery phases of substorms.

maximum expansion. The reason for this behaviour is to be found in the violent time-dependence of the near-Earth reconnection. Fig. 5.11 shows that on the earthward side of the near-Earth reconnection X-line the reconnection process re-establishes the dipolar structure of the geomagnetic field in the tail while at the same time transports the newly reconnected dipole-like flux tubes inward in the form of so-called *bursty bulk flows* as shown in Fig. 5.14. This inward transport of closed magnetic flux tubes leads to a pile up of flux in the inner magnetosphere near the tailward

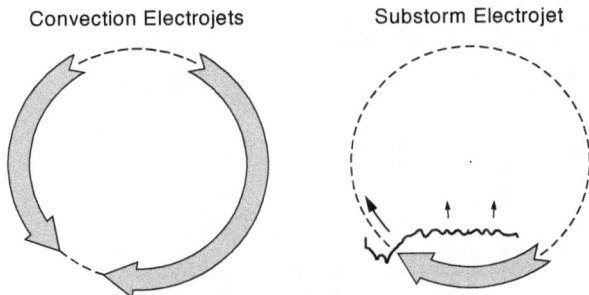

Figure 5.15: Comparison of convection electrojets and the substorm electrojet. During substorms the new electrojet amplifies the westward part of the convection electrojet causing poleward expansion and braking of the duskside eastward electrojet.

plasmaspheric boundary where it is braked and cannot proceed further inward. Piling-up then starts a dipolarization front to move anti-earthward into downtail direction re-establishing the dipolar magnetic field structure in the tail readily reaching the location of the near-Earth neutral line at 20–25 R_E. Since reconnect stops in a dipolar magnetic field configuration, the near-Earth neutral line moves rapidly tailward, still producing some fast earthward flows which however now divert from the radial direction. This is the beginning of the recovery phase of the substorm. Hence, near-Earth reconnection effectively strangles itself after some relatively short time, quenching itself by the successive near-Earth pile-up of the closed dipolar field lines it generates. In doing so, it also pushes the plasmoid tailward until the latter is finally ejected from the magnetotail. Its plasma is lost to the downtail solar wind and the former near-Earth neutral line has become the distant neutral line.

5.7 Substorm Currents

During magnetospheric substorms the ionospheric current flow is affected in two ways. On the one hand, the current flow in the auroral electrojets sketched on the left-hand side of Fig. 5.15 increases along with the enhanced convection. The strengthening of the convection electrojets is already seen during the growth phase and is caused mainly by the increasing convection electric field.

   In addition to the overall growth of the auroral electrojet current, the unloading of magnetic flux previously stored in the magnetotail leads to the formation of a substorm electrojet with strongly enhanced westward current flow in the midnight sector. The *substorm electrojet* is concentrated in the region of active break-up aurora and expands westward during the course of the expansion phase along with the westward traveling surge. In contrast to the convection electrojets, the strength of the substorm electrojet current is mainly determined by the strong increase in ionospheric conductance due to the strong particle precipitation in the bright substorm aurora.

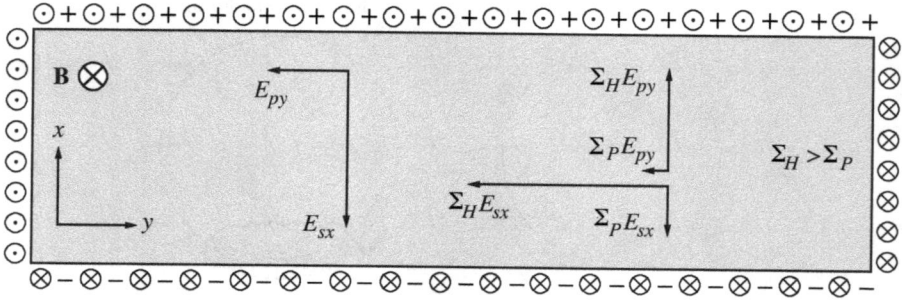

Figure 5.16: Generation of a Cowling channel inside the westward traveling surge.

5.7.1 Substorm Electrojet

Since the substorm electrojet is governed by the strong increase of the conductivities inside the westward traveling surge, the situation is similar to that in the equatorial electrojet described in Section 4.5. However, in the present case the Cowling channel is not perfect since field-aligned currents will remove part of the space charge deposited at the boundaries of the highly conducting channel.

The generation of an imperfect Cowling channel is summarized in Fig. 5.16. Assuming that the conductivities are negligible outside the highly conducting strip, we have the following situation. The primary convection electric field possesses a substantial westward component in the midnight sector (see Fig. 5.10). This primary westward electric field component, E_{py}, drives a primary northward Hall current, $\Sigma_H E_{py}$, across the highly conducting strip. It turns out that only a fraction of this current can be closed via field-aligned currents at the strong conductivity gradients along the northern and southern boundaries (see p. 106). The excess Hall current deposits positive charges at the northern border of the high-conductivity channel while negative charges build up at its southern boundary. These charges give rise to a southward polarization electric field, E_{sx}, which drives a southward Pedersen current, $\Sigma_P E_{sx}$, to balance that part of $\Sigma_H E_{py}$ which is not continued via field-aligned currents.

Introducing the fraction, α_p, of the primary Hall current which is not closed via field-aligned current, we obtain a relation between the primary and secondary electric fields

$$\alpha_p \Sigma_H E_{py} - \Sigma_P E_{sz} = 0 \tag{5.29}$$

which yields for the secondary southward polarization electric field

$$E_{sx} = \alpha_p \frac{\Sigma_H}{\Sigma_P} E_{py} \tag{5.30}$$

The westward currents due to the primary convection and secondary polarization electric field add up to an intense westward current

$$J_y = \Sigma_P E_{py} + \Sigma_H E_{sz} = \left(\Sigma_P + \alpha_p \frac{\Sigma_H^2}{\Sigma_P} \right) E_{py} \tag{5.31}$$

The sum of conductivities appearing on the right-hand side of this expression can be regarded as an imperfect Cowling conductance

$$\Sigma'_C = \Sigma_P + \alpha_p \frac{\Sigma_H^2}{\Sigma_P} \qquad (5.32)$$

For a typical Hall-to-Pedersen conductance ratio of about four and $\alpha_p = 0.5$, the Cowling current is about one order of magnitude stronger than the normal westward Pedersen current.

Also the westward Cowling current encounters conductance gradients at the western end eastern boundary of the channel. But here all current can be closed via field-aligned currents. The strong conductance gradient at the western edge leads to intense localized upward field-aligned currents near the head of the westward traveling surge. Under normal conditions, the density of these upward field-aligned currents would be much higher than what can be carried by the ambient plasma. However, the intense precipitation of 10–30 keV electrons in the same area constitutes an upward current, which is large enough to serve for current continuity. The conductance gradient at the eastern boundary is much more gradual than the gradients at the other borders of the conductance channel. The downward field-aligned currents flowing here are less intense and more wide-spread than those produced at the other boundaries and can be carried by the ambient particles.

Typical values for conductances, fields, and currents inside the westward traveling surge are as follows. Within the active region the Hall conductance reaches peak values of more than 100 S. The convection electric field pattern is distorted by the superposition of a southward polarization electric field with a strength of up to 50 mV/m inside the westward traveling surge. The ionospheric current has sheet current densities of 500–1000 mA/m, comparable with typical westward electrojet values. The westward component of the ionospheric current is connected to very localized and intense (about 5–10 μA/m^2) upward field-aligned currents at the western border, near the head of the surge, and to more wide-spread downward current of lower density (1–2 μA/m^2) in the eastern part. The northward current is connected to field-aligned current sheets of 1–2 μA/m^2 at the southern and northern boundaries of the active region.

5.7.2 Substorm Current Wedge

The substorm electrojet is the ionospheric part of the *substorm current wedge* sketched in Fig. 5.17. The substorm current wedge diverts part of the neutral sheet current along magnetic field lines through the ionosphere. The formation of the current wedge is naturally associated with the formation of the near-Earth neutral line. When the tail magnetic field dipolarizes in the vicinity of a neutral line, the cross-tail neutral sheet current (see Sections 1.3 and 7.4) must be reduced in that region. Indeed the collapse of the tail-like field lines to a dipolar configuration would not occur if the cross-tail current remained at its original level.

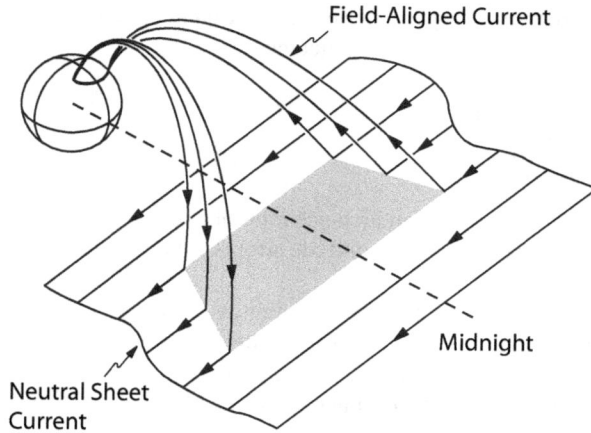

Figure 5.17: Diversion of neutral sheet current through the ionosphere. Current disruption by some unspecified process in the neutral sheet forces the currents to divert from the equatorial plane on the morning side and to flow along the magnetic field into the ionosphere where they close before returning to the magnetosphere along evening side magnetic field lines to continue the neutral sheet current.

However, the question of cause and effect is not quite settled yet. The more common view is that the near-Earth neutral line is the cause of the diversion of the cross-tail current and the formation of the substorm current wedge. Another school of thought believes that the strong enhancement of the westward ionospheric current inside the westward traveling surge and the need to close it via field-aligned currents and transverse current flow in the tail leads to a diversion of the neutral sheet current. In this model the formation of the near-Earth neutral line is a consequence of the diversion of the cross-tail current and the associated dipolarization.

5.8   Concluding Remarks

In the present chapter we have treated the ionosphere as a relatively passive medium with its dynamics governed by the processes that take place in the magnetosphere. The real situation is more complex. The magnetosphere is dominated by a collision-free plasma, while the ionosphere is the region where the effects of collisions of charged particles with neutral particles cannot be neglected and electrical conductivities transverse to the geomagnetic field maximize. The magnetic field connects electrically the ionosphere and the magnetosphere, causing an exchange or coupling of energy and momentum between the two regions. In a sense, the *magnetosphere-ionosphere coupling* is the interaction of different physical processes taking place in either of these two regions. Strong coupling occurs since the two regions are connected by magnetic field lines.

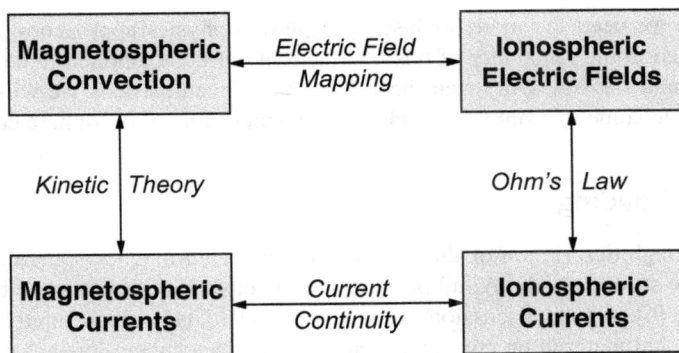

Figure 5.18: The logic of magnetosphere-ionosphere coupling.

A simplified version of the logic governing the entire magnetosphere-ionosphere coupling system is shown in Fig. 5.18. One can understand the physical mechanisms fully only if one examines the coupled system in its entirety, since a change in one of the boxes will imply changes in all other boxes. For example, a variation in magnetospheric convection will modify the electric field mapped to the ionosphere. This will change the ionospheric current flow related to the electric field through Ohm's law and, by the requirement of current continuity, the field-aligned currents and the magnetospheric current distribution. Since the latter is, at least partially, caused by the magnetic drift of energetic particles which may change the magnetospheric electric field distribution due to shielding (see Section 5.2), the whole chain must be readjusted to reach an equilibrium situation.

The situation is further complicated by the fact that all logical connections between the boxes work in both directions. For example, the flow of ionospheric currents is governed by the ionospheric electric field, but the currents may affect the field through polarization charges set up at conductivity gradients, whenever current continuity would require field-aligned current densities higher than what can be carried by the ambient plasma. Subsequently, these ionospheric electric polarization fields will affect the magnetospheric electric field and, hence, the convection of the magnetospheric plasma.

The coupled set of equations represented by the logic diagram of Fig. 5.18 has been solved only recently. For example, the Rice Convection Model can reproduce the large-scale features of the coupled magnetosphere-ionosphere system in near-equilibrium situations. For small-scale features associated with auroral forms or the highly dynamic situations like during substorm onsets only partial solutions exist, linking at most three of the boxes in Fig. 5.18.

Actually, magnetosphere-ionosphere coupling cannot be dealt with using only the single particle approach plus collisions. Especially, the link between magnetospheric convection and magnetospheric currents involves collective plasma effects already for large-scale steady state situations if more than just the ring current is discussed. For small-scale and highly dynamical situations collective effects govern the whole chain, and we need the more sophisticated plasma physics approaches, which will be addressed in the remainder of this book and in our companion volume, *Advanced Space Plasma Physics*. Only near the end of the companion book we will be able to come back to some microphysical aspects of magnetosphere-ionosphere coupling.

Further Reading

A thorough description of the physics of solar wind-magnetosphere coupling is given in the second article. A full proof of the frozen-in flux theorem can be found in monograph [6]. A good discussion about corotation and its limits is found in reference [8]. Additional material on current systems and magnetosphere-ionosphere coupling is found in references [4] and [7]. Readers interested in the aurora should have a look at the third article. Recent results on the behavior of the magnetotail during substorms are presented in the first article. A good summary of our present knowledge about convection and substorms is given in reference [5]. The Rice Convection Model mentioned above is described in the last article.

References

[1] W. Baumjohann, *Space Science Rev.* **64** (1993) 141.

[2] S. W. H. Cowley, *Rev. Geophys. Space Phys.* **30** (1982) 531.

[3] T. J. Hallinen, in *Geomagnetism, Vol. 4*, ed. J. A. Jacobs (Academic Press, London, 1991), p. 741.

[4] Y. Kamide and W. Baumjohann, *Magnetosphere-Ionosphere Coupling* (Springer Verlag, Heidelberg, 1993).

[5] C. F. Kennel, *Convection and Substorms* (Oxford University Press, Oxford, 1995).

[6] D. R. Nicholson, *Introduction to Plasma Theory* (Wiley & Sons Inc., New York, 1983).

[7] J. Untiedt and W. Baumjohann, *Space Science Rev.* **63** (1993) 245.

[8] V. M. Vasyliunas, in *Solar-Terrestrial Physics*, eds. R. L. Carovillano and J. M. Forbes (D. Reidel Publ. Co., Dordrecht, 1983), p. 479.

[9] R. A. Wolf, in *Solar-Terrestrial Physics*, eds. R. L. Carovillano and J. M. Forbes (D. Reidel Publ. Co., Dordrecht, 1983), p. 303.

Problems

Problem 5.1 *Derive the general induction equation (5.2). Which assumptions have to be made in order to arrive at this form?*

Problem 5.2 *If you introduce an arbitrary length L, velocity V, time T which condition must be satisfied in order to reduce the induction equation to the diffusion equation (5.3)?*

Problem 5.3 *Calculate the diffusion coefficient for the solar wind ($n_e \sim 5$ cm^{-3}, $B \sim$ 10 nT, $T_e \sim 50$ eV), plasmasphere ($n_e \sim 100$ cm^{-3}, $B \sim 1000$ nT, $T_e \sim 1$ eV) and magnetosphere ($n_e \sim 0.1$ cm^{-3}, $B \sim 10$ nT, $T_e \sim 1$ keV). What do you conclude?*

Problem 5.4 *What is the physical meaning of the frozen-in theorem? What is its implication for the electric field in plasmas?*

Problem 5.5 *The magnetic Reynolds number. Reynolds discovered his number in discussing viscous fluids. What is the expression for it in that case? What is the difference to the magnetic Reynolds number? What would have to be changed if fluid viscosity would be included into the induction equation?*

Problem 5.6 *Merging:* Figure 5.3 *suggests that antiparallel field lines connect forming an X configuration. Give an argument for the formation of the X. Why do the antiparallel field lines when meeting not annihilate over their whole lengths?*

Problem 5.7 *Let two antiparallel fields of equal magnitude B annihilate in an X-line. What happens to the energy density stored in the magnetic fields? How much total energy is involved in the merging process? Hint: Remember that the magnetic energy density is defined as $B^2/2\mu_0$.*

Problem 5.8 Figure 5.4 *suggests the mixing of plasmas during reconnection without any need for the plasma to flow across the neutral line. How long will it take for plasmas to mix after reconnection over a length of, say, 2 R_E if the plasmas have densities $n_1 = 10$ cm^{-3}, $n_2 = 0.1$ cm^{-3} and parallel temperatures $T_{\|1} = 30$ eV and $T_{\|2} = 1$ keV, not making any difference between electrons and ions? Which effect should be observed on the low temperature side?*

Problem 5.9 Figure 5.5 *is a schematic view of the field line transport during reconnection in the magnetosphere. Why does merging at the front magnetopause necessarily imply merging in the tail? Describe what happens to the geomagnetic field lines on the surface of Earth. How long does external (solar wind) plasma need to reach the night-side equatorial plane (use the numbers of the former problem for medium 1)? At what distance downtail will this happen if the solar wind had speed of 500 km s^{-1}?*

Problem 5.10 *Derive the equation for the magnetospheric convection potential. Why does the longitude ψ appear in this expression? Which condition has the plasma to satisfy in order for the electric equi-potentials being plasma drift paths?*

Problem 5.11 *Derive the expression for the corotation electric field $\mathbf{E}_E$ and potential ϕ_{cr}.*

Problem 5.12 *At which distance for rigid rotation of the magnetic field with the planetary body will corotation not manage anymore to hold the plasma? What distance would this be in the case of Earth? Hint: Rotation implies centrifugal forces.*

Problem 5.13 *Give the explicit expression for the sum of corotation and convection potentials. Describe why the plasmapause is formed. At what distance would the plamapause be found when the Earth rotated 4 times faster?*

Problem 5.14 *What happens with plasma around the stagnation point when the magnetospheric conditions would suddenly change? Say, convection would suddenly become stronger by a factor 2 as it happens during magnetic storms? Hint: Consider the stagnation point distance L_{sp}!*

Problem 5.15 *Which simplification has been made in deriving the height integrated conductivities? When would it not be permitted to make them?*

Problem 5.16 *Consider the expression for the parallel current in the auroral ionosphere. List all the possible causes for generation of parallel currents. Will the parallel currents be distributed smoothly if there is small-scale structure and inhomogeneities in the ionosphere? Which structures would be dominant and how do you believe would they be caused in the auroral ionosphere?*

Problem 5.17 *From merging at the magnetopause we know that there exists a permanent neutral line in the far downstream magnetotail as otherwise the process could not be stationary. Why, during substorms, does an extra neutral line form much closer to Earth? What, knowing the reasons for formation of the distant neutral line, could be an external reason for the near-Earth neutral line? If this reason would be lacking, what else would be required? Hint: Think of any possible nonstationarity of the magnetosphere!*

Problem 5.18 *Derive the expression for the imperfect Cowling conductance Σ'_C and discuss its relevance in auroral physics.*

— 6 —

Elements of Kinetic Theory

Collective behaviour leads to entirely new and otherwise unknown effects. Some of these effects have already been mentioned in Sections 1.1 and 4.1. The collective interactions were hidden behind such terms as the Debye length, plasma frequency, plasma parameter, and Coulomb logarithm. These quantities attribute common average properties to large parts of the plasma consisting of very many particles or to the plasma as such, describing its average behaviour in a global way.

The collective behaviour has its roots in the many-particle character of plasmas. The reason for its appearance is the existence of long-range interparticle interactions between the charged particle components of the plasma due to the electric fields, $\mathbf{E}(\mathbf{x},t)$, connected to each point charge, q, and the magnetic fields, $\mathbf{B}(\mathbf{x},t)$, generated when the charges move at a given velocity, $\mathbf{v}$. The other charged particles in the plasma respond to these fields in a way which leads to momentum and energy exchange between the particles as well as the fields. As a consequence, for a plasma consisting of many particles, with each of them generating its own field and reacting to the microscopic fields of other particles, the actual field configuration is the sum over all the microscopic contributions of the particles to the fields. This average field is of extremely complicated spatial structure and, in addition, varies on a variety of different time scales. At the same time, the motion of the particles in all the microscopic fields is far from the simple motion of single particles discussed before. Consequently, accounting for all the fields and the full particle dynamics is a very complex and almost untreatable task.

The obviously most precise way to obtain more realistic average field and particle configurations than assumed in the previous chapters is to consider all the self-generated microscopic fields, to calculate the trajectories of all particles in these fields by solving their equations of motion and to self-consistently account for their self-generated fields during this motion. Subsequently, one should average over the fastest time scales which one is not interested in when describing average properties of the plasma. It is obvious that such a kind of description immediately runs into enormous computational difficulties which cannot be overcome by even the fastest computers available. Hence, one seeks for a more approximate description. Such a description is necessarily of statistical nature and is found in the so-called *kinetic plasma theory*, the basic elements of which will be developed in this chapter as a preparation for later use and further simplifications.

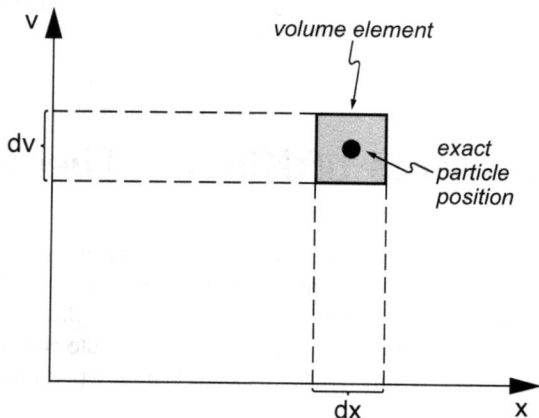

Figure 6.1: Particle position in a phase space volume element.

6.1 Exact Phase Space Density

In contrast to the point of view taken in the previous chapters, where the plasma consisted of single particles, we now assume that it is a *strongly interacting system* of very many particles, each having a time dependent position $\mathbf{x}_i(t)$ and velocity $\mathbf{v}_i(t)$. It is then useful to take these positions and velocities as independent coordinates in a hypothetical six-dimensional space with the coordinate axes $(\mathbf{x}, \mathbf{v})$, called the *phase space*. The particle at a certain time t_0 is then characterized as one point in this space inside a phase space volume element, $d\mathbf{x}d\mathbf{v}$ (see Fig. 6.1), and the particle path at subsequent times $t_1, t_2, \ldots, t_5, \ldots$ is a curve in this phase space as indicated in Fig. 6.2, because of the three space and three velocity coordinates it is actually a six-dimensional curve.

6.1.1 Exact Particle Density

In the presence of very many particles, as in the case of a plasma, one may define an exact number density $\mathscr{F}_i$ of the single i-th particle through

$$\mathscr{F}_i(\mathbf{x}, \mathbf{v}, t) = \delta\big(\mathbf{x} - \mathbf{x}_i(t)\big)\delta\big(\mathbf{v} - \mathbf{v}_i(t)\big) \tag{6.1}$$

where $\delta(\mathbf{x} - \mathbf{x}_i) = \delta(x - x_i)\delta(y - y_i)\delta(z - z_i)$ and $\delta(\mathbf{v} - \mathbf{v}_i)$ are three-dimensional Dirac delta functions and, as usual in particle physics, it is assumed that the particle has zero extension which is valid on all scales of interest in plasma physics. Equation (6.1) then tells that the particle density in phase space is different from zero only at the position and velocity of the i-th particle at time t which is the dynamic path of the particle it performs under the action of all the forces it experiences. For the single i-th particle this density is singular because its position in phase space, $\mathbf{x} = \mathbf{x}_i$ and $\mathbf{v} = \mathbf{v}_i$, is nothing else but the singular point in the small phase space volume of Fig. 6.1. Integrating over the full phase space gives the value one, which means that the particle can be found with certainty somewhere in phase space.

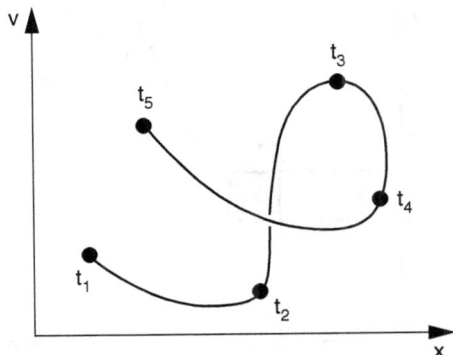

Figure 6.2: The path of a particle in phase space under the action of forces. The particle shifts in time from t_1 to t_5 both in location and velocity in a sequence of being retarded, accelerated, again retarded and reaccelerated ending up neither at the original location nor initial velocity.

Since the history of the particle in phase space is described by the set of all points it assumed between the initial time t_0 and the actual time t, the exact particle density is a function of the phase space coordinates and time, as written explicitly in Eq. (6.1). The total exact particle density function of the plasma is then the sum over all the single exact particle densities given by Eq. (6.1)

$$\mathscr{F}(\mathbf{x}, \mathbf{v}, t) = \sum_i \delta(\mathbf{x} - \mathbf{x}_i(t)) \delta(\mathbf{v} - \mathbf{v}_i(t)) \tag{6.2}$$

If the plasma consists of several components, this equation holds separately for each component. To obtain the total exact phase space density, one has to sum Eq. (6.2) over all particle species.

The geometrical content of the definition Eq. (6.2) is that the phase space volume occupied by the plasma consists of all the phase space points of the single particles or, equivalently, of all the single particle phase space volume elements of Fig. 6.1. Because the particles building up the phase space volume of the plasma are subject to the action of forces, and the forces are different for each of the particles, the phase space volume of the plasma will deform under the action of forces. Because the number of particles does not change, however, the volume remains constant, merely changing its shape. As shown in Fig. 6.3, the simplest kind of change is a mere rotation and stretching of the volume caused by a slight change in the particle velocity, $\mathbf{v}$, and position, $\mathbf{x}$, under the action of a microscopic force, e.g., microscopic electric fields, collisions between particles, anomalous collisions, etc., during the time dt.

6.1.2 Equation of Motion

It is important to remember that $\mathbf{x}$ and $\mathbf{v}$ are independent coordinates in phase space. The particle position itself is determined by its equation of motion under the

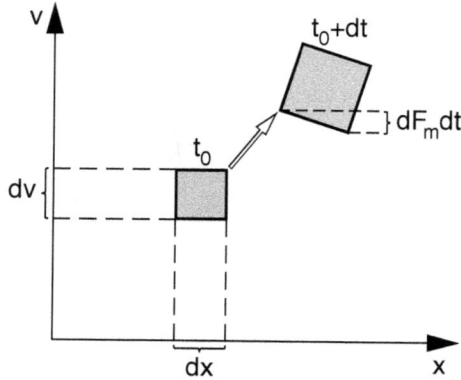

Figure 6.3: Deformation of a phase space element $dx\,dv$ under the action of a microscopic force which stretches and rotates the element.

action of all the microscopic electromagnetic fields, where the instantaneous particle velocity is given by $\mathbf{v}_i(t) = d\mathbf{x}_i(t)/dt$ and d/dt indicates the total derivative with respect to time. Denoting the microscopic fields by the index m, the equation of motion (2.7) reads

$$\frac{d}{dt}\mathbf{v}_i(t) = \frac{q}{m}\left[\mathbf{E}_m(\mathbf{x}_i(t),t) + \mathbf{v}_i(t) \times \mathbf{B}_m(\mathbf{x}_i(t),t)\right] \tag{6.3}$$

The microscopic electric and magnetic fields depend on the particle position, which is a function of time, and also explicitly on time. They are defined as the fields generated by all the particles in the plasma at the exact instantaneous position of the i-th particle and satisfy the microscopic Maxwell equations

$$\nabla \times \mathbf{B}_m(\mathbf{x},t) = \mu_0\mathbf{j}_m(\mathbf{x},t) + \varepsilon_0\mu_0\frac{\partial}{\partial t}\mathbf{E}_m(\mathbf{x},t) \tag{6.4}$$

$$\nabla \times \mathbf{E}_m(\mathbf{x},t) = -\frac{\partial}{\partial t}\mathbf{B}_m(\mathbf{x},t) \tag{6.5}$$

$$\nabla \cdot \mathbf{E}_m(\mathbf{x},t) = \frac{1}{\varepsilon_0}\rho_m(\mathbf{x},t) \tag{6.6}$$

$$\nabla \cdot \mathbf{B}_m(\mathbf{x},t) = 0 \tag{6.7}$$

The coupling of the i-th particle to all other plasma particles proceeds via the microscopic electric space charge, ρ_m, and current densities, $\mathbf{j}_m$, of all particles, which generate the electric and magnetic fields $\mathbf{E}_m$ and $\mathbf{B}_m$, respectively. These charge and current densities are defined as

$$\rho_m(\mathbf{x},t) = \sum_s q_s \int \mathscr{F}_s(\mathbf{x},\mathbf{v},t)\,d^3v \tag{6.8}$$

$$\mathbf{j}_m(\mathbf{x},t) = \sum_s q_s \int \mathscr{F}_s(\mathbf{x},\mathbf{v},t)\,\mathbf{v}\,d^3v \tag{6.9}$$

where $d^3v = d\mathbf{v} = dv_x dv_y dv_z$ and the sum has to be taken over all particle species, electrons, protons and other ions, with exact phase space densities $\mathscr{F}_s$ and charges q_s.

The equation of motion (6.3) together with the microscopic Maxwell equations, which govern the electric and magnetic fields, form a system of equations which is exact and self-consistent. It describes all particle motions and all fields in the plasma in a given phase space. Particle-in-Cell (PIC) computer simulations of plasmas can be based on it. One simply assumes an initial given exact phase space particle density in the simulation box together with an initial given field configuration and then solves the equation of motion for all the particles together with the Maxwell equations for the fields at each point for successive times. The results of such calculations are phase space evolution plots at several fixed successive times consisting of a large number of points. However, due the large particle numbers in a plasma such calculations require enormous amounts of computer time. It is, in fact, still impossible to simulate all the particles in even a small number of Debye spheres each containing some 10^20 particles. Currently at the time of writing this treatise the maximum number of particles in a simulation is about 10^{12}. Simulations therefore refer to 'macro-particles' assuming that each simulated particle represents a very large number of real particles, i.e. plays the role of one particle picked out of a large group of particles which for some unidentified reason have a common dynamics. We do not go either into the technical problems involved nor into the interpretational problems of such simulations.

6.1.3 Klimontovich-Dupree Equation

If no particle is lost from or added to the plasma, the exact phase space density (6.2) is conserved during the dynamic evolution of the plasma. Thus the total time derivative of $\mathscr{F}(\mathbf{x},\mathbf{v},t)$, taken along the dynamic path of all the particles in the plasma, i.e., along the path of the phase space volume occupied by the plasma, must vanish

$$\frac{d}{dt}\mathscr{F}(\mathbf{x},\mathbf{v},t) = 0 \tag{6.10}$$

Since both $\mathbf{x}$ and $\mathbf{v}$ depend on time, we have to use the differential chain rule

$$\frac{d}{dt}f[g(t)] = \frac{df}{dg}\frac{dg}{dt} \tag{6.11}$$

to write the total time derivative in six-dimensional phase space as

$$\frac{d}{dt} = \frac{\partial}{\partial t} + \mathbf{v}\cdot\nabla_{\mathbf{x}} + \frac{d\mathbf{v}}{dt}\cdot\nabla_{\mathbf{v}} \tag{6.12}$$

where the indices of the two ∇ operators indicate differentiation with respect to particle position and velocity. Equation (6.12) is a *convective derivative* since it gives the time derivative of a quantity along a particle trajectory in phase space.

Using the equation of motion (6.3) to replace the derivative of the velocity, one obtains the following expression

$$\frac{\partial \mathscr{F}}{\partial t} + \mathbf{v} \cdot \nabla_{\mathbf{x}} \mathscr{F} + \frac{q}{m}\left(\mathbf{E}_m + \mathbf{v} \times \mathbf{B}_m\right) \cdot \nabla_{\mathbf{v}} \mathscr{F} = 0 \qquad (6.13)$$

Equation (6.13) is the evolution equation of the exact particle density in phase space. It is known under the name *Klimontovich-Dupree equation* and describes the plasma state in phase space at all times. It provides a clear insight into the function of the phase space and its connection to the concept of particle distributions. However, it does not offer a method of calculation of the exact distribution function $\mathscr{F}$ other than through the solution of all the single particle equations of motion. It is thus just a compact description of the particle dynamics in phase space. In order to make some progress in the calculation we will have to make certain approximations on the cost of losing exactness but gaining solutions which can be compared with measurements and observations.

6.2 Average Distribution Function

The Klimontovich-Dupree equation still contains the exact microscopic fields. Solving it is a difficult task; it is impractical if many particles are involved. One therefore seeks for a simpler way which is ultimately found in the method of averaging the above distribution over a larger number of particles, considering them as being statistically correlated in time, space, and velocity through their mutual interactions.

6.2.1 Kinetic Equation

To this end one defines an *ensemble-averaged phase space density*, $\langle \mathscr{F}(\mathbf{x}, \mathbf{v}, t) \rangle = f(\mathbf{x}, \mathbf{v}, t)$, and express the exact phase space density as the sum of its average and a fluctuation, $\delta \mathscr{F}$, which accounts for the deviation of the exact phase space density from the average distribution as

$$\mathscr{F}(\mathbf{x}, \mathbf{v}, t) = f(\mathbf{x}, \mathbf{v}, t) + \delta \mathscr{F}(\mathbf{x}, \mathbf{v}, t) \qquad (6.14)$$

Since the fluctuations should form a statistical ensemble, the ensemble average over the fluctuation is equal to zero, $\langle \delta \mathscr{F} \rangle = 0$. The term ensemble average is in statistical mechanics not well defined; one understands under an ensemble average an average over all the particle properties over all particles in the ensemble. Clearly then, considering one property of the particles, relative to the average of this property in the ensemble there must be equal numbers of deviations of this property in both directions, and the average over these deviations must vanish. This is the content of the above decomposition. In a similar way one decomposes the microscopic fields as sums of averages and fluctuations

$$\mathbf{E}_m(\mathbf{x},\mathbf{v},t) = \mathbf{E}(\mathbf{x},\mathbf{v},t) + \delta\mathbf{E}(\mathbf{x},\mathbf{v},t)$$
$$\mathbf{B}_m(\mathbf{x},\mathbf{v},t) = \mathbf{B}(\mathbf{x},\mathbf{v},t) + \delta\mathbf{B}(\mathbf{x},\mathbf{v},t) \tag{6.15}$$

with $\langle \delta\mathbf{B} \rangle = 0$ and $\langle \delta\mathbf{E} \rangle = 0$. Inserting Eqs. (6.14) and (6.15) into Eq. (6.13) and taking the ensemble average yields the *kinetic equation* for the average phase space density of a plasma

$$\frac{\partial f}{\partial t} + \mathbf{v} \cdot \nabla_{\mathbf{x}} f + \frac{q}{m}(\mathbf{E} + \mathbf{v} \times \mathbf{B}) \cdot \nabla_{\mathbf{v}} f = -\frac{q}{m} \langle (\delta\mathbf{E} + \mathbf{v} \times \delta\mathbf{B}) \cdot \nabla_{\mathbf{v}} \delta\mathscr{F} \rangle \tag{6.16}$$

The kinetic equation describes the evolution of the coarse-grained phase space density $f(\mathbf{x},\mathbf{v},t)$ in time and space under the action of the average fields $\mathbf{E}, \mathbf{B}$. This coarse-grained density is called the (one particle) *distribution function* of the particles in phase space and is interpreted as the probability each single particle must have in order to be found in a certain phase space volume element $d\mathbf{x}d\mathbf{v}$ (see Fig. 6.1).

The fields in Eq. (6.16) are average fields. They are governed by average Maxwell equations, which can be obtained from the microscopic equations by inserting the decomposed fields from Eq. (6.15) and similarly decomposed charge and current densities into the latter and taking the ensemble average. Because the Maywell equations are linear in the fields, the form of the resulting Maxwell equations for the average fields is the same as that of the microscopic equations, but now containing average charge densities and currents.

The kinetic equation has the advantage that both the average distribution $f(\mathbf{x},\mathbf{v},t)$ and the average fields do not depend any more on the single coordinates of all the single particles of a species, but only depend on the phase space coordinates $(\mathbf{x},\mathbf{v},t)$. The ensemble average has smeared out the exact positions of the particles over the phase space volume occupied by the particle group under consideration. Hence, the average particle distribution does not describe any more the exact position of the particles in the volume but instead accounts for the probability to find the ensemble group in question in the interval $\{\mathbf{x},\mathbf{x}+d\mathbf{x}\}, \{\mathbf{v},\mathbf{v}+d\mathbf{v}\}$. The function $f(\mathbf{x},\mathbf{v},t)$ has thus become a probability distribution function for groups of particles with similar properties, and the above equation is the dynamic equation for the evolution of the probability function under the action of the average fields on the particles.

6.2.2 Boltzmann Equation

The term in angular brackets on the right-hand side of Eq. (6.16) contains all the correlations between the fields, particles and their fluctuations. Its calculation poses a very serious problem. Various steps of evaluation can be imagined. Since $f(\mathbf{x},\mathbf{v},t)$ does not distinguish anymore between single particles but accounts only for their dependence on space and velocity, evaluation of the correlation term must provide this information in terms of two-point correlations, three-point correlations, and so on.

One way to simplify the kinetic equation is to neglect the correlations between the fields and to account only for correlations between the particles themselves via collisions. Then Eq. (6.16) can be written as

$$\frac{\partial f}{\partial t} + \mathbf{v} \cdot \nabla_{\mathbf{x}} f + \frac{q}{m}(\mathbf{E} + \mathbf{v} \times \mathbf{B}) \cdot \nabla_{\mathbf{v}} f = \left(\frac{\partial f}{\partial t}\right)_c \qquad (6.17)$$

where the right-hand side is the time rate of change of $f(\mathbf{x}, \mathbf{v}, t)$ due to all kinds of collisions. This equation is the generalized *Boltzmann equation*, which is well-known from statistical mechanics.

In order to solve Eq. (6.17), the exact functional form of its right-hand side has to be specified. For hard-core collisions between particles, the collision term on the right-hand side has been evaluated already by Boltzmann himself at the end of the nineteenth century. In the simplest case of collisions in a plasma that is collisions between charged particles and neutrals in a partially ionized plasma, the collision term can be approximated by the so-called *Krook collision term*

$$\left(\frac{\partial f}{\partial t}\right)_c = v_n(f_n - f) \qquad (6.18)$$

where $f_n(\mathbf{x}, \mathbf{v}, t)$ is the distribution function of the neutral atoms and v_n is the neutral collision frequency defined in Section 4.1. In a fully ionized plasma the long-range Coulomb interactions between the particles complicate the situation considerably. Their inclusion replaces the collision term with the *Landau collision integral* which can be approximated by taking into account the Coulomb collision frequency (see Section 4.1). Since the latter depends on density and temperature, the collision term is further complicated. In the case of fully ionized collisionless dilute plasmas, the collision term vanishes, but the right-hand side of the Boltzmann equation is then dominated by correlations between the particles which are caused by their contributions to the field variations. This correlation term becomes a rather complicated function of the change of velocity of the particles. The equation including this quasi-collision term

$$\left(\frac{\partial f}{\partial t}\right)_c = \nabla_{\mathbf{v}} \cdot (\mathbf{D} \cdot \nabla_{\mathbf{v}} f) \qquad (6.19)$$

is called *Fokker-Planck equation*. Here, $\mathbf{D}(\mathbf{v})$ is a velocity space diffusion coefficient, which is actually a tensor. It is a function of velocity and is derived from the averages over the first- and second-order fluctuations, $\langle \Delta \mathbf{v} \rangle$ and $\langle \Delta \mathbf{v} \Delta \mathbf{v} \rangle$, of the particle velocities. The resulting collisions are not collisions in the usual sense but instead describe changes of the particle velocities under the action of the mutual microscopic particle fields. Thus it is a diffusion in phase space. In a later chapter we will encounter such an equation when considering wave particle interactions in a plasma.

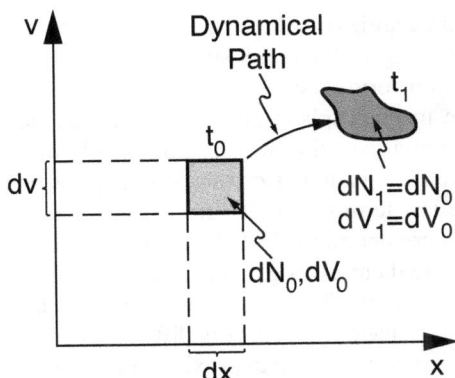

Figure 6.4: Illustration of Liouville's theorem which states that the phase space volume remains constant along the dynamical phase-space particle paths enclosed in the volume while the shape of the volume can be deformed arbitrarily.

6.2.3 Vlasov Equation

Since space plasmas are collisionless, except for the ionosphere, one can often entirely neglect the collision term in the Boltzmann equation. This results in the simplest possible form of kinetic equation of a plasma, the *Vlasov equation*

$$\frac{df}{dt} \equiv \frac{\partial f}{\partial t} + \mathbf{v} \cdot \nabla_{\mathbf{x}} f + \frac{q}{m} \left(\mathbf{E} + \mathbf{v} \times \mathbf{B} \right) \cdot \nabla_{\mathbf{v}} f = 0 \qquad (6.20)$$

In the absence of collisions the phase space density remains constant under the interaction of the particles with the ensemble averaged self-consistent fields in the Lorentz force as it is convected with the particles. This behaviour is the content of *Liouville's theorem*, which states that the phase space volume can be deformed but its density is not changed during the dynamic evolution of the plasma. The Liouville theorem holds exactly for the exact distribution function. In the case of the Vlasov equation, the Liouville theorem holds only if collisions and correlations between the particles and microscopic fields can be neglected. For most applications this approximation is valid.

The Liouville theorem states that a phase space volume element, dV_0, moves under the action of the Lorentz force like an incompressible fluid in phase space, because $\nabla \cdot \mathbf{v} = 0$ holds for the phase space coordinates. This behaviour is visualized in Fig. 6.4. Imagine a phase space element $dx dv$ with density $f(\mathbf{x}_0, \mathbf{v}_0, t_0)$. At time t_0 all particles in this volume element have nearly the same position and velocity. At later time t_1 the particles will have moved to different positions, with their slightly different initial velocities and under the action of the Lorentz forces acting on them. This leads to deformation of the phase space volume element. However, because the number of the particles in the element is conserved, the volume of the phase space element, dV_1, i.e., the total number of points occupied by the dN_0 particles, is conserved. It is constant along the dynamical trajectories of all the particles it contains. Hence, the phase space density, f, is constant along such an orbit such that $f(\mathbf{x}_1, \mathbf{y}_1, t_1) = f(\mathbf{x}_0, \mathbf{v}_0, t_0)$.

The Vlasov equation forms the basis of all kinetic theory in collisionless plasmas like those in the magnetosphere and solar wind. Its mathematical structure is that of a partial differential equation which, though only of first order, is coupled to the full set of Maxwell's equations (for the ensemble averaged fields, charges and currents) through the last term on the left-hand side of Eq. (6.20). In addition, the electromagnetic fields are determined by the charge and current densities which, as will be proved below, are themselves given as integrals over the distribution function. One therefore realizes that the Vlasov equation is in fact a highly nonlinear equation in six-dimensional phase space which is very difficult to solve in full generality. This difficulty forces one to seek for further approximative methods to find solutions under special conditions and in special regimes of the plasma. The Vlasov equation forms the basis for the discussion of plasma processes in this book.

6.2.4   Approximations to the Vlasov Equation

The full electromagnetic Vlasov equation (6.20) is exact in the sense of neglecting particle correlations and using ensemble averaged fields. It represents a total differential along the dynamical orbit of the phase space element. This property can be used to transform the Vlasov equation to other, sometimes more convenient coordinates.

Electrostatic Vlasov Equation

The first reasonable approximation to the Vlasov equation is the assumption of an unmagnetized plasmas in the absence of currents. In this case there is no magnetic field, and the electric field is a pure potential field $\mathbf{E} = -\nabla_{\mathbf{x}}\phi$ with potential ϕ. This simplifies the third term in Eq. (6.20) and frees one from three of the Maxwell equations. One is left with only two scalar equations, the *electrostatic Vlasov equation*

$$\frac{\partial f_s}{\partial t} + \mathbf{v} \cdot \nabla_{\mathbf{x}} f_s - \frac{q}{m}(\nabla_{\mathbf{x}}\phi) \cdot \nabla_{\mathbf{v}} f_s = 0 \tag{6.21}$$

for the distribution function of particle species s, and the Poisson equation for the electrostatic potential as function of the total space charge density $\rho = \sum_s q_s n_s$, with the sum taken over all particle species with particle charge q_s and number density n_s. The Poisson equation reads

$$\nabla_{\mathbf{x}}^2 \phi = -\frac{\rho}{\varepsilon_0} = -\sum_s \frac{q_s}{\varepsilon_0} \int f_s \, d^3 v \tag{6.22}$$

Its right-hand side is the charge density expressed now in terms of the distribution functions f_s of the different species. This form takes care of summing over all particles and their charges in phase space.

The Poisson equation couples all particle species together in their contributions to the electric potential viz. electric field in the plasma. It shows that even in this simple case the particle distributions of the different plasma components are not independent; they are related through the average field that is created by the total charge content of the plasma. This does still hold when the single particle fields are screened in their Debye spheres showing that there can be no absolutely complete screening. We will later see that under completely quasi-neutral conditions when $\rho = 0$ electric fluctuations will be generated in the plasma with non-vanishing potential and oscillations in the distribution function and particle densities.

Since these two equations are scalar they describe the most simple kinetic form of a plasma. However, even in this simple case one observes that the system is highly nonlinear with f_s appearing on the right-hand side of Poisson's equation, and $\nabla_x \phi$ being multiplied with the velocity derivative of f_s in Vlasov's equation. This system will later serve us as starting point for investigating the microscopic theory of waves in a plasma.

Drift-Kinetic Equation

Another option is, for instance, to not use the full particle velocity in the Vlasov equation but to restrict to drift motions which conserve the magnetic moment μ of the particles. Then one may replace the one-particle velocity in the Vlasov equation (6.20) with the general total particle drift velocity

$$\mathbf{v}_d = \frac{v_\parallel \mathbf{B}}{B} + \mathbf{v}_E + \frac{\mathbf{F} \times \mathbf{B}}{qB^2} \tag{6.23}$$

which includes the parallel velocity $v_\parallel$ and has become a function of the generalised perpendicular force

$$\mathbf{F}_\perp = -\mu \nabla_\perp B - mv_\parallel^2 \frac{\mathbf{R}_c}{R_c^2} - m\frac{d\mathbf{v}_E}{dt} \tag{6.24}$$

and is now a function of space and time. Therefore, to be correct one must use the conservation equation form of the Vlasov equation when performing the replacement. Moreover, since the perpendicular electric field has already been absorbed in $\mathbf{v}_E$ in the expression for the drift velocity, the entire force term is to be replaced by the only surviving parallel force

$$F_\parallel = -\mu \nabla_\parallel B + qE_\parallel \tag{6.25}$$

which contains the magnetic moment and the parallel electric field. With these replacements the average drift distribution $f_d(v_\parallel, \mu, \mathbf{x}, t)$ depends on the parallel velocity, $v_\parallel$, the magnetic moment μ, space and time, and evolves according to the *drift kinetic equation*

$$\boxed{\frac{\partial f_d}{\partial t} + \nabla_x \cdot (\mathbf{v}_d f_d) + \frac{\partial}{\partial v_\parallel}\left(\frac{F_\parallel}{m} f_d\right) = 0} \tag{6.26}$$

In this form it is assumed that the gradient scales $L_x \equiv \nabla_x^{-1} \gg r_g$ are large compared with the particle gyro-radius. Thus the equation describes particle mirroring and related effects with time variation of the order of the bounce frequency. Electric, gradient, curvature and polarisation drifts in an inhomogeneous magnetic field and variable electric fields are included in this form of the Vlasov equation which therefore is suitable to investigate the average behaviour of trapped plasmas in the magnetospheric magnetic field as well as the electric convection field and its slow variation conserving the particle magnetic moment.. If one wants to include effects on the scale of the gyro-radius second-order terms in $r_g \nabla_x$ must be included. This can be achieved by expanding the Vlasov equation with respect to the radio r_g / L_x which then leads to an equation which permits to include finite-gyro-radius effects. Such effects are equivalent to viscous interactions. Derivation of the complete equation in second order already is a formidable task.

In the above equation the distribution function is still a function of the magnetic moment which is a constant of motion along phase space orbits. Hence there is no term in the equation which explicitly depends on a derivative with respect to $\mu = m v_\perp^2 / 2B$ because such a term would be multiplied by $d\mu/dt = 0$. When, however, a variation of the magnetic moment would be included, such a term should be added to the drift kinetic equation which takes care of the derivative of the distribution function with respect to the magnetic moment. Assuming that the variation with μ is caused by a variation of the perpendicular energy respectively perpendicular velocity $\mathbf{v}_\perp$ this additional term would be modelled by $\nabla_{v_\perp} \cdot (\mathbf{F}_\perp f_d / m)$. Of the perpendicular force only the term including μ needs to be taken. Then, replacing the derivative with respect to the perpendicular velocity yields the equation

$$\frac{\partial f_d}{\partial t} + \nabla_\mathbf{x} \cdot (\mathbf{v}_d f_d) - \left(\nabla_\perp \sqrt{\frac{2B}{m}} \right) \frac{\partial}{\partial \sqrt{\mu}} (\mu f_d) + \frac{\partial}{\partial v_\parallel} \left(\frac{F_\parallel}{m} f_d \right) = 0 \qquad (6.27)$$

which now takes care of a weak variation of the distribution with the magnetic moment during the evolution of the plasma.

Gyro-Kinetic Equation

One particular set of coordinates are guiding center coordinates $(\mathbf{X}, v_\parallel, \mathbf{v}_\perp, \psi, t)$, where $\mathbf{X}$ is the guiding center position and ψ the particle gyration phase-angle. Introducing such coordinates becomes useful if one considers the behaviour of the distribution function averaged over the gyratory motion, $\langle f(\mathbf{X}, v_\parallel, \mathbf{v}_\perp, \psi, t) \rangle$. The relation between the guiding center and the exact particle coordinates is

$$\mathbf{x} = \mathbf{X} - \frac{\mathbf{v} \times \mathbf{B}}{\omega_g B}$$

$$\mathbf{v}_\perp = v_\perp (\hat{\mathbf{e}}_x \cos \psi - \hat{\mathbf{e}}_y \sin \psi) \qquad (6.28)$$

where the magnetic field direction is thought to be aligned with the z axis and $\hat{\mathbf{e}}_x$ and $\hat{\mathbf{e}}_y$ are unit vectors along the two other axes. Let us for convenience restrict to the electrostatic limit when no magnetic field changes occur ($\partial \mathbf{B}/\partial t \approx 0$). Introducing these expressions into Eq. (6.20), performing the various first-order differentiations and rearranging, one obtains the Vlasov equation expressed in the new coordinates as

$$\frac{\partial f}{\partial t} + \left(\frac{v_{\parallel}\mathbf{B}}{B} + \mathbf{v}_E\right) \cdot \nabla_{\mathbf{X}} f + \frac{qE_{\parallel}}{m}\frac{\partial f}{\partial v_{\parallel}} + \frac{q\mathbf{E}_{\perp} \cdot \mathbf{v}_{\perp}}{mv_{\perp}}\frac{\partial f}{\partial v_{\perp}} + \omega_g\left(1 - \frac{\mathbf{v}_E \cdot \mathbf{v}_{\perp}}{v_{\perp}^2}\right)\frac{\partial f}{\partial \psi} = 0$$

where $\nabla_{\mathbf{X}}$ denotes the vector derivative with respect to guiding center position. In this representation only the electric field drift, $\mathbf{v}_E$, is included while gradient and curvature drifts have been neglected. One must now average over the ensemble gyratory motion of the particles to obtain the evolution of $\langle f \rangle$ in phase space. This average assumes that the distribution is to first-order constant over the gyro-orbit, i.e., the gyroradius is much smaller than the typical scale length over which the density varies. Then $\partial \langle f \rangle/\partial \psi = 0$ and $\langle \mathbf{E}_{\perp} \cdot \mathbf{v}_{\perp}\rangle = 0$, and only non-adiabatic effects contribute to the evolution of $\langle f \rangle$. The equation obtained is the so-called *gyrokinetic equation* in the electrostatic limit

$$\boxed{\frac{\partial \langle f \rangle}{\partial t} + \left(v_{\parallel}\frac{\mathbf{B}}{B} + \mathbf{v}_E\right) \cdot \nabla_{\mathbf{X}}\langle f \rangle + \frac{q}{m}E_{\parallel}\frac{\partial}{\partial v_{\parallel}}\langle f \rangle = 0} \qquad (6.29)$$

where $E_{\parallel}$ is the gyro-averaged parallel electric field which does not depend on particle velocity. One basic difference to the Vlasov equation is that here the coefficient of the second term is a function of the average space coordinate $\mathbf{X}$ because $\mathbf{B}$ and $\mathbf{v}_E$ are both functions of space. Therefore, in contrast to the Vlasov equation, this coefficient cannot be exchanged with the operator $\nabla_{\mathbf{X}}$. The non-commutativity is a result of the averaging procedure over the gyromotion. Writing the gyrokinetic equation in the form of a conservation equation

$$\frac{\partial \langle f \rangle}{\partial t} + \nabla_{\mathbf{X}} \cdot \left(v_{\parallel}\frac{\mathbf{B}}{B} + \mathbf{v}_E\right)\langle f \rangle + \frac{q}{m}\frac{\partial}{\partial v_{\parallel}}(E_{\parallel}\langle f \rangle) = \langle f \rangle \nabla_{\mathbf{X}} \cdot \left(v_{\parallel}\frac{\mathbf{B}}{B} + \mathbf{v}_E\right) \quad (6.30)$$

one recognizes that this average introduces a source term on the right-hand side of the equation which contributes to the evolution of the distribution function $\langle f \rangle$ but is itself a function of $\langle f \rangle$.

Here we just derived the electrostatic gyro-kinetic equation for the evolution of the gyro-orbit averaged particle distribution function $\langle f \rangle$. The general equation of motion of charged particles however depends also on the variations of the magnetic field $\delta \mathbf{B}$. These can be caused by fluctuations of the density $\mathbf{j}$ of any currents flowing in the plasma either along the magnetic field or perpendicular to it. They can also be caused by the spectrum of low-frequency electromagnetic waves flowing on the plasma background. There is a large number of such waves which in a magnetized plasma are allowed to propagate both along and oblique to the magnetic field. Waves of this kind will be introduced and discussed in Chapter 10. They can be divided into low frequency (or fluid) and high frequency (or kinetic) plasma waves and, as discussed in Chapter 11 under certain very frequently realised conditions are excited in plasma if only some extra free energy is available from which they could be fed and which ignites some instability of the plasma. Taking into account such waves and in general magnetic fluctuations then causes an electromagnetic modification of the gyro-kinetic equation transforming it into a more complicated fully electromagnetic gyro-kinetic equation.

6.3 Velocity Distributions

It is impossible to give graphic representations of the six-dimensional phase space distribution function, $f(\mathbf{x}, \mathbf{v}, t)$, as it varies in space, velocity, and time. But, usually, the most interesting property of the distribution function is its dependence on the velocity at a fixed position in configuration space. Many of the characteristic features of plasmas can be understood by knowing this velocity dependence. In the following we discuss a few examples of typical space plasma *velocity distribution functions*, $f(\mathbf{v})$, assuming the plasma to be spatially homogeneous and, in addition, stationary. Under these conditions the plasma does not change in time and does not exhibit spatial variations. In general, such a situation can be realized only when the plasma is in equilibrium; however, there are cases when the velocity distribution is of a form which is far from equilibrium. In such a case it must either be maintained by external means which inhibit relaxation of the distribution to its equilibrium form, or the time the distribution is observed is short with respect to the relaxation time.

6.3.1 Maxwellian Distributions

The general equilibrium velocity distribution function of a collisionless plasma is the *Maxwellian velocity distribution* or simply Maxwellian. A Maxwellian plasma is in thermal equilibrium which implies that it does not contain anymore free energy and, hence, there are no energy exchange processes between the particles in the plasma. It is then clear that the velocities the particles can assume must be distributed randomly around the average velocity. For a plasma at rest, the latter is zero, and the distribution of the velocities follow a simple Gaussian distribution of errors

$$g(\Delta x) = \left(\pi \langle \Delta x \rangle^2\right)^{-1/2} \exp\left(-\frac{(\Delta x)^2}{\langle \Delta x \rangle^2}\right) \tag{6.31}$$

Replacing Δx with one component of the velocity v_x, replacing the variance $\langle \Delta x \rangle$ with the average velocity spread $\langle v_x \rangle$, and multiplying with the average particle density, n, Eq. (6.31) yields the one-dimensional equilibrium velocity distribution function

$$f(v_x) = \frac{n}{\left(\pi \langle v_x \rangle^2\right)^{1/2}} \exp\left(-\frac{v_x^2}{\langle v_x \rangle^2}\right) \tag{6.32}$$

This distribution can be generalized to three dimensions by observing that $v^2 = v_x^2 + v_y^2 + v_z^2$. Hence, multiplying the three one-dimensional Maxwellians in the three

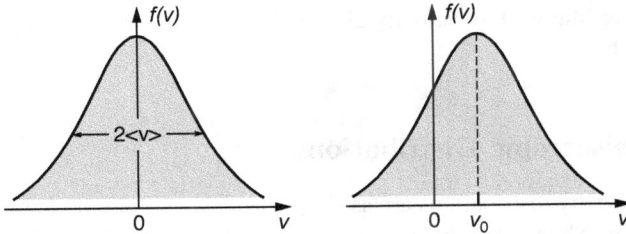

Figure 6.5: *Left*: Maxwellian distribution of velocity spread $\langle v \rangle$. *Right*: The same distribution shifted by the bulk velocity v_0 becomes a drifting Maxwellian velocity distributions.

orthogonal directions and taking care of the normalisation to the average density n one finds the full Maxwellian in an isotropic plasma

$$f(v) = \frac{n}{(\pi \langle v \rangle^2)^{3/2}} \exp\left(-\frac{v^2}{\langle v \rangle^2} \right) \tag{6.33}$$

Sometimes this function is written conveniently in the form

$$\boxed{f(v) = n \left(\frac{m}{2\pi k_B T} \right)^{3/2} \exp\left(-\frac{mv^2}{2k_B T} \right)} \tag{6.34}$$

with m denoting the particle mass and $k_B T$ the average thermal energy. The velocity spread, $\langle v \rangle = (2k_B T/m)^{1/2}$, can be identified as the thermal velocity, a relation which will be derived below in Section 6.5. Using the integral

$$\int_{-\infty}^{\infty} \exp(-x^2)\, dx = \sqrt{\pi} \tag{6.35}$$

it is easy to verify that the integral of the Maxwellian over the whole three-dimensional velocity space is n, the macroscopic number density, thus confirming the correct normalisation of the (probability) distribution. Hence, the Maxwellian velocity distribution tells us how the particle density in equilibrium, at a given point in space and time, is distributed over velocity space, depending on the average thermal energy of the particles. This function is sketched in the left-hand panel of Fig. 6.5. Its functional form is symmetric with respect to the three velocity components and only depends on the magnitude v of the velocity, and its half-width gives the average velocity spread.

A simple further generalization of the equilibrium velocity distribution function is obtained by observing that in a plasma streaming at common velocity, $\mathbf{v_0} = v_0\hat{\mathbf{e}}_x$, in the x direction, the average velocity of the distribution function with respect to v_x is nonzero. Therefore, v_x must be replaced with $v_x - v_0$ or, more generally, $\mathbf{v}$ is replaced by $\mathbf{v} - \mathbf{v_0}$ to obtain

$$f(\mathbf{v}) = n \left(\frac{m}{2\pi k_B T} \right)^{3/2} \exp\left(-\frac{m(\mathbf{v} - \mathbf{v_0})^2}{2k_B T} \right) \tag{6.36}$$

for the drifting Maxwellian velocity distribution, which is sketched in the right-hand panel of Fig. 6.5.

6.3.2 Anisotropic Distributions

Not all velocity distributions are as simple as the isotropic (or drifting) Maxwellian distribution. Already the drifting Maxwellian obeys some kind of asymmetry with respect to the zero point in velocity which is a kind of anisotropy. Especially the presence of magnetic fields introduces an anisotropy because it leads to different particle velocities parallel and perpendicular to the magnetic field. A particularly important case is that of gyrating particles. In this case the velocity distribution is independent of the angle of gyration, depending only on $v_\perp$ and $v_\parallel$. Because these two velocity components are independent, the equilibrium distribution can be modelled as the product of two Maxwellians (6.32) according to

$$f(v_\perp, v_\parallel) = \frac{n}{\pi^{3/2} \langle v_\perp \rangle^2 \langle v_\parallel \rangle} \exp\left(-\frac{v_\perp^2}{\langle v_\perp \rangle^2} - \frac{v_\parallel^2}{\langle v_\parallel \rangle^2}\right) \tag{6.37}$$

The resulting anisotropic distribution function is called *bi-Maxwellian distribution*. It accounts explicitly for the difference in the two average velocities $\langle v_\perp \rangle, \langle v_\parallel \rangle$ in the directions parallel and perpendicular to the magnetic field. Another representation can be based on Eq. (6.34)

$$f(v_\perp, v_\parallel) = \frac{n}{T_\perp T_\parallel^{1/2}} \left(\frac{m}{2\pi k_B}\right)^{3/2} \exp\left(-\frac{mv_\perp^2}{2k_B T_\perp} - \frac{mv_\parallel^2}{2k_B T_\parallel}\right) \tag{6.38}$$

Distributions of the form $f(v_\perp, v_\parallel)$, are the velocity distributions most often found in space plasmas. They are essentially two-dimensional and *gyrotropic velocity distributions*, which do not depend on the phase angle of the gyromotion and are often plotted as contour maps, i.e., curves of $f(v_\perp, v_\parallel) = $ const, or grey-scale plots, where the grey-level encodes the phase space density. The left-hand side of Fig. 6.6 shows a sketch of the bi-Maxwellian (6.37) with $T_\perp > T_\parallel$, i.e., the average thermal energy of the particles perpendicular to the field greater than the average parallel energy. Instead of the circular contours of an isotropic Maxwellian, the contours are deformed into an elliptical shape.

The distribution in the middle of Fig. 6.6, with its circular contours displaced from the origin, is a *drifting Maxwellian* as described by Eq. (6.36). Here all particles drift with the same velocity perpendicular to the field, in addition to their thermal motion. Such a distribution would be representative for a plasma that is in thermal equilibrium, but convects perpendicular to the magnetic field under the action of an external electric field. In cases where the drift velocity is large compared to the thermal velocity

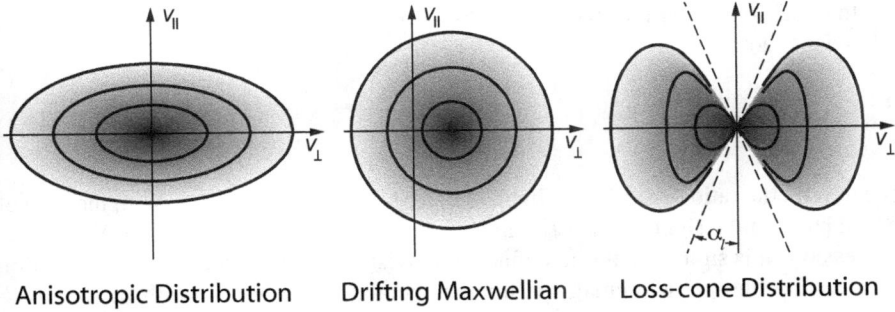

Anisotropic Distribution   Drifting Maxwellian   Loss-cone Distribution

Figure 6.6: Contours of constant phase space distribution f in velocity space $(v_{\parallel}, v_{\perp})$ parallel and perpendicular to an ambient magnetic field **B** for typical anisotropic space plasma velocity distributions.

such a distribution is called a *streaming distribution*. The corresponding perpendicular distribution function is of the same type as Eq. (6.36)

$$f(\mathbf{v}) = \frac{n}{T_{\perp} T_{\parallel}^{1/2}} \left(\frac{m}{2\pi k_B} \right)^{3/2} \exp \left(-\frac{m(\mathbf{v}_{\perp} - \mathbf{v}_{0\perp})^2}{2k_B T_{\perp}} - \frac{mv_{\parallel}^2}{2k_B T_{\parallel}} \right) \tag{6.39}$$

For example, a plasma drift in crossed electric and magnetic fields causes a drift velocity $\mathbf{v}_{0\perp} = \mathbf{v}_E$. In an observer's frame the particle distribution has the form of Eq. (6.39).

In exactly the same way one can model a distribution which drifts along the magnetic field at velocity $v_{0\parallel}$

$$f(v_{\parallel}, v_{\perp}) = \frac{n}{T_{\perp} T_{\parallel}^{1/2}} \left(\frac{m}{2\pi k_B} \right)^{3/2} \exp \left(-\frac{mv_{\perp}^2}{2k_B T_{\perp}} - \frac{m(v_{\parallel} - v_{0\parallel})^2}{2k_B T_{\parallel}} \right) \tag{6.40}$$

Such a distribution function is called a *parallel beam distribution*, a type of distribution function frequently encountered in the auroral magnetosphere, plasma sheet boundary layer, and in the foreshock region in front of the Earth's bow shock wave.

6.3.3   Jüttner Distribution

One may ask whether the relativistic motion of the higher energy particles in plasma (for instance electrons in the Radiation Belts) does reflect itself also in the structure of the Maxwell-Boltzmann distribution. This is indeed the case as reference to the relativity causes a change in the form of the distribution as this can simply be expected by the existence of the speed of light c which poses an upper limit on the achievable speed for any particle massive (like any known particles: leptons and baryons in general) or also massless (like the photon and the gluons in elementary particle theory: the known light bosons).

In relativity the particle energy $\varepsilon = \frac{1}{2}mv^2$ must be replaced by $\varepsilon = m\gamma c^2$ where the relativistic factor

$$\gamma = \left(1 - \frac{v^2}{c^2}\right)^{-1/2} = \sqrt{1 + \left(\frac{p}{mc}\right)^2} \tag{6.41}$$

depends on the ratio of particle to light velocity $\beta = v/c = \sqrt{1 - \gamma^{-2}}$. With the help of β, we have on the right γ expressed in terms of the particle momentum $\mathbf{p}$. With these expressions it is simple matter to define a relativistic Maxwell-Boltzmann distribution by replacing the particle energy in the exponential of Eq. (6.34) with the relativistic expression. Before doing this, let us define the relativistically normalized thermal energy

$$\Theta = k_B T / mc^2 \tag{6.42}$$

Moreover, as in the case of the Maxwell-Boltzmann distribution, the distribution function must again after integration over all the momenta of the particles and space just produce the total particle density n. This requires performing an integration over the momentum dependence of $\gamma(p)$ over all momentum space $d^3 p = 4\pi p^2 dp$, which can be done but leads to the appearance of the modified Bessel function $K_2(\Theta^{-1})$. One finally obtains for the isotropic Jüttner distribution (or relativistic Boltzmann distribution) now as function of the particle momentum the expression

$$f_J(\mathbf{p}) = \frac{n}{4\pi (mc)^3 \Theta K_2(\Theta^{-1})} \exp\left[-\frac{\gamma(p)}{\Theta}\right] \tag{6.43}$$

This is the celebrated *Jüttner distribution* function, an almost trivial though useful extension of Maxwell-Boltzmann's distribution into the relativistic domain. If one wants to express it in terms of the particle energy, then it is more convenient to stay with γ which up to a constant factor is in fact the energy. Then the Jüttner distribution reads

$$\boxed{f_J(\gamma) = \frac{n\gamma^2 \beta(\gamma)}{\Theta K_2(1/\Theta)} \exp\left(-\frac{\gamma}{\Theta}\right)} \tag{6.44}$$

Application of this distribution makes sense whenever one is dealing with the presence of energetic particles in plasma, because it takes care of the high velocity cut-off of the particles which in application of the Maxwell-Boltzmann distribution in such cases always produces some usually neglected effect the over estimation of the presence of such particles in the distribution causes. Such effects are spurious and are indeed suppressed by the Jüttner distribution. Such high energy particles are usually present as particles streams or have other properties than the thermal population. In many cases they obey anisotropies. In this case one needs to consider the anisotropic version of the Jüttner distribution. Its derivation is substantially more complicated than that of the above isotropic result. Here we just provide its expression.

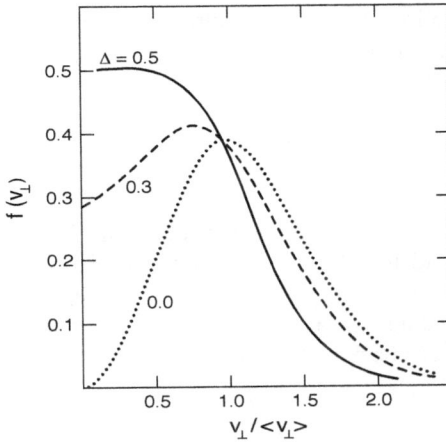

Figure 6.7: The partially filled loss cone distributions as function of the loss-cone filling parameter Δ. Note that increasing Δ not only fills the loss cone but also steepens its slope at larger perpendicular velocities, i.e. cools the distribution in the perpendicular direction.

Define $\Theta_{\perp,\parallel} = k_B T_{\perp,\parallel}$ and temperature anisotropy $A = T_\perp/T_\parallel$, then the anisotropic Jüttner distribution becomes

$$f_J(\mathbf{p}, A) = nC \exp\left\{ -\frac{1}{\Theta_\perp} \sqrt{1 + \frac{p_\perp^2}{m^2 c^2} + A\frac{p_\parallel^2}{m^2 c^2}} \right\} \qquad (6.45)$$

with normalization constant similar to Jüttner's original finding

$$C \equiv \frac{\sqrt{A}}{4\pi (mc)^3 \Theta_\perp K_2(1/\Theta_\perp)} \qquad (6.46)$$

The only difference in this factor is the appearance of the anisotropy A and reference to the perpendicular normalised temperature.

6.3.4 Loss Cone Distributions

In the inner magnetosphere, where mirroring particles with high velocities parallel to the magnetic field may be lost to the ionosphere (see Section 3.2), an originally Maxwellian distribution will loose all particles inside the loss cone, α_ℓ. The right-hand side of Fig. 6.6 shows how such a *loss cone distribution* would look like. The form of the loss cone distribution depends heavily on the processes giving rise to the loss of particles. The simplest form is when the particles at the edge of the loss cone are cut out, but such distributions are quite unrealistic. A reasonable way is to model them via multiplication with some power of the perpendicular velocity $v_\perp^{2j} = v^{2j} \sin^{2j} \alpha$, describing an emptied loss cone near pitch angles $\alpha \approx 0$. Using a symmetric power

$2j$ guarantees that the loss cone is symmetric for pitch angles $\alpha = 0$ and $\alpha = \pi/2$. This way one obtains, after normalization, the so-called *Dory-Guest-Harris loss cone distribution*

$$f(v_\parallel, v_\perp) = \frac{n}{A_j \langle v_\parallel \rangle \langle v_\perp \rangle^{4j}} \left(\frac{v_\perp}{\langle v_\perp \rangle} \right)^{2j} \exp \left(-\frac{v_\parallel^2}{\langle v_\parallel \rangle^2} - \frac{v_\perp^2}{\langle v_\perp \rangle^2} \right) \tag{6.47}$$

and the constant is $A_j = \pi^{2j+1/2}(j!)^{2j}$. This distribution function looks complicated and has a very deep and absolutely empty loss cone. Sometimes the loss cone in the magnetosphere is not empty. In such a case it is more appropriate to use a loss cone distribution which accounts for a partial filling of the loss cone. The *partially-filled loss cone distribution* can be modelled by subtracting simple anisotropic Maxwellians

$$f(v_\parallel, v_\perp) = \frac{n}{\pi^{3/2} \langle v_\parallel \rangle^2 \langle v_\perp \rangle^4} \exp \left(-\frac{v_\parallel^2}{\langle v_\parallel \rangle^2} \right) G(v_\perp, \Delta, \beta) \tag{6.48}$$

The first part of this function is a parallel Maxwellian. The information about the loss cone is contained in the function $G(v_\perp, \Delta, \beta)$

$$G = \Delta \exp \left(-\frac{v_\perp^2}{\langle v_\perp \rangle^2} \right) + \frac{1-\Delta}{1-\beta} \left[\exp \left(-\frac{v_\perp^2}{\langle v_\perp \rangle^2} \right) - \exp \left(-\frac{v_\perp^2}{\beta \langle v_\perp \rangle^2} \right) \right] \tag{6.49}$$

Here Δ and β are parameters chosen to fit the loss cone. In particular, $\Delta = 0$ describes an empty loss cone, the simplest form of the distribution equation (6.47), and $\Delta = 1$ reproduces a simple Maxwellian. Figure 6.7 gives examples of such partially filled loss cones. Since the dependence of the distribution function on $v_\parallel$ is Maxwellian, only the perpendicular distribution, i.e., the part depending on $v_\perp$ is shown. It is obvious from this figure that the loss cone gradually fills up with particles when Δ increases from 0 to 1. The constant β gives another freedom of changing the slope of the distribution inside the loss cone. It changes the average velocity of the subtracted Maxwellian component.

6.3.5 Energy Distributions

Considering the Maxwellian distribution equation (6.34) one realizes that the exponential depends on the ratio of two energies, the kinetic energy of the particles $mv^2/2$, and the average or thermal equilibrium energy $k_B T$. Hence, the equilibrium distribution function can be easily generalized to the inclusion of cases where the particles are localized in an external potential field. For an external electric field or potential, $\mathbf{E} = -\nabla \phi$, the potential energy is given by $U = -q\phi$, with q the charge of the particle. The total energy of the particle is then the sum of its kinetic and potential energies $W = mv^2/2 + U$, and the distribution function becomes

$$f(v) = n \left(\frac{m}{2\pi k_B T} \right)^{3/2} \exp \left(-\frac{W}{k_B T} \right) \tag{6.50}$$

This distribution can be written entirely in terms of the energy W as a variable by observing that the integral over the distribution must reproduce the density n. Hence,

$$f(W) = 2 \left[\frac{2(W - U)}{m} \right]^{1/2} f(v) \tag{6.51}$$

with $f(v)$ given by Eq. (6.50). This distribution function is the *Boltzmann distribution*. It depends only on the particle energy.

We may note that in Eq. (6.44) above we already gave one particular form of energy distribution in the case of the relativistic Maxwell-Boltzmann distribution known as the Jüttner distribution. Its extension to include an external potential energy which however we do not give at this place, is unfortunately non-trivial.

6.3.6 Kappa and Power Law Distributions

All the former distribution functions, which are probability distributions for the particle density in phase space, are based on the fundamental Gaussian (or Gibbsian) distribution which is an exact thermodynamic equilibrium distribution and follows from first fundamental principles like energy and particle number conservation. Energy distributions frequently measured in the collisionless plasma in space do, however, in most cases not resemble Boltzmann distributions. They do not result from some simple equilibrium statistics but include non-equilibrium processes. In those processes microscopic interactions include not just collisions between particles but also interaction with the various fields which penetrate space and are themselves the result of the particle motion. Such processes are called non-linear or by another terminology complex. To account for all of them is in praxis impossible. Thus there is no surprise that observed distribution probability or distribution functions may substantially deviate from Maxwell-Boltzmann in all its different forms. They have more complicated shapes, for instance exhibiting long tails indicating that by some non-equilibrium process a number of particles are scattered away from the idealised central limit, ending up at higher speeds and energies which in the end form some tail on the core distribution. Examples of such distributions are regularly observed in the solar wind, in the energetic particles near the bow shock, in the magnetosphere itself and form the main ingredient of the celebrated cosmic rays. There is no convincing theory yet for their generation and it is not known whether they do indeed represent thermodynamic non-equilibria at all or are just transitional states in the evolution of particle distribution functions towards some kind of thermodynamic equilibrium which in our time integrated measurements appear as about stationary.

Heuristically the tails of those distribution functions can, without calling for a fundamental theory, simply be modelled by *power law distributions*, where the energy distribution function varies like $f(W) \propto (W_0/W)^{-\kappa}$ with κ some constant power. (Similarly it can of course be taken or rewritten in terms of momentum **p** or velocity **v**.) This was proposed by Stan Olbert in the sixties of the past century in an attempt to quantify measured particle distributions in the magnetosphere (see Appendix A).

This functional dependence is an approximation to a more general distribution, called the *kappa distribution*

$$f_\kappa(W) = n\left(\frac{m}{2\pi\kappa W_0}\right)^{3/2} \frac{\Gamma(\kappa+1)}{\Gamma(\kappa-1/2)} \left(1+\frac{W^*}{\kappa W_0}\right)^{-(\kappa+1)} \tag{6.52}$$

which has the simple property of being Boltzmann-Maxwellian at low energies and to include the Boltzmann distribution for $\kappa \to \infty$. Here W_0 is the particle energy at the peak of the distribution which can be related to the average thermal energy by $W_0 = k_B T(1 - 3/2\kappa)$. For $\kappa \gg 1$ the two are identical, and the distribution becomes a simple Boltzmann-Maxwellian. For smaller $\kappa \ll \infty$ this formal distribution exhibits a high-velocity tail. Using $W^* = (\sqrt{W} - \sqrt{W_s})^2$, where W_s is a so-called shift energy, instead of the more simple $W = mv^2/2$ provides an additional parameter by which the distribution can be shifted in energy or velocity space, leaving sufficient freedom to fit measured energy distribution functions.

It should be stressed again, however, that in contrast to the Maxwell-Boltzmannian distribution function this distribution just provides nothing more than an analytical fit to observations and has not been derived from first principles yet even though there is a wide literature of various attempts reaching from guesses of modified thermodynamics, propositions of modified entropies and from most complicated nonlinear models of wave particle interaction. Moreover, in the above form the distribution does not conserve particle number and energy for arbitrary κ thus being severely restricted to the particular observational conditions. It should not be therefore not be taken as fundamental in the spirit of Maxwell-Boltzmann or other distributions which result from the rigorous formulation of statistical or kinetic theory.

In order to guarantee that the distribution is at least a probability, i.e. conserves particle number, the restriction on the 'free' parameter is $\kappa > \frac{1}{2}$. Though sufficient for purely statistical purposes like data representation or mathematical statistics, application in physics, however, imposes more severe conditions. In all cases conservation of energy must be required on the distribution because the tail of the distribution reaches up to infinite energies, even though one could artificially introduce an exponential high-energy cut-off to inhibit divergence. In order to conserve energy and allow for the definition of a temperature (or average energy) the κ-parameter must be further restricted to $\kappa > \frac{5}{2}$ as will be explained in Appendix A.

Observations show that measured κ values are quite large with the lowest values found, for instance in the solar wind, being larger but close to this limit, indicating that κ is not really 'free' but is determined by processes which take care of energy conservation and other requirements by thermodynamics and/or non-equilibrium kinetic theory.

The real physical process which determines κ is not generally known but certainly depends on the different microscopic physics in each application. It is suspected that nonlinear wave-particle interactions and turbulence sign responsible for the generation of tails. Such processes can be very different and depend on the properties of the plasma, the initial and also the boundary conditions, and the only limitation they must satisfy is that they take care of either κ not exceeding the thermodynamically permitted lowest limit or otherwise truncating the distribution function at some highest energy.

In Appendix A we present a brief account of the most advanced theory of the kappa distribution as kind of a general power law distribution based in stationary state quasi-equilibrium theory far from thermodynamic equilibrium.

6.4   Measured Distribution Functions

Distribution functions are probability densities in phase space. As such, the concept of a distribution function looks rather theoretical. However, there is another quantity, which is less theoretical and, more important, easy to measure, namely the particle flux. In the following we are going to relate the measured particle fluxes of charged particles to the distribution function the determination of which is the ultimate goal of the measurement.

6.4.1   Differential Particle Flux

There is a close relationship between $J(W, \alpha, \mathbf{x})$, the *directional differential particle flux* per unit area at a given energy, angle, and position and the particle phase space distribution $f(\mathbf{v}, \mathbf{x})$. A particle flux across a surface is given by the number density times the velocity component normal to the surface. Looked at differentially or, in other words, considering the particles found in a velocity interval dv coming from a solid angle $d\Omega$, the number density of particles with velocity v in a phase space volume element is $dn = f v^2 dv d\Omega$. Multiplying by v, one finds that the differential flux of particles with velocity v is given by

$$J(W, \alpha, \mathbf{x}) dW d\Omega = f(v_\parallel, v_\perp, \mathbf{x}) v^3 dv d\Omega \qquad (6.53)$$

The left-hand side of this expression has been written in terms of the particle energy in the interval dW, simply because it is easier to measure the energy of particles in a certain interval than their individual velocities. Since $dW = mv dv$, the relation between the flux and the distribution function becomes simply

$$\boxed{J(W, \alpha, \mathbf{x}) = \frac{v^2}{m} f(v_\parallel, v_\perp, \mathbf{x})} \qquad (6.54)$$

a very useful formula which directly relates the measured flux in a certain energy interval to the velocity distribution function of the measured particles.

Due to the factor v^2 even simple Maxwellians, which drop monotonically with increasing velocity if displayed as $f(v)$, exhibit a peak if plotted as $J(W)$. The particle energy at the peak of the particle flux, W_0, can be related to the average thermal energy by $W_0 = k_B T$. Figure 6.8 shows an example of the particle flux distribution for a Maxwellian and a kappa distribution.

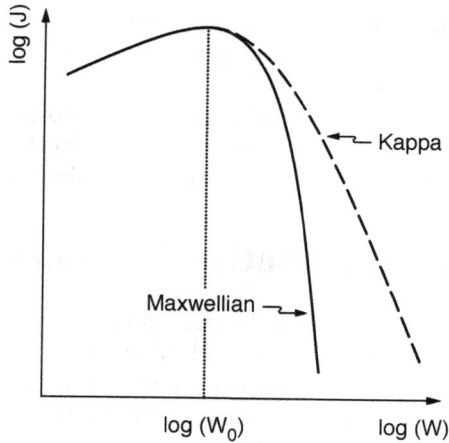

Figure 6.8: Example of a kappa distribution exhibiting a long power law tail. Comparing to the Maxwellian one realises that most of the energy has been shifted from thermal energy into the tail. Thus kappa distributions correspond to acceleration and production of high energy particles. However, for being realistic the power κ must satisfy several severe restrictions.

Another quantity which is often used in space plasma physics is the *differential directional energy flux*, which is defined as the product of particle flux times particle energy. The energy flux drops off even less rapidly than the particle flux and this representation is thus often used to highlight features in the high-energy tail of a particle distribution. Especially in case of isotropic distributions, differential fluxes are often integrated over the solid angle $d\Omega$ and then called *omnidirectional differential flux*.

The differential fluxes form the basis of measurement of velocity distributions in space. Such distributions have been measured since the mid-sixtieth of this century in the solar wind and in the magnetosphere and have been used to obtain information about the plasma state in these regions. The instrumental technique is based on the measurement of a directed particle flux entering the narrow window of the instrument, generally a retarding potential analyzer, from a certain angular direction. One either has a large number of such windows and instruments distributed over a solid angle 4π, or one takes advantage of the rotation of the spacecraft to cover the full solid angle. Ions and electrons can be discriminated by applying positive or negative potentials. The former prevent ions, the latter electrons from penetration. Selection with respect to particle energies takes place by other charged grids which deflect lower energy particles an prevent them from entering the instrument.

6.4.2   Particle Fluxes in Near-Earth Space

Measured velocity distribution functions in space are numerous. One can characterize the different regions in near-Earth space (see Section 1.2) by their characteristic long time averaged distribution functions. A number of such distributions is given in

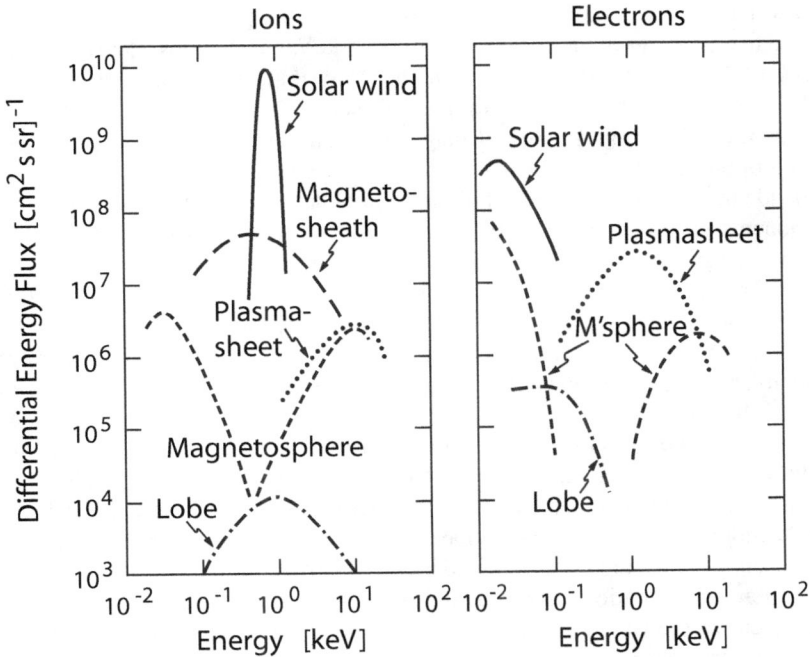

Figure 6.9: Typical omnidirectional differential particle energy fluxes measured in near-Earth space. The opening of the fluxes is a measure of the temperature of the particle component. Temperatures and central energies vary considerably between the different space plasmas and between ions and electrons. One may note the low ion temperature but large kinetic energies of solar wind ions compared with the high temperature and low kinetic energy of the solar wind electrons. To compare, in the plasmasheet the temperatures of both components are comparable, while the electrons have higher energy fluxes than the ions.

Fig. 6.9. This figure shows characteristic average omnidirectional differential ion and electron energy fluxes.

Figure 6.9 exhibits the great variety of particle distributions and energy fluxes in near-Earth space. The variation of the measured maximum fluxes covers about six orders of magnitude for the ions, and three orders of magnitude for electrons. The solar wind has the highest ion fluxes which are distributed over a narrow energy range close to an energy of 1 keV, the typical streaming energy of solar wind protons in the rest frame of the magnetosphere. These high fluxes identify to some extent the solar wind as the main particle and energy source in near-Earth space. Ion fluxes in the magnetosheath have been degraded by two orders of magnitude indicating that the solar wind has passed through the bow shock, slowed down and become heated. Slowing-down and heating is obvious from flux and energy decrease, as well as the increase of the energy spread in the differential flux.

The tail lobe flux distributions found on the open polar cap field lines close to the magnetopause have an average energy of 1 keV, but very low density levels. The energy of the lobe fluxes is slightly higher than that of the magnetosheath fluxes, while the shape of the two distributions is the same. This implies that the lobe plasma is predominantly a tiny fraction of magnetosheath plasma which has gained a small amount of energy when adiabatically entering the open lobe field lines.

Inside the magnetosphere cold plasmaspheric and warm outer ring current plasma components are found. They have similar flux levels, but are clearly separated in their average energy. The plasma sheet fluxes are similar to the warm magnetospheric component. The low energy component has a different origin than the other two. It comes from the plasmasphere in the inner magnetosphere and therefore has no relation to the solar wind. On the other hand, the energetic plasma sheet distribution is clearly related to the magnetosheath plasma. It has a similar shape but lower fluxes and higher energy. Hence, it has been energized by some process to about ten times the typical energy of the magnetosheath plasma during or after entry into the magnetosphere. The magnetospheric energetic component belongs to the energetic ring current and is part of the plasma sheet ions which perform at least a partial drift around the Earth. Clearly it is slightly more energetic than the former and has a lesser energy spread implying that the ring current ions are faster but colder than the plasma sheet plasma.

Similar considerations apply to the electron distributions. The streaming solar wind electrons are warm, considerably warmer than the solar wind ions, as observed from their large energy spread. On the other hand, they have lower streaming energy due to their low mass. When passing through the bow shock, their energy and temperature increases due to processes acting at the bow shock and turbulence in the magnetosheath, but to a lesser extend than found for the ions (not shown in figure).

Again the two magnetospheric components are well separated into cold and dense plasmaspheric electron fluxes and hot outer magnetospheric trapped electrons of solar wind origin which have been energized in the magnetosphere. These fluxes are considerably lower than the low energy plasmaspheric fluxes, indicating the dilute state of the external electron component. The plasma sheet electron fluxes have lower energies than the plasma sheet ions, reflecting both the lower magnetosheath electron temperature and lesser heating during their transfer from the magnetosheath into the plasma sheet. Finally the electron component in the lobe has both low fluxes and low energies, but the fluxes are higher than the lobe ion fluxes. Their origin is less obvious than that of the ions. Escaping polar ionosphere electrons may provide a significant contribution.

In addition to these average flux measurements many different types of particle distributions have been measured at different locations in the near-Earth space. The solar wind generally exhibits streaming Maxwellians, while in front of the bow shock electron and ion beam distributions dominate. In the bow shock one finds some kind of top-flattened electron distribution. On closed field lines in the magnetosphere various types of loss cone distributions of the energetic electron and ion components are found, while the plasma sheet plasma is well approximated by bi-Maxwellian distributions.

Finally, in the lower magnetospheric auroral zone loss cone distributions dominate again but ion and electron beam distributions are sometimes observed.

6.5 Macroscopic Variables

One may ask oneself whether these distribution functions are practical in a more general sense than discussed in the previous section and particularly the last subsection. Measurement of a distribution function does not provide physical quantities as velocities and densities, but gives merely probabilities how many particles are found in a certain velocity or energy interval. Hence the question is how to find such macroscopic measurable quantities from a known distribution function. Fortunately, there is a unique answer to this question which follows from the very definition of a probability distribution. Given a probability distribution, a physical quantity related to the probability is defined as a certain *velocity moment* of this distribution.

6.5.1 Velocity Moments

The idea behind the procedure of calculating moments is simple. The distribution function depends on the velocity, on space, and on time. The physical macroscopic quantities like density, n, bulk flow velocity, $\mathbf{v}_b$, average temperature, T, etc., do not depend on the particle velocities but only on space and time. Hence, to obtain a quantity which does not depend on velocity, one naturally would integrate over all velocities contributing to it. The distribution of particles with velocity is given by the distribution function $f(\mathbf{v},\mathbf{x},t)$. Therefore the corresponding integrals must be weighted by f before performing the integration. To find the i-th moment of the distribution function, the following integral must be calculated

$$\mathcal{M}_i(\mathbf{x},t) = \int f(\mathbf{v},\mathbf{x},t)\,\mathbf{v}^i d^3 v \tag{6.55}$$

where $\mathbf{v}^i$ denotes the i-fold dyadic product, a tensor of rank i (see Appendix B.4). The number of moments which can be calculated from the distribution function is in principle infinite. However only the first few are of physical relevance. Their definition shows the usefulness of the distribution function which permits to calculate macroscopic and more familiar quantities by simple integrations.

Using the above definition, one can calculate the first few moments, $i = 0, 1, 2$, and identify important macroscopic quantities as variants of these moments. The *number density* is given by the zero-order moment

$$\boxed{n = \int f(\mathbf{v})d^3 v} \tag{6.56}$$

The mean or *bulk flow velocity* $\mathbf{v}_b$ is naturally defined by the first-order moment

$$\boxed{\mathbf{v}_b = \frac{1}{n}\int \mathbf{v} f(\mathbf{v})d^3 v} \tag{6.57}$$

The bulk velocity describes the macroscopic flow of the entire particle component in which each particle participates. It is an average flow velocity of the particle species or component under consideration. We have denoted it by the symbol v_b in order to distinguish it from the particle velocities. In later fluid applications where no confusion will be possible we will replace it by the conventional notation v of a velocity.

The *pressure tensor* is defined as the contribution of the fluctuation of the velocities of the ensemble from this mean velocity. Its calculation is based on the second-order moment

$$\mathbf{P} = m \int (\mathbf{v} - \mathbf{v_b})(\mathbf{v} - \mathbf{v_b}) f(\mathbf{v}) d^3 v \tag{6.58}$$

Since the two velocity product appearing in the pressure integral is a dyadic product, the pressure is a tensor.

Sometimes the next higher moment is also used to describe deviations from equilibrium. This moment is called the *heat tensor*

$$\mathbf{Q} = m \int (\mathbf{v} - \mathbf{v_b})(\mathbf{v} - \mathbf{v_b})(\mathbf{v} - \mathbf{v_b}) f(\mathbf{v}) d^3 v \tag{6.59}$$

It is a third rank tensor or dyad. In itself it is not a very useful quantity, but its trace vector $\mathbf{q}$

$$\mathbf{q} = \frac{m}{2} \int (\mathbf{v} - \mathbf{v_b}) \cdot (\mathbf{v} - \mathbf{v_b})(\mathbf{v} - \mathbf{v_b}) f(\mathbf{v}) d^3 v \tag{6.60}$$

is the *heat flux vector* describes the transport of heat into a direction in the plasma which is not necessarily the direction of the mean flow.

6.5.2 Concept of Temperature

The pressure tensor consists of a trace and the traceless off-diagonal part. The former gives, in an isotropic plasma, the isotropic pressure $p = n k_B T$, in an anisotropic plasma the anisotropic pressure. The traceless part contains the stresses in the plasma. The thermal pressure p can be used to define the temperature of the plasma component

$$T = \frac{m}{3 k_B n} \int (\mathbf{v} - \mathbf{v_b}) \cdot (\mathbf{v} - \mathbf{v_b}) f(\mathbf{v}) d^3 v \tag{6.61}$$

a definition which identifies the temperature as a scalar quantity. This temperature is the *kinetic temperature*, a quantity which can formally be calculated for any type of distribution function and therefore is not necessarily a true temperature in the thermodynamic sense, which can only be calculated for plasmas in or close to thermal equilibrium. It rather is a measure of the spread of the particle distribution in velocity space. In addition, because each particle species may have its own distribution function, the kinetic temperatures of the plasma components may differ from each other. Also, in an anisotropic plasma the temperatures parallel and perpendicular to the magnetic field are in general different, because the particle distributions parallel and perpendicular have different shapes (see Section 6.3).

To demonstrate the physical thermodynamic meaning of Eq. (6.61) we calculate the kinetic temperature for the Boltzmann-Maxwellian velocity distribution (6.33)

$$f(v) = \frac{n}{(\pi \langle v \rangle^2)^{3/2}} \exp\left(-\frac{v^2}{\langle v \rangle^2}\right) \tag{6.62}$$

in an isotropic plasma at rest, $\mathbf{v}_b = 0$, knowing that this distribution holds for thermal equilibrium. Performing the integration in Eq. (6.61) one finds that in the isotropic case the volume element becomes $d^3v = 4\pi v^2 dv$. It is shown in Appendix B.7 how the remaining integrals over velocity can be treated. One finds that the density and the factors π cancel, and the result is

$$T = \frac{m \langle v \rangle^2}{2k_B} \tag{6.63}$$

In agreement with thermodynamics the thermal energy of the plasma is $k_B T$. For later convenience, we define the *thermal velocity* as $v_{th}^2 = \langle v \rangle^2 / 2$ or

$$v_{th} = \left(\frac{k_B T}{m}\right)^{1/2} \tag{6.64}$$

Performing a similar calculation for the anisotropic bi-Maxwellian distribution we would have obtained two different temperatures, a parallel temperature, $k_B T_\parallel = m v_{th\parallel}^2$, and a perpendicular temperature, $k_B T_\perp = m v_{th\perp}^2$, instead of the isotropic temperature, $k_B T$, all taken here in energy units. In all these cases the temperature has a well-defined meaning, because the plasma is in thermal equilibrium even when the equilibrium may be anisotropic. For more complicated non-equilibrium conditions, however, the temperature loses its original meaning. As a kinetic temperature it merely contains information about the random part of the average kinetic energy in the plasma.

6.6 Concluding Remarks

In the Newtonian spirit every physical phenomenon can be described by the action of fields on particles. This is the basic idea also of plasma physics. The big problem is, however, that in plasma physics the system does not consist of one or two or three particles for which it would still be possible to solve the equations of motion. Plasma physics is the theory of the physical behaviour of a huge number of particles under the influence of external fields and also of the fields generated by the particles themselves, the electromagnetic fields which have their sources in charge differences and electric current flow.

At a first glance already it is clear that one cannot solve the equations of motion of the extremely large number of particles contained in a volume. Even the best computers are incapable of doing it though numerical simulations are unavoidable when inferring about the small-scale plasma behaviour as we will see at two examples in Chapters 12 and 13, where we will, as an application, review the theory of reconnection and shock waves in plasma. However, numerical simulations take just a small number of particles in place of the large real number, considering each of the particles treated as standing for very many others. This is still an unresolved drawback of the numerical simulations which only slowly is going to be overcome by the future supercomputers.

The other way of dealing with such large numbers is the statistical way. It is based on the assumption that only average properties of the large ensembles are of interest in daily physics with the smaller effects averaging out or, if superposing, lead to a large and observable effect which includes many particles at one time. Such effects are mostly so-called resonances. However, for applying statistical theory one needs to transform the microscopic equations of each particle into equations which describe their larger scale properties.

This has been done by introducing the concept of a statistical distribution of particles in phase space which is then further reduced to velocity space. A procedure of this kind leads to the definition of an equation for the phase-space distribution function which we have derived from the completely basic and valid Klimontovich-Dupree equation. The derivation of the Vlasov equation from it is quite intuitive. It is, however, only one transparent way of constructing the kinetic equations of a plasma. Among the many different approaches there have been very involved diagrammatic methods, operator approaches and various field theoretical procedures. Historically one has gone a different route, starting from the Liouville equation of statistical mechanics and descending from it to the Vlasov equation. This is achieved by the construction of reduced distribution functions, where integration over the individual phase spaces of each particle gets rid of the coordinates of this particle and, hence, its individuality. The two approaches are entirely equivalent, but the Liouville equation approach is mathematically more stringent while being much more complicated. However, because of its greater rigour it may be preferred by the more theoretically oriented readers. For this purpose we sketched this approach in Appendix C.4.

Independent of the method by which an equation of the form of the Vlasov equation has been derived. This equation turns out to very nicely describe many faces of the plasma in space in particular under collisionless conditions. It can thus be considered as the canonically valid fundamental equation of the whole of plasma physics in space and particularly of the plasma in its collisionless state when added to the system of Maxwell equations of electrodynamics including the material equations by defining moments of the Vlasov equation.

We have done both, deriving the Vlasov equation from the more basic Klimontovich-Dupree equation (or also the Liouville equation) and have defined the relevant plasma moments. Unfortunately these include a further approximation as it turns out that there are infinitely man such moments which would be equivalent to solving the

Vlasov-equation exactly. this is of course impossible. Hence, one has to restrict to a small number of moments only assuming that the higher moments play little role for the global plasma dynamics. Fortunately, on the other hand, the material equations of electrodynamics contain only two of these moments, the two lowest ones: number density and current density. Thus restricting to the Vlasov equation and using these two moments still produces a most general precise solution of the processes of the plasma.

This is the most satisfactory way of plasma description. Still it is complicated. Therefore, in the following chapters we will be going to consider the possibility of finding an even simpler though less precise description of plasmas in the assumption of the plasma consisting not of particles but of differently charged fluids.

Further Reading

Kinetic theory is expanded upon in the monographs [7–11] each assuming a completely different point of view. The most basic approach is given in the monograph [7] where the exact (nonrelativistic and relativistic) microscopic theory has been developed. Reference [8] is an early though thoroughly compiled and about complete treatise of the whole of plasma physics, including nuclear fusion from the plasma physics viewpoint while excluding nuclear physics. The original high level theory is found in [9]. This book is of substantial physical insight but difficult to read. The complete kinetic account with derivation of transport coefficients is contained in [10], while the rather formal monograph [11] emphasises linear wave theory.

The general electrodynamic theory is found in [5], including a few aspects of plasma physics and magnetohydrodynamics. This is done mainly from the electrodynamics viewpoint and lacks any application to space physics. The construction of the Vlasov equation using statistical mechanics is described in [7–10]. Each of these volumes takes a different viewpoint and is of different stringency. For the derivation of the Fokker-Planck equation, a particularly good reference is [10], where the expressions for the averages over the fluctuations are explicitly given. These are important in the derivation of transport coefficients, thermalisation time scales, time scales of acceleration and diffusion. The drift-kinetic equation in the form presented here is found in [4]. The full theory of the gyro-kinetic equation is given in [1], including ion viscosity caused by finite Larmor-radius effects. This is a rather involved problem making the paper difficult to read. It requires some sophistication in mathematical analysis and algebra. Loss cone and anisotropic distribution functions are given without derivation in [6]. The rigorous non-trivial derivation of the anisotropic Jüttner distribution has been given only recently [13]. For the measurement principles of distribution functions in space application and examples the interested reader should consult [12]. This latter monograph is particularly valuable for the relation between distribution functions and measurements.

References

[1] A. J. Brizard and T. S. Hahm, Foundations of nonlinear gyrokinetic theory, *Rev. Mod. Phys.* **79**, 421–468, doi: 10.1103—RevModPhys.79.421 (2007).

[2] P. C. Clemmow and J. P. Daugherty, *Electrodynamics of Particles and Plamas* (Addison-Wesley, Raeding MA, 1971).

[3] D. A. Gurnett and A. Bhattacharjee, *Introduction to Plasma Physics* (Cambridge University Press, 2005).

[4] A. Hasegawa, *Plasma Instabilities and Nonlinear Effects* (Springer Verlag, Heidelberg, 1975).

[5] J. D. Jackson, *Classical Electrodynamics* (J. Wiley and Sons, New York, 1975), Chpt. 10, p. 469.

[6] C. F. Kennel and M. Ashour-Abdalla, in *Magnetospheric Plasma Physics*, ed. A. Nishida (D. Reidel Publ. Co., Dordrecht, 1982), p. 245.

[7] Y. L. Klimontovich, *The Statistical Theory of Non-Equilibrium Processes in a Plasma* (MIT Press, Cambridge, 1967).

[8] N. A. Krall and A. W. Trivelpiece, *Principles of Plasma Physics* (McGraw-Hill, New York, 1973).

[9] E. M. Lifschitz and L. P. Pitaevskii, *Physical Kinetics* (Pergamon Press, Oxford, 1981).

[10] D. L. Montgomery and D. A. Tidman, *Plasma Kinetic Theory* (McGraw-Hill, New York, 1964).

[11] D. R. Nicholson, *Introduction to Plasma Theory* (Wiley & Sons Inc., New York, 1983).

[12] M. Schulz and L. J. Lanzerotti, *Particle Diffusion in the Radiation Belts* (Springer Verlag, Heidelberg, 1974).

[13] R. A. Treumann and W. Baumjohann, Anisotropic Jüttner (relativistic Boltzmann) distribution, *Annales Geophysicae* **34**, 737–738, doi: 10.5194/angeo-34-737-738 (2016).

Problems

Problem 6.1 *What is the difference between real configuration space and phase space? What is to be known in order to describe the orbit of a single particle in phase space?*

Problem 6.2 *Consider the harmonic oscillator: A particle oscillates between two points x_1 and x_2 with frequency ω. Write its equation down. What is the phase space orbit of the particle?*

Problem 6.3 *Draw the phase space orbit of an electron gyrating in the magnetic field with cyclotron frequency ω_{ge}. The magnetic field is in z direction. The gyration plane is (x, y).*

Problem 6.4 *An electron (mass m_e, charge $q_e = -e$) is subject acceleration in a constant electric field along a constant magnetic field. What is its phase space orbit?*

Problem 6.5 *Write the Klimontovich-Dupree equation for an electron in terms of the Hamilton function of the particle. Remember that the Hamilton function $\mathcal{H} = K + U$ is the sum of the particle kinetic K and potential U energies. So you need to know the potential energy provided by the electric field $\mathbf{E}_m = -\nabla\phi_m(\mathbf{x})$ at the location $\mathbf{x}$ of the particle as well as its kinetic energy. Remember from classical mechanics the Hamilton relations between particle moment and position. With this knowledge you should be able to bring the Klimontovich-Dupree equation into a very simple form.*

Problem 6.6 *Compare the kinetic equation with the Boltzmann equation. What role plays the average fluctuation term on the right-hand side of the kinetic equation?*

Problem 6.7 *Liouville's theorem holds strictly for the Klimontovich-Dupree equation. Why? Take a number of 10 point-like particles and draw a two-dimensional phase space assigning each particle a location. What is the volume of phase space these particles occupy? Now move the particles around to different locations each. Describe what happens to the phase space volume occupied by the particles. When now going to the kinetic equation, is this picture still correct? For the Vlasov equation, what condition is necessary that the phase space volume remains constant?*

Problem 6.8 *Derive the electrostatic Vlasov equation from the full electromagnetic Vlasov equation.*

Problem 6.9 *Why must the distribution function of particles in phase space be normalised to the number of particles? Show that this implies normalisation of the velocity distribution to particle density. Show this explicitly for the one-dimensional Maxwellian velocity distribution function (6.32). Show that the three-dimensional isotropic Maxwellian velocity distribution is indeed given by Eq. (6.34).*

Problem 6.10 *Derive the anisotropic Maxwellian distribution Eq. (6.38).*

— 7 —

Plasma Magnetohydrodynamics

Magnetohydrodynamics is the theory of electrically conducting fluids. These are not necessarily plasmas. For instance, in the Earth's interior in the outer core, the transition layer between core and mantle and in the Earth's mantle itself the matter is electrically highly conducting. At the same time the matter in the outer core and the transition layer is in the state of a rather slowly moving fluid. In a certain sense the lower mantle can also be considered as being plastic or fluid-like. These regions though being hot, with temperatures around 5000 K in the outer core, are not in the plasma state. At the enormous pressure and density of the Earth's interior they are highly collisional in a state of a solid state fluid. In spite of the high temperature in the Earth's interior they are cold in the physical sense of the word, i.e. their temperature is far below the Fermi temperature T_F which varies between 6,000 K and 4,000 K. For all micro-scale considerations the matter in the Earth's core is in the quantum state. Nevertheless, its macroscopic macro-scale dynamics can be described classically by magneto-hydrodynamic fluid theory which in application to the fluid outer core has succeeded to explain the origin of the geomagnetic field as caused by the dynamo action of the turbulent outer core fluid motion which is driven by a combination of the Earth's rotation and nuclear heating from below.

The precise description of plasmas, on the other hand, requires the evolution of particle distribution functions in phase space. In the previous chapter it has been shown that such distribution functions evolve according to kinetic equations, the Vlasov equation in the special case of a collisionless plasma. In many cases, however, it is not necessary to know the exact evolution of the distribution function but it is sufficient to determine the spatial and temporal development of the macroscopic moments of the distribution, such as densities, velocities and temperatures. In particular, for slow time variations the moments describe the state of the plasma good enough in most cases of interest. Clearly, it will be simpler to investigate their evolution than to determine the exact evolution of the distribution function which for many purposes contains extra information which is unnecessary.

Since the macroscopic moments are quantities which one is already familiar with from fluid and gas dynamics, the resulting theory falls into the domain of fluid theories. The aim of the present chapter is to derive and discuss such a hydrodynamic theory for plasmas. This theory will be called *magnetohydrodynamics* because it is the fluid theory of electrically charged fluids subject to the presence of external and internal magnetic fields in complete analogy to electrically conducting fluids. However, magnetohydrodynamics is already a further approximation to a more general hydrodynamic theory, the *multi-fluid theory* of plasmas (which, for instance, in the

Earth's interior is not applicable). In the following we are progressing stepwise, first going to derive the multi-fluid equations before proceeding to the next more restricting approximation to this theory, i.e. single-fluid magneto-hydrodynamics. The discussion and application of the former will be delayed to later chapters, because of the greater transparency of the magnetohydrodynamic approach when applied to the large-scale slow time variations of the plasma in the magnetosphere and solar wind.

7.1 Multi-Fluid Theory

Fluid theory is looking for evolution equations for the basic macroscopic moments, i.e., number density, $n_s(\mathbf{x},t)$, bulk flow velocity, $\mathbf{v}_{b,s}(\mathbf{x},t)$, pressure tensor, $\mathbf{P}_s(\mathbf{x},t)$, and kinetic temperature, $T_s(\mathbf{x},t)$, of the particle species s in a plasma. For a two-fluid plasma consisting of ions and electrons, we have $s = i, e$. Further, since no distinction is made anymore between the individual particle velocities, we will henceforth drop the bulk flow subscript 'b' and understand that $\mathbf{v}_s$ is the mean flow velocity of species s.

The definition of the moments given in the previous chapter suggests that the evolution equations under question can be derived from the Vlasov equation by performing an appropriate integration with respect to velocity space. For the sake of a better understanding we will explicitly show how such fluid-like equations arise from the kinetic equation of a plasma.

7.1.1 Continuity Equation

In order to demonstrate how this procedure works, let us take the zero-order moment and integrate the Vlasov equation (6.20) over the entire velocity space

$$\int \left[\frac{\partial f_s}{\partial t} + \mathbf{v} \cdot \nabla_{\mathbf{x}} f_s + \frac{q_s}{m_s} (\mathbf{E} + \mathbf{v} \times \mathbf{B}) \cdot \nabla_{\mathbf{v}} f_s \right] d^3 v = 0 \qquad (7.1)$$

Since the velocity-space volume element, $d^3 v$, does not depend on time (remember that the particles are moving in a fixed phase space; what changes is the volume element occupied by the particles, not the phase space volume), the time derivative can be exchanged with the integral to find that the first term in Eq. (7.1) becomes

$$\frac{\partial}{\partial t} \int f_s d^3 v = \frac{\partial n_s}{\partial t} \qquad (7.2)$$

where the definition of the zero-order moment, the particle density of species s, has been used. The second term is evaluated in exactly the same way observing that spatial and velocity coordinates are independent variables and, hence, the integration over $d^3 v$ and the differentiation $\nabla_{\mathbf{x}}$ can be exchanged. Since according to Eq. (6.57) the integral represents the particle flux density, $n_s \mathbf{v}_s$, the whole expression yields the divergence of the particle flux of species s

$$\nabla_{\mathbf{x}} \cdot \int \mathbf{v} f_s d^3 v = \nabla \cdot (n_s \mathbf{v}_s) \qquad (7.3)$$

where we dropped the index $\mathbf{x}$ on ∇ on the right-hand side because the particle flux density depends only on space and time while the velocity dependence has been integrated away. Finally, consider the last term in the integrated Vlasov equation. Applying the velocity gradient $\nabla_{\mathbf{v}}$ to the full integrand, $(\mathbf{E} + \mathbf{v} \times \mathbf{B}) f_s$, makes it a total differential. Integrating over the total differential gives its values at the boundaries in velocity space which non-relativistically are located at infinity. Since no particle has infinite speed, the distribution function is zero here, and this integral vanishes. Now, we are left with

$$\int f_s \nabla_{\mathbf{v}} \cdot (\mathbf{E} + \mathbf{v} \times \mathbf{B}) \, d^3 v \tag{7.4}$$

The triple vector product in this expression can be written as $\mathbf{B} \cdot (\nabla_{\mathbf{v}} \times \mathbf{v}) = 0$ and vanishes, because the magnetic field does not depend on particle velocity. Moreover, the electric field is a function of space only and the first term also vanishes. Hence, the last integral in the integrated Vlasov equation does not contribute. Collecting the surviving terms, we find

$$\frac{\partial n_s}{\partial t} + \nabla \cdot (n_s \mathbf{v}_s) = 0 \tag{7.5}$$

This is the continuity equation of the s-component fluid of particles in the plasma. Its physical meaning is that in the absence of any interaction processes which create or annihilate particles of this species, the particle number density, and also mass and charge densities, are conserved during the motion of the fluid.

7.1.2   Equation of Motion

The continuity equation is the first fluid equation of a multi-fluid plasma; it is the zero-order moment equation of the collisionless Vlasov equation. However, it is not an equation for the density alone but it couples the s-species plasma density to its bulk fluid velocity. Therefore, another equation is required for the velocity $\mathbf{v}_s$ of the plasma. Since the latter is the first moment of the distribution function, this second fluid equation will naturally result from a first moment treatment of the Vlasov equation.

Let us dyadically multiply the Vlasov equation with the phase space velocity variable $\mathbf{v}$ and integrate term by term with respect to the velocity

$$\int \mathbf{v} \left[\frac{\partial f_s}{\partial t} + \mathbf{v} \cdot \nabla_{\mathbf{x}} f_s + \frac{q_s}{m_s} (\mathbf{E} + \mathbf{v} \times \mathbf{B}) \cdot \nabla_{\mathbf{v}} f_s \right] d^3 v = 0 \tag{7.6}$$

Applying the reasoning of the previous paragraph since also the phase space coordinate, $\mathbf{v}$, is independent of time, and again exchanging differentiation and integration, the first integral term results in

$$\frac{\partial}{\partial t} \int \mathbf{v} f_s d^3 v = \frac{\partial}{\partial t} (n_s \mathbf{v}_s) \tag{7.7}$$

which is the temporal variation of the flux density of s-component fluid. Thus identification of the first term is trivial. However, the second term provides more serious difficulties, because it contains a dyadic form, $\mathbf{vv}$. As a first step let us again exchange the differentiation with respect to space and the integration over velocity which is permitted because both are independent coordinates, keeping in mind that because of the symmetry of the dyad $\mathbf{v}(\mathbf{v} \cdot \nabla_{\mathbf{x}}) = \nabla_{\mathbf{x}} \cdot (\mathbf{vv})$. Let us rewrite the dyadic form as

$$\mathbf{vv} = (\mathbf{v} - \mathbf{v}_s)(\mathbf{v} - \mathbf{v}_s) - \mathbf{v}_s\mathbf{v}_s + \mathbf{vv}_s + \mathbf{v}_s\mathbf{v} \tag{7.8}$$

and introduce this into the second integral, with $\nabla_{\mathbf{x}}$ extracted out of the integral. The integral over the term resulting from the first product on the right-hand side of Eq. (7.8) is then after Eq. (6.58) identified as the fluid pressure tensor divided by the mass, $\mathbf{P}_s/m_s$. The integral over the term resulting from the second product on the right-hand side of Eq. (7.8) becomes simply $-n_s\mathbf{v}_s\mathbf{v}_s$ since the fluid bulk velocity $\mathbf{v}_s$ is independent of $\mathbf{v}$. Finally, the two remaining integrals reproduce twice the same value but with positive sign, $+2n_s\mathbf{v}_s\mathbf{v}_s$, so that the total sum is $n_s\mathbf{v}_s\mathbf{v}_s$. Combining all these terms, the second integral becomes

$$\nabla_{\mathbf{x}} \cdot \int \mathbf{vv} f_s d^3v = \nabla \cdot (n_s\mathbf{v}_s\mathbf{v}_s) + \frac{1}{m_s}\nabla \cdot \mathbf{P}_s \tag{7.9}$$

The last integral of the Vlasov equation can be treated by the same method of applying the operator $\nabla_{\mathbf{v}}$ to the full integrand, i.e., transforming it into a total derivative with respect to $\mathbf{v}$, the integral over which vanishes because one integrates over all of velocity space, and subtracting the part which one has added to obtain the total derivative. In the remaining non-vanishing integral the operator $\nabla_{\mathbf{v}}$ is applied only to the velocity, $\nabla_{\mathbf{v}}\mathbf{v}$. The operator is the unit tensor, $\mathbf{I}$. As on p. 158, the mixed vector product vanishes and the electric field does not depend on velocity. Hence, this last integral becomes

$$\int f_s(\nabla_{\mathbf{v}}\mathbf{v}) \cdot (\mathbf{E} + \mathbf{v} \times \mathbf{B})\, d^3v = -n_s(\mathbf{E} + \mathbf{v}_s \times \mathbf{B}) \tag{7.10}$$

We can now add all non-vanishing integrals to obtain our final result

$$\boxed{\frac{\partial(n_s\mathbf{v}_s)}{\partial t} + \nabla \cdot (n_s\mathbf{v}_s\mathbf{v}_s) + \frac{1}{m_s}\nabla \cdot \mathbf{P}_s - \frac{q_s}{m_s}n_s(\mathbf{E} + \mathbf{v}_s \times \mathbf{B}) = 0} \tag{7.11}$$

This equation is the *momentum density conservation equation* of the s-component fluid of the plasma or the equation of motion of this fluid component. It is the equation for the fluid velocity which we looked for. It relates the fluid velocity to density and electromagnetic force acting on the fluid element, but not on the single particles anymore.

The momentum density equation has a close relationship to conventional hydrodynamics where it is known as the Navier-Stokes equation. Hence, the plasma momentum conservation equation is the Navier-Stokes equation including an electromagnetic Lorentz force acting on the charges in the plasma. The appearance of this force in the equation of motion couples the plasma fluid to the full set of electromagnetic equations and makes it very distinct from conventional hydrodynamics where the only forces acting on the fluid are pressure and viscous forces. The appearance of this force also couples all the charged plasma fluid components together. This is obvious from the fact that the electric and magnetic fields in the Lorentz force act on all charged components and, at the same time, all charged components do contribute to the electric and magnetic fields. It is therefore clear that in solving the equations for the motion of one plasma component one is unavoidably confronted with the problem of solving all the equations of motion for all plasma fluid components because neither of them are independent of each other and of the electric and magnetic fields.

7.1.3 Energy Equation

However, as we already expected, the equation of motion does not close the system of equations because the next higher order quantity, the pressure tensor, $\mathbf{P}_s$, appears in it, which again requires another equation determining its evolution. Such an equation is again expected to be found by calculation of the second-order moment equation from the Vlasov equation, i.e., multiplying the Vlasov equation by the second-order dyad $\mathbf{vv}$ and integrating over velocity space. This integration is much more involved than that which we have performed until now. Its result is another equation, the heat transfer or *energy density conservation equation*

$$\boxed{\frac{3}{2}n_s k_B \left(\frac{\partial T_s}{\partial t} + \mathbf{v}_s \cdot \nabla T_s \right) + p_s \nabla \cdot \mathbf{v}_s = -\nabla \cdot \mathbf{q}_s - (\mathbf{P}'_s \cdot \nabla) \cdot \mathbf{v}_s}$$ (7.12)

where T_s is the temperature defined in Section 6.5 and p_s is the scalar pressure, both of which are related by the ideal gas equation, $p_s = n_s k_B T_s$. The quantity $\mathbf{q}_s$ is the heat flux vector and $\mathbf{P}'_s$ denotes the off-diagonal part of the full pressure tensor, $\mathbf{P}_s$, and describes the *shear stress*, e.g., the transfer of y-momentum by motion in the x-direction. Naturally this equation again contains a new undetermined quantity, the heat flux $\mathbf{q}_s$ which is a third-order moment and, hence, requires an additional expression which can be obtained from the third-order equation multiplying the Vlasov equation by $\mathbf{vvv}$ and integrating. This is a formidable task and, as one expects, generates another fourth order quantity. Such it goes until infinity reflecting the fact that the moment procedure generates an infinite chain of moment equations which in order to arrive at a complete solution would have to be solved for all infinitely many moments. This is, however, the equivalent to solving the Vlasov equation for all particles itself. In most cases, however, one can neglect the heat flux and thereby truncate the system of basic equations at the energy conservation equation.

7.2 Equation of State

When calculating moment equations we have found that the fluid equations of a plasma form a hierarchy of ever increasing order where each order contains a next-order quantity which must be determined from the next-order equation. Such a procedure must be closed by truncation of the hierarchy at a certain level. The most common and simplest way is assuming an *equation of state* for the pressure which makes the energy equation obsolete and avoids explicitly taking into account the transport of heat.

The actual form of the equation of state depends on the form of the pressure tensor, e.g., isotropic or anisotropic, and even more generally on the behaviour of the fluid condensed in the number and momentum density equations. The equations of state may differ for the different species s and are indeed often quite different for different fluid components, especially for the electrons and ions because of their enormous mass difference.

7.2.1 Isotropic Pressure

If the pressure is taken to be isotropic, the pressure tensor becomes diagonal

$$\boxed{\mathbf{P}_s = p_s \mathbf{I}} \tag{7.13}$$

which reads in matrix notation

$$\mathbf{P}_s = \begin{pmatrix} p_s & 0 & 0 \\ 0 & p_s & 0 \\ 0 & 0 & p_s \end{pmatrix} \tag{7.14}$$

and only one such equation of state is needed. The actual form of this equation may differ drastically, but here we will give the equations for the two most important cases.

The most simple equation of state is that for the *isothermal* case, $T_s = \text{const}$. Here one simply takes the ideal gas equation

$$p_s = n_s k_B T_s \tag{7.15}$$

Isothermal conditions can be applied when the temporal variations are so slow that the plasma has sufficient time to redistribute energy in order to maintain a constant heat bath temperature. Such conditions frequently apply to the global situation of the magnetosphere. When $T_s = T_{s0}$ is assumed constant, the pressure becomes proportional to the density of the species

$$\boxed{p_s = n_s k_B T_{s0}} \tag{7.16}$$

and the system of equations is truncated to a closed set.

In the other extreme, when the time variations are so fast that no susceptible heat exchange can take place the plasma evolves *adiabatically*. The change in temperature is then related in a simple way to the change in density. Intuitively this is clear because any gas will cool during a fast expansion of the volume and heat up when the volume is compressed. To find the adiabatic relation we set the right-hand side of

Eq. (7.12) to zero. Using the continuity equation (7.5) in order to replace the divergence of the velocity by a time derivative of the density and applying the total time derivative $d/dt = \partial/\partial t + \mathbf{v} \cdot \nabla$, the heat transfer equation (7.12) can be cast into the form

$$\frac{3}{2}\frac{d(n_s k_B T_s)}{dt} - \frac{5}{2}k_B T_s \frac{dn_s}{dt} = 0 \tag{7.17}$$

an equation which is identical to

$$n_s \frac{dT_s}{dt} - \frac{2}{3}T_s \frac{dn_s}{dt} = 0 \tag{7.18}$$

which has the well known adiabatic solution

$$T_s = T_{s0}\left(\frac{n_s}{n_{s0}}\right)^{\gamma-1} \tag{7.19}$$

or, written in terms of the scalar pressure and using Eq. (7.15)

$$\boxed{p_s = p_{s0}\left(\frac{n_s}{n_{s0}}\right)^{\gamma}} \tag{7.20}$$

The *adiabatic index*, $\gamma = c_p/c_v = 5/3$, which is the ratio of the two specific heats at constant pressure and constant volume, is constant in a collisionless ideal isotropic plasma, and as long as there are no further interactions between the different species in the plasma the temperatures and densities of each species evolve according to the adiabatic law. The index γ can also be regarded as a *polytropic index*, comprising not only the adiabatic case, but also the isobaric or constant pressure, $\gamma = 0$, the isothermal or constant temperature, $\gamma = 1$, and the isometric or constant density, $\gamma = \infty$, cases.

7.2.2 Anisotropic Pressure

In anisotropic plasmas, the pressure tensor splits into parallel and perpendicular pressure

$$\boxed{\mathbf{P}_s = p_{s\perp}\mathbf{I} + (p_{s\|} - p_{s\perp})\frac{\mathbf{BB}}{B^2}} \tag{7.21}$$

where the dyad, $\mathbf{BB}$, is proportional to the magnetic pressure tensor; cf. Eq. (7.62). In a coordinate system where the z axis is aligned with the magnetic field direction, it reads

$$\mathbf{P}_s = \begin{pmatrix} p_{s\perp} & 0 & 0 \\ 0 & p_{s\perp} & 0 \\ 0 & 0 & p_{s\|} \end{pmatrix} \tag{7.22}$$

It is then not a priori clear that both parallel and perpendicular pressures evolve according to the same adiabatic laws while it is still a good approximation to use the ideal gas equation for both pressures

$$
\begin{aligned}
p_{s\parallel} &= n_s k_B T_{s\parallel} \\
p_{s\perp} &= n_s k_B T_{s\perp}
\end{aligned}
\tag{7.23}
$$

If the adiabatic approximation is justified, one can use the general definition of the adiabatic index

$$
\gamma = (d+2)/d
\tag{7.24}
$$

where d is the degree of freedoms the particles of the plasma have. In monatomic $3d$ plasmas one recovers $\gamma = 5/3$. In $1d$ plasmas $\gamma = 3$, and in $2d$ plasmas $\gamma = 2$. The parallel pressure has $d = 1$, while the perpendicular has $d = 2$, and the adiabatic equations of state become

$$
\begin{aligned}
p_{s\parallel} &= p_{s\parallel 0} \left(\frac{n_s}{n_{s0}} \right)^3 \\
p_{s\perp} &= p_{s\perp 0} \left(\frac{n_s}{n_{s0}} \right)^2
\end{aligned}
\tag{7.25}
$$

However, the above reasoning clearly neglects the coupling between the two pressure components due to inhomogeneous magnetic fields. Most important, the above equations do not include any dependence on the magnetic field strength, whereas, for example, in a pure mirror geometry (see Section 2.5), the magnetic field strength determines the ratio between parallel and perpendicular pressures. Hence, these equations can only be applied in situations where the magnetic field is homogeneous or of minor importance, i.e. when the magnetic field is weak and the dynamics dominated by the flow.

7.2.3 Double-Adiabatic Invariants

A slightly better approximation to the equation of state of anisotropic plasmas can be found by assuming that the presence of a sufficiently strong magnetic field not only introduces the symmetry breaking between parallel and perpendicular pressures but also some kind of ordering. For instance, for the perpendicular motion of the particles we already know that under some rather weak conditions, when the variation of the plasma is slower than the gyration, the magnetic moment $\mu = mv_\perp^2/2B$ of the particles is conserved. This conservation clearly affects the evolution of the perpendicular energy of the particles and, hence, the perpendicular pressure.

Calculating the perpendicular temperature moment for a Maxwellian anisotropic distribution and dividing by the magnetic field strength, one realizes that the result is the distribution function average over the magnetic moment

$$\langle \mu \rangle = \frac{k_B T_\perp}{B} = \frac{p_\perp}{nB} \tag{7.26}$$

Because the ensemble averaged $\langle \mu \rangle$ must be conserved, the right-hand side of this equation is a constant, and the perpendicular pressure evolves in proportionality to the magnetic field strength. The perpendicular adiabatic law can be read from this behaviour as

$$\frac{d}{dt}\left(\frac{p_\perp}{nB}\right) = 0 \tag{7.27}$$

Since this expression results from magnetic moment conservation, no heat is transferred into the perpendicular direction. No adiabatic exponent enters this equation because the pressure is entirely determined by the magnetic field. Moreover, this relation couples magnetic field, perpendicular temperature and density together so that no simple adiabatic index exists. Instead it depends on either the field or the temperature.

Finding the parallel adiabatic equation of state is more involved because no such simple conservation equation exists as for the magnetic moment. One must rewrite the heat transfer equation for an anisotropic plasma, neglect all dissipative terms and inhibit parallel in addition to perpendicular heat transfer. This requirement is particularly strong for electrons of which we know that they easily escape along the magnetic field lines. Therefore the parallel adiabatic equation will from the very beginning impose a rather strong condition on the applicability to a plasma. One may apply it to ions, but the application to electrons will be severely limited.

One can show that the equation obtained for the parallel pressure $p_\parallel$ under the assumption of suppressed heat flow and no dissipation becomes

$$p_\perp \frac{dp_\parallel}{dt} + 2p_\parallel \frac{dp_\perp}{dt} + 5p_\perp p_\parallel \nabla \cdot \mathbf{v} = 0 \tag{7.28}$$

Replacing the divergence of $\mathbf{v}$ with the help of the continuity equation (7.5) written in a different form

$$-n\nabla \cdot \mathbf{v} = \partial n/\partial t + \mathbf{v} \cdot \nabla n \tag{7.29}$$

and using $d/dt = \partial/\partial t + \mathbf{v} \cdot \nabla$, Eq. (7.28) becomes, after rearranging some terms, a total time derivative

$$\frac{d}{dt}\left(\frac{p_\parallel p_\perp^2}{n^5}\right) = \frac{d}{dt}\left(\frac{p_\parallel B^2}{n^3}\right) = 0 \tag{7.30}$$

These are two versions of the parallel equation of state. Similar to the perpendicular equation of state, the parallel pressure turns out to be a function of the magnetic field and plasma density of the species under consideration.

Combining the two double-adiabatic or *Chew-Goldberger-Low equations* of state with the parallel and perpendicular ideal gas equations (7.23), it is easy to show that in this theory the parallel and perpendicular temperatures depend on the magnetic field strength according to

$$
\begin{aligned}
T_\perp &\propto B \\
T_\parallel &\propto (n/B)^2
\end{aligned}
\tag{7.31}
$$

Hence, for an increasing magnetic field strength the perpendicular temperature will increase while the parallel temperature decreases, unless $n/B = $ const for some reason. A streaming plasma where the magnetic field strength increases along the direction of the stream will therefore exhibit a growing temperature anisotropy with a higher temperature in the perpendicular than in the parallel direction. Such a situation is encountered in the magnetosphere and in the near-Earth magnetosheath.

To conclude this section we briefly discuss the behaviour of the parallel and perpendicular adiabatic indices. Formally it is possible to define such indices in analogy to the adiabatic case by simply writing $p_\parallel \propto n^{\gamma_\parallel}$ and $p_\perp \propto n^{\gamma_\perp}$. Using the double-adiabatic equations of state, one finds that

$$
\begin{aligned}
\gamma_\perp &= 1 + \frac{\ln(B/B_0)}{\ln(n/n_0)} \\
\gamma_\parallel &= 3 - 2\frac{\ln(B/B_0)}{\ln(n/n_0)}
\end{aligned}
\tag{7.32}
$$

are functions of magnetic field and density. However, from Eq. (7.30) one can derive a condition on the parallel and perpendicular adiabatic indices

$$
\gamma_\parallel + 2\gamma_\perp - 5 = 0
\tag{7.33}
$$

which shows that the parallel and perpendicular adiabatic indices in double-adiabatic plasmas are closely related. It is sufficient to know one of the indices in order to determine the other one. Yet these quantities are not constants but spatially varying functions.

7.3 Single-Fluid Theory

Plasmas consist of electrons of mass, m_e, and charge, $q_e = -e$, and ions of mass, m_i, and charge, $q_i = Ze$. Let us for simplicity assume that there is only one ion component present in the plasma. If these are protons, the atomic charge is $Z = 1$. For ease of use, let us repeat the fundamental equations of such a plasma, the continuity equation (7.5) and the equation of motion (7.11) for the s-component fluid

$$\frac{\partial n_s}{\partial t} + \nabla \cdot (n_s \mathbf{v}_s) = 0 \tag{7.34}$$

$$\frac{\partial (n_s \mathbf{v}_s)}{\partial t} + \nabla \cdot (n_s \mathbf{v}_s \mathbf{v}_s) = -\frac{1}{m_s} \nabla \cdot \mathbf{P}_s + \frac{n_s q_s}{m_s} (\mathbf{E} + \mathbf{v}_s \times \mathbf{B}) \tag{7.35}$$

which must be completed with the equations of state for the electron and ion components and with the set of Maxwell's equations, where we define the charges and currents by

$$\rho = e(n_i - n_e) \tag{7.36}$$

$$\mathbf{j} = e(n_i \mathbf{v}_i - n_e \mathbf{v}_e) \tag{7.37}$$

Quasineutrality in such a case is defined by a vanishing electric space charge $\rho = 0$, yielding $n = n_e = n_i$, which implies equal charge densities. For a current-free plasma, $\mathbf{j} = 0$, the particle flux densities must be equal, $n_i \mathbf{v}_i = n_e \mathbf{v}_e$. This is not generally the case, since most plasmas are quasineutral but carry currents.

Sometimes it is convenient to neglect the difference between the particle species in a plasma and to consider the plasma as a conducting fluid carrying magnetic and electric fields and currents. In such a case the fluid field variables are some combinations of the densities and velocities of the single components. The resulting equations are called *magnetohydrodynamic equations* of a plasma. They can be derived from the above two-fluid equations by choosing as center-of-mass variables the following combinations

$$n = \frac{m_e n_e + m_i n_i}{m_e + m_i} \tag{7.38}$$

$$m = m_e + m_i = m_i \left(1 + \frac{m_e}{m_i} \right) \tag{7.39}$$

$$\mathbf{v} = \frac{m_i n_i \mathbf{v}_i + m_e n_e \mathbf{v}_e}{m_e n_e + m_i n_i} \tag{7.40}$$

for the single fluid number density, n, mass, m, and the single fluid velocity, $\mathbf{v}$.

7.3.1 Continuity Equation

With these definitions it is now simple matter to derive the continuity equation for the total fluid. We multiply the two-fluid continuity equation (7.34) for ions by m_i and for electrons by m_e, add the two resulting equations, and make use of the above definitions in Eqs. (7.38) through (7.40) to obtain

$$\boxed{\frac{\partial n}{\partial t} + \nabla \cdot (n \mathbf{v}) = 0} \tag{7.41}$$

This equations represent the usual form of a *fluid continuity equation* and does not anymore discriminate between the different kinds of particles. The physical content of the continuity equation is that in a classical and nonrelativistic plasma flowing with centre of mass velocity, mass is conserved.

7.3.2 Equation of Motion

Constructing the momentum density conservation equation for the total fluid, or *fluid equation of motion*, is more difficult because of the appearance of the nonlinear terms, $n_s v_s v_s$, in Eq. (7.35). We therefore demonstrate explicitly how this equation is obtained. To be even more general, let us for convenience include a simple collisional term in Eq. (7.35), which may account for a momentum transfer between electrons and ions via some kind of friction, either collisional in the classical sense or due to anomalous collisions between the two kinds of particles. As demonstrated in Chapter 4, the presence of collisions gives rise to a non-vanishing electric field and resistive currents. As a consequence an Ohm's law exists in the plasma. The derivation of the momentum conservation equation for the fluid therefore necessarily results also in a generalized Ohm's law as a second material equation for the plasma, coupling the currents in the medium to the electromagnetic field.

The collisional term is defined by $\mathbf{R} = \mathbf{R}_{ei} = -\mathbf{R}_{ie}$. Because this term describes transfer of momentum from ions to electrons in the ion equation and from electrons to ions in the electron equation, conservation of the transferred momentum requires that the two terms are equal in magnitude but have different sign. Hence, the two momentum equations read

$$\frac{\partial(n_e \mathbf{v}_e)}{\partial t} + \nabla \cdot (n_e \mathbf{v}_e \mathbf{v}_e) = -\frac{1}{m_e} \nabla \cdot \mathbf{P}_e - \frac{n_e e}{m_e}(\mathbf{E} + \mathbf{v}_e \times \mathbf{B}) + \frac{\mathbf{R}}{m_e}$$

$$\frac{\partial(n_i \mathbf{v}_i)}{\partial t} + \nabla \cdot (n_i \mathbf{v}_i \mathbf{v}_i) = -\frac{1}{m_i} \nabla \cdot \mathbf{P}_i + \frac{n_i e}{m_i}(\mathbf{E} + \mathbf{v}_i \times \mathbf{B}) - \frac{\mathbf{R}}{m_i} \tag{7.42}$$

The equation of motion of the single-fluid plasma is obtained by multiplying the first equation by m_e, the second by m_i, and adding up. Then the two collisional terms cancel and the right-hand side becomes in center-of-mass variables

$$-\nabla \cdot (\mathbf{P}_e + \mathbf{P}_i) + e(n_i - n_e)\mathbf{E} + e(n_i \mathbf{v}_i - n_e \mathbf{v}_e) \times \mathbf{B} = -\nabla \cdot \mathbf{P} + \rho \mathbf{E} + \mathbf{j} \times \mathbf{B} \tag{7.43}$$

where we have introduced the total pressure tensor, $\mathbf{P} = \mathbf{P}_e + \mathbf{P}_i$, and made use of the definitions of the space charge in Eq. (7.36) and of the current density in Eq. (7.37). The first term on the left-hand side of Eq. (7.42), after adding up and observing the definition of the fluid bulk velocity, $\mathbf{v}$, in Eq. (7.40) becomes

$$\frac{\partial}{\partial t}(m_e n_e \mathbf{v}_e + m_i n_i \mathbf{v}_i) = \frac{\partial}{\partial t}(n m \mathbf{v}) \tag{7.44}$$

In the second nonlinear term of Eq. (7.42) we take advantage of the smallness of the electron mass, $m_e \ll m_i$, and assume that the two densities are nearly equal. In this case $n_i \approx n_e$, and the dominant term in the sum $m_i n_i [\mathbf{v}_i \mathbf{v}_i + \mathbf{v}_e \mathbf{v}_e (m_e n_e / m_i n_i)]$ is the first ion term. With this approximation, which is good for nearly quasineutral plasmas, the equation of motion becomes

$$\frac{\partial (nm\mathbf{v})}{\partial t} + \nabla \cdot (nm\mathbf{v}\mathbf{v}) = -\nabla \cdot \mathbf{P} + \rho \mathbf{E} + \mathbf{j} \times \mathbf{B} \tag{7.45}$$

This is the momentum conservation equation in plasma magnetohydrodynamics.

7.3.3   Generalized Ohm's Law

The momentum conservation equation (7.45) contains the electric current density, $\mathbf{j}$, as a new variable. To close the system of equations, one therefore needs an additional expression for the evolution of $\mathbf{j}$. This equation is the *generalized Ohm's law* of a plasma. It is found by subtracting the two-fluid momentum equations (7.42). For practical reasons it is convenient to multiply the electron equation by m_i and the ion equation by m_e before subtraction. For small current densities, we can neglect quadratic terms in the velocities, i.e., the second terms on the left-hand sides of the momentum equations. These terms do, however contribute to Ohms law when the plasma exhibits strong turbulence causing fluctuations in the streaming velocity or when the flow contains strong gradients in velocity. In the following we neglect these effects even though they may be very important in regions like the magnetosheath or in plasma shear flows. It is then easy to show that the current density satisfies the following relation

$$\begin{aligned}
\frac{m_e}{e} \frac{\partial \mathbf{j}}{\partial t} &= \nabla \cdot \left(\mathbf{P}_e - \frac{m_e}{m_i} \mathbf{P}_i \right) - \left(1 + \frac{m_e}{m_i} \right) \mathbf{R} \\
&\quad + n_e e \left(1 + \frac{m_e n_i}{m_i n_e} \right) \left[\mathbf{E} + \left(\mathbf{v}_e + \frac{m_e n_i}{m_i n_e} \mathbf{v}_i \right) \times \mathbf{B} \right]
\end{aligned} \tag{7.46}$$

The right-hand side of this equation still contains the partial densities, velocities, and pressures. It is, however, possible to replace them by use of Eqs. (7.38) through (7.40). The algebra is simplified when neglecting terms with small mass ratios, $m_e / m_i \ll 1$, and assuming quasineutrality, $n \approx n_i \approx n_e$. Then only the electron pressure plays a role in the first term on the right-hand side of the above equation. Moreover, quasineutrality also simplifies the terms containing $\mathbf{E}$ and $\mathbf{B}$. In particular, it allows to omit the term containing $\mathbf{v}_i$. Hence, in this approximation the above equation can be rewritten as

$$\frac{m_e}{e} \frac{\partial \mathbf{j}}{\partial t} = \nabla \cdot \mathbf{P}_e + ne(\mathbf{E} + \mathbf{v}_e \times \mathbf{B}) - \mathbf{R} \tag{7.47}$$

Before proceeding, we note briefly some interesting implications of this expression. The first one is that in the one-fluid theory thermal effects on the electric current density enter only through the electron pressure, while the macroscopic bulk fluid velocity is also affected by the ion pressure via the full pressure tensor entering Eq. (7.45). It is the changes in the electron partial pressure which modulate the current. Second, the Lorentz term $\mathbf{E} + \mathbf{v}_e \times \mathbf{B}$ on the right-hand side contains only the electron velocity $\mathbf{v}_e$. Hence, even in the one-fluid theory the electron fluid behaves different from the ion fluid and has a greater effect on the current. In particular, in the stationary ideal case the electron fluid is stronger frozen-in to the magnetic field than the ions.

Again omitting small mass ratio terms and assuming quasineutrality, one can find a fluid expression for the electron velocity, since under these assumptions Eq. (7.40) can be written as

$$\mathbf{v}_i = \mathbf{v} \tag{7.48}$$

Inserting this approximation into Eq. (7.37) one obtains

$$\mathbf{v}_e = \mathbf{v} - \frac{\mathbf{j}}{ne} \tag{7.49}$$

Inserting the above expression into Eq. (7.47) yields the following equation

$$\frac{m_e}{e} \frac{\partial \mathbf{j}}{\partial t} = \nabla \cdot \mathbf{P}_e + ne(\mathbf{E} + \mathbf{v} \times \mathbf{B}) - \mathbf{j} \times \mathbf{B} - \mathbf{R} \tag{7.50}$$

Finally, we remember that the friction term, $\mathbf{R}$, is proportional to the velocity difference of the two oppositely charged species, and is quadratric in density. The proportionality factor is the collision frequency, ν_c, times the electron mass

$$\mathbf{R} = m_e n^2 \nu_c (\mathbf{v}_i - \mathbf{v}_e) \tag{7.51}$$

The factor in front of the brackets on the right-hand side includes similar quantities as the plasma resistivity, $\eta = m_e \nu_c / ne^2$, introduced in Section 4.2, thus permitting to write

$$\mathbf{R} = \eta ne \mathbf{j} \tag{7.52}$$

Hence, the generalized though strongly simplified Ohm's law of a single-fluid magnetohydrodynamic plasma becomes after rearrangement of the different terms

$$\boxed{\mathbf{E} + \mathbf{v} \times \mathbf{B} = \eta \mathbf{j} + \frac{1}{ne} \mathbf{j} \times \mathbf{B} - \frac{1}{ne} \nabla \cdot \mathbf{P}_e + \frac{m_e}{ne^2} \frac{\partial \mathbf{j}}{\partial t}} \tag{7.53}$$

This equation is an important expression in several respects. First one recognizes that in a plasma the simple Ohm's law derived in Chapter 4 becomes considerably more complicated. In addition to the resistive term, $\eta \mathbf{j}$, it contains the anisotropic electron pressure term, a Lorentz force term $\mathbf{j} \times \mathbf{B}$ which is often called *Hall term* and even in a collisionless plasma gives rise to electric field components transverse to both the current $\mathbf{j}$ and the magnetic field $\mathbf{B}$. One also recognises the already mentioned continuous presence of a finite Hall resistivity $\eta_H = B/en$ in plasma which is independent of any other resistivity. Its cause is the $\mathbf{E} \times \mathbf{B}$ motion imposed by the Lorentz force. Finally, the generalized Ohm's law turns out to contain the time variation of the current which can be interpreted as the contribution of electron inertia to the current flow. The corresponding *inertial resistivity* η_e of the electrons can be read from this equation to be

$$\eta_e = (\varepsilon_0 \omega_{pe}^2 \tau_j)^{-1} \tag{7.54}$$

where τ_j is the typical time scale of current variation. Clearly, electron inertia plays a role only for very fast current fluctuations of time scales $\tau_j \omega_{pe} \sim 1$. Hence, in applications this term is less important than the other terms in the equation unless the plasma is extremely diluted. Another reason for neglecting this term entirely is that the equations derived for the single fluid plasma are valid only on time scales which are long with respect to the plasma fluctuation time scale. This conclusion shows that in dealing with formally derived expressions one has to be careful when applying them to real problems.

From Eq. (7.53) it also is obvious that in an ideally conducting magnetohydrodynamic fluid with $\eta = 0$ the convective approximation or frozen-in condition

$$\mathbf{E} = -\mathbf{v} \times \mathbf{B} \tag{7.55}$$

requires additional assumptions. Vanishing electron pressure gradients and slow time variations of current density are necessary to neglect the two corresponding terms in the generalized Ohm's law. But even under these conditions neglecting the Lorentz force term, $\mathbf{j} \times \mathbf{B}$, is more difficult to justify. Actually, using Eq. (7.49) in Eq. (7.53), one can readily show that

$$\mathbf{E} = -\mathbf{v}_e \times \mathbf{B} \tag{7.56}$$

Hence, only the electron fluid is frozen to the magnetic field, while the motion of the ion fluid may deviate from the field. Nevertheless, when the transverse currents are small, the ideal magnetohydrodynamic condition is frequently applied to space plasmas. In the solar wind and the magnetosphere for slow variations, negligible pressure gradients and weak currents it is often satisfied very well.

7.3.4   Energy Conservation Equation

For completeness, we add the *energy conservation equation* of magnetohydrodynamics. It follows by multiplying the momentum conservation equation (7.45) by the fluid velocity, $\mathbf{v}$, and manipulating the resulting expression with the help of the continuity equation into the form

$$\frac{\partial}{\partial t} \left[nm \left(\tfrac{1}{2} v^2 + w \right) + \frac{B^2}{2\mu_0} \right] = -\nabla \cdot \mathbf{q} \tag{7.57}$$

The right-hand side of this expression is the divergence of the magnetohydrodynamic energy or heat flux density vector

$$\mathbf{q} = \left(\frac{v^2}{2} + w + \frac{p + B^2/\mu_0}{nm} \right) nm\mathbf{v} - \frac{\mathbf{B}}{\mu_0} \left(\mathbf{v} + \frac{\mathbf{j}}{ne} \right) \cdot \mathbf{B} - \frac{\eta \mathbf{j} \times \mathbf{B}}{\mu_0} + \frac{jB^2}{\mu_0 ne} + \frac{m_e \mathbf{B}}{\mu_0 ne^2} \times \frac{\partial \mathbf{j}}{\partial t} \tag{7.58}$$

Of the dissipative terms in Ohm's law we have taken into account only the resistivity and — for completeness — electron inertia in deriving this form of $\mathbf{q}$. The quantity w is the free internal energy or *enthalpy* density $w = u + p$ of the fluid, p is the isotropic pressure and u the internal energy density. The left-hand side of Eq. (7.57) is the local time derivative of the total energy density which is the sum of the kinetic and internal energy densities to which the pressure of the magnetic field has been added. On the other hand, the energy flux contains the convective losses of kinetic energy, internal energy, and temperature, Joule heating due to resistivity and electron inertia, and magnetic field energy flux related to the divergence of the Lorentz force term. If all these losses vanish, the divergence of $\mathbf{q}$ is zero, and the total energy density including kinetic, inner, and magnetic energy does locally not change in time.

7.3.5   Entropy Equation

In Eq. (7.57) the conservation of energy has been written in the form of a heat conduction equation. This is not the only possible way to express energy conservation. Several other forms have been proposed in the literature. For example, one may introduce the entropy, $\mathscr{S}$, and write Eq. (7.57) as a continuity equation of the entropy

$$\frac{d\mathscr{S}}{dt} = \frac{\partial \mathscr{S}}{\partial t} + \mathbf{v} \cdot \nabla \mathscr{S} = 0 \qquad (7.59)$$

To calculate the entropy change in an ideal isotropic gas from the measurements of temperature and density, one can use

$$\Delta \mathscr{S} = R_0 \ln\left[\frac{T^{1/(\gamma-1)}}{n}\right] \qquad (7.60)$$

where R_0 is the ideal gas constant (cf. Appendix B.1) and γ is the adiabatic index (see Appendix B.6, where also a formula for anisotropic plasmas is given).

Equation (7.59) replaces the energy conservation and heat transfer equations insofar as the entropy is the fundamental state quantity of any thermodynamic system. It is defined differentially as the ratio $d\mathscr{S} = d\mathscr{Q}/T$ of the differential amount of heat $d\mathscr{Q}$ generated in the gas divided by the temperature T. Since the entropy can only grow or stay constant, the differential is always positive or vanishes when no heat is produced. Heat production requires dissipation. Hence, in a dissipationless medium like ideal magnetohydrodynamics the entropy is conserved. This is the content of Eq. (7.59). But if the plasma is dissipative, heat is generated, and the right-hand side of Eq. (7.59) is different from zero. Then one must add a Joule heating term, $\mathbf{j} \cdot \mathbf{E}$, to the right-hand side, where the electric field is being replaced with the help of the general Ohm's law (see p. 168).

7.3.6   Magnetic Tension and Plasma Beta

The Lorentz force or Hall term, $\mathbf{j} \times \mathbf{B}$, appearing in the magnetohydrodynamic equation of motion and the generalized Ohm's law introduces a new effect which is specific for magnetohydrodynamics and which we will discuss separately. It is the effect of the *magnetic tension* on a conducting magnetohydrodynamic fluid.

For slow variations, when the displacement current in the plasma can be neglected, the first Maxwell equation (2.3) can be used to rewrite

$$\mathbf{j} \times \mathbf{B} = -\frac{1}{\mu_0} \mathbf{B} \times (\nabla \times \mathbf{B}) \tag{7.61}$$

Applying some vector algebra to the right-hand side, this expression can be written as

$$\boxed{\mathbf{j} \times \mathbf{B} = -\nabla \left(\frac{B^2}{2\mu_0} \right) + \frac{1}{\mu_0} \nabla \cdot (\mathbf{BB})} \tag{7.62}$$

The first term on the right-hand side of this equation corresponds to a pressure term with the *magnetic pressure* defined as

$$p_B = \frac{B^2}{2\mu_0} \tag{7.63}$$

This pressure simply adds to the thermal pressure of the plasma. The second term is a consequence of the vector product of current and magnetic field and thus of the vector character of the magnetic field. It is the divergence of a *magnetic stress tensor*, $\mathbf{BB}/\mu_0$. The magnetic field introduces a magnetic stress in the plasma, which contributes to tension and torsion in the conducting fluid.

The concept of magnetic pressure can also be used to define another useful quantity. Starting from the fluid equation of motion (7.45) and assuming quasineutrality, which cancels the $\rho \mathbf{E}$ term, equilibrium conditions, $d\mathbf{v}/dt = 0$, and using Eq. (7.61), we are left with

$$\nabla \cdot \mathbf{P} = -\frac{1}{\mu_0} \mathbf{B} \times (\nabla \times \mathbf{B}) \tag{7.64}$$

This equation forms together with those Maxwell's equation describing the divergence and the curl of $\mathbf{B}$ a closed set and is often called *magnetohydrostatic equation*. It tells us that in equilibrium the particle pressure gradient is balanced by magnetic tension.

If we neglect the off-diagonal or stress terms, e.g., for cases where the particle pressure is nearly isotropic and the magnetic field is approximately homogenous, and use Eq. (7.62), we can approximate Eq. (7.64) by

$$\nabla \left(p + \frac{B^2}{2\mu_0} \right) = 0 \tag{7.65}$$

Hence, in an equilibrium, isotropic, and quasineutral plasma the total pressure is a constant. Under these conditions one can define a *plasma beta* parameter as the ratio of thermal and magnetic pressure

$$\beta = \frac{2\mu_0 p}{B^2} \tag{7.66}$$

In anisotropic plasmas where the pressure splits into parallel and perpendicular components one frequently uses parallel and perpendicular plasma beta

$$
\begin{aligned}
\beta_\| &= \frac{2\mu_0 p_\|}{B^2} \\
\beta_\perp &= \frac{2\mu_0 p_\perp}{B^2}
\end{aligned} \tag{7.67}
$$

The β parameters measure the relative importance of particle and magnetic field pressures. A plasma is called a *low-beta plasma* when $\beta \ll 1$ and a *high-beta plasma* for $\beta \approx 1$ and greater. Both types of plasmas are encountered in near-Earth space.

7.3.7   Equation of State

The above equations are not complete until one adds appropriate equations for the pressure tensor components. Neglecting all dissipative effects, except for the electrical resistivity, the pressure tensors may have up to two independent components, $p_\perp$ and $p_\|$, which require additional equations. Under the assumption that the medium behaves like an ideal gas, this pressure equation becomes the ideal gas equation $p = nk_B T$, or two equations of state in the anisotropic case, providing the connection between pressure, temperature, and density.

The equation determining the behaviour of the temperature is the energy or heat conduction equation. For an isotropic magnetohydrodynamic fluid it is sufficient to derive an overall energy conservation equation. But this equation will contain the unknown heat flux and must be truncated. If the heat flux is neglected, the energy equation becomes an equation for the temperature, as has been discussed in Section 7.1. In many cases, however, one can avoid to solve this equation assuming either isothermal conditions with $T = \text{const}$ or adiabaticity or, more general, polytropicity with $p \propto n^\gamma$ (or the various anisotropic equivalents of this expression), thereby closing the full set of magnetohydrodynamic relations.

7.4   Stationarity and Equilibria

Stationarity implies absence of any time variations which mathematically means that partial and sometimes even total time derivatives are set to zero. Stationarity also implies that the state of the plasma persists for long time. It is thus an equilibrium state and one sometimes speaks of plasma equilibria. Finding the equilibrium state of a fluid plasma requires solution of the time-independent fluid equations for the special case under consideration. Under given boundary conditions this may become a formidable task. But it is possible to draw some more general conclusions about the plasma behaviour.

7.4.1   Boltzmann's Law

Let us first treat a rather simple case, which involves only the electron fluid. Consider the stationary electron momentum conservation equation (7.11) for scalar pressure and absence of an external magnetic field. Setting the convective derivative to zero, it can be written as

$$\nabla p_e = -n_e e \mathbf{E} \tag{7.68}$$

demonstrating that the electrons are in equilibrium with the electric field. The electric field can be represented as the gradient of an electric potential, $\mathbf{E} = -\nabla \phi$. Now assuming that the electron temperature is constant, which is reasonable since under stationary conditions one expects that the plasma had sufficient time to achieve an isothermal state, the above equation becomes with $p_e = n_e k_B T_e$

$$\nabla \left(\ln n_e - \frac{e\phi}{k_B T_e} \right) = 0 \tag{7.69}$$

The solution of this equation is the *Boltzmann law* which relates the stationary electron density to the electrostatic potential as

$$\boxed{n_e = n_0 \exp \left(\frac{e\phi}{k_B T_e} \right)} \tag{7.70}$$

where n_0 is the average electron density. The interpretation of this law is that in a stationary electric field, present in a plasma and maintained by external means, the electron density is necessarily spatially inhomogeneous and changes exponentially with the local electrostatic potential. The electron fluid reacts very sensitively to an electric field. According to the attraction exerted on the negative electrons by positive electric potentials the electron density assumes its maximum where the electric potential attains its highest positive value. Application of this law is possible in all cases when electric fields are present in the plasma in the direction parallel to the magnetic field and when the time variations are so slow that electron motions can be neglected.

7.4.2   Diamagnetic Drift

The next conclusion can be obtained in full generality even for the multi-fluid case. Let us return to the fluid momentum conservation equation (7.11) for the s-component fluid. Let us also assume stationary conditions so that the convective derivative terms

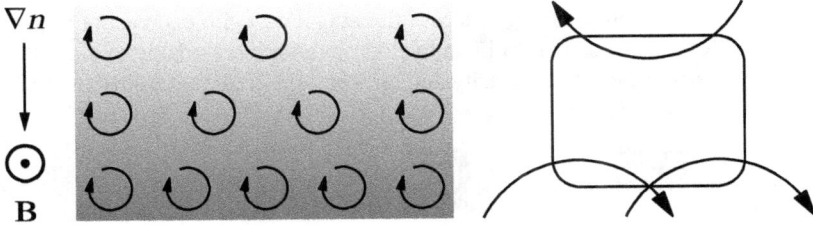

Figure 7.1: The origin of diamagnetic drift in the pressure (or density) gradient of the plasma, assuming positive ions.

can be dropped. For simplicity we also assume that the pressure tensor, $\mathbf{P}_s$, is anisotropic in the form of Eq. (7.21). Then the fluid equation of motion reduces to

$$q_s n_s (\mathbf{E} + \mathbf{v}_s \times \mathbf{B}) = \nabla p_{s\perp} + \nabla \cdot \left[(p_{s\|} - p_{s\perp}) \frac{\mathbf{BB}}{B^2} \right] \qquad (7.71)$$

Taking the cross-product of this equation with $\mathbf{B}/B^2$ and rearranging the different terms, we obtain for the stationary drift velocity of the s-component fluid

$$\mathbf{v}_s = \frac{\mathbf{E} \times \mathbf{B}}{B^2} + \frac{1}{q_s n_s B^2} \mathbf{B} \times \nabla p_{s\perp} + \frac{1}{q_s n_s B^2} \mathbf{B} \times \nabla \cdot \left[(p_{s\|} - p_{s\perp}) \frac{\mathbf{BB}}{B^2} \right] \qquad (7.72)$$

In a manner analogous to single particle theory, this expression defines the drift of the s-component fluid of the plasma across the magnetic field. As one immediately recognizes, the first term on the right hand side is nothing else but the $\mathbf{E} \times \mathbf{B}$ drift of the fluid which is the effect of the Lorentz transformation. The second and third terms are, however, entirely new and did not appear in single particle theory.

The second term describes a drift perpendicular to the magnetic field and to the transverse gradient (because only $\nabla_\perp$ survives the cross-product with the magnetic field) of the perpendicular pressure $p_{s\perp}$. Its dependence on the pressure, which is a moment of the distribution function and thus an average variable of the plasma, indicates that this drift arises due to a collective behaviour of the plasma.

Consider a plasma of gyrating particles of one species. All particles gyrate in the same direction around the field. Consequently, at each point in a homogeneous plasma there would be exactly the same number of particles having exactly same but oppositely directed transverse velocities, resulting from particles which are displaced by just one gyroradius across the magnetic field, so that the average velocity would be zero. In a non-uniform plasma, the change in transverse pressure can be either due to a gradient in density or a gradient in transverse temperature.

The presence of a transverse density gradient introduces an asymmetry since in the direction of decreasing particle density there are less particles gyrating and, hence, not sufficient oppositely directed velocities to average the transverse velocity out. Consequently there is an excess of transverse particle gyration velocity perpendicular to the density gradient which remains and simulates a gross fluid drift motion (see Fig. 7.1).

If the plasma temperature changes across the magnetic field, decreasing temperature implies smaller transverse gyroradii and velocities which are unable to make the average velocity zero. Hence, both density and temperature gradients contribute to a transverse *diamagnetic fluid drift*

$$\mathbf{v}_{dia,s} = \frac{\mathbf{B} \times \nabla_\perp p_{s\perp}}{q_s n_s B^2} \tag{7.73}$$

across the magnetic field. Because this velocity depends on the charge of the fluid particles, differently charged fluid components will drift into opposite directions, thus giving rise to an effective drift current flow in the plasma. In a quasineutral electron-ion plasma this *diamagnetic current* becomes

$$\mathbf{j}_{dia} = \frac{\mathbf{B} \times \nabla_\perp p_\perp}{B^2} \tag{7.74}$$

where $p_\perp = p_{e\perp} + p_{i\perp}$ is the total perpendicular pressure. Any plasma containing transverse density or pressure gradients carries such diamagnetic currents. They are called diamagnetic since they diminish the external magnetic field.

In an isotropic stationary plasma this diamagnetic drift is the only relevant drift term. In the anisotropic case the third term on the right-hand side of Eq. (7.72) comes into play. It is a little more involved to treat this term, since it contains derivatives of the type $\nabla \cdot (\mathbf{BB}/B^2)$. These can be replaced by using the fact that the derivative of a unit vector is equal to the outer (negative sign) normal, $\mathbf{n}$, divided by the radius of curvature, R_c. Hence, the third term in Eq. (7.72) becomes

$$\mathbf{v}_{c,s} = -\frac{(p_{s\|} - p_{s\perp})}{q_s n_s B^2 R_c} \mathbf{B} \times \mathbf{n} \tag{7.75}$$

This is a fluid drift velocity which depends on the pressure difference and is non-zero only in a curved magnetic field. It thus resembles the curvature drift but this time it is caused by the collective effects in the pressure anisotropy. Its direction is the same as the diamagnetic drift. It is interesting to note that this fluid drift may be negative as well as positive depending on the sign of the pressure anisotropy. The anisotropic curvature current resulting from this drift is given by

$$\mathbf{j}_{dia,c} = -\frac{p_\| - p_\perp}{B^2 R_c} \mathbf{B} \times \mathbf{n} \tag{7.76}$$

with $p_\|$ the total parallel pressure. This current, for large parallel pressure, is negative but can change its sign when the perpendicular pressure becomes large. It may therefore amplify or weaken the effect of the isotropic diamagnetic current.

It may be of interest to note at this place that a fundamental theorem by Bohr and van Leeuwen states that diamagnetism is quantum mechanical and there is no classical diamagnetism. Clearly this theorem does not hold for classical high temperature magnetised plasmas in which diamagnetism is naturally generated if only density or temperature gradients exist. It is the diamagnetic currents which cause many interesting effects in space plasmas. Among the notable effects are the storm time ring current which we have already mentioned and, more generally, the generation of the outer boundary of the magnetosphere, the magnetopause which is a thin layer of current flow caused be the gradient in the plasma density between the solar wind and the magnetosphere. Diamagnetic plasma clouds have been observed being occasionally ejected from the solar corona into interplanetary space during coronal mass ejection events. Another magnetospheric example is the neutral sheet current in the tail of the magnetosphere.

7.4.3 Neutral Sheet Current

A typical example of a diamagnetic current is the neutral sheet current in the geomagnetic tail which divides the tail into northern and southern lobes with their stretched magnetic field lines. In the southern lobe the field lines extend from the southern polar cap and point anti-sunward, while in the northern lobe they come from the distant tail pointing sunward and ending in the northern polar cap.

This stretching of the otherwise approximately dipolar terrestrial magnetic field can be accounted for by a diamagnetic current flowing across the magnetospheric tail from dawn to dusk (see Section 1.3 and Fig. 1.6). Such a current transports positive charges from dawn to dusk and negative charges from dusk to dawn across the tail and, because of its stationarity and its macroscopic magnetic effect, cannot be anything else but a diamagnetic current. Its cause must therefore be a gradient in the plasma pressure perpendicular to the current layer pointing from north to south in the upper (northern) half and from south to north in the lower (southern) half of the current layer. Hence, the current layer is a concentration of dense and hot plasma which is called the *neutral sheet* because of the weak magnetic field it contains.

Spacecraft measurements have revealed that the neutral sheet in the geomagnetic tail contains a quasineutral ion-electron plasma of roughly 1–10 keV temperature and a density of about 1 cm^{-3}. The transverse magnetic field in the neutral sheet is not zero but rather weak, of the order of 1–5 nT. Due to the weak magnetic field, the plasma beta parameter has typically rather high values, $\beta \approx 100$. The main current sheet has a typical thickness of 1–2 R_E and the maximum current density is of the order of some nA/m^2. But especially during disturbed times and before substorm onset, the current sheet can be much thinner and the current density much higher.

The width of the current layer must be larger than an ion gyroradius. To find an estimate of the pressure gradient length in the neutral sheet, L_p, we perform a dimensional analysis. Since the neutral sheet current is a diamagnetic current, it is approximated as

$$j = \frac{2p_\perp}{L_p B_L} \tag{7.77}$$

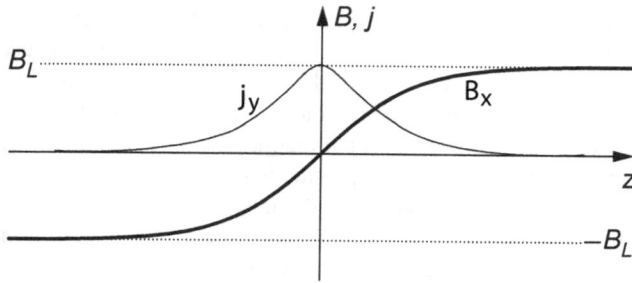

Figure 7.2: Harris model of the neutral sheet current j_y and magnetic field B_x. The magnetic field is of opposite direction to both sides of the sheet. In order to obey equilibrium, the density and current are maximum in the sheet centre.

where $p_\perp$ is the perpendicular plasma pressure in the neutral sheet and B_L is the lobe magnetic field strength. From Ampère's law this current is equal to the curl of the lobe magnetic field

$$j = \frac{B_L}{\mu_0 L_B} \tag{7.78}$$

where L_B is the variation length of the lobe magnetic field. Hence, one finds that

$$\frac{L_p}{L_B} = \frac{2\mu_0 p_\perp}{B_L^2} \tag{7.79}$$

This ratio is something like the plasma beta, but with respect to the lobe magnetic field and the neutral sheet plasma pressure. In equilibrium, when the magnetosphere is undistorted, this ratio is about one. But in the interior of the neutral sheet we have $\beta \gg 1$ and the pressure has nearly flat profile, while outside the neutral sheet, where β is small, the pressure drops steeply towards the lobe.

A very simple theoretical model of the neutral layer, which empirically accounts for these observational properties, is the so-called *Harris sheet*. It assumes that the magnetospheric tail has a simple geometry with the magnetic field pointing in the sunward x direction in the northern lobe and in the $-x$ direction in the southern lobe and varies in the z direction, perpendicular to the plasma sheet equatorial plane, according to a hyperbolic tangent function

$$B_x = B_L \tanh(z/L_B) \tag{7.80}$$

where B_L is again the lobe magnetic field strength and L_B is assumed to be a constant cross scale typical for the width of the Harris current sheet. The current profile of the Harris sheet can be calculated from Ampère's law taking the derivative of B_x. It yields the required current maximum at $z = 0$ from the expression

$$j_y = \frac{B_L}{\mu_0 L_B} \frac{1}{\cosh^2(z/L_B)} \tag{7.81}$$

This model, which is shown in Fig. 7.2, neglects any northward B_z component or inclination of the magnetic field in the neutral sheet. Since the measured magnetic field inside the neutral sheet has also a northward B_z component, typically of the order of 1–5 nT, the Harris model can be taken only as a crude approximation to real tail field. However, it is used as the initial current and field configuration in most of the numerical simulations of reconnection in two-dimensional current sheets separating two antiparallel magnetic fields.

7.4.4 Field-Aligned Currents

The various perpendicular drift currents in a plasma add up to a total perpendicular current density $\mathbf{j}_\perp$ which in an inhomogeneous and possibly time-varying plasma is not necessarily divergence-free. Under slowly variable conditions the requirement of closed current circuits necessarily leads to the generation of field-aligned currents in a way similar to that discussed in connection with the generation of field-aligned currents in the ionosphere (see Section 5.4). Since from Ampére's law under quasi-stationary conditions the divergence of total current must vanish

$$\nabla \cdot \mathbf{j} = \nabla \cdot (\mathbf{j}_\perp + \mathbf{j}_\parallel) = 0 \tag{7.82}$$

$\mathbf{j}_\parallel$ and $\mathbf{j}_\perp$ are not independent. Instead, one obtains a relation between the perpendicular and parallel currents

$$\nabla \cdot \mathbf{j}_\parallel = -\nabla \cdot \mathbf{j}_\perp \tag{7.83}$$

Introducing the coordinate s along the magnetic field, this expression can be rewritten in terms of the scalar parallel current density

$$\frac{\partial}{\partial s} \left(\frac{j_\parallel}{B} \right) = -\frac{\nabla \cdot \mathbf{j}_\perp}{B} \tag{7.84}$$

The divergence of the perpendicular current contributing to the right-hand side, most notably the diamagnetic neutral sheet current, serves as the source of the field-aligned currents in the magnetosphere. They close in the ionosphere as discussed in Sections 5.5 and 5.7.

For $B = B(s)$ Eq. (7.84) can be integrated to obtain a very useful expression for

$$j_\parallel(s) = -\int_0^s \frac{\nabla \cdot \mathbf{j}_\perp(s')}{B(s')} \, ds' \tag{7.85}$$

the field-aligned current density at a given position s along a field line as function of the divergence of the perpendicular current density and magnetic field strength along the field line. Only the part from $s = 0$ to the location s contributes to the field-aligned current density.

7.4.5 Polarization Drifts

We have so far assumed stationarity and thus neglected any time dependence of the fluid velocity. If we assume that such time variations exists, but with a time scale much longer than the gyration period, one can add the term $m_s n_s d\mathbf{v}_s/dt$ to the right-hand side of Eq. (7.71), where the fluid velocity of a particular species, $\mathbf{v}_s$, is now understood as the slow drift velocity of the fluid. The dominant term in this drift velocity is the electric convection fluid drift and one recovers the guiding center *polarization drift* velocity (see Section 2.3)

$$\mathbf{v}_{P,s} = \frac{1}{\omega_{gs}B} \frac{d\mathbf{E}_\perp}{dt} \tag{7.86}$$

In contrast to the gradient and curvature particle drifts, the polarization drift turns out to survive the transition from single particles to fluids. It is one of the important fluid drifts. Since ω_{gs} carries the sign of the charge and also depends on the mass, the polarization drift term also leads to *polarization current* or *inertial current* in the plasma.

In addition to the electric polarization drift, time variations in the pressure gradient term, i.e., the first term on the right-hand side of Eq. (7.71), yield a further contribution to the polarization drift. Assuming that the magnetic field is stationary and that the fluid temperature is constant, we obtain the *density polarization drift* term as

$$\mathbf{v}_{Pn,s} = -\frac{k_B T_s}{q_s \omega_{gs} B} \nabla_\perp \frac{d\ln n_s}{dt} \tag{7.87}$$

This drift is a pure fluid drift and caused by slow variations in the perpendicular gradient of the plasma density. It is a collective effect of the fluid which does not exist for single particles. Since the signs of ω_{gs} and q_s cancel, ions and electrons drift in the same direction. But they drift with different velocities because of their different mass and, possibly, different temperature, and thus add to the polarization current.

Another polarization drift can be obtained for the one-fluid case from the generalized Ohm's law (7.53). This law contains the electron inertial term in the form of a time derivative of the current density $\mathbf{j}$. Vector multiplication of Ohm's law with $\mathbf{B}/B^2$ and solving for the fluid velocity yields the *current polarization drift*

$$\mathbf{v}_{Pc} = \frac{m_e}{ne^2 B^2} \mathbf{B} \times \frac{\partial \mathbf{j}}{\partial t} \tag{7.88}$$

This drift does neither depend on the charge nor the mass of the species and does not contribute to a current. It is a pure plasma flow, where a time variation of the perpendicular current density sets the whole plasma into motion in a direction which is both perpendicular to the magnetic field and to the current.

7.4.6 Force-Free Fields

The ideal magnetohydrodynamic equilibrium equations are obtained by making the transition to the stationary case while at the same time dropping all dissipations.

Would we not drop the dissipative terms, it can be trivially realized that the final long-time stationary state would be the null state with maximum temperature, no motion and no fields because dissipation would have destroyed the field. Hence, in a medium with dissipation only time dependent states of physical significance are of interest. In an ideal magnetohydrodynamic fluid, however, we can consider the conditions for equilibrium by returning to the basic equations. The equation of motion for a quasineutral fluid

$$mn\frac{d\mathbf{v}}{dt} = -\nabla \cdot \mathbf{P} + \mathbf{j} \times \mathbf{B} \tag{7.89}$$

suggests that equilibrium can be established whenever the right-hand side of this equation vanishes and the plasma becomes force-free because all forces cancel.

One particularly interesting case, frequently discussed in magnetohydrodynamic applications to stars as the sun, is the case when the pressure force is neglected. Then the force-free condition reduces to a vanishingly small Lorentz force

$$\mathbf{j} \times \mathbf{B} = 0 \tag{7.90}$$

The vanishing of the Lorentz force implies that the currents in the plasma flow entirely parallel to the magnetic field. The currents are field-aligned and this alignment with the field applies not only to the external field but to the total field. Hence, the current flows parallel to the sum of the external field and the field generated by the current itself. A fully general representation of the same condition on the current would then be to write the current as being proportional to the magnetic field and flowing along the magnetic field lines

$$\mu_0 \mathbf{j} = \alpha_L \mathbf{B} \tag{7.91}$$

where $\alpha_L(\mathbf{x})$ is a scalar to make sure that the current is really parallel to the field. This proportionality factor is called *lapse field*. It may depend on position, but is constant in time. Thus one can use the stationary Ampère's law, $\nabla \times \mathbf{B} = \mu_0 \mathbf{j}$, to find that the condition of force-free magnetohydrodynamics can be written as

$$\mathbf{B} \times (\nabla \times \mathbf{B}) = 0 \tag{7.92}$$

This condition shows that force-free fields must have a complicated helical structure, which is intuitively clear because the parallel current generates a helical field around the external guiding field. Using Eq. (7.91), Ampère's law can be written as

$$\nabla \times \mathbf{B} = \alpha_L \mathbf{B} \tag{7.93}$$

or taking the divergence of both sides of this equation and using $\nabla \cdot \mathbf{B} = 0$ yields

$$\mathbf{B} \cdot \nabla \alpha_L = 0 \tag{7.94}$$

This is a condition on the lapse field, which shows that α_L in a force-free field must be constant along the magnetic field lines and can vary only in the direction perpendicular to the total magnetic field. Hence, the force-free field-aligned current is also proportional to the total self-consistent magnetic field.

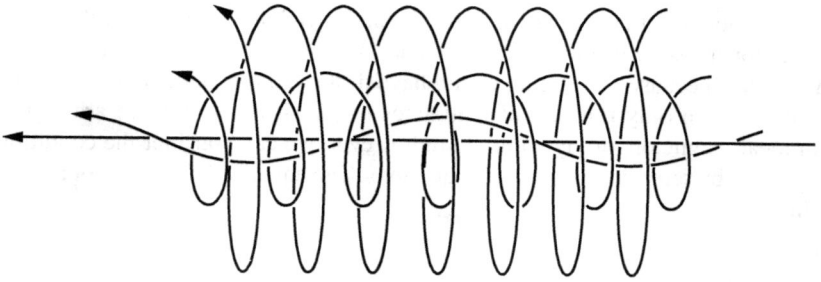

Figure 7.3: Typical force-free field and current configuration.

Figure 7.3 shows a schematic force-free magnetic field configuration which can be imagined schematically. The originally straight magnetic field, when exposed to a sufficiently strong field-aligned current which carries its own solenoidal field, is deformed to become helical. Due to the superposition of the straight and solenoidal fields, the resulting magnetic field configuration winds up into a helical field in order to compensate for the Lorentz force.

Equation (7.94) has a very interesting analogy. It is well-known that in a magnetized plasma which convects with velocity $\mathbf{v}_E = \mathbf{E} \times \mathbf{B}/B^2$ the electric convection potential is constant along the magnetic field and, hence, satisfies the same equation as the lapse function α_L. It is therefore tempting, to identify α_L with an electric convection field potential. This is a quite general interpretation, because the requirement of a force-free field is identical to the requirement that the Lorentz force $\mathbf{E} + \mathbf{v} \times \mathbf{B}$ vanishes. This in turn is just the condition that the plasma convects at velocity $\mathbf{v}_E$.

We may thus interpret α_L as an equivalent electric potential which must be constant along the magnetic field. If α_L itself is constant throughout the whole space this condition is identically satisfied, corresponding to the special case that the electric field in the plasma vanishes. In all other cases of varying α_L the plasma must necessarily undergo convective motion at the velocity

$$\mathbf{v}_L = \frac{\mathbf{B} \times \nabla \alpha_L}{B^2} \tag{7.95}$$

showing that all general force-free field configurations can exist only in convecting plasmas, and a quiescent motionless force-free field can only exist when α_L is uniform. In this case one finds by taking the curl of Eq. (7.93) that the force-free magnetic field satisfies the Helmholtz equation

$$\nabla^2 \mathbf{B} + \alpha_L^2 \mathbf{B} = 0 \tag{7.96}$$

In the near-Earth plasma environment force-free fields are of little importance, since the strong guiding field of the Earth does not permit for force-free conditions. The field-aligned currents flowing in the inner magnetosphere are far too weak to affect the external magnetospheric field effectively. However, force-free fields have been successfully used to model magnetic fields in the distant magnetospheric tail and in the atmospheres of stars as the Sun. In the latter case, in particular, field aligned currents are believed to be very strong in the solar corona, strong enough to deform the original solar guide field until it becomes helical and possibly even of such short screw height that it can undergo internal reconnection.

7.5 Concluding Remarks

The magnetohydrodynamic equations derived in this chapter are simplifications of the more general multi-fluid equations which on their own are moments of the kinetic equation. Each of the steps leading to these sets of equations is based on simplifying assumptions which limit their applicability. In concluding this chapter, it is useful to review some of the conditions which have to be observed when applying the set of magnetohydrodynamic equations summarized below.

7.5.1 Validity of Magnetohydrodynamics in Plasmas

The most basic condition of validity of magnetohydrodynamics is found when remembering that in magnetohydrodynamics no distinction is made between the different components of the plasma. This approximation must be justified. It requires that time-scales of variation of the fluid and fields must be substantially longer than the characteristic time-scales of the heaviest and thus most inert particle component. Hence, the characteristic frequency, ω, of any change in the fields has to be less than the ion cyclotron frequency

$$\omega < \omega_{gi} \tag{7.97}$$

For the same reason, the characteristic length scale, L, where magnetohydrodynamics will become applicable must substantially exceed the ion gyroradius

$$L > r_{gi} \tag{7.98}$$

Magnetohydrodynamics is therefore restricted to very low frequencies, sufficiently long times and to large spatial scales. At such low frequencies one can in most cases ignore the displacement current term in the first Maxwell equation and work in the quasi-stationary limit of electrodynamics where only current flow is to be considered.

The above basic conditions are very important and generally used when deriving low-frequency waves from kinetic theory. One should, however, remark that these conditions are merely derived from the equation of motion of the fluid. Further conditions follow when considering Ohm's law and the induction equation. The equation of state is unaffected by the above conditions because it describes the thermodynamic state. It is thus more precise than magnetohydrodynamics requires.

The *general induction law* of magnetohydrodynamics follows, in the same way as described in Section 5.1, from a combination of Faraday's law and the generalized Ohm's law (7.53), which serves to replace the electric field

$$\frac{\partial \mathbf{B}}{\partial t} = \nabla \times \left(\mathbf{v} \times \mathbf{B} - \frac{1}{ne}\mathbf{j} \times \mathbf{B} + \frac{1}{ne}\nabla \cdot \mathbf{P}_e - \frac{m_e}{ne^2}\frac{\partial \mathbf{j}}{\partial t} - \eta \mathbf{j} \right) \qquad (7.99)$$

As discussed earlier, the first term on the right-hand of this generalized induction equation (7.99) describes frozen-in fields and plasmas. The other terms tend to break the frozen-in state. If they can all be neglected, one ends up with ideal magnetohydrodynamics. For the pressure gradient term to be neglected one compares it with the magnitude of the first term, $|\mathbf{v} \times \mathbf{B}|$, on the right and finds dimensionally

$$\frac{p_e}{neVBL_p} = \frac{n_e k_B T_e}{neVBL_p} = \frac{m_e v_{\text{the}}^2}{2eVBL_p} = \frac{r_{ge}}{L_p}\frac{v_{\text{the}}}{V} \qquad (7.100)$$

where L_p is the gradient scale length of the electron pressure and V and B are average velocity and magnetic field strength, respectively. Hence, when the pressure gradient scale divided by the electron gyroradius is much larger than the ratio of electron to flow velocity, the pressure term can be ignored. This is valid for sufficiently cold plasmas with small $\beta \ll 1$.

The dimensional ratio of the second-last to the last term on the right-hand side of the induction law is obtained by replacing the $\partial/\partial t$ term by the characteristic frequency of change, ω, in the second-last term

$$\frac{m_e \omega}{ne^2 \eta} = \frac{\omega}{v_c} \qquad (7.101)$$

where use has been made of the definition of the resistivity $\eta = m_e v_c/ne^2$. Hence, the inertial term in the induction equation can be neglected whenever the characteristic frequency of change is smaller than the electron-ion collision frequency. This condition is well met in resistive magnetohydrodynamics. However, in time-varying ideal magnetohydrodynamics it poses severe problems and holds strictly only for stationary cases.

A similar problem is encountered when considering the Hall-term. Forming the dimensional ratio of the second to the last term in the induction law, one finds

$$\frac{B}{ne\eta} = \frac{eB}{m_e v_c} = \frac{\omega_{ge}}{v_c} \qquad (7.102)$$

the ratio of electron gyrofrequency to collision frequency. In most of the cases the former is much higher than the latter. It is thus clear that neglecting the Hall term in the ideal magnetohydrodynamic case is rather delicate and is, in general, not permitted, unless it can be justified by additional assumptions. Only for strongly collisional plasmas in weak magnetic fields the Hall term may be dropped. In the light of this discussion the relevant induction equation becomes, when dropping the pressure and resistive terms

$$\frac{\partial \mathbf{B}}{\partial t} = \nabla \times \left(\mathbf{v} \times \mathbf{B} - \frac{1}{ne} \mathbf{j} \times \mathbf{B} - \frac{m_e}{ne^2} \frac{\partial \mathbf{j}}{\partial t} \right) \tag{7.103}$$

Dividing the two last terms yields dimensionally the ratio of the characteristic frequency to the electron gyrofrequency

$$\frac{m_e \omega}{eB} = \frac{\omega}{\omega_{ge}} \tag{7.104}$$

This ratio is always much smaller than one and the inertial term can usually be neglected in comparison to the Hall term. Finally, the dimensional ratio of the Hall and convection terms can be written in the form

$$\frac{j}{neV} = \frac{B}{\mu_0 L_B neV} = \frac{c}{V} \frac{c}{\omega_{pe} L_B} \frac{\omega_{ge}}{\omega_{pe}} \tag{7.105}$$

where L_B is the characteristic length over which the magnetic field varies. The first term, the ratio of the light to the flow velocity is large, typically of the order of 1000. The second ratio is the ratio of the *electron inertial length*, c/ω_{pe}, to the magnetic field scale length, which is a small number. The last ratio is the electron cyclotron to plasma frequency ratio. In strong magnetic fields it is large and the Hall term becomes important. Only in dense plasmas, where this ratio becomes small, the Hall term is negligible against the convection term.

The induction equation (7.103), neglecting the presumably small electron inertia term, allows another interesting conclusion. Because of the relation (7.49) between the current and the flow velocities, the simplified induction equation becomes

$$\frac{\partial \mathbf{B}}{\partial t} = \nabla \times (\mathbf{v}_e \times \mathbf{B}) \tag{7.106}$$

In this form the induction equation shows that in collisionless magnetohydrodynamics with Hall currents not neglected, the electrons are the only plasma component which is frozen to the magnetic field. Hence, the magnetic field is moving with the electron fluid while the plasma flow can, in general, deviate from the motion of the field lines. Only when Hall currents become small or can be ignored, the frozen-in concept of plasma physics applies to the magnetohydrodynamic flow itself.

7.5.2 Summary of Magnetohydrodynamic Equations

In this summary of magnetohydrodynamic equations the time derivative of the electric field is understood to be so slow that the validity of the magnetohydrodynamic approximation is not violated. Then the last Maxwell equation, $\nabla \cdot \mathbf{E} = \rho/\varepsilon_0$, is not needed for closing the system of equations and can be used to determine the space charges maintained in the plasma. The system of equations given below has to be complemented by an energy equation or an equation of state. Usually one assumes adiabatic conditions and uses $p \propto n^\gamma$ with $\gamma = 5/3$. However, as discussed in Section 7.2, other forms of this equation may equally be valid in certain situations.

$$\frac{\partial n}{\partial t} + \nabla \cdot (n\mathbf{v}) = 0$$

$$\frac{\partial (nm\mathbf{v})}{\partial t} + \nabla \cdot (nm\mathbf{v}\mathbf{v}) = -\nabla \cdot \mathbf{P} + \rho \mathbf{E} + \mathbf{j} \times \mathbf{B}$$

$$\mathbf{E} + \mathbf{v} \times \mathbf{B} = \eta \mathbf{j} + \frac{1}{ne}\mathbf{j} \times \mathbf{B} - \frac{1}{ne}\nabla \cdot \mathbf{P}_e + \frac{m_e}{ne^2}\frac{\partial \mathbf{j}}{\partial t}$$

$$\nabla \times \mathbf{B} = \mu_0 \mathbf{j} + \mu_0 \varepsilon_0 \frac{\partial \mathbf{E}}{\partial t}$$

$$\nabla \times \mathbf{E} = -\frac{\partial \mathbf{B}}{\partial t}$$

$$\nabla \cdot \mathbf{B} = 0$$

Further Reading

A thorough description of the equations of state is given in [26]. Magnetohydrodynamics is discussed in many textbooks, but in most cases from the point of view of a conducting fluid, which applies more to the interior of the Earth than to its plasma environment. A general excellent first reference is [19], but this book misses the generalized Ohm's law, which is given in [15]. For other applications of magnetohydrodynamics to space and astrophysics see [23]. The remaining references are presentations of general MHD or applications, some of them being from the early time when MHD was still under development. They are of historical interest mainly today where MHD has in view of plasma application been overthrown by kinetic plasma theory.

References

[1] G. Bateman, *MHD Instabilities* (The MIT Press, Cambridge MA, 1978).

[2] D. Bershader (ed.), *Plasma Hydromagnetics* (Stanford University Press, Stanford CA, 1962).

[3] D. Biskamp, *Nonlinear Magnetohydrodynamics* (Cambridge University Press, Cambridge UK, 1993).

[4] T. J. M. Boyd and J. J. Sanderson, *Plasma Dynamics* (Barnes and Nobel Press, New York, 1969).

[5] S. Chandrasekhar, *Hydrodynamic and Hydromagnetic Stability* (Clarendon Press, Oxford, 1961).

[6] F. F. Chen, *Introduction to Plasma Physics and Controlled Fusion, Vol. 1* (Plenum Press, New York, 1984).

[7] P. C. Clemmow and J. P. Daugherty, *Electrodynamics of Particles and Plamas* (Addison-Wesley, Raeding MA, 1971).

[8] T. G. Cowling, *Magnetohydrodynamics* (Interscience Publ., New York, 1957).

[9] J. L. Delcroix, *Plasma Physics* (J. Wiley, New York, 1965).

[10] S. Gartenhaus, *Elements of Plasma Physics* (Holt-Rinehart-Winston, New York, 1964).

[11] E. H. Holt and R. E. Haskell, *Foundations of Plasma Dynamics* (Macmillan, New York, 1965).

[12] J. D. Jackson, *Classical Electrodynamics* (J. Wiley and Sons, New York, 1975), Chpt. 10, p. 469.

[13] A. Jeffrey and T. Taniuti, *Magnetohydrodynamic Stability and Thermonuclear Containment* (Academic Press, New York, 1966).

[14] N. G. van Kampen and B. U. Felderhof, *Theoretical Methods in Plasma Physics* (North-Holland Publ. Co., Amsterdam, 1967).

[15] N. A. Krall and A. W. Trivelpiece, *Principles of Plasma Physics* (McGraw-Hill, New York, 1973).

[16] P. C. Kendall and C. Plumpton, *Magnetohydrodynamics, Vol. 1* (Pergamon Press, New York, 1964).

[17] F. Krause and K.-H. Rädler, *Mean-Field Magnetohydrodynamics and Dynamo Theory* (Pergamon Press, New York, 1980).

[18] W. B. Kunkel (ed.), *Plasma Physics in Theory and Application* (McGraw-Hill, New York, 1966).

[19] L. D. Landau and E. M. Lifshitz, *Electrodynamics of Continuous Media* (Pergamon Press, Oxford, 1975).

[20] R. K. M. Landshoff (ed.), *Magnetohydrodynamics* (Stanford University Press, Stanford CA, 1957).

[21] M. A. Leontovich (ed.), *REviews of Plasma Physics, Vol. 8* (Consultants Bureau, New York, 1980).

[22] C. L. Longmire, *Elementary Plasma Physics* (Interscience Publ., New York, 1963).

[23] E. N. Parker, *Cosmical Magnetic Fields* (Clarendon Press, Oxford, 1979).

[24] G. Schmidt, *Physics of High Temperature Plasmas* (Academic Press, New York, 1966).

[25] J. A. Shercliff, *A Textbook of Magnetohydrodynamics* (Pergamon Press, Oxford, 1965).

[26] G. L. Siscoe, in *Solar-Terrestrial Physics*, eds. R. L. Carovillano and J. M. Forbes (D. Reidel Publ. Co., Dordrecht, 1983), p. 11.

[27] W. B. Thompson, *An Introduction to Plasma Physics* (Pergamon Press, New York, 1962).

Problems

Problem 7.1 *Derive the continuity equation for plasma species s from the Boltzmann equation given earlier. What difference does it make to the continuity equation derived from the Vlasov equation? Explain the reason for such a difference if you find any.*

Problem 7.2 *Derive the momentum equation for species s. Show the origin of the different terms in this equation, in particular the origin of the pressure term. Why is the pressure a tensor?*

Problem 7.3 *Derive the energy or heat conduction conservation equation of the multifluid plasma. Give the explicit form of the heat flux vector.*

Problem 7.4 *Explicate the conditions under which the pressure tensor becomes an anisotropic diagonal tensor. Show that it can be written in dyadic form.*

Problem 7.5 *Derive the adiabatic relation for the ideal gas density of the sth plasma component. Plot the pressure-density and temperature-density relations for the polytropic cases $\gamma = 3/2, 5/3, 11/3$.*

Problem 7.6 *Show explicitly that the average magnetic particle moment of species s can be written as $\langle \mu_s \rangle = p_{s\perp}/n_s B$. Prove that this moment is conserved under certain conditions and show that it can be used to find the first of the double adiabatic laws.*

Problem 7.7 *Derive the second adiabatic law. Discuss the conditions for its validity. Find some cases when the second adiabatic law is violated.*

Problem 7.8 *In the double adiabatic theory the temperatures evolve differently. Derive the expressions for the parallel and perpendicular temperatures.*

Problem 7.9 *Show that the double adiabatic indices γ are related by the equation $\gamma_{\parallel} + 2\gamma_{\perp} = 5$. Plot this relation. Derive an upper limit on $\gamma_{\perp}$.*

Problem 7.10 *Demonstrate explicitly that the mass conservation law for a single-species fluid is formally the same as that for each sth species. When is this conclusion correct and why? Show that the introduction of the centre of mass quantities $n, m, \mathbf{v}$ is correct. Show that the mass conservation equation remains to be correct even when not dropping the electron-to-ion mass ratio.*

Problem 7.11 *Derive the full one-fluid momentum conservation equation (equation of motion) without making the assumption of small electron-to-ion mass ratio. Discuss the additional terms which appear in the equation.*

Problem 7.12 *Find the generalised Ohm's law without making any of the simplifications we did in the derivation. Discuss the additional terms. What is their meaning and importance?*

Problem 7.13 *Along the same lines find the heat conduction equation and derive the expression for the heat flux density vector.*

Problem 7.14 *Assume a constant parallel electric field $E_\| = E_0$ in the plasmasphere at mid-latitudes above 1000 km altitude along the geomagnetic field. Electrons of density n_0 and temperature T_e located in this field in pressure equilibrium obey Boltzmann's law. Determine the scale height of the electrons in this electric field.*

Problem 7.15 *Determine the orbit of an ion with gyroradius equal to the width of the magnetic neutral layer in the tail. What kind of orbit is an ion performing when only its gyration is considered and it has no velocity component parallel to the magnetic field? Sketch the result.*

Problem 7.16 *Derive the expression for the Harris current profile. Which assumptions have been made? Calculate the expression for the plasma density in the Harris sheet model from your knowledge of the magnetic field and current profiles as given in the text.*

Problem 7.17 *Derive the expression for the field-aligned current. Which are the closure currents in the mid-latitude and auroral ionospheres?*

— 8 —

Flows and Discontinuities

Plasma equilibria frequently arise in plasma flows. The dominant plasma flow in near-Earth space is the solar wind. It interacts with the bodies of the solar system, planets, moons, and comets. This interaction depends on the properties of these bodies. It is particularly strong when they are strongly magnetized as is the case for Earth, Mercury, Jupiter, Saturn, Uranus and Neptune. Such an interaction results in the formation of a *magnetosphere* with a thin outer boundary, a magnetic skin, called *magnetopause* and representing a sudden transition from the solar wind plasma to the planetary magnetic field region. It confines the planetary magnetic field into the magnetosphere which is a region of limited spatial extent obeying its proper dynamics. This dynamics is, to a large degree, controlled by the solar wind stream. Transition layers like the magnetopause are called discontinuities. Under the condition that they are thin, it is possible determine their properties from the equations which govern the interaction. In this chapter we consider the properties of the discontinuities which evolve in the interaction of the solar wind with the geomagnetic field.

8.1 Solar Wind

The *solar wind* is the high-speed particle stream which is continuously blowing out from the solar corona into interplanetary space, extending far beyond the orbit of the Earth and terminating somewhere in interstellar space after having hit the weakly ionized interstellar gaseous medium around 160 AU (1 Astronomical Unit = 1.50×10^{11} m; see Appendix B.1). Near the Earth's orbit at 1 AU, the solar wind velocity typically ranges between 300–1400 km/s. The value of 500 km/s given in Section 1.2 is the most probable value of the solar wind velocity. It corresponds to an about 4-day particle flight from the Sun to the Earth. Streams with velocities of less than 400 km/s are known as low-speed flows, those with velocities exceeding 600 km/s are called high-speed solar wind streams.

Since the solar atmosphere, where the solar wind originates, is a quiescent low-speed region of low temperature near 6000 K, the solar wind must be accelerated by some not completely understood mechanism to its high interplanetary streaming velocities in the *solar corona*. The corona is a hot region with a temperature of about 1.6×10^6 K and a density near 5×10^{17} cm^{-3}. The strength of the magnetic field in the corona is not well known, but at its bottom it might be of the order of some 10^{-2} T, decaying away with increasing distance. Most of the dynamics of the solar wind is determined by this field. When the field is closed with both foot points on the Sun, the highly ionised plasma of the solar atmosphere is trapped. But in regions, where the field lines are stretched out into interplanetary space, the so-called *coronal holes*, which are located normally at higher solar latitudes, the solar atmospheric plasma can flow out. These coronal holes are the source regions for the solar wind.

8.1.1 Solar Coronal Outflow

Neglecting the minor influence of the magnetic field, the expansion of the solar wind from a coronal hole can in the most simple approximation applied by E. N. Parker when neglecting any magnetic fields be treated as a one-dimensional radial hydrodynamic flow problem. Following Parker's approach, radial flux conservation requires

$$\frac{d}{dr}(r^2 nv) = 0 \tag{8.1}$$

where v is the purely radial flow velocity and n is the radially variable coronal plasma density. Stationary momentum conservation yields

$$v\frac{dv}{dr} = -\frac{GM_\odot}{r^2} - \frac{1}{nm_i}\frac{dp}{dr} \tag{8.2}$$

where G is the gravitational constant, $M_\odot$ the solar mass, and p the thermal pressure of the coronal plasma. Instead of an energy law one assumes a simple polytropic relation

$$\frac{d}{dr}\left(\frac{p}{n^\gamma}\right) = 0 \tag{8.3}$$

where adiabatic conditions correspond to $\gamma = \frac{5}{3}$ and isothermal and turbulent flows have $\gamma = 1$. Intermediate values, $1 < \gamma < \frac{5}{3}$, indicate the presence of coronal heating. The latter two equations can be integrated to yield

$$\frac{v^2}{2} + \frac{\gamma}{\gamma-1}\frac{p}{nm_i} - \frac{GM_\odot}{r} = K \tag{8.4}$$

where K is a constant. Identifying the term $p/nm_i = c_s^2$ as the square of the *sound velocity*

$$\boxed{c_s^2 = \frac{\gamma k_B T}{m_i}} \tag{8.5}$$

permits to eliminate the explicit dependence on pressure and radial velocity from Eq. (8.4) by defining the square of the *sonic Mach number*, M_s, as the ratio of twice the flow energy (density) to thermal energy (density)

$$\boxed{M_s^2 = \frac{m_i v^2}{\gamma k_B T} = \left(\frac{v}{c_s}\right)^2} \tag{8.6}$$

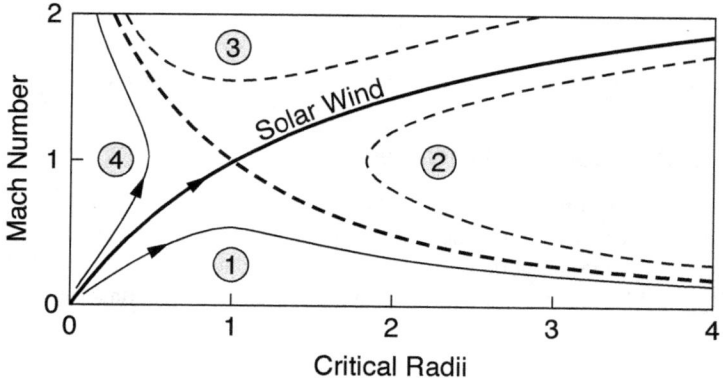

Figure 8.1: The four types of outflow solutions in the (M, R) plane.

Introducing the gravitational potential at the surface of the Sun at radius $R_\odot$

$$\Phi_\odot = GM_\odot / R_\odot \tag{8.7}$$

and integrating Eq. (8.1) with the help of Eqs. (8.4) and (8.6), yields for $R = r/R_\odot$ the Parker solution:

$$\left(K + \frac{\Phi_\odot}{R}\right) \frac{M_s^2 - 1}{m^2} R \frac{dM_s^2}{dR} = \left(1 + (\gamma - 1)\frac{M_s^2}{2}\right)\left(4K + \frac{3\gamma - 5}{\gamma - 1}\frac{\Phi_\odot}{R}\right) \tag{8.8}$$

It is impossible to solve this equation analytically, but one can draw some qualitative conclusions. For $|M_s| = 1$ the left-hand side vanishes. Since the right-hand side must also vanish, the polytropic index must lie in the range $1 < \gamma < 5/3$, which requires coronal heating. For such γ the *sonic point*, $M_s = 1$, lies at the critical radial distance

$$R_c = \frac{\Phi_\odot}{4K} \frac{5 - 3\gamma}{\gamma - 1} \tag{8.9}$$

For the sonic point to be found outside the solar surface $R_c > 1$ and thus

$$K < \frac{\Phi_\odot}{4} \frac{5 - 3\gamma}{\gamma - 1} \tag{8.10}$$

Solutions satisfying this result describe coronal plasma outflows which at $R = R_c$ pass from flow speeds smaller than the sound velocity to speeds larger than the sound velocity. These are very special solutions, which divide the space of possible solutions into four categories.

The two solid curves in Fig. 8.1 correspond to the two solutions passing through the point $M_s(R_c) = 1$. Only two of the four regions in the (M_s, R_c) plane correspond to physical solutions. These are regions (1) and (4). In region (1) the solar wind accelerates from low flow velocity yet subsequently decelerates, never reaching supersonic

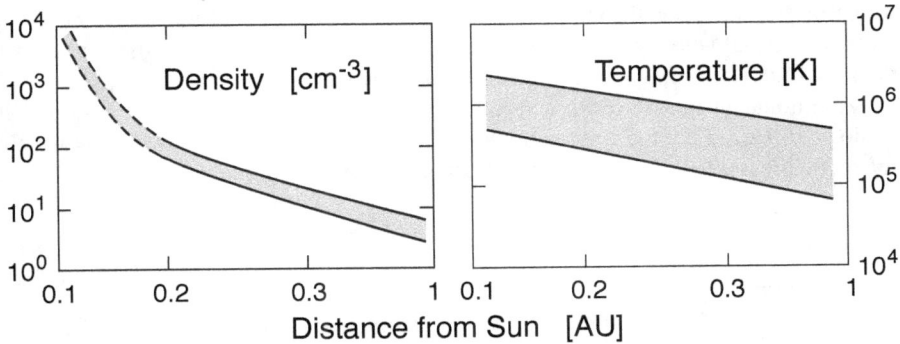

Figure 8.2: Radial variation of solar wind density and temperature.

velocities while escaping from the corona into interplanetary space to large distances. In region (4) it starts at low velocities and increases to supersonic speeds but cannot leave the solar corona falling back into the solar atmosphere. Thus one concludes that since the solar wind is observed to be a high-speed supersonic stream with $M > 1$, it must — in this very simplified fluid theory — necessarily correspond just to the singular critical solution that passes through the sonic point. Clearly this requires some kind of 'fine tuning' of the solar wind acceleration process in the corona which up to this point has not been clarified. In spite of this deficiency, Parker's theory provides a hint on a possible coronal acceleration process the result of which is what is observed as the super-sonic and — as we will see later — super-Alfvénic solar wind flow, the first observed 'stellar wind' and the only one that has been monitored *in situ* by a wealth of spacecraft over now half a century with ever increasing precision.

8.1.2 Properties of the Solar Wind

Once the solar wind has been formed, its plasma expands into the interplanetary space. This expansion is a radial propagation of the solar wind plasma fluid away from the Sun during which it must become diluted and at the same time cool down. Measurements have shown that the solar wind density outside the solar corona up to a distance of a few AU decreases approximately as $n(r) \propto r^{-2}$ (see Fig. 8.2). Close to the Earth, the solar wind is a fully ionized plasma consisting of electrons, protons, and α-particles. Since the abundance of He^{2+} is only about 3% of the total density of about 5 cm^{-3}, their presence can typically be neglected.

The decrease in the temperature is less severe (see Fig. 8.2). It decreases by roughly a factor of 20 from coronal temperatures of a few million Kelvin to electron and proton temperatures at 1 AU of about $T_e \approx 1.4 \times 10^5$ K and $T_i \approx 1.2 \times 10^5$ K, respectively. The large uncertainty of the temperature indicated in Fig. 8.2 is caused by fluctuations in the corona and by a yet unknown solar wind heating mechanism. Although the temperatures of electrons and protons are not very different, they have drastically different thermal velocities at 1 AU, with $v_{the} \approx 1500$ km/s and $v_{thi} \approx 35$ km/s. The electron thermal velocity is typically larger than the flow speed, while the ion thermal velocity is always very small compared to the flow speed.

Additional insight into the nature of the solar wind is obtained by defining, in analogy to Eq. (8.5), the 'magnetic sound' or *Alfvén velocity*, v_A, where twice the magnetic pressure, B^2/μ_0, appears instead of the thermal pressure

$$v_A^2 = \frac{B^2}{\mu_0 n m_i} \tag{8.11}$$

The Alfvén velocity is an important plasma parameter and, as will be shown in Chapter 9, it is the fundamental speed at which magnetic signals in a plasma can be transported by waves. Using the typical densities given above and an interplanetary magnetic field amplitude of about 5 nT, we find that with sound and Alfvén velocities of only about 30–50 km/s, the solar wind is a supersonic and super-Alfvénic flow.

With the help of the Alfvén velocity we can define another Mach number, the Alfvénic Mach number M_A, when we divide the thermal energy density by the magnetic energy density $m_i n v_A^2$. Then we obtain instead of Eq. (8.6) for the square of the Alfvénic Mach number

$$M_A^2 = \frac{m_i n v^2}{m_i n v_A^2} = \left(\frac{v}{v_A}\right)^2 \tag{8.12}$$

This quantity in fast streaming magnetised plasmas like the solar wind or stellar winds turns out of much higher importance than its sonic relative. When both thermal and magnetic pressures become important one can define a more general so-called *magnetosonic* Mach number $M_{ms} = M_A(1+\beta)^{-\frac{1}{2}}$ which is a combination of the sonic and Alfvénic Mach numbers and which we will give later its mathematical expression. It is smaller than any of the other Mach numbers because it refers to both pressure contributions.

Usually, however, it is sufficient to refer just to one of the above Mach numbers depending on which pressure contributes more to the dynamics of the plasma. Surprisingly this will in most except the highly relativistic cases be the Alfvénic Mach number, for the magnetic field causes effects in the plasma which are caused by action at a distance through the presence of the magnetic and electric convection fields while the pure thermal effects require direct interaction which in the collisionless case is practically absent outside the Debye sphere.

8.1.3 Interplanetary Magnetic Field

One important property of the solar wind is its state of magnetization. This state can be described by considering plasma beta, the ratio of thermal to magnetic energy densities given in Eq. (7.66). Near 1 AU the solar wind magnetic field is of the order of

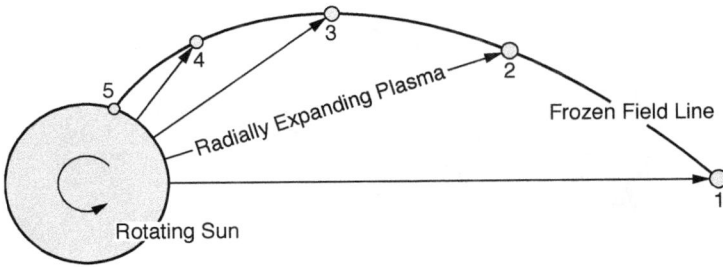

Figure 8.3: Spiral form of interplanetary field line produced in combination of the frozen-in property of the solar coronal magnetic field into the solar wind and the solar rotation.

5 nT, so that $\beta \approx 1 - 30$ is large. The solar wind flow entirely determines the behaviour of the field, which fact has an important consequence for its structure. The radial outflow of the wind from the solar corona transports the field into interplanetary space, while its footpoint remain anchored in the solar atmosphere. Because of the 27-day solar rotation period, the *interplanetary magnetic field* — which is usually abbreviated as IMF — will not maintain the form it had in the solar corona. Figure 8.3 shows schematically what happens to a field line which is frozen into the solar wind plasma flow and is transported radially outward with a constant radial velocity while the Sun rotates. As a result of the combined motion of outflow and rotation the magnetic field line becomes bent into an *Archimedian spiral* form.

At 1 AU this spiral makes an angle of approximately $45°$ to the Earth-Sun line, so that it hits the Earth from the late morning direction. The magnetic field strength at this position is about 5–10 nT. As depicted in Fig. 8.4, the direction of the interplanetary magnetic field in the ecliptic plane is divided into sectors according to solar and antisolar directions. The boundaries between sectors of opposing field direction are regions of zero magnetic field and thus current sheets. To reproduce the direction of the interplanetary magnetic field, the solar wind current sheet must be tilted with respect to the ecliptic plane at the sector boundaries, while high above and below the ecliptic plane it turns into horizontal direction. Close to the sun it looks like a ballerina skirt sketched in Fig. 8.5. At larger distances, however, the interplanetary current sheet becomes an Archimedean double helix winding out from the sun into interplanetary space and crossing the ecliptic periodically from north to south and from south to north with spiral current hills, crests and valleys.

Taking into account of the azimuthal speed of the Earth of about 30 km/s, the solar wind hits the Earth magnetosphere with an *aberration* of typically about $5°$ from the radial direction. This angle decreases the higher the solar wind speed becomes. In occasional high-speed solar wind streams having velocities near and larger 1000 km/s, the solar wind direction is practically radial.

For completeness, we note that the solar wind is by no means a quiet laminar flow. It is subject to many variations in density, flow velocity, temperature, pressure, magnetic field strength and magnetic field direction. These variations do in part originate in

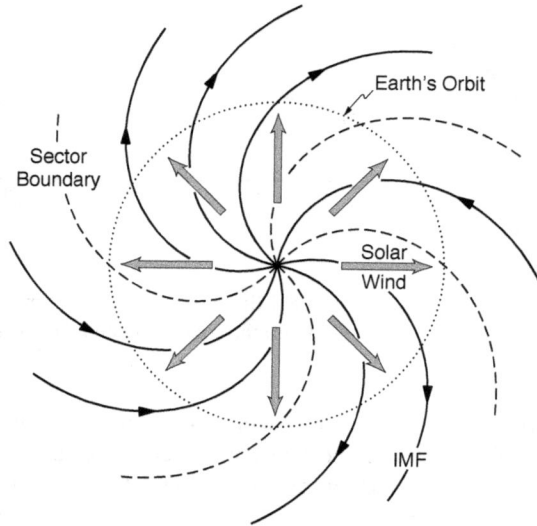

Figure 8.4: Spiral sector structure of interplanetary magnetic field. A picture like this applies to the ecliptic plane out to several AU beyond the Earth's orbit (shown here as a dashed circle). At large distances the interplanetary magnetic field becomes about azimuthal by the solar rotation. The figure also indicates that the interplanetary magnetic field has a four-sector structure with the fields alternating between solar and anti-solar directions. At the sector boundaries, however, the interplanetary field does not really vanish. At these boundaries one has current flow in the solar wind with the interplanetary current strength being largest while undulating from above to below the ecliptic plane and crossing the ecliptic plane just in the sector boundaries. This is shown in the next figure.

the solar atmosphere and are swept along with the solar wind into interplanetary space. However, a large part of them is produced in the solar wind itself. Such variations are either waves which propagate in the solar wind, or they are localized disturbances like narrow boundaries.

8.1.4   Interaction with Obstacles

Because of its supersonic and super-Alfvénic velocity, the solar wind flow is subject to shocking if there is an obstacle in its path. Such obstacles are planets, comets, and asteroids, but the most interesting obstacles are the magnetized planets with their extended magnetospheres. These magnetospheres enhance the cross-section of a planet by a large factor. For the Earth this factor is about 150. One therefore expects that an extended standing *bow shock* exist in front of such planets.

In addition to the creation of a bow shock, the interaction with magnetized planets leads to the formation of another thin boundary, a *magnetopause*. In order to understand the nature of these phenomena we must investigate the various possibilities for

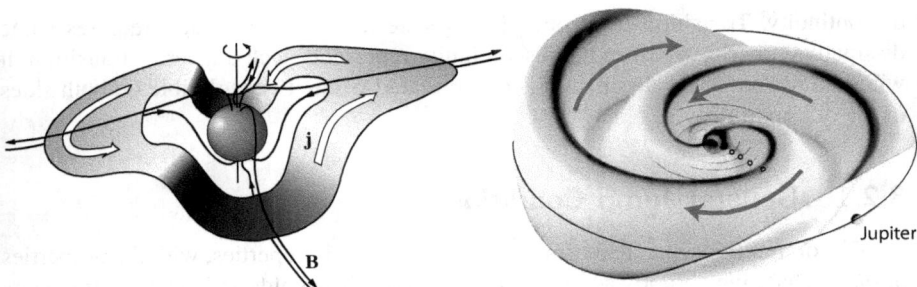

Figure 8.5: Three-dimensional topology of the solar wind current sheet. *Left*: View close to the sun. *Right*: At larger distance the interplanetary current sheet assumes a double spiral form that is winding out from the sun and undulating around the ecliptic plane like shown in this figure up to Jupiter. The small circles indicate the inner planets on their orbits.

the evolution of thin boundaries in a plasma. We will develop the theory of such boundaries in the next section.

8.2 Fluid Boundaries

Equilibria between plasmas of different properties often give rise to the formation of boundaries separating the two adjacent plasma regions. In general these boundaries are not imposed from the outside, but evolve inside the plasma due to the interaction of the plasmas on both of its sides, forming narrow transition layers from one plasma to the adjacent one. These are called *discontinuities*.

Plasmas share the capability for the evolution of discontinuities with ordinary hydrodynamics, yet the much greater richness of plasma processes which include the electromagnetic interactions between charged particles offers more possibilities for the evolution of a discontinuity than in hydrodynamics. Since the properties of an ideal magnetohydrodynamic plasma must be maintained on both sides of the discontinuity, not all kinds of discontinuities can be supported in a given plasma state. Instead the discontinuities will group into different classes. In the following we determine these classes and discuss their properties. We do not go, however, into the structure of a discontinuity as this, by the nature of the formalism and equations, is not contained in the equations but requires a more sophisticated theory which takes care of all the microscopic interactions in the plasma. Such processes, even in the purely collisionless ideal plasma case, include the generation of dissipative processes of one or the other kind by collective interactions between the particles including their self-consistently generated electromagnetic fields.

In order to infer which boundaries can exist in a fluid plasma, it is convenient to start from the ideal one-fluid magneto-hydrodynamic equations, assuming that the plasma flow outside the discontinuity is ideal, in the sense that it is free of dissipation. There is, however, no need for dissipation to be excluded from the interior of the discontinuity. Transition from one plasma state to another typically requires some dissipative process. Because the discontinuity represents such a layer of transition, it will be a region where dissipation is concentrated while it is negligible on both sides of the discontinuity.

8.2.1 General Jump Conditions

Discontinuities separate plasma regions of different properties, with the properties possibly changing abruptly across the discontinuity. The fields and plasma parameters are not independent across the discontinuity but satisfy certain boundary conditions which are imposed on the discontinuity by the requirement that in the adjacent regions the plasma equations should hold. This behaviour is similar to the behaviour of electromagnetic fields at boundaries, where the dielectric properties of optical media change. In all such cases the boundaries are assumed to be infinitesimally thin. In optics the boundary is thin with respect to the wavelength of light and thick with respect to the spacing of the molecules in the crystal structure, while magnetohydrodynamic discontinuities are thin with respect to the scale lengths of the fluid parameters, but thick with respect to Debye length and ion as naturally also the electron gyroradius.

The conditions at the outermost edges of the discontinuities can be derived from ideal one-fluid magnetohydrodynamics. The conventional method to find the boundary conditions is based on an integration of the conservation laws across the discontinuity boundary in order to determine the differences between the field and fluid quantities on both sides of the discontinuity. It is therefore convenient to write the ideal one-fluid equations for a quasi-neutral plasma derived in Section 7.3 in the form of conservation laws

$$\frac{\partial n}{\partial t} + \nabla \cdot (n\mathbf{v}) = 0 \tag{8.13}$$

$$\frac{\partial (nm\mathbf{v})}{\partial t} + \nabla \cdot (nm\mathbf{v}\mathbf{v}) = -\nabla \cdot \left(\mathsf{P} + \frac{B^2}{2\mu_0}\mathsf{I} \right) + \frac{1}{\mu_0}\nabla \cdot (\mathbf{B}\mathbf{B}) \tag{8.14}$$

$$\frac{\partial \mathbf{B}}{\partial t} = \nabla \times (\mathbf{v} \times \mathbf{B}) \tag{8.15}$$

$$\nabla \cdot \mathbf{B} = 0 \tag{8.16}$$

where use has been made of $\mu_0 \mathbf{j} = \nabla \times \mathbf{B}$ for slowly variable fields, the ideal Ohm's law, $\mathbf{E} = -\mathbf{v} \times \mathbf{B}$, has been assumed, and space charges have been neglected, $\rho \mathbf{E} = 0$. These equations have to be complemented by an energy equation and by appropriate equations of state for the independent components of the pressure tensor. A particularly useful form of the energy conservation equation is following from Eq. (7.57) and the first two terms on the right-hand side of Eq. (7.58)

$$\nabla \cdot \left\{ nm\mathbf{v} \left[\tfrac{1}{2}v^2 + w + \frac{1}{nm}\left(p + \frac{B^2}{\mu_0} \right) \right] - \frac{1}{\mu_0}(\mathbf{v} \cdot \mathbf{B})\mathbf{B} \right\} = 0 \tag{8.17}$$

Moreover, we assume in most of the following text that the pressure tensor is isotropic, since an anisotropy introduces only a slight modification of the final results.

For a thin discontinuity in the plasma, the only important changes occur perpendicular to the discontinuity boundary, while parallel to the discontinuity the plasma remains uniform. The discontinuity surface can be described by a two-dimensional function, $S(\mathbf{x}) = 0$. The normal vector of the discontinuity, $\mathbf{n}$, is defined as

$$\mathbf{n} = -\frac{\nabla S}{|\nabla S|} \tag{8.18}$$

It is a unit vector perpendicular to the discontinuity and directed outward taking the discontinuity as if it would be convex. The vector derivative has one single component only in the direction of $\mathbf{n}$, $\nabla = \mathbf{n}(\partial/\partial n)$.

Choosing a certain point on the discontinuity surface, S, and taking the line integral of any field quantity along a rectangular box centered at this point from medium 1 to medium 2 and back to medium 1, the two sides of the box tangential to the surface do not contribute to the integral because the plasma properties do not change (or change very little) along the discontinuity. On the other hand, the two integrations along the normal direction contribute twice due to their opposite senses. Hence, any closed line integral of a quantity X crossing S reduces to

$$\oint_S \frac{dX}{dn}\,dn = 2\int_1^2 \frac{dX}{dn}\,dn = 2(X_2 - X_1) = 2[X] \tag{8.19}$$

The quantity $[X]$ is the jump of X when crossing the boundary. Integrating the conservation laws in the above-prescribed way and observing that an integral over a conservation law vanishes, one can divide by 2 and replace the vector operations by the prescriptions

$$\begin{aligned}
\nabla X &\rightarrow \mathbf{n}[X] \\
\nabla \cdot \mathbf{X} &\rightarrow \mathbf{n} \cdot [\mathbf{X}] \\
\nabla \times \mathbf{X} &\rightarrow \mathbf{n} \times [\mathbf{X}]
\end{aligned} \tag{8.20}$$

Boundaries in a plasma may flow with the medium or even relative to the medium. In both cases it is possible to transform to a system moving with the discontinuity at its local speed $\mathbf{U}$. Because of the Galileian invariance, it is sufficient to replace the time derivative for a stationary flowing discontinuity by

$$\partial/\partial t = -\mathbf{U} \cdot \nabla = -\mathbf{U} \cdot \mathbf{n}(\partial/\partial n) \tag{8.21}$$

while transforming the fluid velocity according to $\mathbf{v}' = \mathbf{v} - \mathbf{U}$. In the co-moving frame the discontinuity is stationary so that the time derivative can be dropped and the conservation laws in Eqs. (8.13) through (8.16) are written in terms of $\mathbf{v}'$. In the following we will drop the prime keeping in mind that viewed from the non-moving frame the velocity is to be replaced by the difference between the plasma velocity and the speed of displacement of the discontinuity.

8.2.2   Rankine-Hugoniot Conditions

With the help of the above definitions for use of the vector derivative, dropping the prime on the velocity, and setting $\mathbf{P} = p\mathbf{I}$, the isotropic conservation laws across the discontinuity can be replaced by *jump conditions* across the discontinuity

$$
\begin{aligned}
\mathbf{n} \cdot [n\mathbf{v}] &= 0 \\
\mathbf{n} \cdot [nm\mathbf{v}\mathbf{v}] + \mathbf{n} \left[p + \frac{B^2}{2\mu_0} \right] - \frac{1}{\mu_0}\mathbf{n} \cdot [\mathbf{B}\mathbf{B}] &= 0 \\
\mathbf{n} \times [\mathbf{v} \times \mathbf{B}] &= 0 \\
\mathbf{n} \cdot [\mathbf{B}] &= 0
\end{aligned}
\tag{8.22}
$$

An additional condition follows from the conservation of energy

$$
\left[nm\mathbf{n} \cdot \mathbf{v} \left\{ \frac{v^2}{2} + w + \frac{1}{nm} \left(p + \frac{B^2}{\mu_0} \right) \right\} - \frac{1}{\mu_0}(\mathbf{v} \cdot \mathbf{B})\mathbf{n} \cdot \mathbf{B} \right] = 0
\tag{8.23}
$$

where the specific internal enthalpy, w, is related to the scalar pressure by

$$
w = \frac{c_v p}{mnk_B}
\tag{8.24}
$$

and the specific heat, c_v, is the typically the same on both sides of the discontinuity.

Paying for the moment no attention to the energy conservation, the last line in the above system of equations (8.22) shows the familiar result that the normal component of the magnetic field is continuous across any surface so that its jump vanishes

$$
[B_n] = 0
\tag{8.25}
$$

Similarly, from the first condition in Eq. (8.22) one finds that the mass flow normal to the discontinuity is always constant and thus

$$
[nv_n] = 0
\tag{8.26}
$$

Splitting between the normal and tangential components of the fields and making use of the last two relations one finds the remaining jump conditions

$$
\begin{aligned}
nmv_n[v_n] &= -\left[p + \frac{B^2}{2\mu_0} \right] \\
nmv_n[\mathbf{v}_t] &= \frac{B_n}{\mu_0}[\mathbf{B}_t] \\
B_n[\mathbf{v}_t] &= [v_n\mathbf{B}_t]
\end{aligned}
\tag{8.27}
$$

where the subscript t denotes the tangential component of the corresponding vector. For a given equation of state these equations are the boundary conditions which must be satisfied across any discontinuity.

These equations are known under the name *Rankine-Hugoniot equations*. They contain all the basic properties of discontinuities in ideal magnetohydrodynamics. This, however, is a severe restriction because these conditions hold only away from the discontinuity under consideration in the region where the plasma behaves ideally and obeys the frozen-in condition. Discontinuities in their interiors violate these condition and may strongly affect their environment. One will have to call for a more precise theory when being interested in the physics involved. This is particularly strongly felt at the magnetopause and at the bow shock though for completely different reasons as we will explain in later chapters. Nevertheless, the jump conditions provide a valuable method of classification of discontinuities.

To determine which classes of discontinuities are possible, the system of nonlinear jump conditions can be rewritten into quasi-linear form by defining averages

$$\langle X \rangle = \tfrac{1}{2}(X_1 + X_2) \tag{8.28}$$

With this definition it is easy to prove that the jump of a product is given by the chain rule

$$[\mathbf{AB}] = [\mathbf{A}]\langle \mathbf{B} \rangle + \langle \mathbf{A} \rangle [\mathbf{B}] \tag{8.29}$$

It is also convenient to define the specific volume, $\mathscr{V} = (nm)^{-1}$, and the constant normal mass flux, $F = nmv_n$ respectively $v_n = F\mathscr{V}$. These definitions allow to rewrite the system of jump conditions when observing from Eq. (8.26) that $[F] = 0$

$$
\begin{aligned}
F[\mathscr{V}] - [v_n] &= 0 \\
F[\mathbf{v}] + \mathbf{n}[p] + \mu_0^{-1}\mathbf{n}\langle \mathbf{B} \rangle \cdot [\mathbf{B}] - \mu_0^{-1}B_n[\mathbf{B}] &= 0 \\
F\langle \mathscr{V} \rangle[\mathbf{B}] + \langle \mathbf{B} \rangle[v_n] - B_n[\mathbf{v}] &= 0 \\
[B_n] &= 0 \\
F\left\{ \tfrac{1}{2}[v^2] + [w] + \langle p/m \rangle[\mathscr{V}] + (4\mu_0)^{-1}[\mathscr{V}][\mathbf{B}_t]^2 \right\} &= 0
\end{aligned} \tag{8.30}
$$

Understanding the average quantities as given and the jumps as the unknowns, these are eight equations for nine jumps. Adding an expression for the enthalpy or an equation of state to relate the pressure to the fields, the system of equations becomes linear and homogeneous in the jumps. Its determinant must vanish to allow for a solution. It is possible, though tedious, to write the determinant in factorized form

$$F\left(F^2 - \frac{B_n^2}{\mu_0\langle \mathscr{V} \rangle} \right)\left\{ F^4 + F^2\left(\frac{[p]}{[\mathscr{V}]} - \frac{\langle \mathbf{B} \rangle^2}{\mu_0\langle \mathscr{V} \rangle} \right) - \frac{B_n^2}{\mu_0\langle \mathscr{V} \rangle}\frac{[p]}{[\mathscr{V}]} \right\} = 0 \tag{8.31}$$

This is a seventh-order equation for the normal mass flux, F, across the discontinuity. It consists of three factors which can be put to zero independently, demonstrating that

in magnetohydrodynamics three different classes of discontinuities may evolve. The
first class of discontinuities is determined by the condition

$$F_{\mathrm{I}} = 0 \qquad (8.32)$$

corresponding to zero mass flow across the discontinuity. We will find that two independent types of discontinuities, *tangential discontinuities* and *contact discontinuities*,
satisfy this condition. The second class attributes a finite value to the mass flux across
the discontinuity

$$F_{\mathrm{II}} = \pm \frac{B_n}{\sqrt{\mu_0 \langle \mathcal{V} \rangle}} \qquad (8.33)$$

This value depends only on the normal component of the magnetic field, which is
continuous, and on the average density of the fluid. The class of discontinuities corresponding to this particular condition is called *rotational discontinuities* and is believed
to describe *stationary magnetic merging* or reconnection (if it exists) in terms of a fluid
theory.

For the third class the expression in the second parenthesis of Eq. (8.31) must
vanish

$$F_{\mathrm{III}}^4 + F_{\mathrm{III}}^2 \left(\frac{[p]}{[\mathcal{V}]} - \frac{\langle \mathbf{B} \rangle^2}{\mu_0 \langle \mathcal{V} \rangle} \right) - \frac{B_n^2}{\mu_0 \langle \mathcal{V} \rangle} \frac{[p]}{[\mathcal{V}]} = 0 \qquad (8.34)$$

This factor contains the jumps of pressure and the specific volume and allows for a
variety of different conditions. It is a biquadratic equation which has two independent
solutions

$$F_{\mathrm{III}}^2 = -\frac{1}{2} \left(\frac{[p]}{[\mathcal{V}]} - \frac{\langle \mathbf{B} \rangle^2}{\mu_0 \langle \mathcal{V} \rangle} \right) \pm \frac{1}{2} \sqrt{\Delta} \qquad (8.35)$$

These solutions for $F_{\mathrm{III}}^2 > 0$ are called *shocks* or *shock waves*. It can be shown that for
any $[p]/[\mathcal{V}]$, positive or negative, the discriminant

$$\Delta = \left(\frac{[p]}{[\mathcal{V}]} - \frac{\langle \mathbf{B} \rangle^2}{\mu_0 \langle \mathcal{V} \rangle} \right)^2 + \frac{4 B_n^2}{\mu_0 \langle \mathcal{V} \rangle} \frac{[p]}{[\mathcal{V}]} > 0 \qquad (8.36)$$

is always positive. Moreover, when $[p]/[\mathcal{V}] < 0$ then the first term in (8.35) is positive
and $\sqrt{\Delta}/2$ is smaller than this term thus yielding shock solution for both signs of
the square root. Shocks principally cause an increase in the pressure $[p] > 0$ across
the shock. For the ratio to be negative, the jump in the specific volume must thus
be negative which implies that the density across the shock increases. The family of
all those shocks satisfying the condition $[p]/[\mathcal{V}] < 0$ is thus *compressive* with $[n] =
n_2 - n_1 > 0$. Thus there exist two types of compressive shocks corresponding to the
two signs of the square root in Eq. (8.35):

$$F_{\mathrm{III}}^{\pm} = \frac{1}{\sqrt{2}} \left(A \pm \sqrt{A^2 - D} \right)^{\frac{1}{2}}, \quad A = \frac{[p]}{|[\mathcal{V}]|} + \frac{\langle \mathbf{B} \rangle^2}{\mu_0 \langle \mathcal{V} \rangle}, \quad D = \frac{4 B_n^2}{\mu_0 \langle \mathcal{V} \rangle} \frac{[p]}{|[\mathcal{V}]|} \qquad (8.37)$$

On the other hand, for $[p]/[\mathscr{V}] > 0$ the first term can be either positive or negative, but in this case $\sqrt{\Delta}/2$ is larger than the first term and when taken with the positive sign of the square root compensates for its negative contribution. Thus even in this case shocks can exist. Across such shocks the pressure increases while the density jump is negative $[n] = n_2 - n_1 < 0$. Such shocks are called *rarefaction shocks*. For them clearly $v_{n2} > v_{n1}$ implying also acceleration. They can only be realised when the temperature across the shock increases sufficiently strongly in order to keep the jump in the pressure positive. This requires that for a rarefaction shock $T_2/T_1 > n_2/n_1$. The last condition can be written

$$\frac{[T]}{\langle T \rangle} > -\frac{[n]}{\langle n \rangle} > 0 \tag{8.38}$$

These three families of shocks in a fluid plasma can be taken as a guide line for the classification of shocks into compressive and rarefaction shocks. It is, however, not practical to work with these very involved expression also because the fluid approximation breaks quickly down for shocks. Below we qualitatively discuss a few simple cases in order to get a feeling for how the parameters in a plasma change across shocks. In later chapter we will more explicitly be going into the physics of shock waves. This can, however, be done only after having developed the theory of plasma waves in both the fluid and kinetic approximations.

8.3 The MHD Discontinuities

The general jump conditions lead to three different families of discontinuities. For each of these an explicit set of jump conditions can be specified.

8.3.1 Contact and Tangential Discontinuities

The first family of discontinuities is characterized by zero normal mass flow and thus $v_n = 0$. Such a situation can be realized in two different cases. In the first the magnetic field has a non-vanishing but continuous component normal to the discontinuity surface, $B_n \neq 0, [B_n] = 0$. In this case the second condition in Eq. (8.27) requires that the tangential magnetic field vector is continuous across the discontinuity. Armed with this knowledge, it is easy to show that the third jump condition (8.27) demands continuity of the tangential velocity vector, while the first jump condition (8.27) reduces to the continuity of the pressure. The only quantity which can experience a change across the discontinuity is the plasma density, while all other quantities are continuous.

This type of discontinuity is called *contact discontinuity*, because two plasmas are attached to each other at the discontinuity and rigidly tied by the normal component of the magnetic field such that they flow together at the same tangential speed. Contact discontinuities therefore satisfy the rather restrictive relations

$$\boxed{[p] = [B_n] = 0 \quad \text{and} \quad [\mathbf{v}_t] = [\mathbf{B}_t] = 0} \tag{8.39}$$

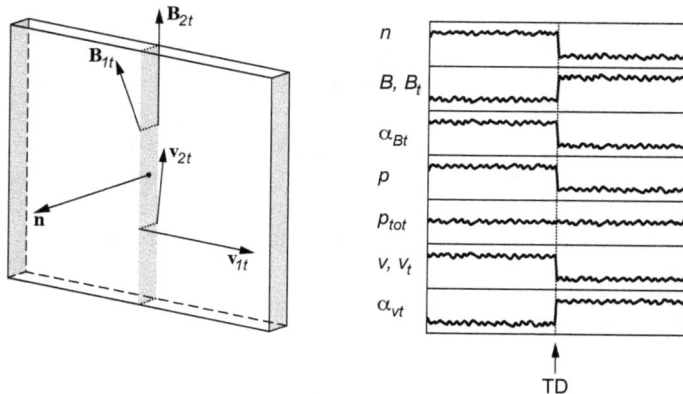

Figure 8.6: Changes of magnetic field and plasma moments across a tangential discontinuity. Since $v_n = B_n = 0$ the tangential velocities and magnetic fields across the discontinuity are independent and can have any direction. The panels on the right show the schematic changes density n, magnetic field $B = B_t$, angle of magnetic field α_{Bt}, pressure p, total pressure p_{tot}, velocity $v = v_t$ and angle of velocity α_{vt} across the tangential discontinuity.

with the vector conditions telling that the tangential velocities and the tangential magnetic fields to both sides of the discontinuity are identical. Thus it is only density and temperature which are allowed to change. However, since the pressure remains constant, any change in density must be balanced by a change in temperature. On the other hand, a temperature difference between both sides of the discontinuity should rapidly be dispersed by electron heat flux along the magnetic field, and thus contact discontinuities should usually not persist for long time but decay if not artificially renewed.

The second and more interesting case has its magnetic field purely tangential to the discontinuity, with zero normal component, $B_n = 0$. Since $v_n = 0$, the second and third of the Rankine-Hugoniot conditions (8.27) are trivially satisfied for any jumps in the tangential velocity and magnetic field. The only nontrivial condition which survives is the first, requiring the continuity of the total pressure across the discontinuity

$$\boxed{\left[p + \frac{B^2}{2\mu_0} \right] = 0} \tag{8.40}$$

Such a discontinuity is a surface of total pressure balance between the two contacting plasmas with no mass or magnetic flux crossing the discontinuity from either side, while all other quantities can experience arbitrary changes. It is called *tangential discontinuity*, because both the plasma flow and the magnetic field are tangential to but discontinuous at the discontinuity. Typical changes of magnetic field, density, pressure, and bulk velocity across a tangential discontinuity are sketched in Fig. 8.6. The left part of the figure shows that the tangential magnetic and velocity vectors may arbitrarily change their magnitudes, B_t, v_t, and directions, α_{Bt}, α_{vt} across the discontinuity.

The right-hand part of the figure shows how plasma and field quantities would change for a spacecraft crossing a tangential discontinuity.

8.3.2 Rotational Discontinuities

Assuming a finite mass flow across the discontinuity, $nv_n \neq 0$, but a continuous normal flow velocity, $[v_n] = 0$, yields discontinuities of family II. Because of the continuity of nv_n there can be no jump in the plasma density across the discontinuity. But the condition $F_{\mathrm{II}} \neq 0$ requires that a non-vanishing normal flow is possible only when the magnetic field also has a non-vanishing normal component, $B_n \neq 0$. The first condition in Eq. (8.27) with vanishing left-hand side requires the continuity of the total pressure, like for tangential discontinuities. However, the second and third condition (8.27) show that the tangential components of the velocity and the magnetic field can only change together. The second condition together with Eq. (8.33) yields an appropriate jump condition for the tangential components. Applying the rule (8.29) to resolve the bracket in the third Rankine-Hugoniot condition (8.27)

$$B_n[\mathbf{v}_t] = v_n[\mathbf{B}_t] \tag{8.41}$$

reveals another interesting property. Since v_n and B_n are constant, the tangential velocity and tangential magnetic field difference vectors are parallel. Magnetic field strength and thermal pressure are each continuous across the discontinuity. Summarizing, discontinuities of the second family obey the following jump conditions

$$
\begin{aligned}
[n] = [p] = [v_n] = [j_n] = [B_n] = [B^2] &= 0 \\
\left[\mathbf{v}_t - \frac{\mathbf{B}_t}{\sqrt{nm\mu_0}} \right] &= 0
\end{aligned}
\tag{8.42}
$$

A discontinuity obeying these boundary conditions is called *rotational discontinuity* and, as sketched in Fig. 8.7, is a region where the tangential flow and magnetic field rotate by some arbitrary angle. The last condition actually relates the jumps, i.e. the vectorial differences in the tangential components of the flow velocity and the Alfvén velocity defined in Eq. (8.11), to each other

$$[\mathbf{v}_t] = \left[\frac{\mathbf{B}_t}{\sqrt{nm\mu_0}} \right] \equiv \frac{[\mathbf{B}_t]}{\sqrt{nm\mu_0}}, \qquad \text{or equivalently} \qquad [\mathbf{j}_t] = \frac{1}{\mu_0} \mathbf{n} \times [\mathbf{B}_t] \tag{8.43}$$

The first form of this equation is often called *Walén relation*. At a rotational discontinuity the difference vector in the tangential flow velocity is exactly equal and directed exactly parallel to that of the tangential Alfvén velocity, a fact which shows that rotational discontinuities are closely related to the transport of magnetic signals across a boundary from one medium to the other. Since the density in rotational discontinuities must be constant, the jump in the tangential Alfvén velocity arises only from the jump in the tangential magnetic field. This is expressed in the second form of the

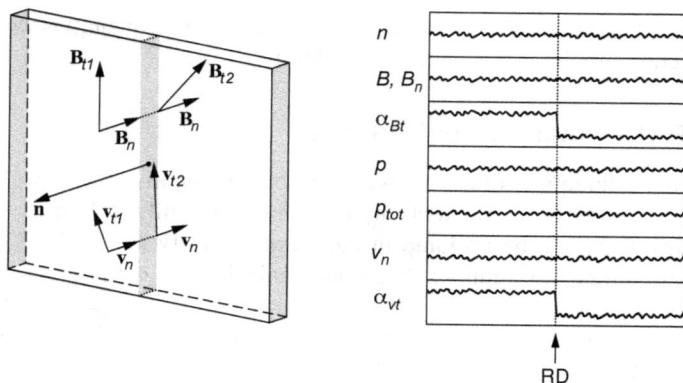

Figure 8.7: Rotational discontinuity. Velocity and magnetic difference vectors are parallel while field and velocity rotate. Plasma flows at velocity $v_n \sim B_n$ across the discontinuity.

above equation where the density just serves as a scaling factor for the two jumps while keeping their directions identical. This implies rotation of the vectors across the discontinuity.

The constant normal component of the flow velocity is naturally related to the normal component of the Alfvén velocity. It is obtained from the constancy of F_{II}, observing that the continuity of the density implies that the specific volume is also continuous

$$v_n = \frac{B_n}{\sqrt{nm\mu_0}} \tag{8.44}$$

This latter equation can be used to determine the normal component of the flow velocity from a measurement of B_n and n. These two quantities are comparably easy to observe while a direct measurement of v_n provides severe difficulties. The presence of non-vanishing normal magnetic component has led to relate rotational discontinuities to stationary reconnection.

Since plasma pressure and density must each be constant across a rotational discontinuity, the temperature of the plasmas on both sides of a rotational discontinuity must also be the same, at least in an isotropic plasma flow. This implies in turn that rotational discontinuities do not lead to an increase in entropy. In the ideal case, they are isentropic discontinuities which in the fluid theory have no internal dynamics. However, this is a very approximative view, because reconnection is necessarily accompanied by non-isentropic processes.

The three discontinuities encountered until now all are reversible. In the case of the former two there was no mixing of the plasma and reversibility was trivial. But in the case of the rotational discontinuity mixing of two different plasmas takes place and, as just mentioned, reversibility comes in as a surprise and, as it turns out, is the artefact of application of ideal fluid plasma theory to a plasma under conditions (like reconnection) where reversibility becomes highly questionable.

8.3.3 Effect of Pressure Anisotropy

So far we have considered cases when the plasma pressure was isotropic. It is straightforward to generalize the results obtained to an anisotropic plasma with the pressure given by Eq. (7.21). The inclusion of pressure anisotropy affects, in the first place, only the equation of motion and the second condition in Eq. (8.22) yielding

$$\mathbf{n} \cdot [nm\mathbf{v}\mathbf{v}] + \mathbf{n} \left[p_\perp + \frac{B^2}{2\mu_0} \right] - \frac{1}{\mu_0} \mathbf{n} \cdot \left[\mathbf{B}\mathbf{B} \left(1 - \frac{\mu_0(p_\parallel - p_\perp)}{B^2} \right) \right] = 0 \qquad (8.45)$$

Since tangential discontinuities have $B_n = 0$, anisotropy has only a minor effect on the conditions for tangential discontinuities. In the pressure jump condition the isotropic pressure is simply replaced by the perpendicular pressure for tangential discontinuities

$$\left[p_\perp + \frac{B^2}{2\mu_0} \right] = 0 \qquad (8.46)$$

It is considerably more involved to determine the anisotropic jump conditions for rotational discontinuities and only a few important points are mentioned here. The Walén relation becomes modified in this case such that the normal component of the flow is equal to a modified normal Alfvén velocity in Eq. (8.44)

$$v_n = \left\{ \left(\frac{B_n^2}{nm\mu_0} \right) \left(1 - \frac{\mu_0(p_\parallel - p_\perp)}{B^2} \right) \right\}^{1/2} \qquad (8.47)$$

on both sides of the discontinuity. Since the tangential magnetic fields can have arbitrary directions in an anisotropic plasma, the two tangential magnetic field vectors and the vector normal to the rotational discontinuity are in general not coplanar anymore. Moreover, the density may also assume a non-vanishing jump, $[n] \neq 0$, and pressure equilibrium holds only for the total pressure, $[p + B^2/2\mu_0] = 0$ where now $p = \frac{1}{3}(p_\parallel + 2p_\perp)$.

8.3.4 Entropy Changes

It is instructive to investigate the behaviour of the entropy at discontinuities. Consider the stationary ideal heat conduction or entropy conservation equation (7.59)

$$\mathbf{v} \cdot \nabla \mathscr{S} = 0 \qquad (8.48)$$

In incompressible media with $\nabla \cdot \mathbf{v} = 0$, this equation can be rewritten as

$$\nabla \cdot (\mathbf{v}\mathscr{S}) = 0 \qquad (8.49)$$

or in the form of a jump condition as

$$[v_n \mathscr{S}] = 0 \qquad (8.50)$$

In the more general case of compressible media, the right-hand side does not vanish, showing that the compressibility of the fluid acts as a source of entropy

$$\nabla \cdot (\mathbf{v}\mathscr{S}) = \mathscr{S}\nabla \cdot \mathbf{v} \tag{8.51}$$

At a one-dimensional discontinuity the last equation can be rewritten as

$$[v_n\mathscr{S}] = \int dn\,\mathscr{S}\frac{dv_n}{dn} = \int \mathscr{S}dv_n \tag{8.52}$$

Hence, discontinuities with constant normal velocity, $[v_n] = 0$, conserve entropy. All ideal single-fluid plasma discontinuities discussed so far belong to this class of *isentropic discontinuities* with no increase of entropy when crossing from the upstream to the downstream regions. Discontinuities in compressible media with a non-vanishing jump in the normal flow speed will necessarily lead to an increase in entropy across the discontinuity because the compression of the plasma requires work which implies the production of entropy, i.e. not leaving the temperature unaffected. They therefore are accompanied by irreversible changes in the state of the plasma across the discontinuity. These types of discontinuities can appear even in an ideal plasma. The non-ideality of the plasma is then confined to the internal transition region of the discontinuity. The extended class of such discontinuities are shocks.

8.4   Shocks

The third discontinuity family is characterized by non-vanishing normal fluxes, $nv_n \neq 0$. For this family the plasma moves across the discontinuity as in the case of rotational discontinuities, but in contrast to rotational discontinuities the density across a shock changes discontinuously.

8.4.1   Intermediate Shocks

We first consider the simplest case of such discontinuities as given by Eq. (8.34) which suggests that $[\mathscr{V}] \neq 0$, because otherwise this condition reduces to

$$[p]\left(F_{\text{III}}^2 - \frac{B_n^2}{\mu_0\langle\mathscr{V}\rangle}\right) = 0 \tag{8.53}$$

The latter equation has two solutions. Either

$$[p] = 0 \tag{8.54}$$

which implies that the pressure is continuous, or

$$\left(F_{\text{III}}^2 - \frac{B_n^2}{\mu_0\langle\mathscr{V}\rangle}\right) = 0 \tag{8.55}$$

showing that under the special condition of continuous density, a family III discontinuity resembles a rotational discontinuity. Indeed, the latter condition is identical with
condition (8.33), which the normal flux satisfies at rotational discontinuities.

In this very special case and with $[p] \neq 0$ one, however, speaks of an *intermediate shock* allowing for a jump in pressure and the related heating (and production of
entropy) inside the discontinuity. On the other hand, when both factors in the above
condition are simultaneously zero, one recovers the ordinary proper rotational discontinuities. Rotational discontinuities thus form a subclass of intermediate shocks with
continuous pressure, no dissipation and heating and thus no increase in entropy. From
this observation we may immediately conclude that the application of proper rotational
discontinuities to reconnection, for instance, would be incorrect. The correct choice
of discontinuity which possibly may sign responsible for stationary reconnection are
the intermediate shocks with $[p] \neq 0$ and $[v_n \mathscr{S}] \neq 0$.

With these conditions, $[B_n] = 0$, and observing that $[\mathscr{V}] = 0$ also implies $\langle \mathscr{V} \rangle = \mathscr{V}$
and $[v_n] = 0$, where $v_n = B_n / \sqrt{\mu_0 mn} = B_n \sqrt{\mathscr{V}/\mu_0} = v_{An}$ is the normal Alfvén speed
based on B_n, the Rankine-Hugoniot conditions yield for the (isotropic) intermediate
shock solution that

$$[p] = -\frac{1}{\mu_0}\langle \mathbf{B}_t \rangle \cdot [\mathbf{B}_t] = -\left[\frac{B_t^2}{2\mu_0}\right] \quad \text{and} \quad [\mathbf{v}_t] = \frac{B_n[\mathbf{B}_t]}{\mu_0 mnv_n} = [\mathbf{v}_{At}] \quad (8.56)$$

The first of these equations is the pressure balance known from the tangential discontinuity. It fixes the total pressure across the intermediate shock. The second is a jump in
the tangential velocity known from the rotational discontinuity. This set of expressions
indicates clearly that the intermediate shock is a mixture of tangential and rotational
discontinuities leaving the freedom to the magnetic fields and normal velocities to
cross the shock transition. The pressure balance takes care of the equilibrium, and the
flow remains isentropic because the change in thermal pressure is compensated by a
change in magnetic pressure. Thus, this intermediate shock is not a true shock; rather
it is a mixed form of cross-flow discontinuity in thermal equilibrium.

However, in dealing with shocks we should have taken into account the energy
equation. Inserting the above expressions for the jumps across the intermediate shock
into the jump condition (8.23) for the energy we find that (after some algebra) all terms
cancel leaving us with the following jump condition

$$v_n[w] = v_n[c_v p] = 0 \quad \text{or, since } v_n \neq 0: \quad [c_v p] = 0 \quad (8.57)$$

With $c_v = \text{const}$ this expression contradicts the requirement that the pressure jump
does not vanish. Thus, naively, we would conclude that intermediate shocks do not
exist. This conclusion is, however, premature. It may not be opportune to assume
that c_v jumps across the intermediate shock like $[c_v] = -[p]\langle c_v \rangle / \langle p \rangle$; in our argument
we may also have used too strong an assumption. This assumption can be found in
the definition of the internal enthalpy w which more generally should be defined as
$w = \varepsilon + p\mathscr{V}$, where $\varepsilon = \mathscr{E}/m$ is the specific internal energy. If using this general

definition, the enthalpy conservation across the intermediate shock in our ideal plasma fluid reduces to

$$[w] = [\mathscr{V}]\langle p \rangle = [\varepsilon + p\mathscr{V}] \tag{8.58}$$

which must be negative. Inserting the condition on $[p]$ we find that the internal energy $\mathscr{E}$ increases across the intermediate shock by the amount

$$[\mathscr{E}] = m\langle \mathscr{V} \rangle \left[\frac{B_t^2}{2\mu_0} \right] = \left[\frac{B_t^2}{2\mu_0 n} \right] \tag{8.59}$$

an increase which is produced by processes that are internal to the intermediate shock transition and are not described by the ideal fluid equations we have used to derive the jumps. Interestingly, this last expression shows that some part of the available magnetic energy difference per particle across the intermediate shock is converted into internal energy of the fluid by the unspecified processes internal to the discontinuity.

8.4.2   True Shocks

In all other cases Eq. (8.34) possesses two pairs of conjugate solutions for non-vanishing jumps in pressure, specific volume, and density, implying that the plasma flow across the discontinuity switches from one thermodynamic state to another. Of these four solutions only the three which yield $F_{\text{III}}^2 > 0$ are physically relevant. These are one solution for $[p]/[\mathscr{V}] > 0$, and two solutions for $[p]/[\mathscr{V}] < 0$. For $[p] > 0$ there is only one solution with $[\mathscr{V}] > 0$ or $[n] < 0$. It has $[v_n] > 0$ and corresponds to a dilutive accelerated transition in a *rarefaction shock*, while there are two solutions with $[n] > 0$ corresponding to compressive transitions. Since the pressure always increases, shocks are always accompanied by an increase in temperature across the shock, indicating that inside the shock transition region the plasma is heated. This is an irreversible process which increases the entropy.

Because the plasma moves across the discontinuity at continuous normal flux, the finite jump in the density implies that the normal velocity changes in the opposite way across the discontinuity. This is most easily observed from the continuity of the normal flow, $[F_{\text{III}}] = [nv_n] = 0$, a condition which can be rewritten using Eq. (8.29)

$$[v_n] = -\frac{\langle v_n \rangle}{\langle n \rangle} [n] \tag{8.60}$$

This a compression $[n] > 0$ implies $v_{n1} > v_{n2}$ corresponding to a slow down in normal speed while a rarefaction $[n] < 0$ implies that $v_{n2} > v_{n1}$ in which case one finds the aforementioned increase in normal speed v_n across the rarefaction shock.

8.4.3 Coplanarity

Knowing that $v_n \neq 0$, the two last conditions of Eq. (8.27) suggest that for shocks with $B_n \neq 0$ the jump in the tangential magnetic field vector is parallel to the jump in the tangential velocity vector. Eliminating $[\mathbf{v}_t]$ from (8.27) one obtains

$$[v_n \mathbf{B}_t] = \frac{B_n^2}{\mu_0 F_{\mathrm{III}}} [\mathbf{B}_t] \tag{8.61}$$

Hence, the cross-product of the right- and left-hand sides must vanish

$$[\mathbf{B}_t] \times [v_n \mathbf{B}_t] = 0 \tag{8.62}$$

When resolving the brackets in the above expression, one obtains

$$(v_{n1} - v_{n2})(\mathbf{B}_{t1} \times \mathbf{B}_{t2}) = 0 \tag{8.63}$$

Since $[v_n] \neq 0$, the upstream and downstream tangential components of the magnetic field on both sides of the shock must be parallel to each other. Hence, the upstream and downstream tangential magnetic field vectors are coplanar with the shock normal vector, they all lie in the same plane normal to the shock. This *coplanarity theorem* implies that the magnetic field across the shock obeys a two-dimensional geometry. The same also holds for the bulk velocity. It is coplanar with the shock normal and has two-dimensional geometry, yet a different one. We should however caution the reader that coplanarity is only warranted in the framework of ideal fluid theory. At real shocks coplanarity fails due to the dominance of kinetic effects on all scales.

8.4.4 Jump Conditions

In order to treat the boundary conditions at shock we need to include the energy transport across the shock, since the shock itself is a region where heat and entropy are generated. Assuming that the plasma behaves like an ideal gas and that all variations proceed so fast that adiabatic conditions can be assumed, the internal enthalpy is

$$w = \frac{p}{nm(\gamma - 1)} \tag{8.64}$$

with γ being a constant polytropic index. With these assumptions in mind, one can bring the conservation of energy (8.23) into the form

$$\left[v_n \left(\frac{nmv^2}{2} + \frac{\gamma p}{\gamma - 1} + \frac{B_t^2}{\mu_0} \right) - \frac{B_n \mathbf{v} \cdot \mathbf{B}}{\mu_0} \right] = 0 \tag{8.65}$$

Using the remaining Rankine-Hugoniot conditions (8.27) and realizing that

$$\left[\left(\mathbf{v}_t - \frac{B_n \mathbf{B}_t}{\mu_0 F_{\mathrm{III}}} \right)^2 \right] = 0 \tag{8.66}$$

the latter expression can be brought into the following form

$$\frac{[v_n^2]}{2} + \frac{\gamma[pv_n]}{(\gamma-1)F_{\mathrm{III}}} + \frac{[v_n B_t^2]}{\mu_0 F_{\mathrm{III}}} - \frac{B_n^2}{\mu_0 F_{\mathrm{III}}^2} \frac{[B_t^2]}{2\mu_0} = 0 \tag{8.67}$$

or, when splitting the jumps of the products into products of jumps and averages,

$$[p]\langle v_n \rangle + [v_n]\gamma\langle p \rangle + \frac{\gamma-1}{4\mu_0}[v_n][B_t^2] = 0 \tag{8.68}$$

Since bulk velocity and magnetic field each obey the coplanarity theorem, one can write Eq. (8.61) in scalar form

$$[v_n]\langle B_t \rangle + [B_t]\langle v_n \rangle = \frac{B_n^2}{\mu_0 F_{\mathrm{III}}}[B_t] \tag{8.69}$$

and find from momentum conservation, i.e., the second Rankine-Hugoniot condition (8.27),

$$\begin{aligned} B_n[v_t] &= \frac{B_n^2}{nm\mu_0}\frac{[B_t]}{\langle v_n \rangle} \\[2ex] F_{\mathrm{III}}[v_n] &= -\left[p + \frac{B_t^2}{2\mu_0}\right] \end{aligned} \tag{8.70}$$

where $B_n^2/nm\mu_0$ is the square of the normal component of the Alfvén velocity given by the Walén relation (8.44). Equations (8.68) through (8.70) form a closed set of equations for the shock jump conditions and are the general Rankine-Hugoniot conditions for shocks, which are valid in an isotropic and adiabatic one-fluid plasma.

8.4.5 Fast and Slow Shocks

We can obtain some general results when eliminating the jump in the normal velocity $[v_n]$ from Eqs. (8.68) and (8.70). This way we obtain a relation between the jumps in plasma and magnetic pressures

$$\left(\frac{\langle v_n \rangle}{\gamma-1} - H\right)[p] = \frac{H}{\mu_0 F_{\mathrm{III}}}[B_t^2] \tag{8.71}$$

where the quantity H is defined as

$$F_{\mathrm{III}}H = \frac{[B_t^2]}{4\mu_0} + \frac{\gamma\langle p \rangle}{\gamma-1} \tag{8.72}$$

Since the pressure always increases across a shock transition from the undisturbed to the disturbed and heated plasma behind the shock, one has

$$[p] > 0 \tag{8.73}$$

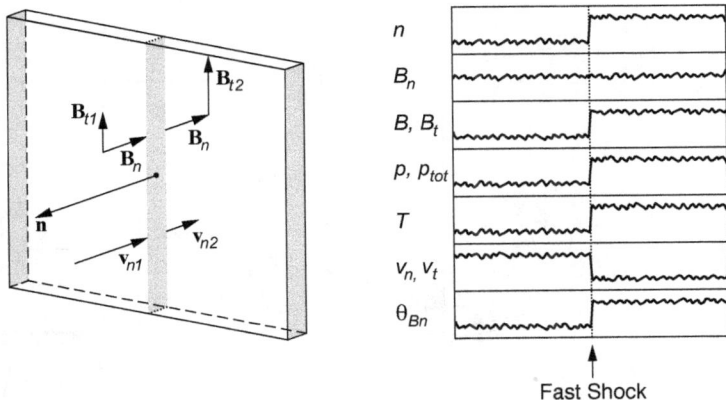

Figure 8.8: Changes of magnetic field and plasma moments across a fast shock.

One can therefore distinguish between two different cases of shock waves. The first type of shocks is characterized by an increase in the Alfvén speed or, correspondingly, an increase in the magnetic pressure

$$[B_t^2] > 0 \tag{8.74}$$

Such shocks satisfy the condition

$$\langle v_n \rangle > (\gamma - 1)H \tag{8.75}$$

and are called *fast shocks* (see Fig. 8.8). The second type experiences a decrease in magnetic pressure when passing through the shock from the undisturbed to the disturbed medium

$$[B_t^2] < 0 \tag{8.76}$$

Shocks of this type satisfy the opposite condition for the normal average velocity

$$\langle v_n \rangle < (\gamma - 1)H \tag{8.77}$$

and are called *slow shocks* (see Fig. 8.9).

As can be recognized by comparing Figs. 8.8 and 8.9 or the two panels on the right-hand side of Fig. 8.11, across fast shocks the magnetic field increases and is tilted toward the shock surface, while across slow shocks it decreases and bends toward the shock normal. The average normal velocity points always downstream, because plasma is flowing across the shock from the undisturbed into the disturbed and heated medium. Hence, slow shocks can exist only when the average pressure is large enough to satisfy

$$\langle p \rangle + \frac{\gamma - 1}{4\gamma\mu_0} \left([B_t^2] - B_{t1} B_{t2} \right) > 0 \tag{8.78}$$

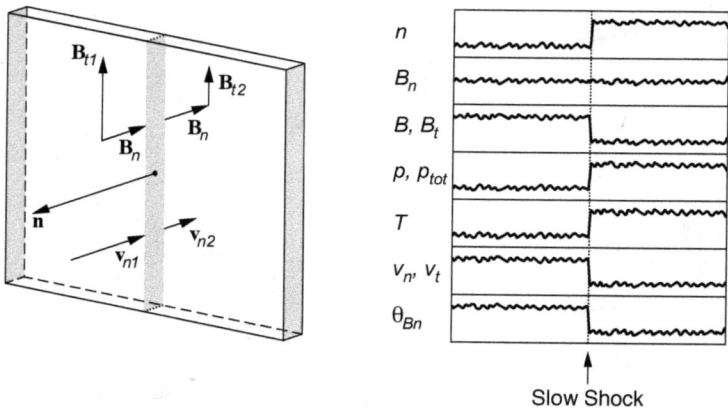

Figure 8.9: Changes of magnetic field and plasma moments across a slow shock.

The two types of shocks can be distinguished by the behaviour of the tangential magnetic field pressures across the shock (see Figs. 8.8 and 8.9). In addition, one finds that fast flows will cause fast shocks to evolve, while slow flows may give rise to slow shocks.

8.4.6 Shock Mach Numbers

So far we have considered the jump boundary conditions which have to be satisfied across a shock transition. These conditions allowed for a classification of shocks it they exist. However, none of these conditions tells under which circumstances in the plasma a shock would develop. This condition is, however, very simply given by the Mach number which we have already defined earlier. Mach numbers are not frame invariant. In the proper frame of a fluid flowing with the flow velocity the Mach number vanishes. However, seen from another frame or putting an obstacle into the flow the Mach number seen by the obstacle is non-zero. A shock may then develop when the fluid hits the obstacle with velocity which velocity exceeds the Alfvénic or more generally (see Section 8.1.2) the *magnetosonic speed*

$$\boxed{c_{ms}^2 = c_s^2 + v_A^2} \qquad (8.79)$$

in the fluid, which replaces the sound velocity (8.5) in an ordinary fluid. In this case the plasma flow is called super-magnetosonic (or when the thermal velocity is much less than the Alfvén speed, super-Alfvénic) and, in analogy to ordinary fluids, a non-moving obstacle will give rise to the evolution of a plasma shock. One defines a *magnetosonic Mach number*, M_{ms}, by relating the fluid velocity to the magnetosonic speed

$$\boxed{M_{ms} = \frac{v}{c_{ms}}} \qquad (8.80)$$

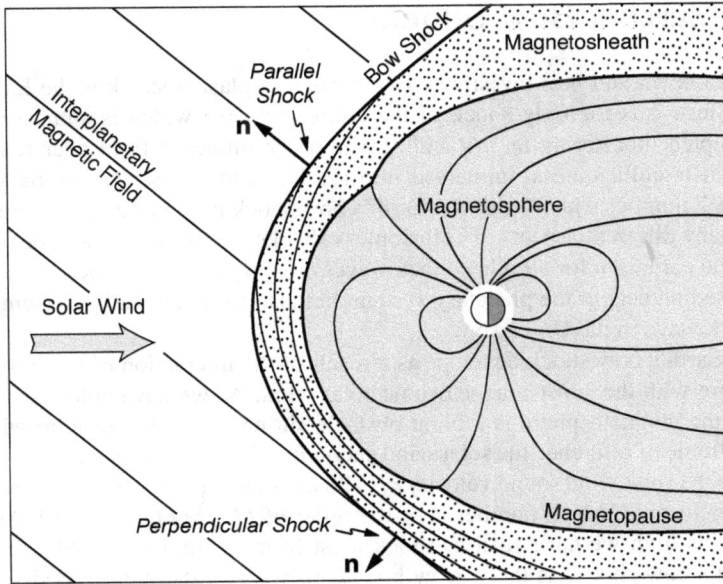

Figure 8.10: The Earth's bow shock is a highly curved shock standing in the solar wind. For a given direction of the interplanetary field the shock normal **n** rotates along the shock by nearly an angle of 135°. For this reason the bow shocks steadily offer all kinds of shock configurations reaching from strictly perpendicular over quasi-perpendicular to quasi-parallel and even strictly parallel shocks when passing along the shock surface from one side of the magnetosphere to the other.

The condition for the evolution of a shock wave expressed in terms of the magnetosonic Mach number in a plasma then becomes that

$$M_{ms} > 1 \tag{8.81}$$

Whenever this condition is satisfied and the plasma flow is distorted by a non-moving obstacle, a shock front develops across which the fluid quantities jump discontinuously. The super-magnetosonic flow is retarded across the shock from upstream to downstream to a sub-magnetosonic flow.

This is, however, not the only condition under which shocks develop. A more general statement is that the nonlinear growth of a wave in the plasma fluid can, under certain circumstances, also lead to the formation of a shock even in the complete absence of an obstacle. Such processes are not treatable in the framework of simple fluid theory (for such processes see Chapter 13 and our companion volume *Advanced Space Plasma Physics*). We mention them here for completeness only.

8.5   Earth's Bow Shock

The best known and best investigated example of a plasma shock is the Earth's bow shock. Since it is the only shock in the entire Universe which is subject to almost uninterrupted monitoring *in situ* and also in a certain sense from near remote, and because it is collisionless, immersed into a dilute and temperate plasma stream of high-Mach number which lends the Earth's bow shock its very complex nature which unites many different aspects of collisionless shocks, the Earth's bow shock serves as the unique paradigm for all other shock waves not only as a bow shock ahead of other magnetised planets in the planetary system, but also for other stellar systems and for any shock wave in the Universe.

The Earth's bow shock develops as a result of the interaction of the Earth's magnetosphere with the super-magnetosonic solar wind. As we have noted several times already, the magnetosphere is a blunt obstacle sitting at rest in the solar wind. Seen from its frame of reference the solar wind encounters it with average velocity $\sim$500 km/s far above the solar wind sound velocity $c_s \sim 20$ km/s and Alfvén velocity $v_A \sim 50$ km/s. Its (magnetosonic) Mach number is of the order of $M_{ms} \approx M_A \sim 8 - 10$. At a Mach number this high we have seen as shock must form in front of the Magnetosphere, the Earth's bow shock, as the narrow boundary between the undisturbed solar wind flow and the disturbed flow region between the bow shock and the obstacle magnetosphere. Since the solar wind cannot simply cross the Magnetosphere it is braked by its presence and turned around it. The solar wind becomes informed bout the presence of the obstacle, retards and deviates from its original direction to flow around the obstacle magnetosphere. Figure 8.10 shows the about hyperbolically shaped surface of the bow shock, across which the solar wind velocity decreases from super-magnetosonic to sub-magnetosonic flow and Mach number. It divides the solar wind flow into two regions, the undisturbed solar wind upstream of the bow shock and the disturbed *magnetosheath* flow on the downstream side of the bow shock.

The shock exists only over a limited region of space in front of the Earth because the Mach number is defined by the solar wind velocity component normal to the shock, $v_n = v_{sw} \cos \theta$. The condition $M_{ms} > 1$ for a bent shock (for instance like a parabola) is satisfied only as long as the angle $\theta < \arccos M_{ms}^{-1}$. For $M_{ms} \approx 8$ the maximum angle between the solar wind velocity and the shock normal up to which the bow shock exists is $\theta_{max} \approx 80°$. Hence, in this case the bow shock forms a spatially restricted shield in front of the magnetosphere and undergoes a transition from a high-Mach number shock at its nose to a low Mach number shock at its flanks. Otherwise when it is of hyperbolic shape it bents out up to a limiting angle only and at larger distances from the ecliptic starts out flattering and resolving.

The bow shock is an ideal object for studying the properties of all kinds of shocks *in situ*. This may, at the first glance, be surprising. But a second glance shows that the bow shock is not a simple and uniform entity but instead offers a number of regions with completely different properties. As a first property we may note that, since the solar wind is a high-Mach number stream with $M_{ms} \approx 8$, the bow shock is a *fast magnetosonic* shock. The density and the magnetic field both increase when crossing from

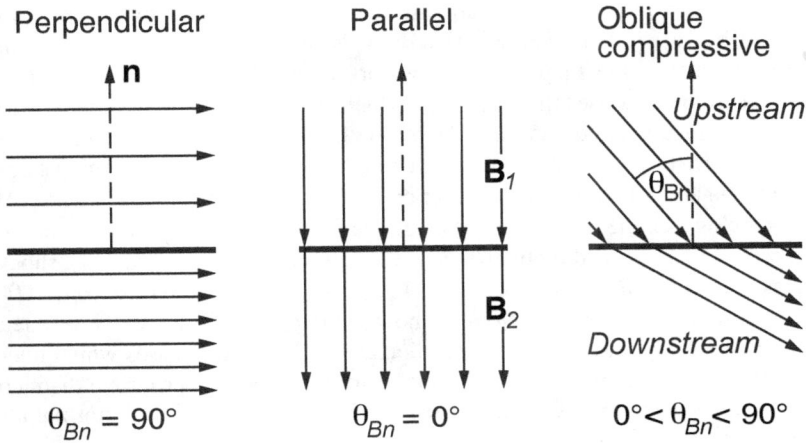

Figure 8.11: Three highly schematic possible average geometries of shock normal and magnetic field. This simplified classification distinguishes between strictly perpendicular, strictly parallel and oblique shocks. The case of a oblique shock is too general, however. The bow shock is a fast shock, and the term oblique is rather imprecise when describing its properties. It splits into two more appropriate terms, a quasi-perpendicular shock for the angle between shock normal and upstream magnetic field $\theta_{Bn} > 45°$ and a quasi-parallel shock for $\theta_{Bn} > 45°$, both with completely different properties as described in the text. In an oblique shock the downstream magnetic field turns always away from the shock normal because of the constancy of B_n and the compression of the tangential field component.

the solar wind into the magnetosheath. As determined experimentally, both quantities jump by about a factor of $\lesssim 4$.

Such high-Mach number shocks having $M_{ms} > M_c$ are called *supercritical*. We will in a later chapter explain what the property of super-criticality physically implies for the properties of a shock. Here we just note that super-critical shocks behave differently from low Mach number shocks with $M_{ms} < M_c$, which are *subcritical*. The *critical Mach number*, M_c, is conventionally defined as the Mach number for which the flow velocity downstream of the shock equals the downstream sound velocity so that the downstream magnetosonic Mach number is equal to unity. Solving the shock jump conditions under this restriction and under the assumption that the magnetic field is tangential to the shock yields a value of $M_c = 2.7$. This value decreases, however, for oblique magnetic field directions. At the bow shock it has been found that an average critical Mach number $1 < M_c < 2$ is more appropriate than the above theoretical value. Hence, the vast majority of observed bow shock transitions are supercritical.

8.5.1 Parallel and Perpendicular Shocks

Another distinctive and most important difference between different parts of the bow shock can be realized from Fig. 8.10. It is controlled by the direction of the upstream (solar wind) magnetic field with respect to the outer shock normal $\mathbf{n}$ which is defined by Eq. (8.18) when $S(r, \theta, \phi) = 0$ is understood as the two-dimensional function describing the shape of the bow shock (in spherical coordinates) in space. As a measure one chooses the angle Θ_{Bn} which is the angle between the (solar wind) upstream magnetic field direction and the (bow) shock normal. For the ideal Archimedian spiral form of the interplanetary field, the shock normal on the morning side of the bow shock is about parallel to the direction of the interplanetary magnetic field while on the evening side the interplanetary magnetic field and the shock normal are close to being orthogonal. Depending on the value of the *shock normal angle*, Θ_{Bn}, shocks can be classified as *parallel shocks* ($\Theta_{Bn} = 0°$), as *perpendicular shocks* ($\theta_{Bn} = 90°$) or as *oblique shocks* ($0° < \Theta_{Bn} < 90°$) as shown in Fig. 8.11. However, since one easily realises that strictly parallel and perpendicular shocks are rare cases which may exist just in one singular point at the real shock surface, a more appropriate distinction is between *quasi-perpendicular* and *quasi-parallel* shocks if the shock normal angle is either $45° < \Theta_{Bn} < 90°$ or $0° < \Theta_{Bn} < 45°$, respectively.

The distinction between the two shock directions is physically relevant. Strictly parallel shocks have their magnetic field directed along the shock normal and since B_n must be continuous, the magnetic field is not affected by the presence of the shock. But this case is never realized in real systems. Realistic parallel shocks are always quasi-parallel and react also magnetically. Any small deviation of the magnetic field direction from being perpendicular to the shock front results in a strong effect on the magnetic field, since the magnetic field is rotated out of coplanarity by plasma waves radiated inside the shock into all directions tangential to the shock front. Such a distortion causes local disturbances which result in short wavelength oscillations of the magnetic field. The shock becomes turbulent. In addition, the generation of the new out-of coplanarity magnetic component turns a parallel shock into a quasi-perpendicular one close to the shock ramp. A magnetized shock always manages to turn the magnetic field locally about quasi-perpendicular, even if far upstream of the shock the magnetic field was parallel.

Figure 8.12 shows characteristic magnetic shock profiles. The typical ideal (sub-critical) perpendicular shock profile consists of upstream and downstream regions connected by a steep *shock ramp*. Quasi-perpendicular shocks usually possess a *shock foot* region in front of the shock ramp which is already part of the shock transition itself, where the magnetic field gradually rises into the shock ramp. In addition, the shock ramp itself generally shows a magnetic *shock overshoot* before settling at the average magnetic field strength behind the shock. Quasi-parallel shocks obey a more complicated behaviour. The quasi-parallel shock transition starts exhibiting strong oscillatory behaviour, which gradually becomes turbulent. Quasi-parallel shocks are highly oscillatory or even turbulent up to large distances in front of the shock. This distorted upstream shock region is called the *foreshock*, since here the upstream medium becomes notified of the shock's presence and thus in a more strict sense of the word already belongs to the shock. These properties can be describes in more detail when distinguishing between sub-critical and super-critical shocks. Sub-critical quasi-perpendicular shocks are about laminar with no turbulence developing in front of the

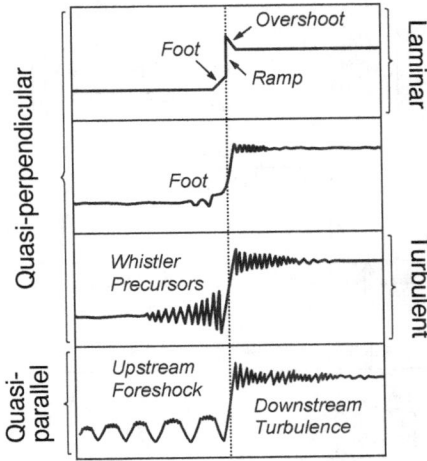

Figure 8.12: Typical schematic magnetic shock profiles. The upper three panels refer to quasi-perpendicular shocks, the lower to parallel shocks. The laminar shock is sub-critical, develops a small foot and shock overshoot in the steep ramp. The super-critical quasi-perpendicular shock is turbulent. It develops an extended foot with whistler precursor waves and is turbulent behind the shock. It also has an overshoot. Quasi-parallel shocks are always turbulent with extended foreshock regions containing long wave-length waves, less well expressed ramps and a highly turbulent downstream region (for further discussion see Chapter 12 on Shock Waves).

shock as shown in the upper part of the Fig. 8.12. Quasi-parallel sub-critical shocks, however, are always oscillatory. On the other hand, super-critical shocks (called turbulent in the figure) exhibit strong precursor oscillations in the quasi-perpendicular case and highly turbulent foreshocks in the quasi-parallel case. This behaviour of shocks is related to a purely kinetic and not anymore fluid-like property of super-critical shocks, i.e. the capability of reflection of a substantial umber of upstream particles back upstream into the undisturbed (solar wind) flow where these particles cause a wealth of kinetic effects. The reasons of this kind of particle reflection can be found in the impossibility for a super-critical shock to dissipate the excess upstream kinetic energy of the flow in the shock transition and to convert it into compression, heating and entropy during the brief passage time of the flow across the shock transition.

8.5.2 Shock Currents

The jump $[B_t] \neq 0$ in the tangential magnetic field across the shock indicates that the bow shock itself is a current layer with an internal surface current density, j_{sh}, which accounts for the change in the magnetic field. It can be estimated from

$$j_{sh} = \frac{[B_t]}{\mu_0 d_{sh}} \tag{8.82}$$

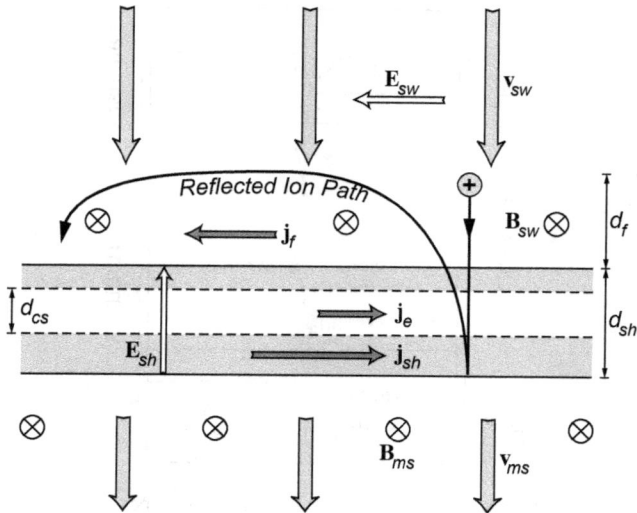

Figure 8.13: Ion reflection and acceleration at an ideally perpendicular shock. The impacting and reflected ions simply change sign of their normal velocities and return into the upstream solar wind where they form the shock foot adjacent to the shock over width d_f. Flowing along the shock they constitute a shock-parallel current $\mathbf{j}_f$ whose magnetic field causes the foot magnetic field of the quasi-perpendicular shock. Inside the shock ramp charge separation between ions and electrons causes a shock electric field $\mathbf{E}_{sh}$ which is responsible for the ion reflection. This field together with the shocked magnetic field also causes an E×B drift of the electrons in the shock transition tangential to the shock causing a purely electron-carried current $\mathbf{j}_e$ along the shock ramp. Its magnetic field generates the shock magnetic overshoot.

where d_{sh} is the shock width. This current increases the magnetic field strength in the shock ramp itself and over a short behind the shock. It should, in principle, partially cancel the magnetic field in front of the shock, but this has not been observed. Instead one finds a slight increase in magnetic field strength in the shock foot region, as indicated in Fig. 8.12, due to the appearance of reflected ions in front of the shock. These ions form themselves a current, and it is this current which completely compensates for the shock ramp current diminishing of the upstream field. All these processes are, however, of kinetic nature having no equivalent in the shock fluid theory of this chapter. They are genuinely plasma effects which we will re-consider only after having developed the kinetic theory of plasma waves. In noting them here, we are already leaving the shock fluid theory.

As for a first contact with the non-fluid theory of a shock we consider here a simple theory of the reflection of solar wind ions from the shock. Solar wind ions and electrons encountering the compressed magnetic field perpendicular shock will have different gyroradii. The ions can penetrate deeper into the field than the electrons. This difference in penetration depth will generate a charge separation electric field in the shock normal direction, pointing toward the sun. Such a field will reflect a number of ions back into the solar wind, while it attracts and captures electrons. It is given by

$$\varepsilon_0 E_{sh} = e(n_{i,sh} + n_{e,sh})d_{cs} \qquad (8.83)$$

where d_{cs} is the width of the charge separation layer and the densities are the densities of the particles inside the shock. The ion density is given as $n_i = n_e - n_{ir}$, with n_{ir} the density of the reflected ions

$$E_{sh} = \frac{en_{e,sh}d_{cs}}{\varepsilon_0}\left(1 - \frac{n_{ir}}{n_{e,sh}}\right) \qquad (8.84)$$

Hence, all ions having (directed) energies perpendicular to the shock less than the electric energy in the potential drop across the shock, $e\phi = eE_{sh}d_{cs}$, are potentially reflected from the shock. They after reflection return into the solar wind in front of the shock and perform another gyration in the solar wind magnetic field as shown in Fig. 8.13. This reflection can, in the simplest case, be assumed to be specular which means that the normal velocity of the particles in the reflection process simply turns by $180°$ and thus changes sign. Clearly such an assumption if highly simplistic as it does not take into account any particle energy losses inside the shock or even translation of the particle along the shock which in the quasi-perpendicular case is unavoidable due to any finite parallel velocity of the particles.

Because the solar wind in the shock frame carries a convection electric field, which lies in the gyration plane of the reflected ions, the reflected ions will be accelerated in this electric field to about twice the solar wind velocity. These reflected and accelerated ions carry a current in the foot region, j_f. Only a fraction of solar wind ions is actually reflected, but this fraction carries a current which closes the shock current in front of the shock and over-compensates the decrease in the magnetic field caused by the shock current.

The magnetic overshoot in the shock profile is related to another current layer inside the shock transition region. This current is a pure electron drift current (see Fig. 8.13). It results from the presence of the charge separation field, E_{sh}, inside the shock. As argued above this field is restricted to a narrow layer, narrower than the ion gyroradius in the compressed shock magnetic field, but wider than the electron gyroradius. Therefore the electrons may perform an electric $E \times B$ drift motion in the crossed electric and magnetic fields within this layer while the ions are not affected. This drift gives rise to an electron current, j_e, flowing in the same direction as the shock current, j_{sh}, and amplifying it locally, thereby causing the magnetic overshoot.

At this place we do not go into further detail of the shock properties and structures. As pointed out, fluid theory is incapable of describing any collisionless shock properties which go into the real shock structure and shock physics. Even in the above we have gone beyond any information which fluid theory can provide. One needs to develop the plasma kinetic theory and apply it to shocks in order to arrive at a state where one can compare theory with observation and interpret the behaviour of a collisionless shock like the Earth's bow shock wave. In the next section we instead describe briefly the properties of another most important discontinuity near the Earth, the Earth's magnetopause. But also there we run into the same difficulty that fluid theory just allows for a classification and determination of the shape but not for any description of the physics and properties of the magnetopause. Inference requires a fully kinetic approach to collisionless plasmas and ultimately numerical simulation techniques.

8.6 Earth's Magnetopause

Similar to Earth's bow shock, Earth's magnetopause is the rare case of a collisionless tangential discontinuity that can and has been monitored for many years and has led to a large amount of insight into the physical behaviour of tangential discontinuities. In this chapter we will, similar to the section on the bow shock, discuss only a few of those properties, just the ones which can be described by simple collisionless one-fluid theory. As described in Section 5.1, the fully ionized and magnetized solar wind plasma cannot mix with the terrestrial magnetic flux tubes. Instead, it will deviate from its original direction and will, by its dynamical pressure, compress the terrestrial field and confine it into a small region of space, the magnetosphere. During this interaction the *magnetopause* evolves as a narrow boundary layer between the solar wind and the geomagnetic field which is confined inside the magnetosphere. This layer is a discontinuity which must be different from the bow shock because the plasma flow behind the bow shock is already sub-magnetosonic (or sub-Alfvénic). To first order, i.e. in the framework of ideal single fluid theory the magnetopause can be regarded as a tangential discontinuity, again the only persistently available to investigation tangential discontinuity in the Universe.

8.6.1 The Shape of the Magnetopause

As a tangential discontinuity the magnetopause is a surface of total pressure equilibrium between the solar wind-magnetosheath plasma and the geomagnetic field confined in the magnetosphere. The weakness of the solar wind magnetic field allows, in a first approximation, to neglect the contribution of the interplanetary field to the pressure balance. In addition, since the main energy of the solar wind flow is stored in the bulk flow of the ions and not in the thermal pressure, it is sufficient to take into account only the solar wind dynamic ram pressure

$$p_{dyn} = n_{sw} m_i v_{sw}^2 \tag{8.85}$$

This equation is valid for ideal specular reflection of the oncoming solar wind particles at the magnetopause boundary as shown in Fig. 8.14. The dynamic pressure exerted on the terrestrial field is proportional to the number density and the total change in energy of the particles during their turn-around. The latter is twice the dynamic solar wind ion energy, since the tiny contribution of the electrons can be neglected. If the particles are not really specularly reflected, one must include an efficiency coefficient

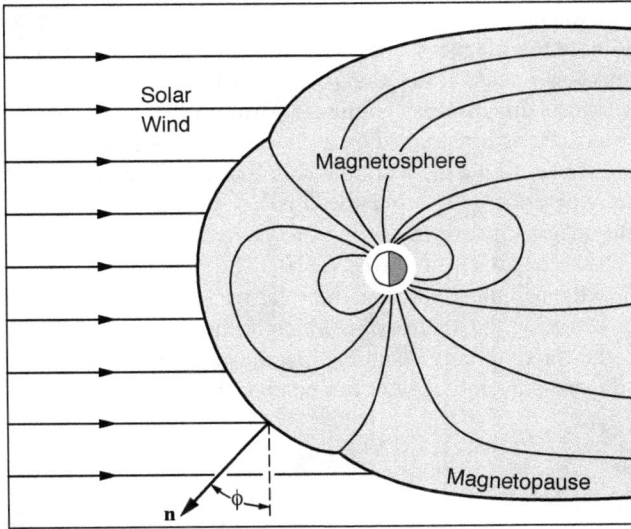

Figure 8.14: The global geometry of the Earth's magnetopause.

κ on the right-hand side of the above expression. On the other hand, inside the magnetosphere the plasma thermal and dynamic pressures can be neglected when compared with the pressure of the geomagnetic field. With these simplifications the tangential discontinuity condition (8.40) becomes

$$2\mu_0 \kappa n_{sw} m_i (\mathbf{n} \cdot \mathbf{v}_{sw})^2 = (\mathbf{n} \times \mathbf{B})^2 \tag{8.86}$$

Here $\mathbf{n}(r, \theta, \phi)$ is the outer normal to the magnetopause surface. It depends on all three spatial directions, because the magnetopause is a complicated curved surface. The left-hand side of Eq. (8.86) selects the normal solar wind velocity component as the only relevant component for the interaction, while the right-hand side takes into account that the magnetospheric field has no component perpendicular to the magnetopause.

Denoting the function describing the magnetopause surface in spherical coordinates

$$S_{mp}(r, \theta, \varphi) = 0 \tag{8.87}$$

the outer normal to the magnetopause (cf. Fig. 8.14) can be expressed as the normalized negative gradient vector of the surface function

$$\mathbf{n}(r, \theta, \varphi) = -\frac{\nabla S_{mp}(r, \theta, \varphi)}{|\nabla S_{mp}(r, \theta, \varphi)|} \tag{8.88}$$

Inserting into Eq. (8.86) yields

$$2\mu_0 \kappa n_{sw} m_i \left(\frac{\nabla S_{mp}}{|\nabla S_{mp}|} \cdot \mathbf{v}_{sw} \right)^2 = \left(\frac{\nabla S_{mp}}{|\nabla S_{mp}|} \times \mathbf{B} \right)^2 \tag{8.89}$$

The above equation contains the complicated global structure of the magnetospheric magnetic field near the magnetopause. It also contains the three-dimensional derivatives of the unknown surface function, S_{mp}, and their second powers. Hence, it is a second-order three-dimensional nonlinear partial differential equation for S_{mp} the solution of which can be constructed by numerical methods only.

One can, however, find a most simple solution for the distance of the nose of the magnetopause, where the solar wind speed reduces to zero, the so-called *stagnation point*. Here the magnetopause is assumed to be symmetrical in the angular coordinates so that all the angular derivatives vanish. The magnetospheric field is assumed perpendicular to the ecliptic plane, and the solar wind velocity is assumed to be in the ecliptic plane. Denoting the *stand-off distance* of the magnetopause from the Earth's center by R_{mp} and for simplicity assuming that the magnetospheric field is dipolar (see Section 3.1), the pressure equilibrium can be expressed as

$$n_{sw}m_i v_{sw}^2 = \frac{KB_E^2}{2\mu_0 R_{mp}^6} \tag{8.90}$$

Here B_E is the magnetic field at the surface of Earth, and the constant K accounts for both κ and the deviation of the magnetic field from its dipolar value at R_{mp}. Rewriting this equation yields the stand-off distance in Earth radii of the solar wind stagnation point at the nose of the magnetopause from the centre of Earth

$$\boxed{R_{mp} = \left(\frac{KB_E^2}{2\mu_0 n_{sw}m_i v_{sw}^2} \right)^{\frac{1}{6}}} \tag{8.91}$$

as the sixth root of the ratio of the magnetic dipole energy at the Earth's surface to the dynamic solar wind energy density. Taking $n_{sw} = 5$ cm^{-3}, $v_{sw} = 400$ km/s, $B_E = 3.1 \times 10^4$ nT, and assuming $K = 2$, one finds $R_{mp} = 9.9\,R_E$. Since the stand-off distance changes as the sixth root of the values involved, it is not very sensitive to variations in the solar wind dynamic pressure. Under quiet magnetospheric conditions the magnetospheric nose (or solar wind stagnation point) lies indeed at $\sim 10\,R_E$ geocentric distance.

Equation (8.91) is independent of the specific terrestrial situation. It is valid for any dipolar magnetic field which interacts with a weakly magnetized plasma stream. It can therefore be applied to any other magnetosphere, ranging from the magnetospheres of the planets to magnetospheres of stars and pulsars interacting with stellar winds or interplanetary gases. Table 8.1 collects the theoretical stagnation point distances for the magnetized planets in our solar system.

A similar conclusion as for the nose distance of the magnetopause can be drawn for the distance of the magnetopause at its flanks. At the flanks the solar wind flow is tangential to the magnetopause, $v_n = 0$, and the ram pressure of the solar wind vanishes. In pressure equilibrium between the non-magnetized solar wind and the dipolar magnetospheric field the so far neglected thermal pressure, $p_{sw} = \gamma n_{sw} k_B T_{sw}$, comes into play at this point. In addition, however, one has to include the magnetic pressure

Planet	Mercury	Earth	Jupiter	Saturn	Uranus	Neptune
R_{mp}/R_P	1.4	10	75	20	20	25

Table 8.1: Stagnation point distances (in planetary radii R_P) of the magnetised planets. These distances are determined by the surface magnetic field strengths and the solar wind dynamic pressure at planetary orbit. The latter decays with density as r^{-2} because of the expansion of the solar wind.

of the interplanetary magnetic field $p_{IMF} = B_{IMF}^2/2\mu_0$ which is approximately of the same order as the thermal pressure. Thus with the total pressure $p = p_{sw} + p_{IMF}$ one obtains

$$R_{mpf} = \left(\frac{KB_E^2}{2\mu_0 p_{IMF}(1 + \beta_{sw})} \right)^{\frac{1}{6}} = \left(\frac{n_{sw}m_i v_{sw}^2}{p_{IMF}(1 + \beta_{sw})} \right)^{\frac{1}{6}} R_{mp} \qquad (8.92)$$

for the theoretical maximum distance of the magnetopause flanks in units of Earth radii from the Sun-Earth line, where $\beta_{sw} = 2\mu_0 \gamma n_{sw} k_B T_{sw}/B_{IMF}^2$ is the solar wind plasma beta. For a solar wind temperature of about 1.3×10^5 K, $\gamma = 5/3$, and the values used above at 1 AU, $n_{sw} = 5$ cm^{-3}, $B_{IMF} = 10$ nT, this distance becomes about $R_{mpf} \approx 1.6 R_{mp}$, roughly one and a half times the distance of the subsolar point. Observations of the shape of the magnetopause have shown that in the average the magnetopause at the dawn-dusk meridian is found at about a distance of 14 R_E to 15 R_E, slightly less than its theoretical distance.

In addition the observations show that the dawn and dusk magnetopauses still experience a non-vanishing normal flow velocity component, $v_n \neq 0$ caused by different mixing effects. In other words, the magnetosphere at dawn and dusk is still inflating and the radius of the magnetosphere still increases, when going from the dayside through the dawn-dusk meridian to the nightside magnetosphere. Only much farther downstream the tail of the magnetospheric boundary becomes approximately parallel to the flow. The reason for such a behaviour can be found in the global magnetospheric current system.

Equation (8.89) has been solved numerically to obtain the shape of the magnetopause. Surprisingly, in the meridional plane there is no continuous solution connecting the dayside magnetopause to the nightside magnetopause. Figure 8.15 shows the calculated magnetopause cross-section in the equatorial and meridional planes. While in the equatorial plane the magnetopause is a smooth curve extending from the dayside into an open tail, at high latitudes the tangent to the magnetopause is discontinuous at one location. This point has been identified as the *polar cusp* and arises from the special geometry of the dipolar geomagnetic field. From this point onward the magnetic field lines are turned around to the tail as a consequence of their interaction with the solar wind. However, the field lines do not experience any discontinuity at the polar

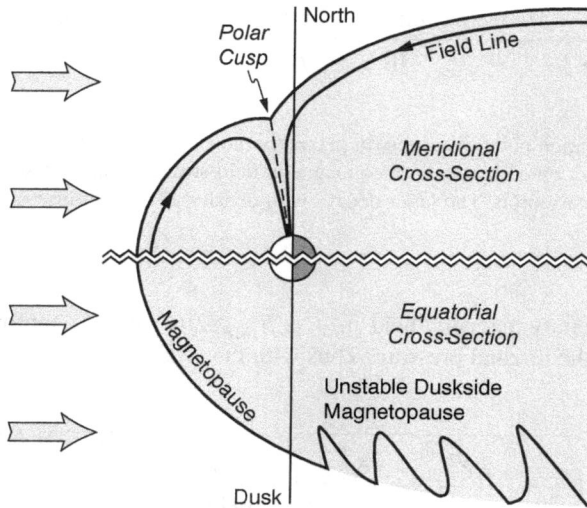

Figure 8.15: Magnetopause cross-sections in the main meridional (upper part) and equatorial ecliptic (lower part) solar-ecliptic coordinate planes. The meridional cross-section exhibits the closed dayside magnetosphere plus the (northern) polar cusp also located on the dayside at high latitudes with the further inflating magnetosphere in solar wind-downstream direction. The equatorial cross-section shows the smooth average inflation of the magnetosphere towards evening and night while also indicating the vulnerability of the dusk side magnetosphere to instability of its boundary causing large-scale wavy Kelvin-Helmholtz tongue structures and plasma mixing. The dawn side equatorial magnetopause (not shown) does not exhibit such waves but is smooth.

cusp, but simply change their topology from dayside-like to tail-like as indicated by the field line included in the figure.

The lower part of the figure shows the dusk side of the equatorial cross section of the magnetosphere. The shape of the equatorial magnetopause obtained numerically from Eq. (8.89) is completely symmetric for both the dawn and dusk sides being of the form of the shaded undisturbed magnetopause shown in the figure. It still opens to the tail becoming tangential to the flow only farther downstream. The wavelike patterns of the actual magnetopause drawn indicate an interesting effect which occurs almost only on the dusk side of the magnetospheric flanks. Here the magnetopause becomes unstable and evolves into such intrusions where solar wind magnetosheath plasma mixes with magnetospheric plasma and the two magnetic fields also interconnect. This interesting effect is not contained in the theory of the magnetopause shape. It can, however, be described in an extension of fluid theory by taking into account the magnetosheath and magnetospheric flows which cause instability (named Kelvin-Helmholtz instability after its discoverers in fluids Lord Kelvin and Hermann von Helmholtz) of the magnetopause and lead it to exhibit a temporary and spatially variable structure of the kind shown in the figure.

8.6.2 The Magnetopause Current System

Separating the solar wind from the magnetospheric magnetic field and being a surface across which the magnetic field strength jumps from its low interplanetary value to the high magnetospheric field strength respectively turns from solar wind-magnetosheath to magnetospheric direction thus causing $\nabla \times \mathbf{B} \neq 0$ across the magnetopause, represents a surface current layer. The origin of this current can be understood from the left-hand side of Fig. 8.16.

Specularly reflected ions and electrons hitting the magnetospheric field inside the magnetopause boundary will perform half a gyro-orbit inside the magnetic field before escaping with reversed normal velocity from the magnetopause back into the magnetosheath. The thickness of the solar wind-magnetosphere transition layer under such idealized conditions becomes of the order of the ion gyroradius, $r_{gi} = v_{sw}/\omega_{gi}$. Electrons also perform half gyro-orbits, but with much smaller gyroradii. The sense of gyration inside the boundary is opposite for both kinds of particles leading to the generation of a narrow surface current layer. This current provides the additional magnetic field, which compresses the magnetospheric field into the magnetosphere and at the same time annihilates its external part. It is a *diamagnetic current* caused by the perpendicular density gradient at the magnetopause.

The current density inside the magnetopause can be estimated from the jump in the magnetic field across the magnetopause and the average thickness of the magnetopause to about 10^{-6} Am^{-2}. The total the current flowing in the magnetopause is of the order of 10^7 A. This magnetopause current forms two quasi-circular current systems on the two hemispheric magnetopauses. However, not all the currents can topologically close in themselves. In the equatorial plane the magnetopause currents being amplified due to the superposition of their northern and southern hemispheric parts which thus naively would form a magnetopause jet current stream surrounding the magnetosphere along the magnetopause. This stream does, however, not exist. The current instead crosses the magnetospheric tail from dawn to dusk, as shown schematically on the right in Fig. 8.16 (see also Fig. 1.6) becoming the neutral sheet current. This neutral sheet current splits into northern and southern parts on both flanks of the magnetopause and creates the magnetospheric lobes. In the equatorial plane it weakens the magnetic field at the tail magnetopause thereby allowing the solar wind convection-electric field to penetrate across the magnetopause into the neutral sheet and across the magnetosphere. It is this partially un-screened electric dawn-to-dusk electric field which drives the neutral sheet current and accelerates the ions from dawn to dusk and the electrons from dusk to dawn across the tail.

8.6.3 Magnetopause Reconstruction

So far we have considered all the discontinuities to behave like plane one-dimensional narrow sheets with a invariable normal along the sheet separating two plasmas of different properties. When allowing for a variation in the direction tangential to the sheet, the normal will not point anymore in the same direction, and the definition of

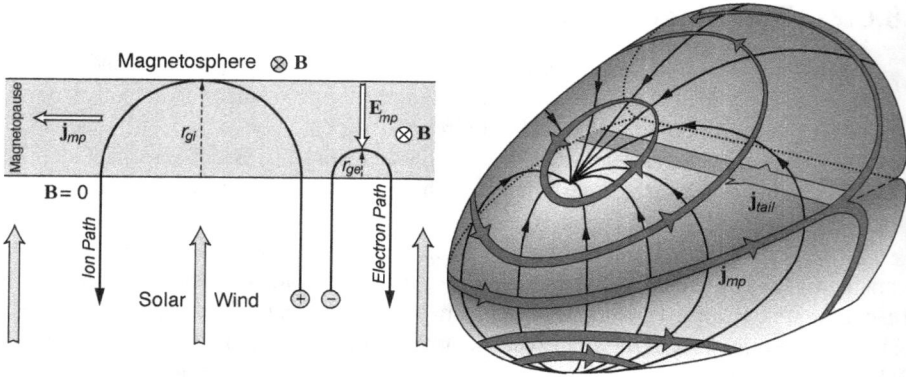

Figure 8.16: Magnetopause current system. *Left:* The mechanism of generation of the diamagnetic magnetopause current system by specular reflection of solar wind particles from the magnetospheric magnetic field when impinging. The magnetic field 'selects' the particle type by the sign of charge and mass. Ions penetrate deeper than electron thereby giving rise to an internal to the magnetopause outward charge-separation electric field component E_{mp}. Ions turn left, while electrons turn right thereby creating a net tangential internal diamagnetic magnetopause current component J_{mp} flowing to the left in the ion direction. *Right:* The closed three-dimensional geometry of magnetopause current system consists of the diamagnetic magnetopause currents closed by the neutral sheet current in the deep tail across the magnetosphere. This current is necessary for closure reasons. It is caused by the solar wind convection electric field potential difference across the tail from dawn to dusk.

jump conditions in the simple way as done in this chapter runs into difficulties. The jumps become non-local in this case.

On the other hand, observations of the magnetopause indicate that the magnetopause undergoes oscillations in one or the other direction. Such oscillations let the magnetopause normal vary. If they are caused by temporal effects one can attribute them to surface waves running along the magnetopause as for instance shown in Fig. 8.15. A different class of magnetopause variations are stationary structures interior to the magnetopause. In such a case a spacecraft passes across such structures and monitors the variation in the magnetic field and other plasma quantities. For the particular case of the magnetopause being a tangential discontinuity in such a case one can reconstruct the structure of the transition from the magnetosheath to the magnetosphere along the cross-section of the magnetopause which has been traversed by the spacecraft. This is made possible by the simple pressure balance condition at the magnetopause which for the tangential discontinuity is written

$$\nabla p = \mathbf{j} \times \mathbf{B} \tag{8.93}$$

Still assuming a two-dimensional state where the variation along the magnetopause boundary has a preferred direction which does not change during the crossing of the spacecraft, two components of the magnetic field can be expressed as the curl of a vector potential $\mathbf{A}(x,y)$ which has only one component $A(x,y)$ such that the magnetic field becomes

$$\mathbf{B} = [\partial A/\partial y, -\partial A/\partial x, B_z(x,y)] \tag{8.94}$$

Now, defining a partial total pressure which includes the thermal pressure and the remaining component B_z as

$$P(A) = p(A) + B_z^2(A)/2\mu_0 \tag{8.95}$$

where in the two-dimensional case p and B_z depend only through A on the coordinates, the above pressure equilibrium condition can be written as

$$\frac{\partial^2 A}{\partial x^2} + \frac{\partial^2 A}{\partial y^2} = -\mu_0 \frac{dP(A)}{dA} = -\mu_0 j_z(A) \tag{8.96}$$

This equation, which is an inhomogeneous Laplace equation for A, is the so-called Grad-Shafranov equation. With given right-hand side, i.e. given current or total pressure profile, which can be measured along the spacecraft orbit across the magnetopause, this equation can be solved numerically in the environment of the spacecraft orbit such that the interior structure of the tangential discontinuity in stationary pressure equilibrium can be reconstructed along the spacecraft orbit by integration. In this way it becomes sometimes possible to infer the internal structure of the magnetic field across the magnetopause as a tangential discontinuity as shown in Fig. 8.17.

8.6.4 Magnetosheath Flow

Knowing the three-dimensional shape of the magnetopause and bow shock, one is in the position to calculate the properties of the flow of the compressed and heated plasma in the magnetosheath which surrounds the magnetosphere in the downstream direction. This can be done in several degrees of sophistication. The simplest one is to entirely neglect the contribution and effect of the also shock-compressed turbulent magnetic field in the magnetosheath, solving the jump conditions across the bow shock and calculating the flow in the magnetosheath by assuming ideal hydrodynamic conditions and the condition of completely tangential flow. Though this is not fully realistic, it for a simple gasdynamic shock simplifies the Rankine-Hugoniot conditions considerably. In such a case the magnetic field in the magnetosheath behaves passively, does not affect the location of the shock and, in particular, allows for a simple estimate of the approximate shock distance from the blunt magnetospheric body, for such a shock satisfies the extremely simple condition

$$\boxed{R_{bs} = \left(1 + 1.1 \frac{n_{sw}}{n_{bs}}\right) R_{mp}} \tag{8.97}$$

where n_{bs} is the magnetosheath density adjacent to the shock ramp. From gasdynamic shock theory it follows that this density is at its maximum about $n_{bs} \approx 4 n_{sw}$, yielding a distance of about $R_{bs} \approx 1.3 R_{mp}$, both in good agreement with observations.

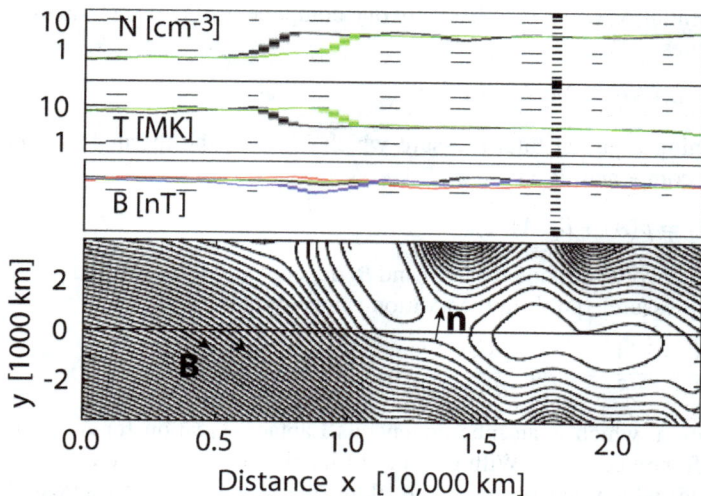

Figure 8.17: Reconstruction of the interior magnetopause structure based on the Grad-Shafranov equation and 3 Cluster spacecraft observations during a magnetopause crossing when the magnetopause was believed to be a tangential discontinuity. *Top*: The three panels show density, temperature and magnetic field observed along the straight horizontal line (spacecraft path) in the lower panel across the magnetopause seen as the kink in the density and temperature and oscillation in the magnetic field. *Bottom*: The reconstructed field line structure around one spacecraft path inside the magnetopause. Note the magnetic islands. The direction of the local normal is shown in one place. The two arrows labeled by the magnetic field show the directions of the observed magnetic field at two points along the spacecraft path which has been used as input into the reconstruction.

The results of such gasdynamic calculations depend on the polytropic index of the solar wind plasma and on its Mach number. As displayed in Fig. 8.18, for weak magnetic fields (in these calculations, which are simple gasdynamical, any magnetic fields have actually been completely neglected calculating the flow in the prescribed magnetospheric, magnetopause and bow shock geometries which have been taken from the boundary conditions as described before) the magnetosphere bends the fluid flow lines into an azimuthal direction, with the flow lines closer together at the magnetopause. At a certain distance from the subsolar point the nozzle effect of the magnetosheath causes the flow to again make the transition from subsonic to supersonic flow. The isodensity contours indicate compression of the magnetosheath plasma in a region close to the stagnation point at the nose of the magnetosphere. Outside this region the plasma is still compressed, but gradually becomes more dilute toward the flanks of the magnetopause. The temperature behind the shock is enhanced showing the generation of entropy in the course of the solar wind shocking process. Close to the stagnation point this enhancement is more than a factor of 20 and is still significant near the flanks of the magnetopause where the flow has cooled adiabatically.

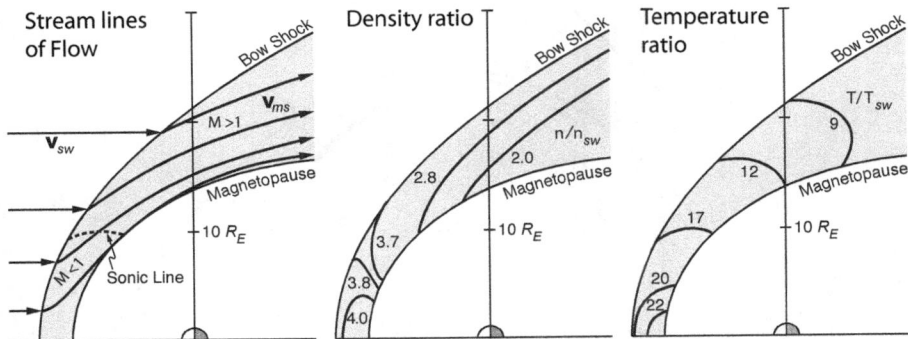

Figure 8.18: Dayside ecliptic equatorial magnetosheath flow stream lines and normalised density and temperature iso-contours for a solar wind Mach (sonic) number $M_s = 8$. The normalisation is to the unmagnetised solar wind density n_{sw} and temperature T_{sw}. One should note that above a certain latitude outside the stagnation point the magnetosheath flow accelerates again to become supersonic as the effect of the off-stagnation point expansion of the magnetosheath plasma.

Neglecting the magnetic field can be justified only if it is so small that it merely reacts passively to the interaction of the flow with the magnetosphere. If this is the case, which is approximately true only for $\beta \gg 1$, the magnetic field is simply convected along the magnetosheath flow in a manner that it stays tangential to the magnetopause and satisfies the ideal magnetohydrodynamic conditions

$$\nabla \cdot \mathbf{B}_{ms} = 0$$
$$\nabla \times (\mathbf{v}_{ms} \times \mathbf{B}_{ms}) = 0 \tag{8.98}$$

The flow then drapes the solar wind interplanetary magnetic frozen-in field lines around the magnetopause (see Fig. 8.19) and at the same time transports it downstream to the nightside. There is some compression of the field in the magnetosheath similar to the closer positioning of the streamlines of the flow (see Fig. 8.18), but there is no reaction of the field on the flow in this extremely simplified model.

Interestingly, for a more parallel direction of the interplanetary magnetic field to the flow the draping occurs only on that side of the magnetosphere where the bow shock is quasi-perpendicular. Since the bow shock is a fast shock the magnetic field lines in the magnetosheath are refracted away from the shock normal. Behind the quasi-parallel part of the bow shock this refraction pulls the field lines away from the stagnation point thereby generating a region of lower magnetic field strength in the magnetosheath between the bow shock and the magnetopause. Usually this region is found on the early morning side of the magnetosheath.

The magnetic field draping has two effects which have been neglected earlier. Firstly, due to the compression of the field related to the draping the magnetic field pressure increases. Ultimately this effect will lead to a breakdown of the gasdynamic model. The enhanced magnetic field pressure inside a compressed magnetosheath

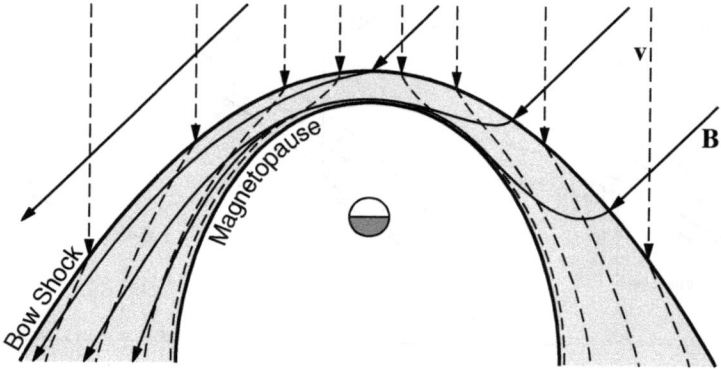

Figure 8.19: Magnetic field draping around the magnetosphere in the magnetosheath as shown in the ecliptic equatorial plane. Flow lines are drawn as dashed lines here. The direction of draping depends on the relative relation between flow and solar wind magnetic field. Since the flow direction does not change much, it is the direction of the interplanetary solar wind magnetic field which determines the kind of draping.

flux tube near the stagnation point will squeeze the magnetosheath plasma out of this tube into the flank-side magnetosheath. This effect has been observed and was called *plasma depletion*. It effectively dilutes the inner magnetosheath plasma near the nose below its theoretical gasdynamic density by pressure balance when the magnetic pressure is included and squeezes the plasma out of the compressed flux tubes in the direction along the field, the only direction where it can escape without violating the frozen-in condition and particle adiabatic invariants. This plasma depletion effect is mainly observed when the magnetic fields in the magnetosheath and magnetosphere are nearly parallel to each other because then the pressure adds up when the magnetosphere appears as an impenetrable and rigid object for the flow which can just be compressed a little more until reaching pressure equilibrium. Otherwise, when the fields are inclined or even anti-parallel a new effect sets on. This is reconnection or field line merging which we have already discussed at an earlier place in this volume.

8.6.5   Reconnection at the Magnetopause

Plasma depletion by enhanced pressure near the magnetopause stagnation point is an effect which could be understood by a simple logical extension of the fluid theory though it is not contained in the equations for the discontinuities but belongs to the physics of the transition layer between the bow shock and the magnetopause, the magnetosheath. Another effect which goes substantially beyond the simple jump conditions is magnetic merging or reconnection. Its importance has already been sketched in Chapter 5 where the reader was referred to the schematic of field line transport by merging of a southward interplanetary and a northward magnetospheric magnetic field line shown in Fig. 5.5. It was also stressed there that merging/reconnection causes plasma mixing. Clearly, none of these effects is covered by the simple pressure balance

condition imposed by the interpretation of the magnetopause as a tangential discontinuity. This pressure balance enabled us to determine the average shape of the magnetopause in three dimensions. It failed in describing the instability of the evening-side
magnetopause, however, and it fails completely when one attempts to understand or
describe reconnection at the magnetopause. The reason is that pressure balance does
not allow for the magnetic field — and thus also the plasma flow — to penetrate
the magnetopause, independent of the direction of the magnetosheath magnetic field
and the angle between the interplanetary and the magnetospheric field. The diamagnetic currents produced by the pressure balance condition exclude the field from the
interior of the magnetopause boundary if it is assumed to behave like a tangential
discontinuity.

Whatsoever the attraction of such a simple model is, however, its validity must
be questioned by the reality of observation. These observations have shown unambiguously that thermal plasma can indeed penetrate across the magnetopause deep
into the inner magnetosphere. This is possible only under two conditions: either the
interaction between the magnetopause and the magnetosphere includes some finite
and substantial viscosity, or the interplanetary and magnetospheric fields merge at the
magnetopause building a magnetic connection which bridges the gap between the two
plasmas, the distorted solar wind in the magnetosheath and the magnetosphere. Otherwise magnetosheath plasma will be unable to enter the magnetosphere across the
rigid magnetopause boundary. In the sixties of the past century a model of the magnetosphere had been proposed which took advantage of the assumption of viscous
interaction. Generation of viscosity is a slow and not very efficient process, however,
as it requires either breaking the adiabatic invariance of the particles or the inclusion of
finite-Larmor radius effects as have been noted earlier in this book. Though the latter
cannot be excluded, their importance has been found to be rather limited causing this
model of a closed teardrop magnetosphere to be abandoned readily after the proposal
of the reconnection model of the magnetosphere shown in Fig. 5.5. But, while this
reconnection model nicely covers all the observed effects and is highly appreciated,
justifying it in the framework of ideal plasma fluid theory readily runs into severe
difficulties.

The problem lies in the fact that the shape of the magnetopause is well described
by the tangential discontinuity while for merging, reconnection and understanding
of the magnetospheric convection rotational discontinuities have to be called for. In
the strict sense of the word both are, however, incompatible. Above we have given
an argument that in the frame of fluid theory the magnetopause might be understood
as the particular case of an intermediate shock which obeys both, pressure balance
and thus the correct description of the global average shape of the magnetopause,
and finite normal magnetic fields and flows. Still, as we have noted above, problems
remain with the energy budget and entropy generation. Thus the problem of describing reconnection in the fluid picture has not been settled yet and principally cannot be
solved without switching to a kinetic approach to solar wind-magnetosphere interaction. Reconnection should in such an approach come out as the dominant plasma and
magnetic transport process. However, the scales of the global interaction which lead

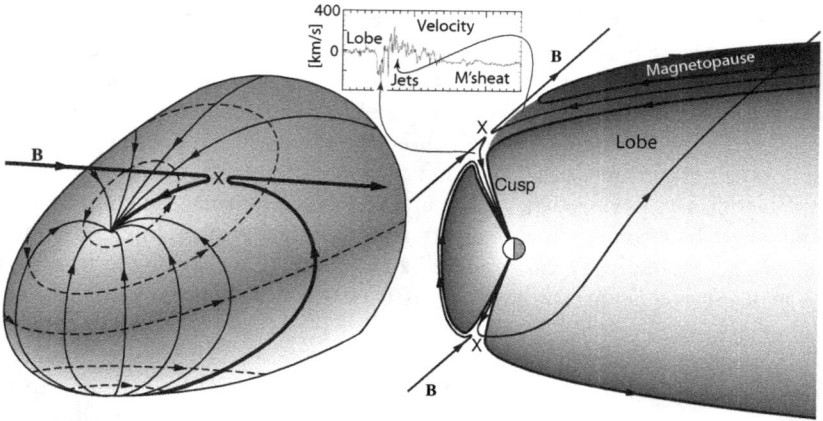

Figure 8.20: High-latitude merging for ecliptic and cusp merging for non-southward interplanetary fields. For the latter case, the inflation of the magnetopause in the northern lobe with displacement of the magnetopause and its compression on the dayside are indicated. Observations of bulk velocities by Cluster 1 during a passage in 2003 have been included. The two reconnection jets generated in the northern cusp by cusp reconnection can be recognised in the velocity data from the switch from negative to positive velocities when passing from south to the north of the cusp reconnection X point. In the southern cusp the magnetosphere will deflate in this case becoming strongly asymmetric in the reconnection process. This is indicated by the long reconnected southern lobe field line which is draped by the solar wind flow over the magnetosphere. High latitude merging will occur only when, in addition, the magnetosphere inflates such that a sufficiently strong normal flow component exists which pushed the two antiparallel fields against each other.

to the formation of the magnetopause and the microscopic scale of reconnection are so enormously separate that it is impossible to treat both processes on the same level. For this reason reconnection itself is always considered separately and only its global consequences are referred to when treating the global macroscopic behaviour of the magnetospheric plasma.

Globally speaking, dayside reconnection occurs when the compressed solar wind magnetic field in magnetosheath possesses a southward component, i.e. a component which is antiparallel to the magnetospheric magnetic field at the dayside magnetopause. As introduced in Sections 5.1 and 5.2, in such a case *reconnection* or *merging* sets on at the magnetopause between the contacting antiparallel magnetosheath and magnetospheric magnetic field lines by some not completely understood local microscopic mechanism. The three-dimensional topology of the magnetopause almost always, i.e. for any interplanetary magnetic field direction, offers some limited area on its surface where the interplanetary magnetic field evolves a component antiparallel to the magnetospheric field. In many such cases reconnection will be going on then if only the flow direction is favourable for reconnection, i.e. if the flow has a finite velocity component normal to the magnetopause in this region which is sufficiently

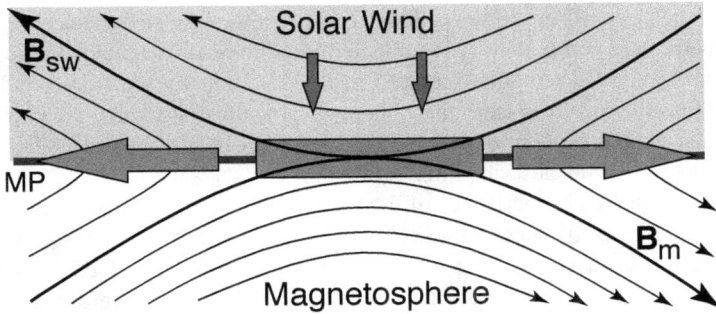

Figure 8.21: The X-point configuration at the dayside magnetopause during magnetopause reconnection. The solar wind flowing in from above (short downward arrows) with its southward (pointed to the left) magnetic field $\mathbf{B}_{sw}$ encounters the magnetopause MP at the short distance of the dark box where it merges with the stronger magnetospheric magnetic field $\mathbf{B}_m$ and subsequently convects to both sides of the X-point along the magnetopause. The merging of the fields causes the incoming solar wind plasma to symmetrically jet out (large arrows) to north (right) and south (left) of the reconnection X-point site. Not shown is the inflowing solar wind plasma which can now stream in from the solar wind into the magnetosphere along the connected magnetic field lines on both sides having access to the magnetosphere through the gate which has been opened by reconnection in the tiny reconnection box. One should note that this inflow is possible all along the magnetopause. The separatrix field lines have been highlighted.

large to push the solar wind magnetic field against the magnetopause. Two such cases are shown in Fig. 8.20. The different types of reconnection may happen simultaneously at different places on the magnetopause causing loosing its global character of a tangential discontinuity and becoming a surface, which in many places has become perforated by reconnection and magnetically connected to interplanetary space.

The by far most important case is, however, when reconnection occurs at the dayside near the nose of the magnetopause. This is the canonical case of the southward solar wind magnetic field in merging with the northward magnetospheric field which we referred to in Fig. 5.5. For northward interplanetary field reconnection occurs preferably at the polar sides of the northern and southern polar cusps as is obvious from the right side of Fig. 8.20. In all these cases anti-parallel magnetic field collide and merge in an X-point or along an X-line (see Fig. 5.3). In Chapter 12 we will describe the small-scale physics of reconnection in more detail.

The global effect of reconnection sketched in Fig. 5.5 implies that magnetosheath field lines are glued to the magnetospheric field lines which are convectively transported tailward by the solar wind flow. At the reconnection site the nature and topology of the magnetopause change fundamentally. While the magnetopause still maintains to be a surface of total pressure equilibrium including kinetic and magnetic pressures, it looses the property of a tangential discontinuity at the X-line by evolving a non-vanishing normal magnetic component, $B_n \neq 0$, and it keeps this property all over the long path the magnetospheric filed line describes during the tailward advection. In such a case the normal flux of matter is also non-zero, $F_{\text{II}} \neq 0$. Plasma from the magnetosheath can all the time get free access to the magnetosphere along the normal magnetic field component being injected into the outer magnetospheric region as long as the field lines remain connected in the transport by the solar wind over the polar cap into the tail. It is clear that this process is most attractive for plasma injection into the magnetosphere during magnetospheric disturbances like substorms and magnetospheric storms described in Section 5.6.

The macro-scale consequences of the micro-scale reconnection process turn thus out to be enormous. Without any need for further generation of viscous, diffusive or resistive transport coefficients all over the vast global size of the magnetopause surface, the merging of field flux tubes in a narrow microscopic region on the dayside or polar cusp magnetopause suffices for the restoration of the outer magnetospheric plasma population. Magnetic reconnection at the magnetopause thus is a paradigm for the huge global consequences of a microscopic physical process that takes place in a miniscule region of space somewhere on the huge surface of the magnetopause.

This is shown in Fig. 8.21 which picks out the tiny region of reconnection on the magnetopause. Though this figure visualises the main macroscopic features of reconnection, showing the connection of the antiparallel solar wind and magnetospheric field lines at the magnetopause, the deflection of the solar wind into to symmetric oppositely directed reconnection jets and the connected field lines which are convected by the flow to both sides along the magnetopause and suggests that plasma from the solar wind can now freely flow in along these connected field lines, it does not give any explanation for what happens at the X point. It can, presumably, be understood only referring to nonlinear kinetic theory and with the help of numerical simulations. This will be the subject of Chapter 12.

8.7 Concluding Remarks

The general Rankine-Hugoniot conditions obtained in this chapter are valid in a plasma as long as the typical scales of interest are much larger than the width of the discontinuity respectively the discontinuous transition region. In such a case one divides the plasma into the large-scale region outside the discontinuity, where the jump conditions hold, and into its interior, where the complicated dissipative processes take place.

There may, however, exist cases when the dissipative processes inside the discontinuity affect the behaviour of the plasma outside the transition region so strongly that even part of the outside region cannot be considered as ideal. This is generally the case for shocks where the character of the upstream region depends on the Mach number and on the magnetic field direction with respect to the shock normal. Strictly speaking, the shock foot and precursor wave region of quasi-perpendicular shocks, and the extended foreshock of quasi-parallel shocks are such disturbed zones which, in principle, must be added to the shock transition. Also the entire downstream turbulence is a property of the shock and should thus be considered as part of the shock transition. If this is all taken into account, then the Rankine-Hugoniot conditions can be applied to outside the upstream and downstream boundaries of the shock transition in both ideal media neglecting all processes inside the transition itself.

It is clear that in the majority of cases the shock transition is by no means narrow and may have become of the scale of the processes of interest. When this happens, as for example at the bow shock where the downstream disturbed region fills the entire space between the bow shock and the dayside magnetopause, then the Rankine-Hugoniot conditions cease to be applicable and one is forced to treat the entire transition not as a discontinuity but by the more precise multi-fluid or even by kinetic theory. This will be the subject of Chapter 13.

Similar arguments do also apply to the magnetopause which at first glance is a narrow tangential transition from the turbulent magnetosheath to the magnetosphere while in reality causing the extended plasma depletion region, the reconnection X point with its two separatrices and the reconnection jets to both of its sides, and allows for the plasma inflow from the magnetosheath and plasma outflow from the magnetosphere along the interconnected magnetic field lines. All these processes act to widen the magnetopause transition until the scales become large.

Still, one can determine the global properties of both bow shock and magnetopause, their locations and global shapes, from the jumps, but the real shock and magnetopause physics and its consequences are barely described by simple fluid theory. In the next chapters we progress to a more precise plasma theory by considering wave propagation in fluid and kinetic approximations. These next steps are still approximative only, accounting for linear disturbances in plasma. Chapter 11 provides the linear theory for wave growth from available free energy, i.e. the theory of plasma instability.

Plasmas are, however, highly nonlinear media allowing for a wealth of interactions through self-generated fields the effects of which on the original settings, particle dynamics and initial fields, may become the dominant effects which reform and completely determine the intermediate and final plasma states. Cases where this happens are the processes of reconnection and the formation of structure at the bow shock, both paradigmatic and of vital importance in space physics. The related nonlinear effects will not be considered in detail except for the brief overviews given on kinetic reconnection and shock physics in Chapters 12 and 13.

Further Reading

The classical treatment of the ideal magnetohydrodynamic jump conditions can be found in [2]. A thorough description of the equations of state is given in the tutorial article [3]. A comprehensive tutorial of the gasdynamic theory of the flow around the magnetosphere has been developed in [5]. Figure 8.18 is based on calculations presented in that publication. A lot of useful information about magnetic reconnection can be found in [1], and the physics of the magnetopause is exhaustively treated in [4]. Finally, a good introduction on bow shock physics is given in [6].

References

[1] E. W. Hones, Jr. (ed.), *Magnetic Reconnection in Space and Laboratory Plasmas* (American Geophysical Union, Washington, 1984).

[2] L. D. Landau and E. M. Lifshitz, *Electrodynamics of Continuous Media* (Pergamon Press, Oxford, 1975).

[3] G. L. Siscoe, in *Solar-Terrestrial Physics*, eds. R. L. Carovillano and J. M. Forbes (D. Reidel Publ. Co., Dordrecht, 1983), p. 11.

[4] B. U. Ö Sonnerup, M. Thomson, and P. Song (eds.), *Physics of the Magnetopause* (American Geophysical Union, Washington, 1995).

[5] J. R. Spreiter, A. Y. Alskne, and A. L. Summers, in *Physics of the Magnetosphere*, eds. R. L. Carovillano, J. F. McClay, and H. R. Radoski (D. Reidel Publ. Co., Dordrecht, 1968), p. 301.

[6] R. G. Stone and B. T. Tsurutani (eds.), *Collisionless Shocks in the Heliosphere: A Tutorial Review* (American Geophysical Union, Washington, 1985).

Problems

Problem 8.1 *Confirm the three expressions for the Mach numbers, sonic, Alfvénic, magneto-sonic. Think about their definitions: either as velocity or energy ratios. Which one is more fundamental and thus more general?*

Problem 8.2 *Explain why the interplanetary magnetic field at large distances from the sun becomes close to azimuthal? Why does it decay like r^{-1} while the density decays like r^{-2}? Explain both laws.*

Problem 8.3 *If you put a slice along the Earth-Sun line far out in space, what kind of curve would the centre of the interplanetary current sheet describe along this line?*

Problem 8.4 *Derive the set of Rankine-Hugoniot equations for the jumps. Invest some effort to bring it into the factorised form given in this chapter and confirm the three families of solutions.*

Problem 8.5 *Prove the coplanarity theorem for shocks.*

Problem 8.6 *Estimate the perpendicular shock cross shock charge separation potential for a Mach number $M = 8$ solar wind flow (normal solar wind temperature, density and magnetic field) consisting of protons and electrons. Calculate the cross shock electric field. What is its dependence on the hock normal angle? How strong is the electron current in the shock transition and what is the overshoot magnetic field caused by it?*

Problem 8.7 *The magnetopause is a diamagnetic current sheet. What about the bow shock?*

Problem 8.8 *Write the equation which determines the shape of the magnetopause surface in spherical coordinates and confirm that it is a nonlinear partial differential equation. Hint: Normalise the radius to R_E. Rewrite the formal expression of a surface $S = 0$ in spherical coordinates and determine the normal vector.*

Problem 8.9 *Is the diamagnetic effect present in a homogeneous and extended plasma? Would anyone feel it here? Why does it occur at the magnetopause?*

Problem 8.10 *Estimate the charge separation field inside the magnetopause for a field difference between the magnetosheath and the magnetosphere of $\Delta B = 30$ nT, quasi-neutral sheath density of $n_i = n_e = 20$ cm^{-3}, ion velocity of $v_i = 250$ km/s and electron temperature $k_B T_e = 30$ eV. Explain the result.*

Problem 8.11 *Why do the magnetopause currents outside the ecliptic form closed curves around the cusps and not the poles?*

Problem 8.12 *Estimate the cross magnetosphere electric potential difference in the neutral sheet which accelerates the neutral sheet closure current of the magnetopause current system. Which direction does it point? Assume a solar wind flow of 500 km/s, a solar wind magnetic field of 10 nT, and an ecliptic equatorial tail diameter of 30 R_E. What should the solar wind ion energy become when it enters the magnetospheric neutral sheet on one side and leaves it on the other after becoming accelerated in the electric potential difference across the tail? Compare it with the real energy the ions achieve in the neutral sheet as was given in an earlier chapter. What do you conclude about the efficiency of electric field penetration? What should be observed adjacent to the magnetopause on the side where the neutral sheet accelerated ions leave the magnetosphere and merge into the solar wind?*

Problem 8.13 *Which stagnation point distance would the magnetopause assume if the thermal pressure would confine the magnetospheric field instead of the kinetic pressure? What is the effect of an isotropic solar wind thermal pressure on the magnetopause shape if one would not neglect it?*

Problem 8.14 *We have mentioned that the ecliptic equatorial magnetopause exhibits a strong asymmetry between dawn and dusk being smooth at dawn and strongly 'tongued' (in addition to varying with time) at dusk. What reason could you propose for this difference if you are asked to explain this observation? Which other asymmetries come to your mind in this connection? Hint: Think of the magnetospheric drift paths in Figs. 3.8 and 5.7!*

Problem 8.15 *Prove the formula for the maximum flank magnetopause distance from the Sun-Earth line.*

Problem 8.16 *Estimate the strength of the dayside magnetopause current assuming a magnetic field twice as strong as the geomagnetic dipole field at the magnetopause, i.e. maximum equilibrium compression field.*

— 9 —

Waves in Plasma Fluids

In a plasma there are many reasons for the evolution of time-dependent effects. The high temperatures required to produce a plasma imply that the plasma particles are in fast motion. Such motions generate microscopic charge separations and currents and therefore temporally changing electric and magnetic fields. Hence, it is quite natural to expect that electric and magnetic fluctuations are typical for a plasma, even in its stationary state. Absolutely quiescent plasmas do not exist. Just due to the thermal motion of the particles in a plasma every plasma in equilibrium contains a certain level of fluctuations, which depends entirely on the temperature of the plasma and is therefore called *thermal fluctuation level*. The thermal spectrum of a plasma can be calculated as the balance between the generation of the thermal fluctuations and the reabsorption and dissipation of these fluctuations, but calculations of this kind require quantum theoretical methods which are outside the scope of this book.

In addition to these unavoidable fluctuations, any plasma will react to a violent distortion of its state imposed by outer means. All such disturbances may be thought of as a superposition of linear waves onto the quiescent plasma state which propagate across the plasma in order to transport the energy of the distortion and to communicate it to the entire plasma volume. Such plasma waves have been measured in many different frequency ranges. Figure 9.1 indicates that their frequencies may be as low as several Millihertz and as high as several tens of Kilohertz. Conventionally, this range is subdivided into ultra-low (ULF), extremely-low (ELF) and very-low frequency (VLF) waves.

But plasma waves are not generated at random. In order to exist, any disturbance must satisfy at least two conditions. First, it must be a solution of the appropriate equations of the plasma. Therefore the number of modes propagating in the plasma will not be continuous but discrete. Secondly, we can speak of a wave only if its amplitude exceeds the level of the thermal fluctuations always present in a plasma. The second condition sets a limit on the initial disturbance causing the waves. If it has an amplitude lower than the thermal noise level and if no mechanism acts to amplify the disturbance in the plasma, this disturbance does not affect the plasma and there is no wave.

In the present and the following chapter we investigate the consequences of the first condition, i.e., we consider the discrete modes which can propagate in a plasma. Since there are several different plasma models available, the number and properties of the wave modes depends on the chosen approximation to the kinetic plasma model. The present chapter investigates wave propagation in plasma fluids. In neglecting the second condition we automatically assume that the thermal noise level is much smaller

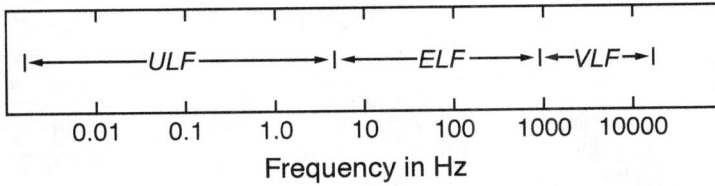

Figure 9.1: Ranges of ultra-low, extremely-low, and very-low frequency waves.

than the wave amplitude. Hence, the plasma is assumed to be sufficiently cold. On the other hand, we will not deal with nonlinear effects in this chapter. Therefore the wave amplitudes are assumed to be small enough to allow any disturbance to be represented as superposition of plane waves.

With these remarks in mind, we can represent any wave disturbance, $\mathbf{A}(\mathbf{x}, t)$, in the plasma by plane waves, i.e., by its Fourier components. If the disturbance itself is a plane wave, it consists only of one Fourier component

$$\mathbf{A}(\mathbf{x}, t) = \mathbf{A}(\mathbf{k}, \omega) \exp(i\mathbf{k} \cdot \mathbf{x} - i\omega t) \tag{9.1}$$

where the amplitude, $\mathbf{A}(\mathbf{k}, \omega)$, is a function of the wave vector, $\mathbf{k}$, and the frequency, ω. This representation allows to define the phase and the group velocity of the wave

$$\mathbf{v}_{ph} = \omega \mathbf{k}/k^2 \tag{9.2}$$
$$\mathbf{v}_{gr} = \partial \omega / \partial \mathbf{k} \tag{9.3}$$

The phase velocity is always parallel to the wave vector, $\mathbf{k}$, and shows the direction of wave propagation. The group velocity may deviate from this direction and describes the speed and direction of the energy flow in the wave.

9.1 Waves in Unmagnetized Fluids

As a first example and to introduce the concept of plasma waves let us consider an unmagnetized plasma consisting of equal numbers of electrons and ions. Two kinds of waves can propagate in such a plasma. The first kind are electromagnetic waves similar to waves in vacuum. Due to the presence of charges which respond to the electric and magnetic field of the waves, the properties of these waves will be modified. The second kind of waves are internal plasma oscillations, which do not exist in the vacuum, but are a specific property of the plasma. We will treat the second type of waves first.

9.1.1 Langmuir Oscillations

Consider a plasma where the ions are fixed while the electrons may undergo small translations relative to the ions. Such an assumption is reasonable if the timescale of the electron translation is so short that the ions cannot follow the electron motion

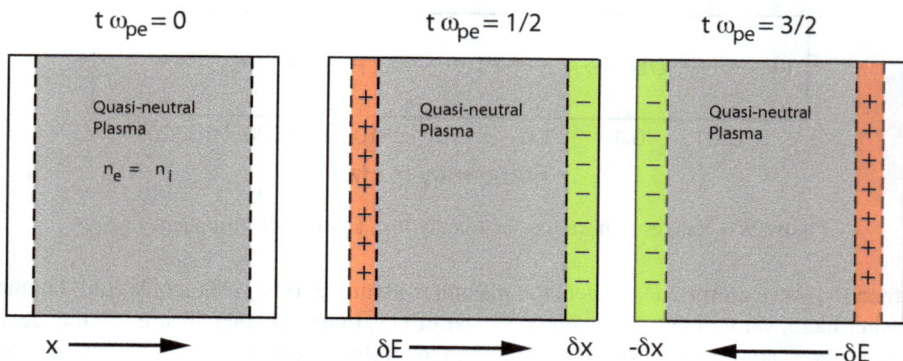

Figure 9.2: Oscillation of a column of electrons at the plasma frequency. The ions are immobile. Displacement of the electrons to the right by the small distance δx causes a negative charge layer on the right, a positive charge layer on the left with a positive electric field δE pointing in x-direction. Elastic restoration by the electric force $-e\delta E$ lets the electrons oscillate to the left exceeding the equilibrium position now by a distance $-\delta x$. Thus the process repeats itself but now in the opposite direction. In the absence of friction this causes an undamped oscillation of electron plasma frequency ω_{pe} around quasi-neutral equilibrium.

because of their large inertia. In other words we consider high-frequency electron oscillations in which the ions do not participate.

Now take a column of electrons and displace this column with respect to the ions by a short distance, δx, in the x direction (see Fig. 9.2). Such a displacement causes an electric field, δE, also pointing in x direction and exerting a force, $-e\delta E$, on each electron which tries to pull the electron back to its mother ion in order to preserve quasineutrality. For the whole column of density n_e this means that the time variation of the density distortion, δn, will be given by the electron fluid continuity equation

$$\frac{\partial \delta n}{\partial t} = -n_e \frac{\partial \delta v_{e,x}}{\partial x} \tag{9.4}$$

as the spatial derivative of the electron velocity disturbance, $\delta v_{e,x}$. The distortion of the velocity is found from the electron momentum conservation as

$$\frac{\partial \delta v_{e,x}}{\partial t} = -\frac{e}{m_e} \delta E \tag{9.5}$$

and the electric field caused by all the displaced electrons satisfies Poisson's law

$$\frac{\partial \delta E}{\partial x} = -\frac{e}{\varepsilon_0} \delta n \tag{9.6}$$

It is now simple to derive an equation for the disturbance of the density. Take the time derivative of the first of the above equations, replace the time derivative of the velocity disturbance in the resulting expression by the second of the above equation, and eliminate the spatial derivative with the help of the third equation. The result is

$$\frac{\partial^2 \delta n}{\partial t^2} + \frac{n_e e^2}{m_e \varepsilon_0} \delta n = 0 \tag{9.7}$$

This is a linear equation for the variation of the density, which has the form of a linear oscillator equation. Clearly, the coefficient of the second term must have the dimension of an inverse time squared. This time is proportional to the characteristic period of the oscillation of the electron column around the equilibrium position of the ion column. The solution of the above equation is found by taking $\delta n \propto \exp(-i\omega t)$, where $\omega = \omega_{pe}$ is the angular frequency of the oscillation

$$\boxed{\omega_{pe}^2 = \frac{n_e e^2}{m_e \varepsilon_0}} \tag{9.8}$$

Hence, the electrons will perform an oscillation around the position of the ions with the electron *plasma frequency*, ω_{pe}, already given in Section 1.1.

9.1.2 Langmuir Waves

The plasma oscillation is somewhat artificial, since the electrons are not at rest but have different velocities and will react differently to the attempt to displace them from their instantaneous positions. To account for this effect one must include the adiabatic variation of the electron thermal pressure, $\delta p_e = \gamma_e k_B T_e \delta n_e$, into the electron momentum conservation equation. Let us, for simplicity, assume that the electron temperature is constant (which for fast adiabatic changes is not completely correct). Then the linearized equation of motion of the displaced electron fluid

$$\frac{\partial \delta v_{e,x}}{\partial t} = -\frac{e}{m_e} \delta E - \frac{\gamma_e k_B T_e}{m_e n_e} \frac{\partial \delta n}{\partial x} \tag{9.9}$$

replaces Eq. (9.5). Eliminating once more δE and $\delta v_{e,x}$ now yields another more precise equation for the variation of density

$$\frac{\partial^2 \delta n}{\partial t^2} - \frac{\gamma_e k_B T_e}{m_e} \frac{\partial^2 \delta n}{\partial x^2} + \omega_{pe}^2 \delta n = 0 \tag{9.10}$$

This equation differs from the former one in the appearance of the second partial derivative with respect to x. Therefore it is of the form of a wave equation and can be solved by introducing the plane wave ansatz for the variation of the electron density, $\delta n \propto \exp(-i\omega t + ikx)$, into Eq. (9.10), which yields a relation between the angular frequency, ω, and the wavenumber, k

$$\boxed{\omega_l^2 = \omega_{pe}^2 + k^2 \gamma_e v_{the}^2} \tag{9.11}$$

where in the one-dimensional case the adiabatic index is $\gamma_e = 3$ and we used the electron thermal velocity, $v_{the} = (k_B T_e / m_e)^{1/2}$ defined in Eq. (6.64). This is the *Langmuir wave dispersion relation*. It determines the dependence of the frequency of the Langmuir waves on wavenumber.

The interesting point is that the thermal motion of the electrons leads to a dispersion of the electron plasma oscillations by introducing a wavenumber dependence into the wave frequency. This dependence drops out only if the electrons have zero temperature or for zero wavenumber, $k = 0$. In both cases one recovers the plasma oscillations. However, for finite temperatures or $k \neq 0$ the oscillations start propagating across the plasma and turn into travelling electrostatic waves which are oscillations of the electric potential field propagating at finite phase velocity ω_l/k through the plasma.

The limit of vanishing wavenumber is of particular interest. Because k is inversely related to the wavelength, $\lambda = 2\pi/k$, the wavenumber becomes zero for infinitely long waves. Langmuir oscillations are thus Langmuir waves of very long wavelength. Here 'very long' means very long compared with some electron scale. This scale can easily be identified as the Debye length λ_D when rewriting the Langmuir wave dispersion relation as the electron plasma frequency multiplied by a thermal correction. It can then be given two versions:

$$\omega_l^2 = \omega_{pe}^2 \left(1 + k^2 \gamma_e \frac{v_{the}^2}{\omega_{pe}^2} \right) = \omega_{pe}^2 \left(1 + \gamma_e k^2 \lambda_D^2 \right) \tag{9.12}$$

of which the second contains the ratio of the Debye to Langmuir wavelengths. The thermal correction is thus identified as becoming important only at wavelengths of the order of λ_D. For sufficiently longer λ one expands the parentheses and obtains the Langmuir-wave dispersion relation in the conventional form

$$\omega_l \approx \omega_{pe} \left(1 + \frac{3}{2} k^2 \lambda_D^2 \right) \tag{9.13}$$

Langmuir waves are one of the fundamental electrostatic waves in plasmas. Figure 9.3 shows two observations of Langmuir waves in space, on the left a measurement *in situ* the solar wind, on the right an observation in the near-Earth auroral magnetosphere at an altitude of a few 1000 km. In looking at the time scales one realises that both observations are very different even though belonging to the same type of waves, i.e. high frequency oscillations of the electric potential field. The observations in the solar wind cut out one hour out of a persistent Langmuir wave band at frequency of about $f_l \approx 5$ kHz showing an emission of very variable bandwidth and electric field amplitude oscillating between ± 0.1 mV/m and ± 1 mV/m thus being a weak wave. The auroral zone observations are at frequency $f_l \approx 350$ kHz cutting out just 1 ms of Langmuir oscillations of very fast large amplitude variation. The wave field here reaches ± 1500 mV/m up to a factor of 10^4 stronger than in the solar wind. Since, by the above formulae, the frequency maps the plasma density, the observation of such waves serves to a most precise determination of the density and density variations.

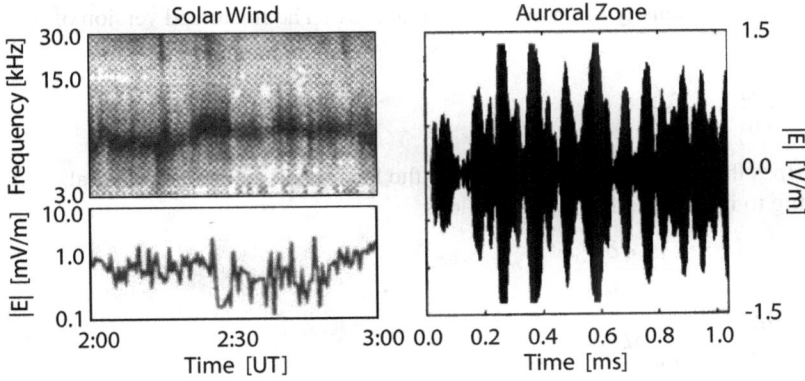

Figure 9.3: Two different observations of Langmuir waves in space. *Left*: Continuous Langmuir wave chains in the solar wind over times much longer than 1 hour. The central frequency of the wave band varies between 5 kHz and 10 kHz. *Right*: Millisecond scale Langmuir wave packets in the auroral zone near Earth where the central frequency is ~ 350 kHz.

9.1.3 Ion-Acoustic Waves

The other fundamental plasma wave is the in-acoustic wave. So far we completely neglected the contribution of the ions and have considered just the very-high frequency electron oscillations. At lower frequencies ion inertia cannot be neglected anymore, and it becomes necessary to take into account the ion equation of motion, in addition to the electron equations. On the other hand, in a first approach one can now safely neglect the electron inertia against that of the ions because the ion plasma frequency

$$\omega_{pi} = \left(\frac{n_i Z^2 e^2}{m_i \varepsilon_0} \right)^{1/2} \tag{9.14}$$

is, for protons with $Z = 1$ and quasineutrality, $n_i \approx n_e$, by a factor of $(m_e/m_i)^{1/2} = 43$ smaller than ω_{pe}. At such low frequencies the electrons react almost without any inertia to the change in the electric field. Electron dynamics then reduces to a simple balance between electron pressure and the electric force (note that $\partial n_0/\partial x = 0$)

$$e\delta E = -\gamma_e k_B T_e \frac{\partial \ln n_e}{\partial x} \tag{9.15}$$

where $n_e = n_0 + \delta n_e$. When introducing the electric potential, $\delta E = -\partial \delta \phi/\partial x$, the above equation reduces to a Boltzmann-like dependence

$$n_e = n_0 \exp \left(\frac{e\delta \phi}{\gamma_e k_B T_e} \right) \tag{9.16}$$

of the electron density n_e on electric potential $\delta\phi$. The linearized version of this equation

$$\frac{\delta n_e}{n_0} = \frac{e\delta\phi}{\gamma_e k_B T_e}$$

(9.17)

describes the linear electron response to the low-frequency wave potential oscillation. Adding to it the linearized ion equations

$$\frac{\partial \delta n_i}{\partial t} = -n_i \frac{\partial \delta v_{i,x}}{\partial x}$$

$$\frac{\partial \delta v_{i,x}}{\partial t} = \frac{e}{m_i}\delta E$$

(9.18)

where we have neglected the ion pressure term, supposing that the ions are much colder than the electrons, and assumed charge neutrality, $\delta n_e = \delta n_i = \delta n$, one arrives at

$$\frac{\partial^2 \delta n}{\partial t^2} - \frac{\gamma_e k_B T_e}{m_i}\frac{\partial^2 \delta n}{\partial x^2} = 0$$

(9.19)

as the ionic equivalent of Eq. (9.10). For plane waves its solution yields

$$\omega_{ia}^2 = \frac{\gamma_e k_B T_e}{m_i}k^2$$

(9.20)

which is the dispersion relation of *ion-acoustic waves*. These waves are called acoustic, since they have the same properties as sound waves in a gaseous medium. Both waves have linear dispersion, $\omega \propto k$, and are pure density fluctuations.

Dividing both sides of Eq. (9.20) by k^2, one finds the phase velocity of ion-acoustic waves as $v_{ph,ia} = c_{ia}$, with the *ion-acoustic speed*

$$c_{ia} = \left(\frac{\gamma_e k_B T_e}{m_i}\right)^{1/2}$$

(9.21)

The latter is given by the square root of the ratio of electron temperature and ion mass and, for protons, is a factor of 43 smaller than the electron thermal velocity. The linear dispersion of ion-acoustic waves implies also that their group velocity is equal to the phase velocity, $v_{gr,ia} = c_{ia}$.

In the above derivation of ion-acoustic waves we have neglected the contribution of ion pressure. Correcting for this imprecision requires the replacement of $\gamma_e T_e$ in the above expressions by the sum of the electron and ion contributions, $\gamma_e T_e + \gamma_i T_i$. Hence, for high ion temperatures the ion sound speed becomes the ion thermal velocity, and the contribution of the electrons to sound waves is lost.

The second approximation used above was the assumption of quasineutrality even for the fluctuating quantities. At higher frequencies close to ω_{pi} this assumption is

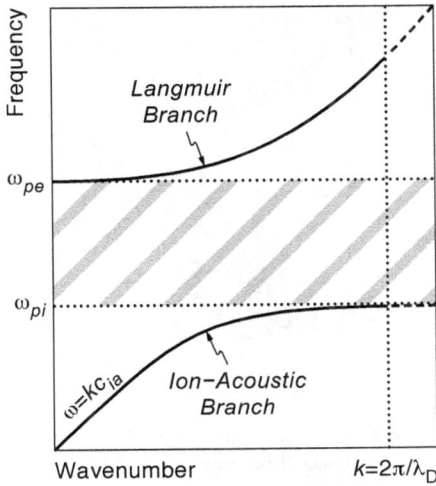

Figure 9.4: The Langmuir and ion-acoustic wave branches of the electrostatic dispersion relation in a non-magnetised plasma showing the cut-off of Langmuir waves at the electron plasma frequency ω_{pe} and of ion-acoustic waves at the ion plasma frequency ω_{pi}. The two branches are separated by a broad band where none of these waves exist. This separation if also seen in observations like those of Fig. 9.5.

incorrect because the electron and ion motions in the wave field become uncorrelated. We therefore replace the condition $\delta n_e = \delta n_i$ with Poisson's equation

$$\frac{\partial^2 \delta \phi}{\partial x^2} = \frac{e n_0}{\varepsilon_0} \left(\frac{\delta n_e}{n_0} - \frac{\delta n_i}{n_0} \right) \tag{9.22}$$

where we assumed quasineutrality of the undisturbed state, $n_e = n_i = n_0$. The electron and ion equations (9.17) and (9.18) can now be used to manipulate Poisson's equation (9.22) into the following form

$$\left(\frac{\partial^2}{\partial t^2} - c_{ia}^2 \frac{\partial^2}{\partial x^2} \right) \delta \phi = \frac{c_{ia}^2}{\omega_{pi}^2} \frac{\partial^4 \delta \phi}{\partial t^2 \partial x^2} \tag{9.23}$$

This is again a linear equation for the electric potential of the wave, $\delta \phi$, and one can apply the plane wave ansatz to obtain the more precise dispersion relation

$$\omega_{ia}^2 = \frac{k^2 c_{ia}^2}{1 + k^2 c_{ia}^2/\omega_{pi}^2} = \frac{k^2 c_{ia}^2}{1 + \gamma_e k^2 \lambda_D^2} \tag{9.24}$$

This expression shows that ω becomes a linear function of k only for long wavelengths or small k. Here the wave has the character of a sound wave. But for short wavelengths comparable to the Debye length introduced in Eq. (1.3) the character of

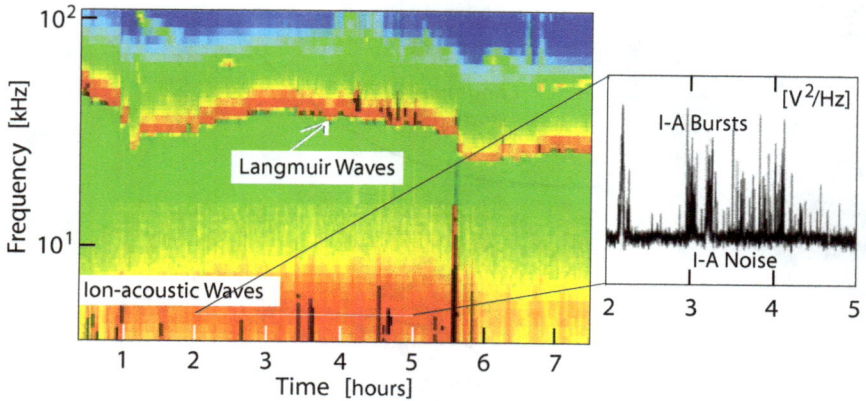

Figure 9.5: Ion-acoustic waves in the solar wind. *Left*: Dynamical spectrogram between 3 kHz and 100 kHz for 7 hours observation time with the wind spacecraft in the solar wind. The intense emission line at high frequencies is the Langmuir mode oscillation in the solar wind. The broadband noise at frequencies below 10 kHz is the ion acoustic wave band in the solar wind. Both Langmuir and ion acoustic waves are rather variable. *Right*: Three hours of wave power flux in the electric field amplitude selected at 5 kHz (along the white line in the left part of the figure) showing that ion acoustic waves are basically noise but irregularly consist of spiky intense bursts.

the sound wave is destroyed. The wave frequency becomes about constant and for very short wavelengths approaches the ion plasma frequency, ω_{pi}, where phase and group velocities both vanish.

Figure 9.4 shows the schematic behavior of the dispersion curves of the two electrostatic waves that exist in an unmagnetized plasma. At high frequencies the Langmuir branch starts at the electron plasma frequency. At low frequencies the ion-acoustic branch starts at zero frequency and approaches the ion plasma frequency. Between the two plasma frequencies, ω_{pi} and ω_{pe}, no electrostatic wave mode can propagate in an unmagnetized plasma. This is nicely seen from observations of the Wind spacecraft in the solar wind shown in Fig. 9.5. The broad band of intense emissions below 10 kHz in this figure on the left are ion acoustic waves. Between them there opens a broad gap until the Langmuir wave band starts above roughly 30 kHz. The right part of the figure shows the electric wave spectral power flux density taken at the white horizontal line on the left over three hour time. The wave power is seen to be quite steady at a relatively low level indicating the presence of continuous electric noise in the solar wind. However, from time to time this noise intensifies briefly to become very spiky bursts of ion acoustic waves. We will in a later chapter discuss some reasons for this kind of intensification (see also our companion volume *Advanced Space Plasma Physics*).

Our claim that Langmuir waves occupy a narrow band while ion-acoustic waves are broad-band is of course to take as a relative statement. It refers to the central frequency of the waves. Langmuir waves are very high frequency waves just above ω_{pe} with bandwidth given by the thermal correction term $\Delta\omega \sim \frac{1}{2}\gamma_e k^2 v_{the}^2 \ll \omega_{pe}$.

Ion-coustic waves are at much lower frequency, have linear dispersion for long wave length and bandwidth roughly $\Delta\omega \sim \omega_{pi}$. In this sense the ion-acoustic waves are broad-band waves of frequency $\omega < \omega_{pi}$ less than their bandwidth. In fact, inspection of Fig. 9.5 suggests that they have even larger bandwidth and are weaker at the lowest frequencies respectively longest wavelengths. The reasons for such a behaviour are manifold. The weakness of the waves at the lowest frequencies is partially an instrumental effect resulting from the restriction of the instrumentation to detect very slow electric oscillation; the much broader extension to high frequencies lies in the spiky nature of the ion acoustic waves which, when taking the Fourier transform of the wave generates spectra which reach far above the theoretical cut-off of the waves. Always observed unexpectedly broad spectra indicate that the waves are spiky.

9.1.4 Debye Length

To our surprise we have encountered an old acquaintance when discussing the dispersion relations for ion sound waves, the Debye length. That it determines the properties of the fundamental electrostatic waves suggests that it arises from thermal charge separation effects at short wavelengths. Since we have now accumulated sufficient knowledge about the dynamics of particles in a plasma we will give a derivation of this quantity.

Assume that a heavy, motionless ion is immersed into the quasineutral plasma. This ion will cause an electric charge separation field to arise in its vicinity that will attract electrons to charge-neutralize the ion. Because the electrons are highly mobile, they will be accelerated toward the ion, pass around the ion, and subsequently escape into the ambient plasma, but in the average there will be more electrons near the ion than outside at large distances. This poses the question of up to what distance the electron density will be slightly distorted due to the presence of the ion.

In the region where the density is distorted, charge neutrality becomes violated and a non-vanishing electric potential, $\phi(r)$, will arise which must satisfy Poisson's equation for an electron-proton plasma

$$\nabla^2\phi = -\frac{e}{\varepsilon_0}(n_i - n_e) \tag{9.25}$$

The ion density is the quasineutral density of the ambient plasma, $n_i = n_0$, while the electron density includes the distortion by the presence of the test ion. For an equilibrium between electron thermal motion and electric force the electrons are Maxwellian and their density in obeys Boltzmann's law

$$n_e(r) = n_0 \exp\left[\frac{e\phi(r)}{k_B T_e}\right] \tag{9.26}$$

For weak potentials, $|e\phi| \ll k_B T_e$, this expression can be expanded into a Taylor series and inserted into Poisson's equation to obtain

$$\nabla^2\phi = \frac{e^2 n_0 \phi}{\varepsilon_0 k_B T_e} \tag{9.27}$$

The problem will be spherically symmetric with radius r centered on the position of the ion, and the potential, $\phi(r)$, must diverge as $1/r$ for $r \rightarrow 0$. Dimensionally the left-hand side of the last equation is equal to the electrostatic potential divided by a characteristic length squared, ϕ/λ_D^2. Comparing the two sides of the equation yields for this length

$$\lambda_D = \left(\frac{\varepsilon_0 k_B T_e}{n_0 e^2} \right)^{1/2}$$

(9.28)

which is the *Debye length* postulated in Section 1.1. The Debye length is thus the typical average screening distance of the electrostatic field of an ion charge in a quasineutral plasma by electrons of temperature T_e. It indicates that around each ion there is a cloud of excess electrons which screen the ion field from the plasma. The sphere of radius λ_D around the ion is called the *Debye sphere*, and the number of electrons within this sphere, roughly $\frac{4}{3}\pi n_0 \lambda_D^3$, is the *Debye number*, or approximately the plasma parameter. Inside the Debye sphere the potential of the ion is not screened. As a consequence, on distances λ_D in a plasma quasineutrality is distorted. Over these distances the electrostatic field experiences relatively strong fluctuations, which are the reason for the change in the dispersive properties of the waves at wavelengths comparable to the Debye length. Such short wavelength waves contain contributions from the unscreened electric fluctuations caused by the electron motions across the Debye sphere around each ion in the plasma.

The electron motions in the Debye sphere are thermal and do always exist. In thermal equilibrium between electrons and electrostatic fluctuations, the energy of the fluctuations is equal to the mean electron energy, $k_B T_e$, and the energy density of the fluctuations is this energy divided by the volume of the Debye sphere. Hence, the energy density contained in the thermal fluctuation is equivalent to the thermal energy of one electron per Debye sphere

$$W_{tf} \approx \frac{k_B T_e}{\lambda_D^3}$$

(9.29)

We have so far neglected the screening of electron charges by the ions. If one takes into account this often minor effect, the effective Debye length becomes

$$\lambda_{D,\mathrm{eff}}^{-2} = \lambda_D^{-2} + \lambda_{Di}^{-2}$$

(9.30)

where

$$\lambda_{Di} = \left(\frac{\varepsilon_0 k_B T_i}{n_0 e^2} \right)^{1/2}$$

(9.31)

is the ion Debye length. In an isothermal plasma with similar electron and ion temperatures both Debye lengths contribute equally to the effective Debye length.

The Debye length can be written as the ratio of electron thermal velocity to electron plasma frequency, $\lambda_D = v_{the}/\omega_{pe}$. As we have shown, this allows to write the dispersion relation of Langmuir waves in the form of Eq. (9.12) repeated here:

$$\omega_l^2 = \omega_{pe}^2 \left(1 + \gamma_e k^2 \lambda_D^2\right) \qquad (9.32)$$

9.1.5 Ordinary Electromagnetic Waves

The appearance of purely electrostatic disturbances which propagate as wave modes is a very particular property of plasmas, which is connected with the presence of free charges in a plasma. Moving charges can, however, contribute also to oscillating plasma currents which should become sources of electromagnetic waves. A large number of such electromagnetic wave modes may propagate in a magnetized plasma. In this introductory section, where we deal with an unmagnetized plasma, we will only consider the most familiar electromagnetic mode, the free-space electromagnetic wave.

An electromagnetic wave of frequency ω in the plasma will set the electrons into motion to generate a linear electron current

$$\delta \mathbf{j}_{em} = -e n_0 \delta \mathbf{v}_e \qquad (9.33)$$

Only the disturbance of the electron velocity contributes to the current, since the plasma was initially at rest. This disturbance can be calculated from the electron equation of motion in the electromagnetic plane wave field, $\delta \mathbf{E}$

$$\delta \mathbf{v}_e = -\frac{ie}{\omega m_e} \delta \mathbf{E} \qquad (9.34)$$

Inserting this into the above expression for the current, one finds that the induced current $\delta \mathbf{j}_{em} = \sigma_{em} \delta \mathbf{E}$ is proportional to the wave electric field in direct analogy to Ohm's law. The constant of proportionality is the electromagnetic wave conductivity

$$\sigma_{em} = \frac{i \varepsilon_0 \omega_{pe}^2}{\omega} \qquad (9.35)$$

It depends on the wave frequency and on the electron plasma frequency and, for real frequency ω, is a purely imaginary quantity (like the inverse inductive resistance). It is zero for very-high frequencies and vanishes in the absence of a plasma. In both cases the electromagnetic wave will become an ordinary electromagnetic free-space wave. The dispersion relation of a free-space electromagnetic wave is

$$N^2 = \frac{k^2 c^2}{\omega^2} \qquad (9.36)$$

where N is the *index of refraction*. In a vacuum $N^2 = 1$. In an unmagnetized plasma, one may replace it by the dielectric function $\in(\omega, \mathbf{k})$ of the plasma to obtain the dispersion relation of the electromagnetic wave as

$$\frac{k^2 c^2}{\omega^2} = \in(\omega, \mathbf{k}) \tag{9.37}$$

We will show in the next section that there is a unique relation between $\in(\omega, \mathbf{k})$ and the wave conductivity, $\sigma(\omega, \mathbf{k})$, which in our special case can be written as

$$\in(\omega) = 1 + \frac{i\sigma_{em}(\omega)}{\varepsilon_0 \omega} = 1 - \frac{\omega_{pe}^2}{\omega^2} \tag{9.38}$$

With the help of this expression, the dispersion relation of the free-space electromagnetic wave in the presence of a plasma becomes

$$\boxed{\omega_{om}^2 = \omega_{pe}^2 + c^2 k^2} \tag{9.39}$$

The wave described by this dispersion relation is called the *ordinary mode*, because it has the same dispersion as the free-space wave when the plasma frequency is discarded, i.e. when there is no plasma but vacuum. The main difference to the free-space wave is that for frequencies below the electron plasma frequency there is no real solution for the wavenumber, and the wave ceases to exist. In other words, the wavenumber of the ordinary mode vanishes at the plasma frequency, ω_{pe}, which is a *cut-off* for the ordinary mode. Physically spoken, the electromagnetic wave coming from the outside is unable to penetrate into the plasma deeper than up to the level where the wave frequency matches the local plasma frequency. At this point its wavenumber vanishes.

A cut-off is a point in the dispersion diagram as well as in real space, where the wavenumber turns zero and, hence, the direction of propagation of the wave reverses. Cut-offs are *wave reflection* points which are related simply to the refractive properties of the plasma and wave propagation across the plasma. The practical implication of the ordinary mode cut-off is that in an unmagnetized plasma ordinary electromagnetic waves cannot propagate below ω_{pe}.

This reflection has important practical consequences. It is responsible for the trapping of long wavelength radio waves with frequencies below the plasma frequency of the dense ionosphere between the Earth's surface and the ionosphere. Moreover, it is responsible for the screening of low-frequency radio waves generated in the topside auroral ionosphere from reaching the ground. This radiation, the *auroral kilometric radiation*, and the radio emission from the radiation belts, the *trapped radiation*, has frequencies well below the ionospheric plasma frequency and is reflected from the topside ionosphere so that it becomes unobservable from the ground while cosmic high-frequency radio waves, for instance solar radio bursts, can propagate without any serious attenuation down to the Earth's surface. On the other hand, launching low-frequency radio waves from the Earth into the ionosphere and recording the reflected wave provides a tool for sounding the ionospheric density profile. This technique is used in ionosondes. Figure 9.6 summarizes the reflection of radio waves at the ionosphere.

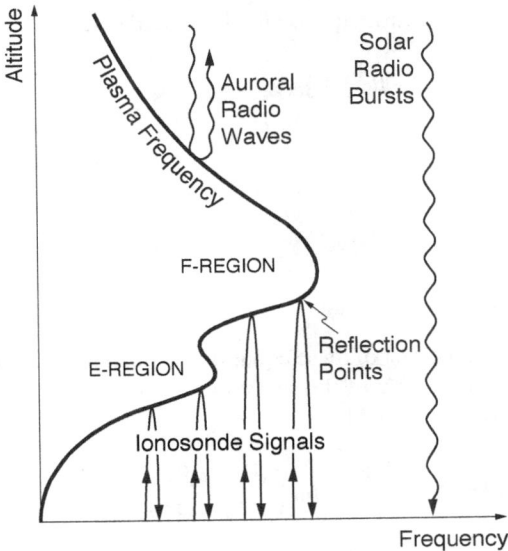

Figure 9.6: Ionospheric reflection of radio waves from ground and space. Ionosonde electromagnetic waves are reflected from the different ionospheric layers where their frequencies match the local plasma frequency. The ionosonde technique of measuring the altitude of the layers is based on sweeping an electromagnetic wave in frequency, detecting the reflected wave and the lapsed time. Auroral kilometric radiation of frequency lower than the F-region plasma frequency cannot pass to the ground whereas high-frequency solar radio bursts reach the Earth's surface.

9.2 General Dispersion Relation

The examples of the previous section have shown that waves can propagate in a plasma. The three simplest wave modes in an unmagnetized plasma have been identified as the electrostatic Langmuir and ion-acoustic waves and the electromagnetic ordinary wave. However, the presence of a magnetic field introduces a large variety of other possible wave modes. On the other hand, the complexity of the interactions between the many particles and fields in the plasma allows only for a finite number of waves to propagate. These waves are the linear eigenmodes of the plasma. To determine their possible branches one can develop a general procedure based on the so-called *general dispersion relation* of a plasma.

9.2.1 General Wave Equation

Let us write Maxwell's equation taking into account a selfconsistent current, $\mathbf{j}$, which consists of the various contributions of the moving plasma particles. Similarly, the selfconsistent plasma charge density in the plasma is ρ. In addition to these selfconsistent sources of the field, there may be external currents and charges, $\mathbf{j}_{ex}, \rho_{ex}$, which are the sources of electromagnetic fields applied externally to the plasma. Then

$$\nabla \times \mathbf{B} = \varepsilon_0 \mu_0 \frac{\partial \mathbf{E}}{\partial t} + \mu_0 (\mathbf{j} + \mathbf{j}_{ex}) \tag{9.40}$$

$$\nabla \times \mathbf{E} = -\frac{\partial \mathbf{B}}{\partial t} \tag{9.41}$$

$$\nabla \cdot \mathbf{B} = 0 \tag{9.42}$$

$$\nabla \cdot \mathbf{E} = \frac{1}{\varepsilon_0} (\rho + \rho_{ex}) \tag{9.43}$$

Taking the derivative of the first of these equations with respect to time and eliminating the magnetic field with the help of the second equation, one finds an inhomogeneous wave equation for the electric field

$$\nabla^2 \mathbf{E} - \nabla(\nabla \cdot \mathbf{E}) - \varepsilon_0 \mu_0 \frac{\partial^2 \mathbf{E}}{\partial t^2} = \mu_0 \left(\frac{\partial \mathbf{j}}{\partial t} + \frac{\partial \mathbf{j}_{ex}}{\partial t} \right) \tag{9.44}$$

We have explicitly included external charges and currents which may be imposed into the plasma by outer means. But because Maxwell's equations are linear in the charges and fields, their contribution can be added subsequently. As the magnetic field can be eliminated from the internal current, $\mathbf{j}$, with the help of Maxwell's equations, it is sufficient to retain only the dependence of the current on the electric field as

$$\mathbf{j} = \int d^3 x' \int_{-\infty}^{t} dt' \sigma(\mathbf{x}, \mathbf{x}', t, t', \mathbf{E}) \cdot \mathbf{E} \tag{9.45}$$

This dependence is not necessarily linear, but can be quite complicated. However, for small perturbations of the fields and plasma properties it can be approximated by a linear relation between the current and the field corresponding to a time-varying Ohm's law

$$\mathbf{j} = \int d^3 x' \int_{-\infty}^{t} dt' \sigma(\mathbf{x}, \mathbf{x}', t, t') \cdot \mathbf{E} \tag{9.46}$$

The integration from $-\infty$ to t contains the concept of causality, where the history of the plasma contributes to its response at time t, while the future behavior is determined by the solution of Maxwell's equations.

The condition (9.46) closes the above system of equations, if the tensor σ is assumed to be known. This tensor contains all of the properties of the plasma. Since the wave current, $\mathbf{j}$, is the sum over all particle motions generated by the wave disturbance, it is clear that the process of finding the wave conductivity, σ, involves the plasma dynamics and thus depends on the choice of the plasma model. However, assuming that the plasma responds linearly to the presence of the wave disturbance, the wave conductivity becomes independent of the wave amplitude. Hence, the general linear dispersion relation can be derived by using a form of the wave conductivity that depends on relative position and time only, $\sigma(\mathbf{x} - \mathbf{x}_0, t - t_0)$. Under these conditions one may linearize the general wave equation (9.44) with $\mathbf{E}_0 = 0$ or $\mathbf{E}(t, \mathbf{x}) = \delta\mathbf{E}(t, \mathbf{x})$ to obtain

$$\nabla^2 \delta\mathbf{E} - \nabla(\nabla \cdot \delta\mathbf{E}) - \varepsilon_0 \mu_0 \frac{\partial^2 \delta\mathbf{E}}{\partial t^2} = \mu_0 \frac{\partial \mathbf{j}}{\partial t} \tag{9.47}$$

with Ohm's law given by

$$\mathbf{j}(t, \mathbf{x}) = \int d^3x' \int_{-\infty}^{t} dt' \sigma(\mathbf{x} - \mathbf{x}', t - t') \cdot \delta\mathbf{E} \tag{9.48}$$

Equation (9.47) is the general linear wave equation. It is applicable to any medium with a linear response to an applied field fluctuation. Its left-hand side represents the purely electromagnetic part independent on the presence of any medium. The response of the medium is entirely included in the fluctuating current term on the right-hand side and, since this current is proportional to the fluctuating field, is contained only in the fluctuating conductivity tensor, σ.

9.2.2 General Wave Dispersion Relation

Interpreting the electric field fluctuations according to Eq. (9.1) as plane waves

$$\delta\mathbf{E}(\omega, \mathbf{k}) = \delta\mathbf{E}_0(\omega, \mathbf{k}) \exp(i\mathbf{k} \cdot \mathbf{x} - i\omega t) \tag{9.49}$$

the dependence of the conductivity tensor on relative spatial and temporal distances turns the integral on the right-hand side of Eq. (9.48) into a folding integral. This fact considerably simplifies the solution of Eq. (9.47) and reduces it to

$$\left[\left(k^2 - \frac{\omega^2}{c^2} \right) \mathbf{I} - \mathbf{kk} - i\omega\mu_0\sigma(\omega, \mathbf{k}) \right] \cdot \delta\mathbf{E}_0(\omega, \mathbf{k}) = 0 \tag{9.50}$$

for the constant wave amplitude, $\delta\mathbf{E}_0(\omega, \mathbf{k})$. Hereby, the field and conductivity satisfy the symmetry conditions

$$\begin{aligned} \delta\mathbf{E}^*(\mathbf{k}, \omega) &= \delta\mathbf{E}(-\mathbf{k}, -\omega) \\ \sigma^*(\omega, \mathbf{k}) &= \sigma(-\omega, -\mathbf{k}) \end{aligned} \tag{9.51}$$

which result from the requirement of real wave field amplitudes. Equation (9.50) is a dyadic (or tensor) equation. Its nontrivial solution requires that the determinant of the expression in brackets vanishes, thereby yielding the general dispersion relation

$$D(\omega, \mathbf{k}) = \text{Det}\left[\left(k^2 - \frac{\omega^2}{c^2} \right) \mathbf{I} - \mathbf{kk} - i\omega\mu_0\sigma(\omega, \mathbf{k}) \right] = 0 \tag{9.52}$$

Sometimes it is more convenient to include the current in Ampère's law into the electric field term on the left-hand side by defining the electric induction

$$\delta \mathbf{D} = \in \cdot \delta \mathbf{E} \tag{9.53}$$

where $\in$ is the *dielectric tensor*. It satisfies the symmetry relations of Eq. (9.51)

$$\in^*(\omega, \mathbf{k}) = \in(-\omega, -\mathbf{k}) \tag{9.54}$$

With its help the current density assumes the representation

$$\delta \mathbf{j}(\omega, \mathbf{k}) = -i\omega\varepsilon_0 \left[\in(\omega, \mathbf{k}) - \mathsf{I}\right] \cdot \delta \mathbf{E}(\omega, \mathbf{k}) \tag{9.55}$$

Using this version of Ohm's law to replace the current by the wave electric field, the dielectric tensor is defined as

$$\boxed{\in(\omega, \mathbf{k}) = \mathsf{I} + \frac{i}{\omega\varepsilon_0}\sigma(\omega, \mathbf{k})} \tag{9.56}$$

This definition allows to rewrite the dispersion relation (9.52) as

$$\boxed{\mathrm{Det}\left[\frac{k^2c^2}{\omega^2}\left(\frac{\mathbf{kk}}{k^2} - \mathsf{I}\right) + \in(\omega, \mathbf{k})\right] = 0} \tag{9.57}$$

Equation (9.57) is the general dispersion relation of any active medium. Its solutions describe propagating linear waves of frequency $\omega = \omega(\mathbf{k})$. Since it is an eigenvalue equation, it has only a finite number of discrete solutions. To determine these solutions, one must first find the plasma dielectric tensor, which itself implies solving the linear dynamic plasma equations. This step depends on the choice of the plasma model.

9.2.3 Isotropic Plasma

The dielectric tensor, $\in(\omega, \mathbf{k})$, contains all the relevant linear properties of the plasma. In general it is an anisotropic tensor. However, in the absence of an external magnetic field, when the plasma is unmagnetized, it becomes isotropic. In this case the dispersion relation can be simplified, since the only particular direction in a plasma is then given by the direction of the wave vector, $\mathbf{k}$. It allows to construct a tensor

$$\mathsf{I}_\mathrm{L} = \frac{\mathbf{kk}}{k^2} \tag{9.58}$$

which is the *longitudinal unit tensor* prescribing the direction of the electrostatic fluctuations. Subtracting it from the unit tensor, I, we obtain the transverse unit tensor

$$\mathsf{I}_\mathrm{T} = \mathsf{I} - \frac{\mathbf{kk}}{k^2} \tag{9.59}$$

which in contrast accounts only for the electromagnetic directions of propagation. These two tensors permit to decompose the dielectric tensor into longitudinal and transverse components

$$\in(\omega,\mathbf{k}) = \in_L(\omega,k)\mathbf{l}_L + \in_T(\omega,k)\mathbf{l}_T \tag{9.60}$$

In this representation the coefficients of the longitudinal and transverse unit tensors are scalar functions, which depend only on frequency and wavenumber, but not on the direction of the latter. Knowing the isotropic dielectric tensor of the plasma, these two dielectric functions can be calculated by multiplying $\in$ from both sides with $\mathbf{k}$

$$
\begin{aligned}
\in_L(\omega,k) &= \frac{\mathbf{k}\cdot\in(\omega,\mathbf{k})\cdot\mathbf{k}}{k^2} \\
\in_T(\omega,k) &= \frac{\mathrm{tr}\in(\omega,\mathbf{k}) - \in_L(\omega,k)}{2}
\end{aligned}
\tag{9.61}
$$

The symbol 'tr' denotes the trace (of the dielectric tensor). In an isotropic plasma the dispersion tensor becomes very simple

$$\mathrm{Det}\left[\in_L(\omega,k)\mathbf{l}_L + \left(\in_T(\omega,k) - \frac{k^2 c^2}{\omega^2}\right)\mathbf{l}_T\right] = 0 \tag{9.62}$$

Obviously, the two tensors are linearly independent, and the two dispersion relations of an isotropic plasma reduce to the two decoupled scalar equations

$$
\begin{aligned}
\in_L(\omega,k) &= 0 \\
\in_T(\omega,k) - \frac{k^2 c^2}{\omega^2} &= 0
\end{aligned}
\tag{9.63}
$$

for the longitudinal electrostatic and transverse electromagnetic waves which can propagate in the plasma.

9.2.4 Dielectric Response Function

There is a close relation between the longitudinal dielectric function in Eq. (9.63) and the dielectric of an anisotropic plasma, the *dielectric response function*

$$\boxed{\in(\omega,\mathbf{k}) = \frac{\mathbf{k}\cdot\in(\omega,\mathbf{k})\cdot\mathbf{k}}{k^2}} \tag{9.64}$$

The former describes longitudinal waves in an unmagnetized plasma and is independent of the direction of the wave vector, while the latter governs the linear response of a plasma to a disturbance in density or electric field and depends on the full wave vector. In this sense the latter function is not restricted to an isotropic plasma but applies to a much wider range of plasma models.

To derive Eq. (9.64), we take advantage of the fact that external charges or currents can be added to the linear electromagnetic equations. Consider an external disturbance ρ_{ex} of the electric space charge in the plasma, for instance a number of test particles added to the plasma from outside. The total disturbance of the charge written in $(\omega, \mathbf{k})$ space is the sum of both linear disturbances

$$\rho_{tot}(\omega, \mathbf{k}) = \rho_{ex}(\omega, \mathbf{k}) + \rho(\omega, \mathbf{k}) \tag{9.65}$$

where $\rho_{tot}(\omega, \mathbf{k})$ as the total disturbance of the charge density is defined to be related to the disturbance $\rho_{ex}(\omega, \mathbf{k})$ via the response function

$$\rho_{tot}(\omega, \mathbf{k}) = \frac{\rho_{ex}(\omega, \mathbf{k})}{\in(\omega, \mathbf{k})} \tag{9.66}$$

From Poisson's equation it then follows for the electric field disturbance, $\delta\mathbf{E}(\omega, \mathbf{k})$, that

$$\delta\mathbf{E}(\omega, \mathbf{k}) = -i\frac{\rho_{tot}(\omega, \mathbf{k})\mathbf{k}}{k^2 \varepsilon_0} \tag{9.67}$$

Now, using the linear Ohm's law, $\mathbf{j} = \sigma \cdot \delta\mathbf{E}$, one arrives for an intermediate step at

$$\mathbf{j} = -i\frac{\rho_{tot}\sigma \cdot \mathbf{k}}{k^2 \varepsilon_0} \tag{9.68}$$

This current causes the induced space-charge disturbance, ρ. We can determine it from the charge-current continuity equation as

$$\rho(\omega, \mathbf{k}) = -\frac{i}{\omega\varepsilon_0}\frac{\mathbf{k}\cdot\sigma\cdot\mathbf{k}}{k^2} \tag{9.69}$$

Inserting this expression into Eq. (9.65) and comparing with Eq. (9.66) leads to

$$\in(\omega, \mathbf{k}) = 1 + \frac{i}{\omega\varepsilon_0}\frac{\mathbf{k}\cdot\sigma(\omega, \mathbf{k})\cdot\mathbf{k}}{k^2} \tag{9.70}$$

and with Eq. (9.56) to the expression for the dielectric response function in Eq. (9.64).

9.3 Plasma Wave Energy

All waves contain energy even though the wave in the average of the amplitude taken over more than one wave train does not exist. This energy is the energy of the fluctuations of the electric and magnetic fields in the wave which is transported across the plasma at the wave group velocity.

9.3.1 Langmuir Plasmons

A simple example shows that waves are carrier of energy. Consider for instance the dispersion relation of Langmuir waves

$$\omega^2 = \omega_{pe}^2(1 + \gamma_e k^2 \lambda_D^2) \tag{9.71}$$

From quantum mechanics we know that any finite frequency implies the existence of a quantum of energy, $\hbar\omega$. Hence, multiplying the above relation by $\hbar^2$ yields the square of the energy contained in one single Langmuir wave packet, a Langmuir *plasmon*. Let us stress the analogy a bit further. The energy of any particle of rest mass m_0 can, for low velocities, be expressed as

$$W = m_0 c^2 + \frac{\mathbf{p}^2}{2m_0} \tag{9.72}$$

Because the momentum of the particle, $\mathbf{p} = -i\hbar\nabla$, is nothing else but the gradient, it can be written as $\mathbf{p} = \hbar\mathbf{k}$. The above dispersion relation can thus be rewritten as

$$\hbar\omega = \hbar\omega_{pe}\left(1 + \frac{\gamma_e \mathbf{p}_l^2 \lambda_D^2}{2\hbar^2}\right) \tag{9.73}$$

where we have expanded the square root and introduced the plasmon momentum, $\mathbf{p}_l$. Comparison with the expression for the energy of a particle shows that the second term in the bracket on the right-hand side of the Langmuir wave dispersion relation corresponds to the kinetic energy of the plasmon, $\hbar\gamma_e k^2 v_e^2/2$, while the first term corresponds to either a potential energy provided by the plasma or to the rest mass energy of the plasmon. Adopting the second interpretation, the mass is given by $m_{0l} = \hbar\omega_{pe}/c^2$ and is typically very small. For instance, in the solar wind where $f_{pe} = \omega_{pe}/2\pi \approx 10$ kHz, its value is $m_{0l} \approx 10^{-46}$ kg.

9.3.2 Average Energy

Plasma waves consist of many plasmons. By their transport of energy across the plasma they contribute to the redistribution of energy and information. It is of considerable interest to find a macroscopic expression for their average energy content as it is of interest to know how much energy is contained in any other plasma wave. In electrodynamics the flux of energy in an electromagnetic wave is given by the *Poynting vector*

$$\boxed{\mu_0\mathbf{P} = \delta\mathbf{E} \times \delta\mathbf{B}} \tag{9.74}$$

Obviously, the Poynting vector, $\mathbf{P}$, is a nonlinear quantity. Being the energy flux, its divergence must balance the leakage of wave energy and the decrease of wave energy density, W_w, in a given volume. The latter is the sum of the changes in the electric and magnetic field energy densities. Calculating $\nabla \cdot \mathbf{P}$ from Eq. (9.74) results in

$$-\mu_0\nabla \cdot \mathbf{P} = \delta\mathbf{E} \cdot (\nabla \times \delta\mathbf{B}) - \delta\mathbf{B} \cdot (\nabla \times \delta\mathbf{E}) \tag{9.75}$$

The curls in this formula can be replaced with the help of Maxwell's equations in order to obtain the conservation law of wave energy density

$$\frac{\partial W_w}{\partial t} + \delta\mathbf{j} \cdot \delta\mathbf{E} = -\nabla \cdot \mathbf{P} \tag{9.76}$$

where W_w is defined as

$$W_w = \varepsilon_0 \delta\mathbf{E}^* \cdot \in \cdot \delta\mathbf{E} + \frac{|\delta\mathbf{B}|^2}{2\mu_0} \tag{9.77}$$

The first term on the right-hand side of Eq. (9.77) is easily recognized as the electrostatic energy density stored in the wave, the second is its magnetic energy density. In a plasma the dielectric tensor contributes to storage of electric field energy. In other words, electric field energy can be stored in the interaction of the plasma particles which leads to the polarization of the plasma. However, the dielectric tensor contributes to the electric energy density only through its product from the right and from the left with the electric wave field. Thereby it reduces to the dielectric response function, $\in(\omega, \mathbf{k})$, in Eq. (9.64). Finally, the term containing the electric current density in Eq. (9.76) is that part of the wave energy which is dissipated by Joule heating or Ohmic losses. This term vanishes if there is no dissipation which holds in general for collisionless plasmas.

We are interested in the two contributions of waves to the energy. Energy is a real quantity. Hence, for a complex wave only the products of wave amplitudes and their conjugates contribute to energy. Calculation of the magnetic energy becomes trivial. Assuming that the wave field is complex one simply obtains $|\delta\mathbf{B}|^2/2\mu_0$.

The calculation of the electrostatic part is more involved. The wave field changes in time and space with phase $\varphi = \omega t - \mathbf{k} \cdot \mathbf{x}$. For complex frequencies, $\omega = \omega_r + i\gamma$, the phase itself is a complex quantity. To simplify the procedure let us assume that the wave amplitude is small so that we are dealing with linear waves. Let us further temporarily abbreviate the product of the dielectric response function and the wave electric field to its right in Eq. (9.77) as the dielectric displacement, $\delta\mathbf{D} = \in(\omega, \mathbf{k})\delta\mathbf{E}(\omega, \mathbf{k})$, where we have implicitly represented the electric wave field by a Fourier integral

$$\begin{aligned}
\delta\mathbf{E}(t, \mathbf{x}) &= \frac{1}{16\pi^4} \int d^3k \int_{-\infty}^{\infty} d\omega\, \delta\mathbf{E}(\omega, \mathbf{k}) \exp(i\varphi) \\
\delta\mathbf{D}(t, \mathbf{x}) &= \frac{1}{16\pi^4} \int d^3k \int_{-\infty}^{\infty} d\omega \in(\omega, \mathbf{k})\delta\mathbf{E}(\omega, \mathbf{k}) \exp(i\varphi)
\end{aligned} \tag{9.78}$$

We are not interested in the instantaneous energy carried by the waves at a particular time, but in their average energy. This average energy is defined as the change in electrostatic energy averaged over volume and time and over the whole ensemble of waves present in the wave mode

$$\langle W_w \rangle = \varepsilon_0 \int_{-\infty}^{t} dt' \int d^3x \left\langle \delta \mathbf{E}(t',\mathbf{x}) \cdot \frac{\partial \delta \mathbf{D}(t',\mathbf{x})}{\partial t'} \right\rangle \tag{9.79}$$

When we now substitute from Eq. (9.78) into the last expression (remember that one now needs two different integration variables ω and ω'), replace the time derivative with $-i\omega$, and make the resulting expression symmetric, we find

$$\langle W_w \rangle = -\frac{i\varepsilon_0}{64\pi^5} \int_{-\infty}^{t} dt' \int d^3k \int_{-\infty}^{\infty} d\omega \, d\omega'$$

$$[\omega \in (\omega,\mathbf{k}) - \omega' \in (-\omega',-\mathbf{k})] \langle \delta \mathbf{E}(\omega,\mathbf{k}) \cdot \delta \mathbf{E}(-\omega',-\mathbf{k}) \rangle \exp[i(\omega'-\omega)t'] \tag{9.80}$$

The terms with the dielectric response function can be simplified if the latter is expanded around the point where the two frequencies coincide ($\omega = \omega'$)

$$\omega' \in (-\omega',-\mathbf{k}) = \omega \in (-\omega,-\mathbf{k}) + (\omega'-\omega) \frac{\partial [\omega \in (-\omega,-\mathbf{k})]}{\partial \omega} \tag{9.81}$$

Since we have assumed that dissipation is small, the symmetry of the response function requires simply that $\in (-\omega,-\mathbf{k}) = \in (\omega,\mathbf{k})$. Integrating over t' we obtain

$$\langle W_w \rangle = \frac{\varepsilon_0}{64\pi^5} \int d^3k \int_{-\infty}^{\infty} d\omega \, d\omega'$$

$$\langle \delta \mathbf{E}(\omega,\mathbf{k}) \cdot \delta \mathbf{E}(-\omega',-\mathbf{k}) \rangle \frac{\partial [\omega \in (\omega,\mathbf{k})]}{\partial \omega} \exp[i(\omega'-\omega)t] \tag{9.82}$$

The term in angular brackets under the integral sign is the spectral energy density function of the electric field

$$\langle \delta \mathbf{E}(\omega,\mathbf{k}) \cdot \delta \mathbf{E}(-\omega',-\mathbf{k}) \rangle = 4\pi^2 \langle |\delta \mathbf{E}(\omega,\mathbf{k})|^2 \rangle \delta(\omega'-\omega) \tag{9.83}$$

With its help we find ultimately for the total wave energy

$$\langle W_w \rangle = \frac{\varepsilon_0}{16\pi^3} \int d^3k \int_{-\infty}^{\infty} \langle |\delta \mathbf{E}(\omega,\mathbf{k})|^2 \rangle \frac{\partial [\omega \in (\omega,\mathbf{k})]}{\partial \omega} d\omega \tag{9.84}$$

Since the energy and the spectral electric energy are real, only the real part of the dielectric response function enters into Eq. (9.84). However, we are not so much interested in the total energy of the electrostatic waves, but in their spectral energy density and its dependence on the wave frequency and the dispersive properties of the plasma waves. This can be directly read from Eq. (9.84) when we write

$$\langle W_w \rangle = \frac{1}{8\pi^3} \int d^3k \int_{-\infty}^{\infty} W_w(\omega,\mathbf{k}) \, d\omega \tag{9.85}$$

Comparing the latter two expressions, we obtain for the spectral energy density

$$W_w(\omega, \mathbf{k}) = \frac{\varepsilon_0}{2} \left\langle |\delta \mathbf{E}(\omega, \mathbf{k})|^2 \right\rangle \frac{\partial [\omega \in (\omega, \mathbf{k})]}{\partial \omega} \tag{9.86}$$

Again, only the real part of the dielectric response function enters the last expression.

It is very important to note once more that it is not the spectral function of the electric field alone (as it is in the case of the magnetic fluctuations) which contributes to the spectral energy density of the wave. The polarization of the plasma in response to the disturbance contributes as well. This contribution is contained in the derivative of the dielectric function with respect to frequency in Eq. (9.86). The actual electrostatic energy stored in the waves can therefore be considerably different from the measured spectral fluctuation of the electric field. To find the correct energy, knowledge of the relevant wave mode and its dispersive properties is required.

We can now use the above formulae to calculate the energy density of Langmuir waves and ion acoustic waves. The Langmuir wave dielectric function is

$$\in_l(\omega, k) = 1 - \frac{\omega_{pe}^2}{\omega^2}(1 + \gamma_e k^2 \lambda_D^2) \tag{9.87}$$

Multiplying by ω and differentiating with respect to ω yields a Langmuir wave energy density of $W_l = 2W_E$, just twice the electric field energy density

$$W_E = \frac{\varepsilon_0}{2}|\delta E|^2 \tag{9.88}$$

Hence, half of the wave energy is stored in the thermal electron motion providing the polarization of the plasma. The same holds for ion acoustic waves, both in the long- and short-wavelength domains.

9.4   Magnetohydrodynamic Waves

In general plasmas are magnetised. We have shown that the simplest magnetised plasma model is the magneto-hydrodynamic fluid. Waves in an ideal magneto-hydrodynamic fluid can be treated on the basis of the previous section by calculating the wave conductivities and dielectric functions. But because of the relative simplicity of the ideal magneto-hydrodynamic fluid equations derived in Section 7.3 it is much simpler to return to these and linearize them directly.

9.4.1   Magnetohydrodynamic Dispersion Relation

We assume stationary ideal homogeneous conditions as the initial state of the single-fluid plasma with vanishing average velocity and electric fields, overall pressure equilibrium, and vanishing magnetic stresses

$$\mathbf{v}_0 = 0$$
$$\mathbf{E}_0 = 0$$
$$\nabla\left(p_0 + B_0^2/2\mu_0\right) = 0 \qquad (9.89)$$
$$(\mathbf{B}_0 \cdot \nabla)\mathbf{B}_0 = 0$$

Plasma density, velocity, magnetic field and electric field are then decomposed as the sums of their initial values and a space and time dependent fluctuation according to

$$n = n_0 + \delta n$$
$$\mathbf{v} = \delta\mathbf{v}$$
$$\mathbf{E} = \delta\mathbf{E} \qquad (9.90)$$
$$\mathbf{B} = \mathbf{B}_0 + \delta\mathbf{B}$$

Because the magnetohydrodynamic equations contain nonlinear terms, the fluctuations must be small for application of the linear approximation. This assumption has to be justified, because some of the initial average quantities are zero. The idea is that we want to arrive at a homogeneous set of linear equations which will lead us to a dispersion relation for the eigenmodes of the plasma. In such a set all variables can be expressed through one single variable, whose value remains free. The assumption of smallness can thus be reduced to the assumption that this one remaining variable will be small compared with its initial value. Hence, if the ambient magnetic field is sufficiently strong, as is usually the case in magneto-hydrodynamics of plasmas, one can assume that the fluctuation amplitude of the magnetic field is much weaker than the stationary magnetic field

$$|\delta\mathbf{B}| \ll B_0 \qquad (9.91)$$

With these assumptions the continuity equation becomes

$$\frac{\partial \delta n}{\partial t} + n_0 \nabla \cdot \delta\mathbf{v} = 0 \qquad (9.92)$$

Similarly, the magnetohydrodynamic momentum conservation equation reduces to

$$m_i n_0 \frac{\partial \delta\mathbf{v}}{\partial t} = -\nabla\left(\delta p + \frac{1}{\mu_0}\mathbf{B}_0 \cdot \delta\mathbf{B}\right) + \frac{1}{\mu_0}(\mathbf{B}_0 \cdot \nabla)\delta\mathbf{B} \qquad (9.93)$$

Since the plasma is typically unable to extinguish the fast temperature variations caused by the fluctuations, one can use the adiabatic pressure law, and the variation of the pressure becomes

$$\frac{\partial \delta p}{\partial t} = m_i c_s^2 \frac{\partial \delta n}{\partial t} = -m_i n_0 c_s^2 \nabla \cdot \delta\mathbf{v} \qquad (9.94)$$

where in the second part of the equation we made use of Eq. (9.92). The quantity $c_s^2 = \gamma p_0 / m_i n_0$ is the square of the sound velocity introduced in Section 8.1. The only remaining equation is Faraday's induction law which after linearization becomes

$$\frac{\partial \delta \mathbf{B}}{\partial t} = (\mathbf{B}_0 \cdot \nabla) \delta \mathbf{v} - \mathbf{B}_0 (\nabla \cdot \delta \mathbf{v}) \tag{9.95}$$

Equations (9.92) through (9.95) represent the desired linear and homogeneous system of equations for δn, $\delta \mathbf{v}$ and $\delta \mathbf{B}$. Because we assumed a uniform plasma with straight magnetic field lines, the direction of the ambient magnetic field is the only direction of symmetry. Hence, we can choose the magnetic field to be aligned with the z axis of our orthogonal system of coordinates, $\mathbf{B}_0 = B_0 \hat{\mathbf{e}}_\parallel$. With these conventions Eqs. (9.93) and (9.95) can be written as

$$\frac{\partial \delta \mathbf{v}}{\partial t} = v_A^2 \nabla_\parallel \left(\frac{\delta \mathbf{B}_\perp}{B_0} \right) - v_A^2 \nabla_\perp \left(\frac{\delta B_\parallel}{B_0} \right) - \nabla \left(\frac{\delta p}{m_i n_0} \right)$$
$$\frac{\partial}{\partial t} \left(\frac{\delta \mathbf{B}}{B_0} \right) = \nabla_\parallel \delta \mathbf{v}_\perp - \hat{\mathbf{e}}_\parallel (\nabla_\perp \cdot \delta \mathbf{v}_\perp) \tag{9.96}$$

where v_A is the Alfvén velocity, introduced in Eq. (8.11). These two equations together with Eq. (9.94) form a closed system of first-order differential equations for the fluctuating components of the magnetic field, pressure and velocity in the plasma. We have now the choice to either directly operate on it with the plane wave ansatz or to first derive one second-order wave equation for one of the field variables. In order to make the physics more transparent, we decide for the second choice. Taking the time derivative of the first equation and eliminating the variation of the magnetic field and pressure we obtain

$$\frac{\partial^2 \delta \mathbf{v}}{\partial t^2} = c_{ms}^2 \nabla (\nabla \cdot \delta \mathbf{v}) + v_A^2 \left(\nabla_\parallel^2 \delta \mathbf{v} - \nabla \nabla_\parallel \delta v_\parallel - \hat{\mathbf{e}}_\parallel \nabla_\parallel \nabla \cdot \delta \mathbf{v} \right) \tag{9.97}$$

which is second-order in all derivative terms and therefore describes travelling waves. Its solution is found by introducing $\delta \mathbf{v} = \delta \mathbf{v}_0 \exp(i \mathbf{k} \cdot \mathbf{x} - i \omega t)$ where $\delta \mathbf{v}_0$ is an arbitrary constant amplitude of the velocity field. This yields

$$\left[(\omega^2 - k_\parallel^2 v_A^2) \mathbf{I} - c_{ms}^2 \mathbf{k} \mathbf{k} + (k \hat{\mathbf{e}}_\parallel + \hat{\mathbf{e}}_\parallel \mathbf{k}) k_\parallel v_A^2 \right] \cdot \delta \mathbf{v}_0 = 0 \tag{9.98}$$

A meaningful solution of this equation is obtained only if $\delta \mathbf{v}_0 \neq 0$, and thus the determinant of the tensor in square brackets must vanish. If we choose a right-handed system where the perpendicular component of the wave vector is parallel to the x axis so that $\mathbf{k} = k_\parallel \hat{\mathbf{e}}_\parallel + k_\perp \hat{\mathbf{e}}_x$, the last equation can be written in the form

$$\begin{bmatrix} \omega^2 - v_A^2 k_\parallel^2 - c_{ms}^2 k_\perp^2 & 0 & -c_s^2 k_\parallel k_\perp \\ 0 & \omega^2 - v_A^2 k_\parallel^2 & 0 \\ -c_s^2 k_\parallel k_\perp & 0 & \omega^2 - c_s^2 k_\parallel^2 \end{bmatrix} \begin{bmatrix} \delta v_{0x} \\ \delta v_{0y} \\ \delta v_{0\parallel} \end{bmatrix} = 0 \tag{9.99}$$

where c_{ms} is the magnetosonic speed introduced in Eq. (8.79).

9.4.2 Shear Alfvén Wave

The above system shows that the velocity fluctuation in the y direction decouples from all other fields, representing a wave with linear dispersion

$$\boxed{\omega_A = \pm k_\parallel v_A} \tag{9.100}$$

This wave propagates parallel to the ambient field and is purely transverse. It is an electromagnetic wave which is called *shear Alfvén wave*. According to Eq. (9.96), the magnetic component of this wave is parallel to the velocity component, $\delta B_y/B_0 = -\delta v_y/v_A$, and thus the wave has zero electric fluctuation field in the direction of the ambient magnetic field, $\mathbf{B}_0$. The wave electric field $\delta E_x = \delta B_y/v_A$ points in the x direction. In addition, because $\delta B_y \ll B_0$, we find that the velocity fluctuation is small compared with the Alfvén speed, $\delta v_y \ll v_A$.

The frequency of the shear Alfvén wave depends linearly on the wavenumber. The shear Alfvén wave is thus non-dispersive and its wave energy flows along $\mathbf{B}_0$, as is recognized from its group velocity, $v_{gr,A\parallel} = v_A$, $v_{gr,A\perp} = 0$. It represents simple string-like oscillations of the magnetic field lines.

9.4.3 Magnetosonic Waves

The remaining four matrix elements couple the parallel velocity component, $\delta v_\parallel$, to the other transverse velocity fluctuation, δv_x. The dispersion relation of this wave is obtained from the vanishing of their determinant

$$\omega^4 - \omega^2 c_{ms}^2 k^2 + c_s^2 v_A^2 k^2 k_\parallel^2 = 0 \tag{9.101}$$

The two roots of this relation are

$$\omega_{ms}^2 = \frac{k^2}{2} \left\{ c_{ms}^2 \pm \left[\left(v_A^2 - c_s^2 \right)^2 + 4 v_A^2 c_s^2 \frac{k_\perp^2}{k^2} \right]^{1/2} \right\} \tag{9.102}$$

The expressions in the curly brackets are the phase velocities of the two *magnetosonic wave* modes described by this dispersion relation. They depend only on the angle θ between the magnetic field and the wave vector through $k_\perp^2/k^2 = \sin^2\theta$. The root with the positive sign is called the *fast magnetosonic wave*, the root with the negative sign is the *slow magnetosonic wave*. We have encountered these waves already in Section 8.4, in connection with our discussion of fast and slow shocks. In fact, fast and slow shocks are the final states of fast and slow magnetosonic waves when evolving to large amplitudes.

Inspection of the dispersion relations of the two magnetosonic waves reveals that for $k = k_\perp$ the root on the right-hand side of the dispersion relation becomes trivial and the dispersion relation can be written as

$$\omega^2 = \tfrac{1}{2} k^2 (c_{ms}^2 \pm c_{ms}^2) \tag{9.103}$$

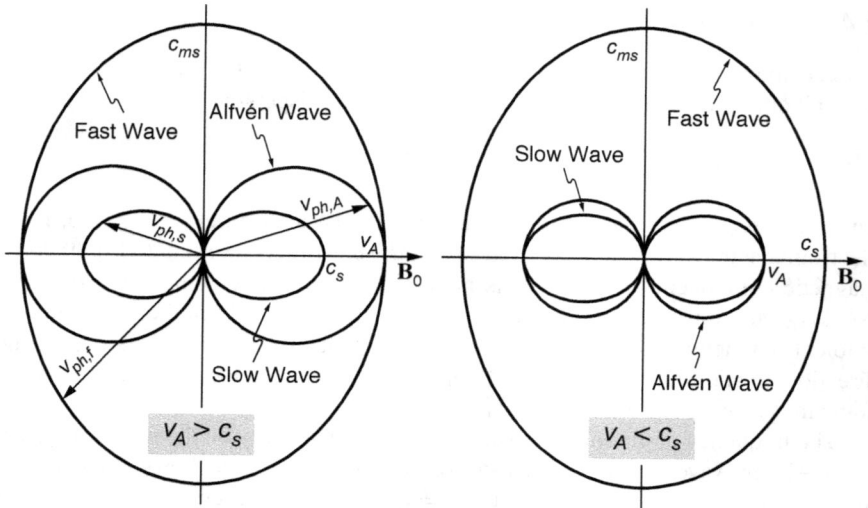

Figure 9.7: Phase velocity diagrams of the three MHD wave modes.

The fast mode (positive sign) propagates into the perpendicular direction with phase velocity $v_{ph,f\perp} = c_{ms}$, while the slow mode (negative sign) does not propagate. Since the shear Alfvén wave does not propagate into the perpendicular direction, too, we find that the only perpendicular magnetohydrodynamic wave is the fast magnetosonic mode.

In the parallel direction the dispersion relation reduces to

$$\omega^2 = \tfrac{1}{2}k^2 \left[c_s^2 + v_A^2 \pm (c_s^2 - v_A^2) \right] \tag{9.104}$$

The character of the waves depends on whether v_A or c_s is higher. For $v_A > c_s$ the parallel phase velocity of the slow mode becomes $v_{ph,s\parallel} = c_s$. The parallel slow magnetosonic mode is a simple sound wave, while the fast mode propagates at Alfvén speed, $v_{ph,f\parallel} = v_A$. In the opposite case, when $c_s > v_A$, the slow mode approaches the Alfvén speed, and the fast mode has sound velocity. This behavior of the phase velocities of the three magnetohydrodynamic modes is illustrated in Figs. 9.7 and 9.8.

Figure 9.7 is the phase velocity diagram of the three magnetohydrodynamic wave modes. In this diagram the ambient magnetic field, $\mathbf{B}_0$, points in the x direction. The arrows are the vectors of the phase velocities for the different modes drawn as functions of their angle with respect to the magnetic field. Their absolute length is the value of the phase velocity at the particular angle. The figure can be interpreted as the instantaneous form of a wave front at unit time having started from the origin and propagating in all directions. This wave front will have become displaced farthest at the angle where the phase velocity is maximum for the mode under consideration.

Figure 9.8 shows the variation of the phase velocities of the three magnetohydrodynamic wave modes as function of the angle between their wave vectors and the direction of the ambient magnetic field for the two cases $v_A > c_s$ and $v_A < c_s$. One

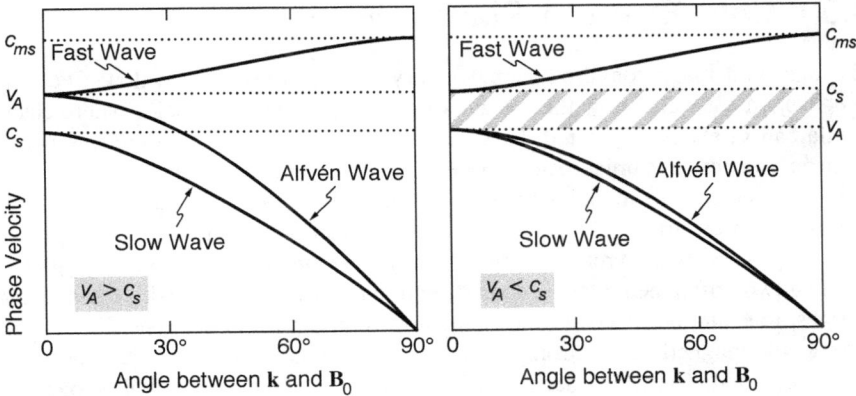

Figure 9.8: Dependence of MHD wave phase velocities on the angle between **k** and **B**₀.

recognizes the increase in the fast wave speed from parallel to perpendicular direction and the corresponding decreases in the Alfvén and slow wave velocities.

For the Alfvén wave we found that the components $\delta v_y, \delta B_y$, and δE_x build up the shear Alfvén mode. Correspondingly $\delta B_x, \delta B_\parallel, \delta v_x, \delta v_\parallel, \delta p$, and δn belong to the two magnetosonic wave modes. In order to understand what happens physically, let us return to the equation of motion and write it in components

$$\omega \delta \mathbf{v} = \frac{\mathbf{k}}{m_i n_0}\left(\delta p + \frac{1}{\mu_0}\mathbf{B}_0 \cdot \delta \mathbf{B}\right) - \frac{\mathbf{k} \cdot \mathbf{B}_0}{\mu_0 m_i n_0}\delta \mathbf{B} \tag{9.105}$$

In the parallel direction this equation reduces to the simple expression

$$\omega v_\parallel = \frac{k_\parallel \delta p}{m_i n_0} \tag{9.106}$$

showing that parallel pressure variations cause variations in the parallel flow but not in the magnetic field. Moreover, dotting the equation of motion with **k** and using $\mathbf{k} \cdot \delta \mathbf{B} = k_\parallel \delta B_\parallel + k_\perp \delta B_x$ yields

$$\omega(k_\parallel v_\parallel + k_\perp v_x) = \frac{k^2 \delta p_{\text{tot}}}{m_i n_0} \tag{9.107}$$

where the total pressure $p_{\text{tot}} = p + B^2/2\mu_0$ is the sum of the isotropic thermal and magnetic pressures.

We already know that $v_\parallel$ is generated by pressure fluctuations only. Hence, we conclude that perpendicular fluid motions are connected with variations in the total pressure and in particular with variations in the magnetic field strength. Because these variations can be in-phase or out-of-phase they either amplify the force on the plasma in which case the wave becomes the fast (accelerated) magnetosonic mode. In the opposite case the force is weakened by the out-of-phase total pressure variation, with the result that the magnetosonic mode is retarded and becomes the slow mode.

9.5 Cold Electron Plasma Waves

The one-fluid magnetohydrodynamic theory is valid only at very low frequencies $\omega \ll (\omega_{gi}, \omega_{pi})$, well below the ion-cyclotron and plasma frequencies where electron inertia can safely be neglected. Near both of these frequencies differences between electron and ion dynamics begin to become important and magnetohydrodynamic waves transform into modes which are not contained in the apparatus of one-fluid theory. When electron and ion inertia are to be taken into account in wave propagation one possibility is to relax the one-fluid assumption and to analyse wave propagation in a two-component plasma consisting of equal numbers of positive and negative charges, ions and electrons. In the following we proceed in two steps. First we consider a cold magnetized electron plasma where the ions constitute merely a neutralizing background. The waves derived will thus also have frequencies well above all ion frequencies, $\omega \gg (\omega_{gi}, \omega_{pi})$. In the next section we consider waves with frequencies intermediate between the magnetohydrodynamic and high-frequency electron waves in a magnetized plasma. Only in this case ion dynamics will explicitly be taken into account.

9.5.1 The Cold Electron Plasma Dispersion Relation

Cold electron dynamics is governed by the single-particle electron motion in a strong magnetic field as given by Eqs. (2.8) and (2.16). For our purposes where the electrons are 'cold', the magnetic field of any wave is not affected by the electron motion. In other words, in the Lorentz force we retain only the linear term $\delta \mathbf{v} \times \mathbf{B}_0$. Since $\mathbf{B}_0$ is a constant vector we can extract it from the product and include it into the electron gyrofrequency vector, $\omega_{ge} = -e\mathbf{B}_0/m_e$, this time also including the sign of the electron charge. Then the parallel and perpendicular equations of motion become

$$\frac{d\delta v_\parallel}{dt} = -\frac{e}{m_e} \delta E_\parallel$$
$$\frac{d\delta \mathbf{v}_\perp}{dt} = -\frac{e}{m_e} \delta \mathbf{E}_\perp - \omega_{ge} \times \delta \mathbf{v}_\perp \tag{9.108}$$

Differentiating this equation with respect to time, substituting for the cross-product from the undifferentiated equation, and rearranging, we obtain the well-known ordinary oscillator equation

$$\frac{\partial^2 \delta \mathbf{v}_\perp}{\partial t^2} + \omega_{ge}^2 \delta \mathbf{v}_\perp = -\frac{e}{m_e} \left(\frac{\partial \delta \mathbf{E}_\perp}{\partial t} - \omega_{ge} \times \delta \mathbf{E}_\perp \right) \tag{9.109}$$

for the velocity variation $\delta \mathbf{v}_\perp$ in the presence of a driving term on the right-hand side. It is our goal to obtain an expression for the linear wave conductivity, $\sigma(\omega, \mathbf{k})$, which we will use later to find the dielectric tensor and the cold plasma dielectric response function.

Recall that we are dealing here with a cold electron plasma of zero pressure where the ions serve as an immobile cold charge-neutralising background which is not to be taken into account. As a consequence the thermal gradient force drops out from the momentum conservation equation and density fluctuations do not contribute to the wave. This is the reason why we are allowed to use the single-electron equation of motion and can ignore the continuity equation. The condition for the validity of the cold model is clearly that the thermal velocity of the plasma electrons is much less than the wave phase velocity.

The relation between the wave velocity and current is given by the sum over all particle momenta. But since in a cold plasma of zero temperature all electrons have the same velocity caused by the wave electric field fluctuation, the sum over all momenta reduces to the product of the undisturbed particle density, n_0, and the disturbance of the electron velocity, δv. Hence, the current density becomes

$$\delta \mathbf{j} = -e n_0 \delta \mathbf{v} = \sigma \cdot \delta \mathbf{E} \tag{9.110}$$

From Eq. (9.109) one realizes that for $\delta \mathbf{E}_\perp = 0$ the electron motion becomes a pure gyration with velocity $\mathbf{v}_{\perp 0}$. The cyclotron motion is the homogeneous solution of Eq. (9.109). We are interested in the particular solution of the inhomogeneous equation

$$\delta \mathbf{v}_\perp - \mathbf{v}_{\perp 0} = -\frac{1}{e n_0} \sigma \cdot \delta \mathbf{E} \tag{9.111}$$

which is a periodic oscillation of frequency ω according to $\delta \mathbf{v} \propto \exp(-i\omega t)$. Inserting into Eq. (9.109) yields (note that here and from now on we disregard the sign of the electron charge and take ω_{ge} as a positive number)

$$\sigma(\omega) = \varepsilon_0 \omega_{pe}^2 \begin{bmatrix} \dfrac{i\omega}{\omega^2 - \omega_{ge}^2} & \dfrac{\omega_{ge}}{\omega^2 - \omega_{ge}^2} & 0 \\[2ex] -\dfrac{\omega_{ge}}{\omega^2 - \omega_{ge}^2} & \dfrac{i\omega}{\omega^2 - \omega_{ge}^2} & 0 \\[2ex] 0 & 0 & \dfrac{i}{\omega} \end{bmatrix} \tag{9.112}$$

The last line in this tensor arises from the parallel component of the electron equation of motion which is independent of the magnetic field. The conductivity tensor of cold plasma waves depends only on the wave frequency. Going back to the definition in Eq. (9.56) it is easy to find the dielectric tensor of the cold plasma

$$\mathcal{E}_{cold}(\omega) = \begin{bmatrix} 1 + \dfrac{\omega_{pe}^2}{\omega_{ge}^2 - \omega^2} & -\dfrac{i\omega_{ge}}{\omega} \dfrac{\omega_{pe}^2}{\omega_{ge}^2 - \omega^2} & 0 \\[2ex] \dfrac{i\omega_{ge}}{\omega} \dfrac{\omega_{pe}^2}{\omega_{ge}^2 - \omega^2} & 1 + \dfrac{\omega_{pe}^2}{\omega_{ge}^2 - \omega^2} & 0 \\[2ex] 0 & 0 & 1 - \dfrac{\omega_{pe}^2}{\omega^2} \end{bmatrix} \tag{9.113}$$

Inserting this into Eq. (9.50) yields the cold electron plasma dispersion relation

$$
\text{Det} \left[\frac{k^2 c^2}{\omega^2} \left(\mathbf{1} - \frac{\mathbf{kk}}{k^2} \right) - \epsilon_{\text{cold}} \right] = 0
\tag{9.114}
$$

The cold dielectric tensor is independent of the wave vector, which enters only through the electrodynamic equations. From this fact we conclude that all electrostatic variations in a cold electron plasma will be mere oscillations while the propagating waves are pure electromagnetic waves. The only oscillation in a cold electron plasma is the Langmuir oscillation with frequency $\omega = \pm \omega_{pe}$ and infinite wavenumber. We found this oscillation already in Section 9.1. Equation (9.114) therefore turns out to be the basic dispersion relation of high-frequency electromagnetic waves.

9.5.2 High-Frequency Electromagnetic Waves

We can write the cold plasma dielectric tensor in Eq. (9.113) in a shorthand version

$$
\epsilon = \begin{bmatrix} \epsilon_1 & -i\epsilon_2 & 0 \\ i\epsilon_2 & \epsilon_1 & 0 \\ 0 & 0 & \epsilon_3 \end{bmatrix}
\tag{9.115}
$$

where the matrix components have been defined as

$$
\begin{aligned}
\epsilon_1 &= 1 - \frac{\omega_{pe}^2}{\omega^2 - \omega_{ge}^2} \\
\epsilon_2 &= -\frac{\omega_{ge}}{\omega} \frac{\omega_{pe}^2}{\omega^2 - \omega_{ge}^2} \\
\epsilon_3 &= 1 - \frac{\omega_{pe}^2}{\omega^2}
\end{aligned}
\tag{9.116}
$$

The electron-cyclotron frequency is taken to be independent of the sign of the electron charge. Then defining the vectorial refractive index, $\mathbf{N} = \mathbf{k}c/\omega$, with $N^2 = N_\perp^2 + N_\parallel^2$, and assuming without any restriction of generality that the wave vector, $\mathbf{k}$, is in the (x, z) plane ($k_y = 0$), the cold plasma dispersion relation (9.114) can be written as

$$
\text{Det} \begin{bmatrix} N_\parallel^2 - \epsilon_1 & i\epsilon_2 & -N_\parallel N_\perp \\ -i\epsilon_2 & N^2 - \epsilon_1 & 0 \\ -N_\parallel N_\perp & 0 & N_\perp^2 - \epsilon_3 \end{bmatrix} = 0
\tag{9.117}
$$

This is the basic dispersion relation of a charge-compensated electron plasma of zero temperature which serves as the starting point of a discussion of high-frequency electromagnetic wave propagation in a plasma. One should, however, keep in mind that the cold plasma approximation is a very crude approximation which is valid only at wavelengths much larger than the gyroradii of the particles and for high phase velocities much larger than the thermal velocity of the electrons. In the following we distinguish between propagation parallel and perpendicular to the magnetic field.

9.5.3 Parallel Electromagnetic Wave Propagation

For parallel propagation we have $N_\perp = 0, N_\parallel = N$. Since the wave vector is parallel to the magnetic field, $\mathbf{B}_0$, the dispersion relation splits into the parallel dispersion relation, $\in_3 = 0$, and into a transverse dispersion relation. The former yields the already well-known plasma oscillations, $\omega^2 = \omega_{pe}^2$, belonging to an electrostatic parallel electric wave field, $\delta E_\parallel \| k_\parallel$. The transverse dispersion relation contains the two perpendicular electric field components, $\delta E_x, \delta E_y$. The cylindrical geometry of this wave suggests the introduction of the following combination of electric wave field components

$$\sqrt{2}\,\delta E_{R,L} = (\delta E_x \mp i\delta E_y) \tag{9.118}$$

These new wave fields describe right-hand (R) and left-hand (L) circularly polarized waves. This is easily realized by taking the ratio of the Cartesian field components

$$(\delta E_y / \delta E_x)_{R,L} = \pm i \tag{9.119}$$

showing that the electric field vector of the R-wave rotates in positive y direction while that of the L-wave rotates in negative y direction.

The transformation from $\delta E_{x,y}$ to $\delta E_{R,L}$ does not change the electric field. It is a unitary matrix satisfying $\mathbf{U}\cdot\mathbf{U}^\dagger = \mathbf{U}^\dagger\cdot\mathbf{U} = 1$ and is defined as

$$\mathbf{U} = \begin{bmatrix} 1/\sqrt{2} & -i/\sqrt{2} & 0 \\ -i/\sqrt{2} & 1/\sqrt{2} & 0 \\ 0 & 0 & 1 \end{bmatrix} \tag{9.120}$$

Accordingly the dielectric tensor transforms as $\mathbf{U}\cdot\in\cdot\mathbf{U}^\dagger$. The result of this procedure is the diagonal dielectric matrix

$$\mathbf{U}\cdot\in\cdot\mathbf{U}^\dagger = \begin{bmatrix} \in_R & 0 & 0 \\ 0 & \in_L & 0 \\ 0 & 0 & \in_3 \end{bmatrix} \tag{9.121}$$

Its transverse components are given by $\in_{R,L} = \in_1 \pm \in_2$ or explicitly as

$$\in_{R,L} = 1 - \frac{\omega_{pe}^2}{\omega(\omega \mp \omega_{ge})} \tag{9.122}$$

The diagonalized dielectric tensor separates the transverse dispersion relation into two independent dispersion relations for the right- and left-circular polarized parallel modes

$$N^2 = \frac{k^2 c^2}{\omega^2} = \in_{R,L} \tag{9.123}$$

The right-hand side of these dispersion relations is independent of the wavenumber, indicating that the dispersion of the transverse modes is entirely due to the refraction index, N. Hence, Eq. (9.123) is already in its explicit form for the parallel wavenumber.

The right-circular polarized electromagnetic wave has refraction index

$$\frac{k^2 c^2}{\omega^2} = 1 - \frac{\omega_{pe}^2}{\omega(\omega - \omega_{ge})} \tag{9.124}$$

For $\omega \to 0$ as well as for $\omega \to \omega_{ge}$ the refraction index diverges. In the latter case this implies that the wavenumber diverges, $k \to \infty$. Hence, $\omega - \omega_{ge} \to -0$ and

$$\omega_{R,res} = \omega_{ge} \tag{9.125}$$

is the *electron-cyclotron resonance frequency* for the right-circular polarized mode.

Resonances are quite complex physical phenomena requiring deep insight into the physical processes of the interaction of the wave with the plasma itself. The observation that $k \to \infty$ at a resonance implies that the resonant wavelength becomes very short, $\lambda = 2\pi/k \to 0$, microscopically small indeed. In addition, because the frequency of the wave is constant, the phase velocity of the wave in the plasma frame becomes zero, while the wave plasmon momentum, $\hbar \mathbf{k}$, increases. In such a case the interaction between the cold plasma particles and the wave must necessarily become very strong, while over the length of one wavelength the wave starts 'resolving' single particles and violently affects their orbits. During this strong interaction the wave will either give energy to the particles and will become damped, or it will extract energy and momentum from the particles in resonance and will grow. Either case is possible so that one can distinguish between *resonant absorption* of wave energy or *resonant amplification* of the wave.

At sufficiently low frequencies one can neglect the one in the dispersion relation of the low-frequency slow mode and obtain

$$\omega = \frac{\omega_{ge}}{1 + \omega_{pe}^2/k^2 c^2} \tag{9.126}$$

In the long-wavelength limit of small k this dispersion relation simplifies to

$$\omega_w = k^2 c^2 \omega_{ge} \omega_{pe}^{-2} \tag{9.127}$$

This is the dispersion relation of the *electron whistlers* frequently observed on the ground and in the Earth's magnetosphere and often excited by lightnings. They propagate nearly strictly along the magnetic field lines from one hemisphere to the other. Their phase and group velocities are both proportional to k and thus $v_{ph}^w \sim v_g^w \sim \sqrt{\omega_w}$, implying that higher-frequency waves have higher group and phase velocities. Hence, the high-frequency part of a whistler injected from, say, lightning in the southern hemisphere will reach the northern hemispheric footpoint of the magnetic field line earlier

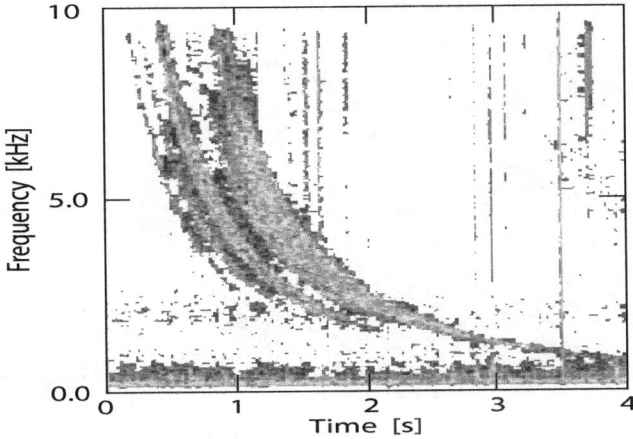

Figure 9.9: Sonogram of real whistlers recorded on the ground in Antarctica showing experimentally that high frequencies of a whistler arrive first at ground, the typical property of a whistler which is the consequence of the proportionality of the whistler phase and group velocities to the wavenumber and thus to $\sqrt{\omega_W}$. The recorded whistler obviously comes from the northern hemisphere and propagates along the connected magnetic field line to the ground station.

than its low-frequency part. The whistler will appear as a falling tone in a frequency-time sonogram like the one shown in Fig. 9.9, a property which gave it its name.

For large $\omega > \omega_{ge}$ the refraction index, and thus the wavenumber, vanishes at the *right-hand cut-off* frequency

$$\omega_{R,co} = \tfrac{1}{2}\left[\omega_{ge} + (\omega_{ge}^2 + 4\omega_{pe}^2)^{1/2}\right] \tag{9.128}$$

which is found by setting the right-hand side of Eq. (9.124) to zero.

Let us now turn to the left-circular wave mode. Its dispersion relation reads

$$\boxed{\frac{k^2 c^2}{\omega^2} = 1 - \frac{\omega_{pe}^2}{\omega(\omega + \omega_{ge})}} \tag{9.129}$$

One immediately recognizes that this dispersion relation does not have any resonances. Moreover, since $N^2 < 1$, left-circular waves have phase velocities larger than c. They are high-frequency waves and have a low-frequency cut-off at

$$\omega_{L,co} = \tfrac{1}{2}\left[(\omega_{ge}^2 + 4\omega_{pe}^2)^{1/2} - \omega_{ge}\right] \tag{9.130}$$

This cut-off is below the R-wave cut-off. In our approximation, where the effects of the ions are neglected, the left-circular wave exists only at frequencies above $\omega_{L,co}$.

Figure 9.10 shows the dependence of the refraction index on frequency. There is no wave propagation for negative values of N^2. The dispersion branches of the R

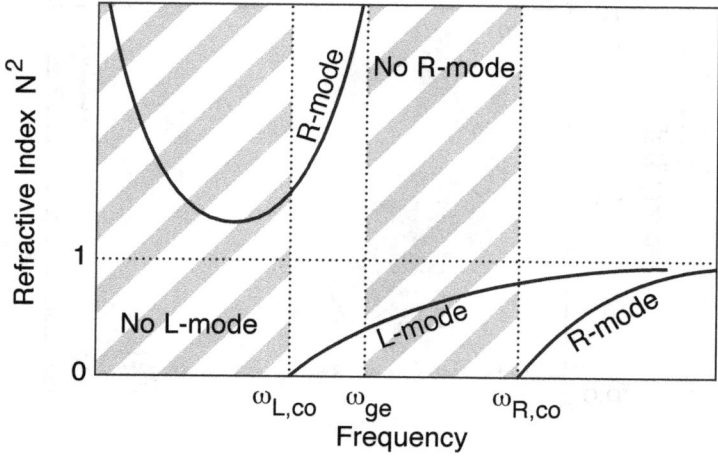

Figure 9.10: Refractive indices for parallel propagating R- and L-waves. For propagating waves one has $N^2 > 0$ which causes the cut-offs $\omega_{L,co}, \omega_{R,co}$ on the $N^2 < 1$ large phase velocity $v_{ph} \le c$ free-space branches of both modes. At low phase velocities $v_{ph} < c$ only the R-mode can exist below the electron cyclotron frequency. The regions where the linear modes cannot propagate are shown hatched. Both L- and R-modes propagate above the R-mode cut-off and in the narrow region between the L-mode cut-off and the electron cyclotron frequency. But the character of the modes is different here. The L-mode is a free space mode while the R-mode is a confined plasma mode.

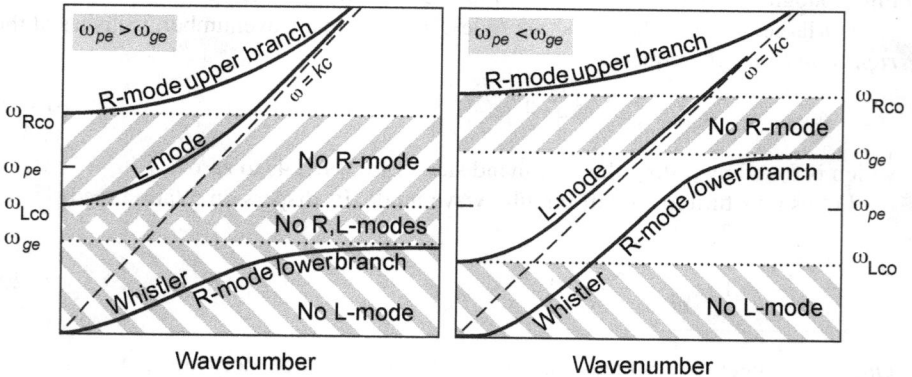

Figure 9.11: Dispersion branches for parallel propagating R- and L-waves shown for the two cases $\omega_{pe} > \omega_{ge}$ and $\omega_{pe} < \omega_{ge}$. The main difference is that in the second case no stop band exists where neither of the modes can propagate. Instead, the confined plasma R- and free space L-modes overlap in the intermediate region.

and L modes for a dense, $\omega_{pe} > \omega_{ge}$, and a dilute, $\omega_{pe} < \omega_{ge}$, plasma are shown in Fig. 9.11. At high frequencies the L-mode and R-mode upper branches become free-space modes of light velocity. It is easily recognized from the tangents to all the curves that the group velocities of the waves are smaller than c. One also observes that there is no connection between the two branches of the R-mode. Both branches are separated by a stop band. High-frequency R-modes cannot penetrate into a plasma beyond the point where their frequency matches the R-mode cut-off. Similarly L-modes cannot penetrate beyond the L-mode cut-off. In a dilute plasma the low-frequency R- and L-mode branches overlap in the interval between the cyclotron frequency and the L-mode cut-off.

9.5.4 Perpendicular Electromagnetic Wave Propagation

The other limiting case are waves propagating perpendicular to the magnetic field with a wave vector $\mathbf{k} = \mathbf{k}_\perp$. In a homogeneous plasma the direction of the wave vector in the perpendicular plane is arbitrary and we can chose it to be parallel to the x axis. For $k_y = k_\parallel = 0$ the general cold plasma dispersion relation (9.117) can be reduced to

$$\text{Det} \begin{bmatrix} -\epsilon_1 & i\epsilon_2 & 0 \\ -i\epsilon_2 & N_\perp^2 - \epsilon_1 & 0 \\ 0 & 0 & N_\perp^2 - \epsilon_3 \end{bmatrix} = 0 \tag{9.131}$$

Let us first consider the third line of the matrix in Eq. (9.131). It tells us that the wave with $\delta E_\parallel$ decouples from the two transverse field components. Its dispersion relation is $N_\perp^2 = \epsilon_3$ or explicitly using the definition of ϵ_3 in Eq. (9.116)

$$\boxed{\omega_{om}^2 = \omega_{pe}^2 + k_\perp^2 c^2} \tag{9.132}$$

Because the wave vector and wave electric field of this mode are perpendicular to each other, this wave is a transverse electromagnetic wave. For $\omega_{pe} \to 0$ it becomes the usual free-space electromagnetic wave. We have recovered the *ordinary wave*, abbreviated as *O-mode*, in a magnetized plasma which we already met in Eq. (9.39) for an unmagnetized plasma. As shown there, it is easy to recognize that this mode has a low-frequency cut-off at the local plasma frequency. The ordinary wave smoothly connects to the high-frequency L-mode when its wave vector turns parallel to the magnetic field.

The dispersion relation of the remaining wave is obtained by solving the determinant of the perpendicular part of Eq. (9.131)

$$\epsilon_2^2 + \epsilon_1(N_\perp^2 - \epsilon_1) = 0 \tag{9.133}$$

which, when using Eq. (9.116), is rewritten as

$$\boxed{k_\perp^2 c^2 = \frac{(\omega^2 - \omega_{R,co}^2)(\omega^2 - \omega_{L,co}^2)}{\omega^2 - \omega_{ge}^2 - \omega_{pe}^2}} \tag{9.134}$$

We have factorized the right-hand side of this expression in a convenient way, making use of the previously defined cut-off frequencies of the R- and L-modes, which turn out to be cut-off frequencies of the new perpendicular mode, too. Since this is topologically impossible with one single branch, the mode described by the above dispersion relation will have two different branches, one being cut off at the R-mode cut-off, the other at the L-mode cut-off. We call this mode the *extraordinary mode*, abbreviated as *X-mode*, to distinguish it from the O-mode.

Before discussing its dispersive properties, we note that its electric field components are in the (x, y) plane. Since $k_\perp = k_x$ by definition, the extraordinary mode has an electric field component parallel to the wave vector. This wave is thus of mixed electrostatic and electromagnetic polarization. It propagates perpendicular to the magnetic field, but does not possess any electric field component parallel to the ambient magnetic field. On the other hand, because being a transverse wave, its wave vector is directed perpendicular to the wave magnetic field. Therefore the extraordinary mode possesses magnetic components in y direction as well as parallel to the ambient magnetic field.

Inspection of the dispersion relation (9.133) and its equivalent form (9.134) for the wave number of the extraordinary mode shows that the X-mode has a resonance at the *upper-hybrid frequency* which is a combination of the electron plasma and cyclotron frequencies:

$$\boxed{\omega_{uh}^2 = \omega_{ge}^2 + \omega_{pe}^2} \tag{9.135}$$

Here the plasma and cyclotron properties of the electrons combine. Since the upper-hybrid frequency is higher than the L-mode cut-off, the lower-frequency branch of the X-mode is resonant at the upper-hybrid frequency. This branch can thus propagate between its cut-off at $\omega_{L,co}$ and the resonance at ω_{uh}. On the other hand, the R-mode cut-off is higher than the upper-hybrid frequency and the upper branch of the X-mode, which for high frequencies becomes a free-space mode propagating above $\omega_{R,co}$. Thus there is a stop-band for the X-mode between $\omega_{R,co} > \omega > \omega_{uh}$.

Figure 9.12 shows the transverse mode branches for perpendicular propagation in dense and dilute plasmas. A comparison with Fig. 9.11 shows that the two branches of the perpendicular extraordinary wave become the two branches of the right-circular R-mode when the wave vector turns into parallel propagation.

9.6 Two-Fluid Plasma Waves

At very low frequencies comparable to the ion-cyclotron frequency, $\omega \approx \omega_{gi}$, one cannot anymore neglect the ion dynamics. In linear theory it is easy to include the ion terms into the dispersion relation, observing that the contribution of the plasma is contained only in the linear conductivity and thus in the dielectric tensor in an additive way. This is clear when one remembers that both the space charges and linear currents are additive quantities containing the sum of the electron and ion contributions.

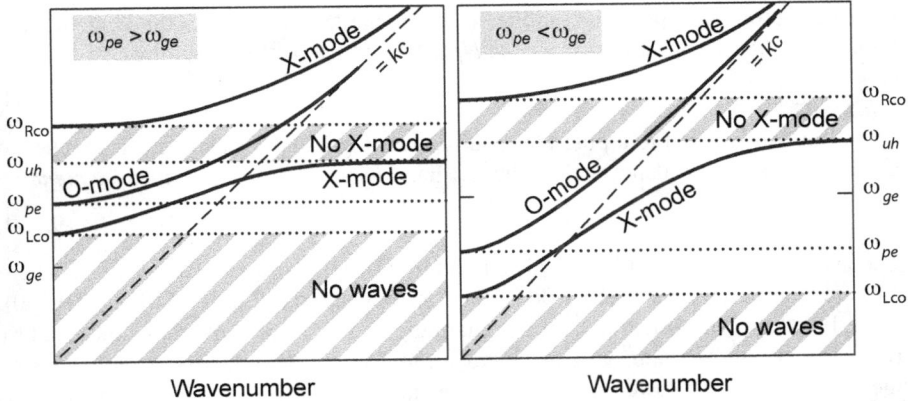

Figure 9.12: Dispersion branches for perpendicular propagating O- and X-modes shown for the two cases $\omega_{pe} > \omega_{ge}$ and $\omega_{pe} < \omega_{ge}$. One may note that the O-mode never becomes a confined plasma mode. It always is a free space mode while the X-mode always splits into free space and confined plasma modes. At low frequencies neither of these modes exists for perpendicular propagation.

9.6.1 Effect of Ion Dynamics

The total linear conductivity is the sum of the electron and ion conductivities. The dielectric does, in addition, contain a unit tensor which results from the field equations and is independent of the plasma. Hence, one must subtract the unit terms in the diagonal terms of the dielectric tensor to obtain the plasma contribution. Doing this and adding the ion terms to the components of the cold dielectric tensor components in Eq. (9.116) one obtains the following tensor elements for low-frequency waves

$$
\begin{aligned}
\in_1 &= 1 - \frac{\omega_{pe}^2}{\omega^2 - \omega_{ge}^2} - \frac{\omega_{pi}^2}{\omega^2 - \omega_{gi}^2} \\
\in_2 &= -\frac{\omega_{ge}}{\omega}\frac{\omega_{pe}^2}{\omega^2 - \omega_{ge}^2} + \frac{\omega_{gi}}{\omega}\frac{\omega_{pi}^2}{\omega^2 - \omega_{gi}^2} \\
\in_3 &= 1 - \frac{\omega_{pe}^2}{\omega^2} - \frac{\omega_{pi}^2}{\omega^2}
\end{aligned}
\tag{9.136}
$$

9.6.2 Parallel Wave Propagation

For parallel propagation, $N_\perp = 0$, the dispersion relation including electron and ion dynamics becomes

$$
N_{\parallel R,L}^2 = 1 - \frac{\omega_{pe}^2}{\omega(\omega \mp \omega_{ge})} - \frac{\omega_{pi}^2}{\omega(\omega \pm \omega_{gi})}
\tag{9.137}
$$

where as before the upper sign applies to the R-mode, the lower to the L-mode. The R-mode has the well-known resonance $N_{\parallel R} \to \infty$ at the electron-cyclotron frequency

$$\omega_{R,res} = \omega_{ge} \tag{9.138}$$

while now, in contrast to a pure electron plasma, the L-mode has also a resonance at the ion-cyclotron frequency, called the left-hand *ion-cyclotron resonance frequency*

$$\omega_{L,res} = \omega_{gi} \tag{9.139}$$

The above dispersion relation (9.137) has cut-offs for $N_{\parallel} \to 0$. These cut-off frequencies are nearly the same as those derived before and given in Eqs. (9.128) and (9.130), with slight corrections due to the ion dynamics. These corrections can be included by replacing $\omega_{pe}^2 \to \omega_{pe}^2 + \omega_{pi}^2$ and $\omega_{ge} \to \omega_{ge} \pm \omega_{gi}$, where the upper sign applies inside the square roots, the lower sign outside the square roots. Because of the small mass ratio these corrections are of little importance.

Solving for the low-frequency R-mode one also finds that the whistler dispersion relation (9.127) is changed by the inclusion of the ion correction

$$\omega = \frac{\omega_{ge}}{2} \left(1 + \frac{\omega_{pe}^2}{k^2 c^2}\right)^{-1} \left[\left(1 + \frac{4\omega_{pi}^2}{k^2 c^2}\right)^{1/2} + 1\right] \tag{9.140}$$

For long wavelengths, $k^2 \ll \omega_{pi}^2/c^2$, where the ion term in the square root dominates, this dispersion relation reduces to the Alfvén wave, $\omega_A = \pm k_{\parallel} v_A$, so that in a quasineutral electron-ion plasma the long-wavelength, low-frequency limit of the R- or whistler mode is a right-handed Alfvén wave with vanishing frequency at infinitely long wavelengths. The Alfvén wave at low frequencies therefore has a right-handed and a left-handed component, which both are dominated by ion inertia. In general, Alfvén waves will have an elliptical polarization, but they can become linearly polarized when the two oppositely polarized modes have about equal amplitudes.

At shorter wavelengths, where the ion term can be neglected, the R-mode with the dispersion relation (9.140) becomes a whistler, whose parallel wavenumber satisfies

$$4\omega_{pi}^2 \ll k^2 c^2 \ll \omega_{ge}^2 \qquad \text{or equivalently} \qquad k^2 \lambda_e^2 \ll 1 \ll \tfrac{1}{4} k^2 \lambda_i^2 \tag{9.141}$$

showing that the whistler wavelength is much longer than the *electron inertial length*, $\lambda_w \gg \lambda_e = c/\omega_{pe}$, but still much shorter than the *ion inertial length*, $\lambda_w \ll \lambda_i = c/\omega_{pi}$.

The dispersion relation of the L-mode at low frequencies can be written down in full analogy to that of the R-mode as

$$\omega = \frac{\omega_{ge}}{2} \left(1 + \frac{\omega_{pe}^2}{k^2 c^2}\right)^{-1} \left[\left(1 + \frac{4\omega_{pi}^2}{k^2 c^2}\right)^{1/2} - 1\right] \tag{9.142}$$

The long-wavelength limit is again an Alfvén wave, this time the left-handed mode of Eq. (9.100). At slightly shorter wavelengths the L-mode wave has a dispersion similar to the electron whistler and becomes an *ion whistler* which runs into the L-mode ion cyclotron resonance, $\omega_{L,res}$, when the wavenumber increases further.

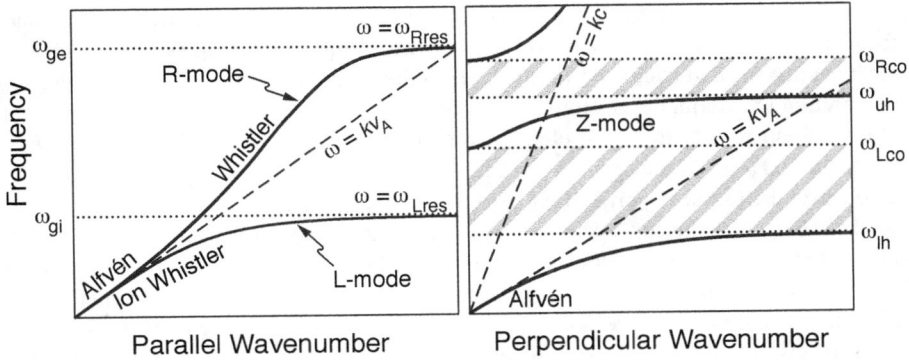

Figure 9.13: Low-frequency dispersion branches for parallel and perpendicular propagation.

9.6.3 Perpendicular Wave Propagation

For perpendicular propagation, the dispersion relation (9.133) can be written as

$$N_\perp^2 = \epsilon_1 - \epsilon_2^2/\epsilon_1 = 0 \tag{9.143}$$

In the extremely-low frequency limit, $\omega \to 0$, we find for $\omega_{ge} \gg \omega_{gi}$

$$\lim_{\omega \to 0} \epsilon_1 = \lim_{\omega \to 0} \epsilon_2 = 1 + \frac{c^2}{v_A^2} \tag{9.144}$$

as the low-frequency dielectric constant for the extraordinary wave. This is the relevant low-frequency dielectric constant of a plasma, because at such low frequencies below the plasma frequency no ordinary wave mode can propagate. Instead the dispersion relation becomes that of the two-fluid Alfvén wave, $k^2 c^2/\omega^2 = \epsilon_1$, or

$$\omega = \pm k v_A \left(1 + \frac{v_A^2}{c^2} \right)^{-1/2} \tag{9.145}$$

whose phase velocity is reduced. As claimed earlier, the X-mode at very low frequencies close to zero frequency becomes an Alfvén wave. Since its wavenumber vanishes, too, it turns out to be a long wavelength mode. Its actual phase velocity is slightly slower than the Alfvén velocity, however. Hence, the L-X-mode very-low frequency branch is an Alfvén wave which propagates parallel as well as perpendicular to the magnetic field. Thus this two-fluid Alfvén mode is nearly isotropic.

Using Eq. (9.136) and assuming $\omega \ll \omega_{ge}$, one finds from Eq. (9.143) that for $\epsilon_1 \to 0$ another resonance appears at

$$\omega_{lh}^2 = \frac{\omega_{pi}^2 + \omega_{gi}^2}{1 + \omega_{pe}^2/\omega_{ge}^2} \tag{9.146}$$

It is called the *lower-hybrid resonance*. The numerator of the lower-hybrid resonance frequency resembles the upper-hybrid resonance frequency in Eq. (9.135) in that the electron plasma and cyclotron frequencies in the latter are replaced here with the corresponding ion frequencies. But it also contains an electron contribution in its denominator which results from the electron response to the low-frequency electric wave field, $\delta E_\perp$. The cold electrons respond to this transverse electric field by performing an electric drift motion perpendicular to the magnetic field.

In most cases the ion-cyclotron frequency can be neglected in the numerator of Eq. (9.146), if the Alfvén velocity is slower than the speed of light, $v_A^2/c^2 = \omega_{gi}^2/\omega_{pi}^2 \ll 1$. This condition may be violated in diluted plasmas or in very strong magnetic fields in which case the ion-cyclotron frequency contributes to the lower-hybrid resonance.

It is easy to show that under the condition $m_e/m_i \ll v_A^2/c^2 \ll 1$ the lower-hybrid frequency is equal to the ion plasma frequency, $\omega_{lh} = \omega_{pi}$, while for $v_A^2/c^2 > 1$ it is equal to the ion-cyclotron frequency, $\omega_{lh} = \omega_{gi}$. In the opposite case of dense plasmas the ratio of electron plasma to cyclotron frequencies in the denominator of Eq. (9.146) dominates unity. In this limit the lower-hybrid frequency approaches

$$\omega_{lh} = (\omega_{ge}\omega_{gi})^{1/2} \tag{9.147}$$

This expression finds wide application deep in the inner magnetosphere and plasmasphere where the density is high and the approximations made in deriving this limit are valid. However, in the outer magnetosphere outside the plasma sheet and in the solar wind the full expression for the lower hybrid frequency must be used.

We have sketched the low-frequency dispersion curves of the two-fluid waves in Fig. 9.13. For parallel propagation the dispersion curves of the R- and L-modes separate from the Alfvén branch at frequencies close to the ion-cyclotron resonance. For perpendicular propagation the low-frequency branch of the X-mode is an Alfvén wave which turns into the resonance at the lower-hybrid frequency, while the two upper branches remain unaffected by the ion dynamics because of their high frequencies. The non-escaping mode between the lower cut-off and the upper hybrid resonance is called *Z-mode*. The Z-mode is a high frequency perpendicular branch of the X-mode which is trapped in the plasma bouncing back and forth between its cut-offs and going into resonance where its frequency comes close to the upper hybrid frequency. It may thus play a role in energy distribution in a plasma preferentially heating the electrons in resonance. The Z- and X-modes are the main constituents of the so-called *auroral kilometric radiation*.

9.6.4   Oblique Propagation

The above theory can be extended to arbitrary directions of propagation. In this case we return to the general dispersion relation of a cold plasma in Eq. (9.114), but to include the ion contribution, so that the components of the dielectric tensor are given by Eq. (9.136). The right-hand and left-hand components of the dielectric tensor in

Eq. (9.122) are similarly redefined. These redefinitions permit us to rewrite the cold plasma dispersion relation (9.117) as function of k^2 and the angle θ between $\mathbf{B}_0$ and $\mathbf{k}$

$$\text{Det} \begin{bmatrix} \left(\dfrac{kc}{\omega}\right)^2 \cos^2\theta - \epsilon_1 & i\epsilon_2 & -\left(\dfrac{kc}{\omega}\right)^2 \sin\theta\cos\theta \\[2mm] -i\epsilon_2 & \left(\dfrac{kc}{\omega}\right)^2 - \epsilon_1 & 0 \\[2mm] -\left(\dfrac{kc}{\omega}\right)^2 \sin\theta\cos\theta & 0 & \left(\dfrac{kc}{\omega}\right)^2 \sin^2\theta - \epsilon_3 \end{bmatrix} = 0 \quad (9.148)$$

This equation can be solved either for the wavenumber, k^2, or for the angle, θ. Writing it as an equation for k^2, it becomes a biquadratic equation in $N^2 = k^2 c^2 / \omega^2$

$$AN^4 - BN^2 + C = 0 \tag{9.149}$$

The coefficients of this equation are

$$
\begin{aligned}
A &= \epsilon_1 \sin^2\theta + \epsilon_3 \cos^2\theta \\
B &= \epsilon_R \epsilon_L \sin^2\theta + \epsilon_1 \epsilon_3 (1 + \cos^2\theta) \\
C &= \epsilon_3 \epsilon_R \epsilon_L
\end{aligned}
\tag{9.150}
$$

The solution of Eq. (9.149) is the *Appleton-Hartree equation*

$$N^2 = \frac{B \pm \sqrt{B^2 - 4AC}}{2A} \tag{9.151}$$

which shows that there are two principal propagating wave solutions, which may have different branches depending on the sign of the expression under the square root. For parallel and perpendicular propagation we have identified these modes as the (R,L)- and (O,X)-modes, respectively. For oblique angles these modes are smoothly connected.

Solving for the angular function, one finds after accordingly rearranging the Eq. (9.151) and use of the definitions of the above coefficients A, B, C that

$$\tan^2\theta = -\frac{\epsilon_3 (N^2 - \epsilon_R)(N^2 - \epsilon_L)}{(\epsilon_1 N^2 - \epsilon_R \epsilon_L)(N^2 - \epsilon_3)} \tag{9.152}$$

From which the angular dependence of the resonances, $k \to \infty$, may be determined as

$$\tan^2\theta_{res} = -\epsilon_3 / \epsilon_1 \tag{9.153}$$

For a pure electron plasma this condition can be rewritten into a biquadratic equation

$$\omega_{res}^4 - \omega_{uh}^2 \omega_{res}^2 + \omega_{ge}^2 \omega_{pe}^2 \cos^2\theta = 0 \tag{9.154}$$

It solutions yield two resonance frequencies, whose variation with θ is plotted in Fig. 9.14 for a dense, $\omega_{pe} > \omega_{ge}$, and a dilute, $\omega_{ge} > \omega_{pe}$, plasma. In the first case

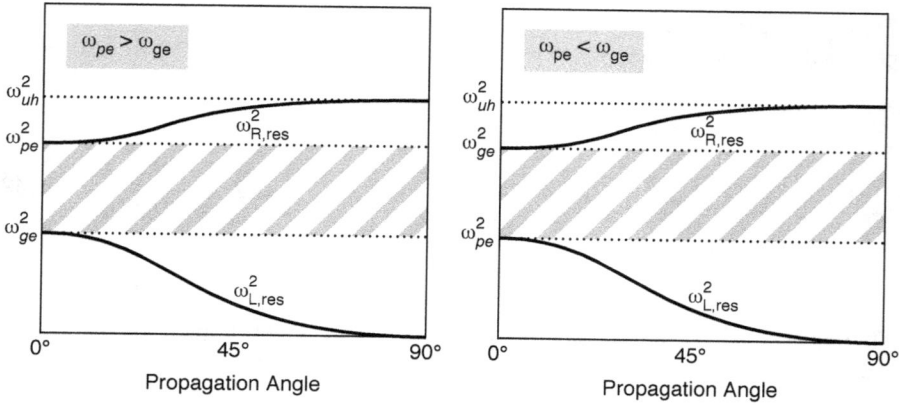

Figure 9.14: Angular variation of cut-offs and resonances.

the upper-hybrid resonance at perpendicular propagation moves to a resonance at the plasma frequency for parallel propagation, while the cyclotron resonance decreases from ω_{ge} at $0°$ to zero at $90°$. In the second case the cyclotron resonance increases from the cyclotron frequency at parallel propagation to the upper-hybrid frequency for perpendicular propagation, while the other resonance moves from zero at $90°$ and to ω_{pe} at $0°$.

Including the ion contribution leads only to minor changes in the angular dependence of the higher frequency resonances. However, the lower-hybrid resonance becomes weakly sensitive on angle. If we approximate Eq. (9.153) for low frequencies, $\omega_{gi}^2 \ll \omega^2 \ll \omega_{ge}^2$, and solve for the lower-hybrid frequency we find

$$\omega_{lh}^2 = \frac{\omega_{pi}^2 + \omega_{pe}^2 \cot^2 \theta}{1 + \omega_{pe}^2/\omega_{ge}^2} \tag{9.155}$$

For perpendicular propagation we have $\cot \theta = 0$ and Eq. (9.155) becomes identical with the corresponding approximation to Eq. (9.146), while for parallel propagation this expression breaks down due to our approximation. Writing (9.153) as

$$\epsilon_3 + \epsilon_1 \tan^2 \theta = 0 \tag{9.156}$$

we find that for $\theta = 0$ this condition implies that $\epsilon_3 = 0$ which means that for parallel propagation the resonance approaches the value $\omega_{res,\|} = \omega_{pe}(1 + m_e/m_i)^{1/2}$. The lower-hybrid resonance thus smoothly approaches a frequency close to the plasma frequency. The general expression for the cut-offs follows from setting $C = 0$ in Eq. (9.151). This is equivalent to putting $\epsilon_3 \epsilon_R \epsilon_L = 0$. The latter does not depend on angle and, hence, the cut-off frequencies do not change with the angle of propagation.

As a final example, we consider the change in the whistler dispersion relation when including the angle of propagation. In this case Eq. (9.124) becomes

$$\frac{k^2 c^2}{\omega^2} = 1 + \frac{\omega_{pe}^2}{\omega(\omega_{ge} \cos \theta - \omega)} \tag{9.157}$$

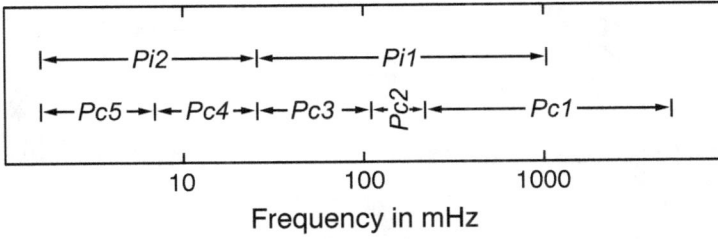

Figure 9.15: Frequency ranges of ULF continuous (Pc) and irregular (Pi) magnetic pulsations.

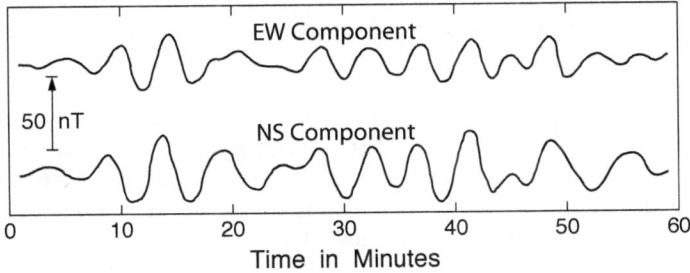

Figure 9.16: Ground magnetic disturbance of a Pc5 pulsation.

The whistler resonance frequency decreases from the cyclotron frequency at parallel propagation to a certain angle, $\cos \theta_{res} = \omega / \omega_{ge}$, beyond which the whistler does not propagate anymore. This is the *whistler resonance angle*. A whistler with a given frequency is confined to a certain cone of propagation around the field line.

Investigation of wave propagation in a cold plasma has turned out to be a complicated play with the dispersive properties of the different wave modes. An exhaustive encyclopedia is beyond the scope of this text, but we note that the *CMA-diagram* given in Appendix C.5 describes a useful graphical tool for two-fluid plasma wave analysis.

9.7   Geomagnetic Pulsations

As an example of the application of the above theory to the Earth's environment we turn to the discussion of a phenomenon which is known in geomagnetism since more than one century. This is the existence of fast fluctuations of the Earth's surface magnetic field in the frequency range from a few Millihertz up to a few Hertz, corresponding to oscillation periods from several hundred seconds to a fraction of a second. Here we are in the ultra-low frequency range which is conventionally divided into five intervals, Pc1–Pc5, for *continuous pulsations* and into two intervals, Pi1 and Pi2, for *irregular pulsations* (see Fig. 9.15). In many cases, the pulsating disturbance fields observed are associated with Alfvén waves

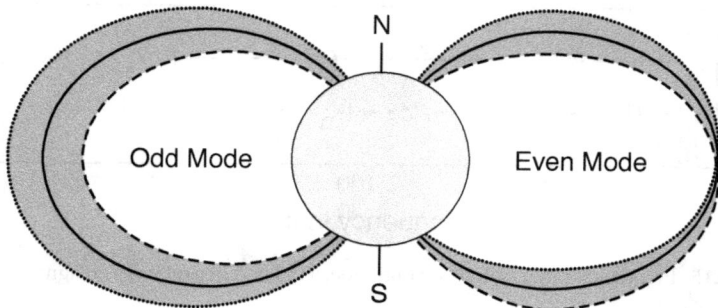

Figure 9.17: Fundamental poloidal field line resonances.

The class of continuous pulsations covers quasi-sinusoidal oscillations of narrow spectral bandwidth, like shown in Fig. 9.16. They may have a comparably long duration from several minutes up to hours. Pc pulsations can generally be observed over a wide latitudinal and longitudinal range on the Earth's surface and in the magnetosphere, but their frequencies and amplitudes often exhibit a latitudinal variation.

The irregular pulsations, in contrast, are shorter-lived, each pulsation sometimes being composed of only a few oscillations decaying in time and neither being of sinusoidal form nor having a well-expressed spectral peak. Rather they have a broad spectrum. Pi pulsations are usually also more localized, both in latitude and longitude and have similar spectrum over the region where they are observed.

9.7.1   Field Line Resonance

The Pc5 pulsations shown in Fig. 9.16 are caused by oscillations of and entire magnetospheric field lines being of long wavelength. The simplest wave mode which can be called for the explanation of standing magnetic field line oscillations is the single-fluid shear Alfvén wave. For a standing wave the length of the field line between the two reflection points, ℓ, must be a multiple of half the parallel wavelength or $\nu_h \lambda_\parallel = 2\ell$, where $\nu_h = 1, 2, 3, \ldots$. From the dispersion relation of the shear Alfvén wave, $\omega = k_\parallel v_A$, one finds for the possible oscillation frequencies of standing Pc oscillations

$$\omega_h = \frac{\nu_h \pi \langle v_A \rangle}{\ell} \tag{9.158}$$

where $\langle v_A \rangle$ is the average Alfvén velocity along the field line. This approximate formula shows that each particular field line has a number of distinct Alfvénic resonances. Since the length of the field lines increases with latitude, the resonance frequency decreases with latitude. For an average Alfvén velocity of 1000 km/s, the fundamental resonance frequency on closed field lines, outside the polar cap, ranges between 1 and 100 mHz and thus falls into the Pc3 to Pc5 range.

Figure 9.17 shows schematically how the dipolar field configuration changes for two fundamental types of field line resonances. The footpoints of the field lines are fixed on the Earth while the dipole field lines may either perform a breathing motion (fundamental odd mode) or a wobbling motion (even mode). In addition to these poloidal modes, the field lines can also exert toroidal oscillations, in which case the elongation of the field line and the plasma bulk flow are purely azimuthal. Pc pulsations often are a mixture of poloidal and toroidal oscillations.

The Alfvénic oscillations of the magnetospheric field lines are, of course, associated with oscillations of the plasma bulk flow velocity. Figure 9.18 shows a spacecraft measurement of the perpendicular magnetospheric plasma drift at geostationary altitude, decomposed into azimuthal, v_ϕ, and radial, v_r, components for the same Pc5 pulsation event shown in Fig. 9.16. Both, on the ground and in space the phase difference between the components indicates elliptic polarization. In particular the plasma drift velocity exhibits a large phase shift of about $180°$ with v_ϕ being ahead of v_r.

9.7.2 Sources of Pc Pulsations

One can imagine the field line resonances to occur as modes which are excited on the entire magnetosphere. Some periodic disturbance of frequency ω_{ex} arriving at the magnetopause may set the field line with resonance frequency $\omega_{res} = \omega_{ex}$ into oscillation while all other field lines, whose resonance frequencies do not match ω_{ex}, are only marginally excited and do not contribute to the pulsation. Since ω_{res} is a function of space, one particular field line or L-shell is excited.

The left-hand lower panel of Fig. 9.18 provides a schematic view of how the magnetospheric flux tubes, i.e., the field lines and their plasma content, respond to the excitation at a particular frequency, ω_{ex}. Since the field line eigenperiod varies with L, the external frequency picks out the resonant field line, where the amplitude of the flux tube motion maximizes. Interestingly, the polarization of the oscillation changes when crossing the resonance, as indicated in the figure. It is nearly linear at the resonant shell, L_{res}.

The second part of the lower panel in Fig. 9.18 shows a ground-based radar measurement of the latitudinal dependence of a Pc5 field line resonance region in high latitudes. This measurement shows the latitudinal resonance width being of about $1°$ which indicates that the source wave which fell into resonance may not have been monochromatic but has resonantly excited an entire $1°$ in latitude wide geomagnetic flux tube of about 1000 km radial width in the magnetosphere.

Concerning the initial wave source which excites the field line resonance, there are two possible scenarios. Surface waves excited via the Kelvin-Helmholtz instability by the flow of the solar wind around the magnetopause (left part of Fig. 9.19) can set the inner magnetosphere into oscillation and become resonant at the resonant L-shell, L_{res}. For Pc5 pulsations this mechanism seems to work and the polarization pattern is in rough agreement with the polarization pattern of the surface wave showing left-handed (LH) oscillations at dawn and right-handed (RH) oscillations at dusk. Other possibilities are compressional waves which enter the magnetosphere at its nose or

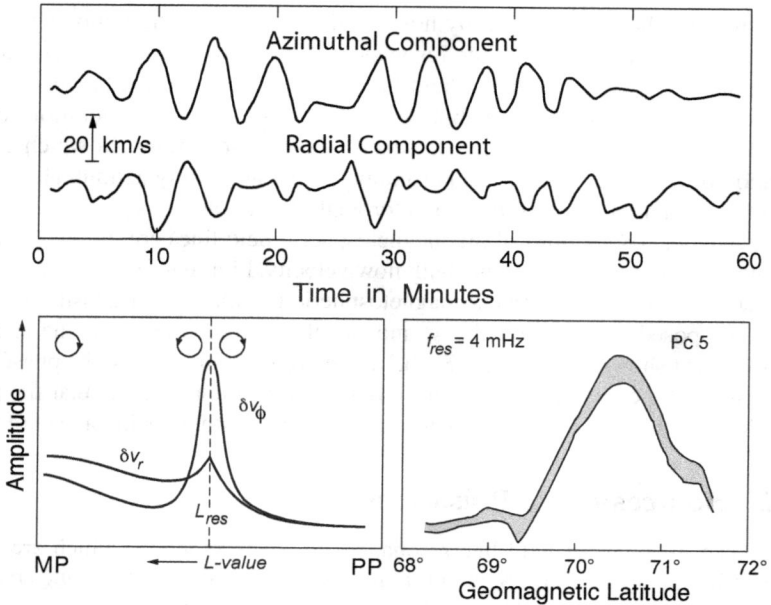

Figure 9.18: *Top*: Response of the magnetospheric plasma drift velocity during a Pc5 pulsation event for both, the azimuthal and radial velocity components. These are about out of phase. *Bottom*: *L*-dependence (left) and azimuthal dependence (right) of the amplitude of Pc5 field line resonance. In the two velocity components the resonance peaks at the resonant field line. The change polarisation across the resonance is indicated by circles. Maximum resonance at 4 mHz occurs between 70° to 71° latitude.

turbulence from the magnetosheath which may enter directly through the cusps and leak into the magnetosphere (right-hand panel of Fig. 9.19).

9.7.3 Pi2 Pulsations

The short-period irregular Pi2 pulsations are associated with the development of the substorm current wedge described in Section 5.7. Whenever field-aligned currents are suddenly switched on somewhere in the magnetosphere, they must be transported to the ionosphere via shear Alfvén waves. Only this transverse magnetohydrodynamic wave mode can carry field-aligned current. Launched in the magnetosphere, the shear Alfvén waves are then reflected back and forth between the ionosphere and the current generator in the tail until a stationary equilibrium is reached.

Figure 9.20 shows qualitatively the development of the magnetic disturbance field and thus the field-aligned current flow after the sudden switch-on of a hypothetical current generator in the magnetotail with the current being allowed to decay with given time constant. At $t = 0$ an Alfvén wave is launched which carries a current corresponding to the generator current and thus has a magnetic disturbance field, ΔB_1,

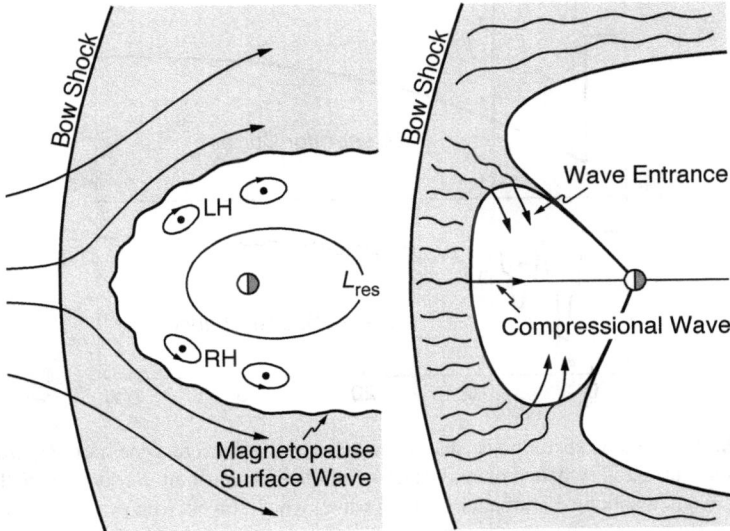

Figure 9.19: Schematic of magnetopause surface waves (for instance excited by the Kelvin-Helmholtz resonance at the magnetopause) and magnetosheath turbulence as sources of pulsations. Kelvin-Helmholtz waves though should exhibit a dawn dusk asymmetry, however. Magnetosheath turbulence may enter the magnetosphere either through reconnection at the magnetopause along the field lines, through the cusps, or by compressive transmission of momentum to the magnetospheric magnetic flux tubes setting them into oscillation.

corresponding to the magnetic disturbance caused by the generator current, $\Delta B_1 = \Delta B_g(0)$. This wave reaches the ionosphere at $t = \tau_A$, the Alfvén wave travel time between magnetosphere and ionosphere of some 30 s. Here about 80% of its amplitude is reflected with the magnetic field of the reflected wave, ΔB_2, adding to the primary disturbance field, ΔB_1.

At $t = 2\tau_A$ the reflected wave comes back to the generator and launches a third wave, whose magnetic disturbance, ΔB_3, is determined by the requirement that the total disturbance matches that caused by the current of the generator at time $t = 2\tau_A$, namely $\Delta B_g(2\tau_A) = \Delta B_1 + \Delta B_2 + \Delta B_3$. Since typically $\Delta B_1 + \Delta B_2 > \Delta B_g(0) > \Delta B_g(2\tau_A)$, the magnetic disturbance of the third wave must be of opposite polarity and thus decrease the total wave magnetic field. Multiple bounces of the wave lead to magnetic field disturbances which oscillate with a period of $4\tau_A$ until they finally converge to match the generator current. In particular, the magnetic oscillations with periods of some 100 s are readily observable as Pi2 pulsations. They are often used as a good indicator for the onset of substorms.

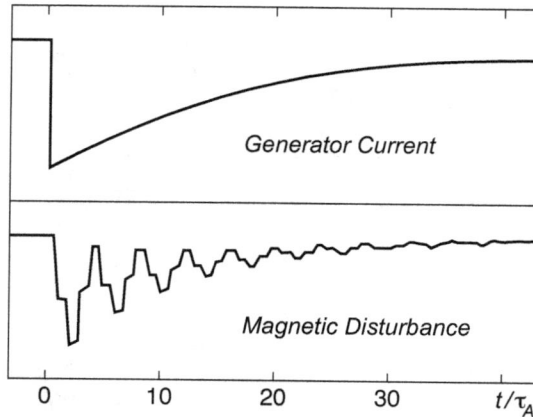

Figure 9.20: Magnetic disturbance due to switch-on of a current generator. *Top panel*: The hypothetical current is switched on and decays with its time constant. *Bottom panel*: The corresponding magnetospheric disturbance (Alfvén wave) which travels with period τ_A several times back and forth between the current source and the ionosphere along the field line. The wave is partially reflected from the ionosphere while the decaying current feeds or damps it each time it arrives (see the description in the text).

9.8   Concluding Remarks

The fluid theory of plasma waves is only the first step in a more sophisticated approach to plasma wave generation and propagation. Since it is based on the fluid model, it neglects all microscopic particle correlations. Such an approach is justified on relatively large scales, if the waves have sufficiently long wavelengths and sufficiently low frequencies. Moreover, dissipative effects must be negligible to justify the ideal plasma model.

The first restriction does not provide any problem as long as the three basic magnetohydrodynamic modes are considered. Neglecting dissipation for these waves is also justified, except in the ionosphere and the lower solar corona. It is, however, not a big problem to include resistive dissipation into the discussion of these waves. Since in the linear treatment the conductivity does not become infinite in a resistive plasma and is independent of the electric field, a simple replacement of the wave frequency by $\omega \to \omega - i\nu_c$, where ν_c is the classical or anomalous collision frequency, will account for dissipative effects. Inclusion of collision frequencies or dissipation simply leads to *wave damping*. As a result the amplitude of the wave decays with time and the wave energy turns into heat, which causes an increase of the plasma temperature and ultimately violates the assumption of a cold plasma. Temperature effects can be taken into account in a fluid approach by including the pressure term into the basic equations, but leads to further splitting of the fundamental wave modes.

The fluid approach to high-frequency electromagnetic waves is justified because for such frequencies the plasma reacts passively like a dielectric medium.

Nevertheless, in the following chapters we will encounter some important microscopic effects on electromagnetic radiation as well.

Further Reading

An exhaustive elementary description of linear fluid plasma waves in one- and two-fluid theory is given in [1]. A compendium of wave propagation in the magnetosphere and ionosphere can be found in [2] also introducing magneto-ionic wave theory. For more elaborate theories of wave phenomena consult [6, 7]. In particular, monograph [6] is a classic among the plasma wave books. However, as its starting point it takes the kinetic theory which will we will develop in the next chapter. Though its approach is intended to be systematic, it is not very useful as a textbook. A comparable, very complete and systematic classical monograph which appeared in Russian even earlier is [3]. Rather one may take it of a (now out of date) reference volume. If one is interested in whistlers, the old monograph [5] is still very useful. Further information about pulsations is contained in [4] and the references therein.

References

[1] J. A. Bittencourt, *Fundamentals of Plasma Physics* (Pergamon Press, Oxford, 1986).

[2] K. G. Budden, *Radio Waves in the Ionosphere* (Cambridge University Press, Cambridge, 1961).

[3] V. L. Ginzburg, *The Propagation of Electromagnetic Waves in Plasmas* 2nd ed. of Engl. translation (Pergamon Press, Oxford, 1970).

[4] K. H. Glaßmeier, in *Handbook of Atmospheric Electrodynamics, Vol. 2*, ed. H. Volland (CRC Press, Boca Raton, 1995), p. 463.

[5] R. A. Helliwell, *Whistlers and Related Ionospheric Phenomena* (Stanford University Press, Stanford, 1965).

[6] D. G. Swanson, *Plasma Waves* (Academic Press, Boston, 1989).

[7] T. H. Stix, *Waves in Plasma* (American Institute of Physics, New York, 1992).

Problems

Problem 9.1 *The Langmuir wave is the simplest high-frequency electrostatic wave in a plasma. Show that this also holds for magnetised plasmas along the magnetic field. What happens in a direction oblique to the field?*

Problem 9.2 *When deriving the dispersion relation of Langmuir waves we have set the temperature as constant. This is barely justified because the oscillations are so fast that no thermal exchange can take place. Hence the temperature must follow an adiabatic law. Take the adiabatic variation of the temperature into account and find the modified dispersion relation. What is the condition for neglecting the temperature variation?*

Problem 9.3 *From the Langmuir wave dispersion relation calculate the Langmuir wave phase and group velocities. Show that for large wavenumbers the phase velocity approaches the electron thermal velocity while the group velocity stays always smaller than it.*

Problem 9.4 *Figure 9.3 shows two observations of Langmuir waves at very different frequencies. What was the plasma density in these both cases? At what distance from the Sun has the solar wind measurement been made? Hint: Refer to the canonical dependence of the solar wind density on radial distance from the Sun.*

Problem 9.5 *Justify the second version of the ion-acoustic wave dispersion relation Eq. (9.24). This version of the ion-acoustic wave is most appropriate at long-wavelengths. Show that the dispersion relation can also be given another form which is appropriate near the ion plasma frequency ω_{pi}.*

Problem 9.6 *Compare the bandwidths of Langmuir and ion-acoustic waves in Fig. 9.3 in order to get a feeling for which of the waves is narrow and which one is broadband.*

Problem 9.7 *Calculate the phase and group velocities for Langmuir and ion-acoustic waves. Calculate the energy contained in both wave types.*

Problem 9.8 *Prove that the dielectric tensor of the plasma which enters the general plasma dispersion relation must satisfy the symmetry relation Eq. (9.54).*

Problem 9.9 *Calculate the equivalent mass of an ion-acoustic plasmon. Hint: For this you will need the form of the ion-acoustic dispersion relation you found in Problem 9.5.*

Problem 9.10 *Derive the dispersion relation of the single-fluid magnetosonic waves. Discuss its angular dependence and the asymmetry of its wave front.*

Problem 9.11 *Show explicitly that the cold plasma dispersion relation takes the for of Eq. (9.117).*

Problem 9.12 *Explain what is meant when speaking of polarisation of a wave. Use the two expressions (9.118) and (9.119).*

Problem 9.13 *Derive the expressions for the R- and L-mode in circular polarisation.*

Problem 9.14 *Give an interpretation of the dispersion of the whistlers seen in Fig. 9.9.*

Problem 9.15 *Prove the expression for the X-mode Eq. (9.134).*

Problem 9.16 *We derived the lower-hybrid frequency as a resonance frequency. It depends on the ratio of plasma to cyclotron frequencies. Give this expression another form. What do you conclude? Also discuss its angular dependence for oblique propagation.*

Problem 9.17 *What would be the reason for a surface wave if excited and travelling on the equatorial magnetopause surface to correspond to a right-hand polarisation on the dusk side and a left-hand polarisation on the inner edge of the magnetopause? Would one expect to find a reason in the magnetosphere for supporting or braking such an effect?*

Problem 9.18 *The magnetopause as a tangential discontinuity does not allow for matter to cross from one side to the other. How can on the closed dayside magnetopause a compressional wave enter into the magnetosphere across the magnetopause? Compare with reconnection.*

— 10 —

Wave Kinetic Theory

The most general theory of plasma waves makes use of the kinetic theory of a plasma. As has been demonstrated in Chapter 6, the only approximations underlying the kinetic theory are the statistical assumption and the assumption that higher order particle correlations can be neglected in a statistical theory based on a perturbation approach. But in contrast to the fluid theory of plasma waves, the wave kinetic theory takes explicitly care of the properties of the particle distribution function and its variations and of the correlations between particles and fields. Hence, entirely new effects will appear in this theory which cannot be covered by the fluid approach to a plasma.

Because in wave kinetic theory we are dealing with distribution functions and their evolution, the set of mass, momentum and energy conservation equations of fluid theory is replaced by the set of Vlasov equations for the different components of the plasma while the field equations remain the same. This implies that for the investigation of linear waves the formal structure of the general linear wave dispersion relation in Eq. (9.57) remains unchanged. The only quantity to be replaced is the dielectric tensor, $\in(\omega, \mathbf{k})$, because it contains the particle dynamics through the relation between the current density and the electric field, Ohm's law, and through Poisson's equation for the charge density. The former we have already needed extensively while the latter assumes a more important role in the kinetic theory. Thus the explicit calculation of the dielectric tensor from the particle distributions and particle dynamics is instrumental to the subsequent investigation of the kinetic properties of the plasma waves.

10.1   Landau-Laplace Procedure

In kinetic plasma wave theory the microscopic electric charge separation fields and particle currents become important. The dispersive properties of the plasma will therefore look rather complicated. Because of this reason, we start by neglecting any magnetic fields and thus considering the plasma as unmagnetized and the waves as purely electrostatic. One should, however, be aware that these assumptions are valid only when there are no oscillating microscopic currents (or fast particle motions) in the plasma, which is a poorly satisfied assumption when considering high-frequency electron oscillations against ions at rest. But in a plasma where the electron temperature is sufficiently small compared to the electron rest energy, i.e., in a nonrelativistic plasma with $v_e \ll c$, the electromagnetic part of the waves is usually weak.

Let us initially consider only the one-dimensional case. Under these assumptions the Vlasov equation Eq. (6.20) for electrons and ions simplifies considerably

$$\frac{\partial f_{e,i}(v,x,t)}{\partial t} + v\frac{\partial f_{e,i}(v,x,t)}{\partial x} \pm \frac{e}{m_{e,i}}E(x,t)\frac{\partial f_{e,i}(v,x,t)}{\partial v} = 0 \tag{10.1}$$

The electric field is purely electrostatic and satisfies the Poisson equation

$$\frac{\partial E(x,t)}{\partial x} = \frac{e}{\varepsilon_0}\int\limits_{-\infty}^{\infty} dv\,[f_i(v,x,t) - f_e(v,x,t)] \tag{10.2}$$

where on the right-hand side we have expressed the particle densities through the moments of the corresponding distribution functions. Although the Poisson equation is a linear equation, the system of Eqs. (10.1) and (10.2) is nonlinear, because eliminating the electric field from the Vlasov equation with the help of Poisson's equation produces a product between the partial derivative of the distribution function with respect to the velocity in the last term of the Vlasov equation and an integral over the distribution function. In addition, the Poisson equation couples the two particle species together.

Considering waves on an otherwise quiet background requires a separation of the distribution functions and the electric field into undisturbed parts, $f_{e,i0}, E_0$, and perturbations, $\delta f_{e,i}, \delta E$, according to

$$f_{e,i} = f_{e,i0} + \delta f_{e,i} \tag{10.3}$$
$$E = E_0 + \delta E \tag{10.4}$$

We now assume that the perturbations of the distribution function are linear, i.e., they are, for all times and at all positions in the plasma, much smaller then the local undisturbed value of the distribution, $|\delta f| \ll f_0$. The physical content of this assumption is that the probability to find a given number of particles at any time and position in the equilibrium state is much larger than to find them in an excited and perturbed state. Hence, the unperturbed particle density will be much larger than the density of particles participating in the oscillation. This assumption permits us to linearize the Vlasov-Poisson system of equations in order to obtain the linear set

$$\frac{\partial \delta f_{e,i}}{\partial t} + v\frac{\partial \delta f_{e,i}}{\partial x} \pm \frac{eE_0}{m_{e,i}}\frac{\partial \delta f_{e,i}}{\partial v} = \mp\frac{e\delta E}{m_{e,i}}\frac{\partial f_{e,i0}}{\partial v} \tag{10.5}$$

$$\frac{\partial \delta E}{\partial x} = \frac{e}{\varepsilon_0}\int\limits_{-\infty}^{\infty} dv\,[\delta f_i - \delta f_e] \tag{10.6}$$

for $\delta f_{e,i}(v,x,t)$ and $\delta E(x,t)$. Yet these equations are still very general, but we can simplify them if we suppose that the undisturbed distribution function varies very slowly in time and space compared to the perturbation. In such a case the equilibrium distribution, $f_0(v)$, becomes a function of velocity only. In addition the undisturbed plasma should be in a quasineutral state. Then $E_0 = 0$ and the third term on the left-hand side of the linearized Vlasov equation (10.5) vanishes. The remaining three coupled equations constitute a linear but inhomogeneous system of partial integro-differential equations in the three variables v, x, t with constant coefficients for the perturbations of the ion and electron distribution functions. After solving them, the electric perturbation field can be calculated separately.

Before proceeding, remember that we wanted to calculate the dielectric function of the plasma. Referring to its definition in Eq. (9.56) this requires to find the plasma wave conductivity, $\sigma(\omega, \mathbf{k})$, which in our one-dimensional and isotropic case is a scalar and is determined from the perturbed current density

$$\delta j = e n_0 (\delta v_i - \delta v_e) = \sigma \delta E \tag{10.7}$$

by comparing the coefficients of δE, after having expressed the perturbations of the particle velocities calculated from the perturbed distribution functions through the electric perturbation field.

10.1.1   Langmuir Waves

In order to demonstrate the method of how to solve the Vlasov-Poisson system we restrict ourselves to the case of high-frequency perturbations applied to the plasma. For frequencies far above the ion plasma frequency the ions can then be considered as an immobile stationary charge-neutralizing background. The Vlasov-Poisson system reduces to the two equations

$$\frac{\partial \delta f(v,x,t)}{\partial t} + v \frac{\partial \delta f(v,x,t)}{\partial x} \;=\; \frac{e}{m_e} \delta E(x,t) \frac{\partial f_0(v,x,t)}{\partial v} \tag{10.8}$$

$$\frac{\partial \delta E(x,t)}{\partial x} \;=\; -\frac{e}{\varepsilon_0} \int_{-\infty}^{\infty} dv\, \delta f(v,x,t) \tag{10.9}$$

We have dropped the index e on the electron distribution function here in order to simplify notation. We also note that for the purely electrostatic longitudinal perturbations the electric field can be written as the gradient of an electrostatic potential

$$\delta E = -\frac{\partial \phi}{\partial x} \tag{10.10}$$

There are two ways to solve the above linear system of Vlasov-Poisson equations. Because the system is linear, one may use the plane wave ansatz which corresponds to a solution by Fourier transformation both in space and time. This method implicitly assumes that the plasma is periodic in space and time. However, time evolves on a linear scale and assuming periodicity is physically not entirely justified. To make physics more transparent we therefore choose the more involved method of Fourier transforming in space, while solving the time-dependent equations in a different way.

Fourier transformation in space yields the following representation of the perturbations of the distribution function and electric field

$$[\delta f(v,x,t), \delta E(x,t)] = (2\pi)^{-1/2} \int_{-\infty}^{\infty} dk \, [\delta f(v,k,t), \delta E(k,t)] \, e^{ikx} \qquad (10.11)$$

where the Fourier transform of the electric field can also be expressed by the Fourier
transformed potential as $\delta E(k,t) = -ik\phi(k,t)$, and the linearized Poisson equation
(10.9) yields

$$\delta E(k,t) = \frac{ie}{\varepsilon_0 k} \int_{-\infty}^{\infty} dv \, \delta f(v,x,t) \qquad (10.12)$$

while from the Vlasov equation (10.8) we obtain the differential equation

$$\frac{\partial \delta f(k,v,t)}{\partial t} + ikv \, \delta f(k,v,t) - \frac{e}{m_e} \delta E(k,t) \frac{\partial f_0(k,v,t)}{\partial v} = 0 \qquad (10.13)$$

for the Fourier transformed perturbation of the distribution function.

10.1.2 Landau-Laplace Procedure

We solve this equation by taking the Laplace transform of the two unknown functions

$$[\delta f(k,v,p), \delta E(k,p)] = \int_{0}^{\infty} dt \, [\delta f(k,v,t), \delta E(k,t)] \, e^{-pt} \qquad (10.14)$$

under the assumption that this integral converges. This assumption requires that the
real part of p is sufficiently large and positive, larger than any possible negative real
part of any time-dependent exponential factor contained in $\delta E, \delta f$. The Laplace transform explicitly takes into account that the perturbation evolves in time. The Laplace
transformed Eqs. (10.12) and (10.13) are

$$(p+ikv)\delta f(k,v,p) \quad - \quad \frac{e}{m_e} \delta E(k,p) \frac{\partial f_0(v)}{\partial v} = g(k,v) \qquad (10.15)$$

$$\delta E(k,p) \quad = \quad \frac{ie}{\varepsilon_0 k} \int_{-\infty}^{\infty} dv \, \delta f(k,v,p) \qquad (10.16)$$

The inhomogeneity appearing on the right-hand side of the upper equation is the initial
value of the perturbed distribution, $g(k,v) = \delta f(k,v,t=0)$. It contains the asymmetric time-behavior of the perturbation. The inhomogeneous set of algebraic equations

(10.15) and (10.16) has a well-defined unique solution for the evolution of the disturbances of the electron distribution function and the electric field. The Fourier-Laplace transform of this solution is

$$\delta f(k, v, p) = (p + ikv)^{-1} \left[\frac{e}{m_e} \delta E(k, p) \frac{\partial f_0(v)}{\partial v} + g(k, v) \right] \tag{10.17}$$

$$\delta E(k, p) = \frac{ie}{\varepsilon_0 k \in (k, p)} \int_{-\infty}^{\infty} dv \frac{g(k, v)}{p + ikv} \tag{10.18}$$

The new term $\in (k, p)$ appearing in the denominator of the solution for $\delta E(k, p)$ and, as a consequence, also in the first term on the right-hand side of the solution for the perturbed distribution function is given by

$$\in (k, p) = 1 - \frac{i\omega_{pe}^2}{n_0 k} \int_{-\infty}^{\infty} dv \frac{\partial f_0(k, v, p)/\partial v}{p + ikv} \tag{10.19}$$

where we used the definition of the electron plasma frequency, $\omega_{pe}^2 = e^2 n_0 / \varepsilon_0 m_e$. To find the physically relevant solution one must transform back to real time. Inversion of the Laplace transform of the electric perturbation field gives

$$\delta E(k, t) = \frac{1}{2\pi i} \int_{a-i\infty}^{a+i\infty} dp\, e^{pt} \delta E(k, p) \tag{10.20}$$

Here a is a real 'large enough' constant, and the contour of integration is a line which is parallel to the imaginary axis in the complex p plane to the right of all singularities of the integrand in Eq. (10.18) in order to warrant convergence

$$\delta E(k, t) = \frac{e}{2\pi \varepsilon_0 k} \int_{a-i\infty}^{a+i\infty} dp\, e^{pt} \left[\frac{\displaystyle\int_{-\infty}^{\infty} dv \frac{g(k, v)}{p + ikv}}{1 - \frac{i\omega_{pe}^2}{n_0 k} \displaystyle\int_{-\infty}^{\infty} dv \frac{\partial f_0(k, v, p)/\partial v}{p + ikv}} \right] \tag{10.21}$$

In this expression the integrals in the numerator and denominator contain poles at $v = ip/k$ even for the analytic functions $g(k, v), \partial f_0(v)/\partial v$. In addition the Laplace integral may has singularities at all points where $\in (k, p) = 0$. Let us therefore assume that at least in a physically acceptable solution, $g(k, v)$ and $\partial f_0(v)/\partial v$ are analytic functions, so that the integrals over v can be analytically continued. In this case we can apply the simple analytic theory of Appendix B.7. Let us further assume without restriction of generality that $k > 0$. Then the two integrals over v become entire functions. The only possible singularities of the integral in Eq. (10.21) are the poles of its denominator

$$\in (k, p) = 0 \tag{10.22}$$

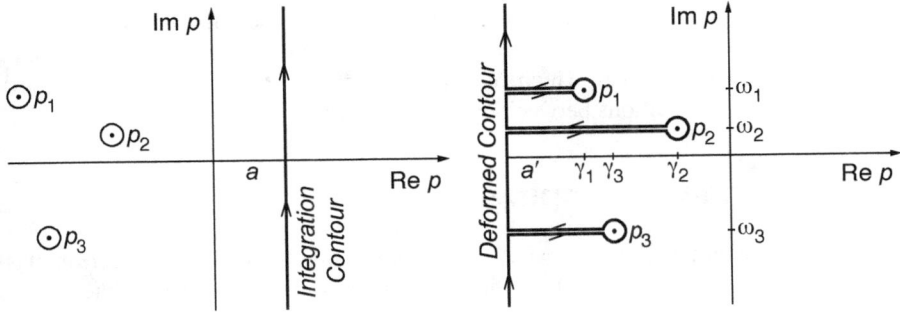

Figure 10.1: Integration contours and poles of Eq. (10.21). The integration path is right-handed parallel to the imaginary axis at negative real distance a' to the left of all poles. To include the poles on its right the integration must be deformed, encircle the pole and return to the path. The two straight excursions parallel to the real axis cancel leaving only the contribution of the pole.

Equation (10.22) may have a finite number of solutions, $p_i(k)$. These are the poles of the integral in Eq. (10.21). Let us choose the position, a, of the path of integration in Eq. (10.21) so that it is larger than the largest of the real parts of the $p_i(k) = \gamma_i(k) - i\omega_i(k)$, where we split p_i into real and imaginary parts. Integrating along a line at $a = $ const parallel to the imaginary axis (as shown on the left-hand side of Fig. 10.1) and assuming that for physical solutions $E(k, p) \to 0$ for $|p| \to \infty$, the integral will exist, and we can deform the contour of integration by pulling it into the negative direction of the real p axis to a position $a = a'$, far beyond all the poles. It will consist of small circles around the poles and a piecewise continuous path parallel to the imaginary p axis with horizontal connections to the poles along which the contributions to the integral cancel (see right-hand side of Fig. 10.1). The integral becomes the sum of all the residua, $r_i(p_i)$, at the poles, $p_i(k)$, plus the integral along the piecewise continuous path parallel to the imaginary axis

$$\delta E(k,t) = \sum_i r_i(p_i) \exp[p_i(k)t] + (2\pi i)^{-1} \int_{a'-i\infty}^{a'+i\infty} dp\, e^{pt}\, \delta E(k,p) \qquad (10.23)$$

The integral taken along $a' = $ const vanishes in the long-time limit, $t \to \infty$, as

$$\lim_{t \to \infty} \exp(-|a'|t) \to 0 \qquad (10.24)$$

so that only the sum of the residua survives. Moreover, of these residua only the one with the smallest negative real part is, in the long-time limit, of interest. For this particular pole, $p_l(k) = \gamma_l(k) - i\omega_l(k)$, the electric perturbation field assumes the asymptotic form

$$\delta E(k,t) \propto \exp[\gamma_l(k)t - i\omega_l(k)t] \qquad (10.25)$$

Hence, the linear solution of the Vlasov-Poisson set of equations is a damped oscillation of frequency, $\omega_l(k)$, and negative damping decrement, $\gamma_l(k) < 0$. As we will demonstrate below, this frequency is the Langmuir plasma oscillation frequency found in Chapter 9. But the damping decrement is entirely new. It is the so-called *Landau damping* of Langmuir waves, which does not come from collisions between the particles, but from decorrelations between particles and waves.

10.2   Landau Damping

Considering Langmuir wave propagation in a pure charge-compensated electron plasma from the point of view of kinetic theory thus leads to a number of surprises. The first surprise is that kinetic theory introduces a larger number of wave modes and in addition shows that harmonic waves do exist only in the time-asymptotic limit when the contribution of the integral at $a' = $ const vanishes fast enough. The second surprise is the appearance of a *collisionless damping*, which cannot be avoided and which is a purely kinetic effect. Actually, it appears as a damping only as long as $\gamma_i(k) < 0$. When this condition is violated, the asymptotic time-limit breaks down and one speaks of an *instability*. But under equilibrium conditions all waves will tend to damp out. In the final state of the equilibrium only thermal fluctuations survive, where the thermal particle motion causes small-scale electric fluctuations which are immediately damped out by Landau damping. It is the equilibrium between damping and the thermal motions which causes the thermal fluctuation level given in Eq. (9.29).

10.2.1   Thermal Plasma

Having clarified the general properties of the solution, let us proceed to one special equilibrium distribution function. Assuming that the disturbance is introduced into a plasma in thermal equilibrium, we choose the undisturbed distribution

$$f_0(v) = n_0 \left(\frac{m_e}{2\pi k_B T_e} \right)^{1/2} \exp\left(-\frac{mv^2}{2k_B T_e} \right) \tag{10.26}$$

to be a Maxwellian of electron temperature T_e. Its derivative with respect to v is always negative, $\partial f_0/\partial v = -(v/k_B T_e)f_0$. Therefore, for $p > 0$ and $k \neq 0$, the function $\in (k, p)$ has no zeros, as is obvious from its definition Eq. (10.19). All the poles of $\in (k, p)$ have negative real parts of p, and the wave perturbation will be damped.

In order to determine p, one must calculate $\in (k, p)$ explicitly and solve the dispersion relation in Eq. (10.22) which gives the real and imaginary parts of p at its poles. We write Eq. (10.19) in the two equivalent forms

$$
\begin{aligned}
\in(k,p) &= 1 - \frac{\omega_{pe}^2}{n_0 k^2} \int_{-\infty}^{\infty} dv \frac{\partial f_0(v)/\partial v}{v - ip/k} \\
&= 1 + \frac{\omega_{pe}^2}{n_0} \int_{-\infty}^{\infty} dv \frac{f_0(v)}{(p + ikv)^2}
\end{aligned}
\tag{10.27}
$$

where the second form arises from partial integration. The real and imaginary parts of the pole, $p = \gamma - i\omega$, are found by performing the v-integration in Eq. (10.27). Now let us assume that the real part of p at the relevant pole is small, so that in agreement with our former discussion of the surviving disturbance the damping will be small. The pole will be close to the imaginary p axis. The integration in $\in (k, p)$ is along the real v axis with v having a pole at $v = (\omega + i\gamma)/k$. Remember that $\gamma < 0$ and small. The integral in Eq. (10.27) then consists of the contribution of the pole and the principal value taken along the real axis (see Appendix B.7). The paths of integration for the three possible cases are shown in Fig. 10.2 and the integral in Eq. (10.27) can be represented as

$$\int_{-\infty}^{\infty} dv \frac{\partial f_0(v)/\partial v}{(v - ip/k)} = \begin{cases} \displaystyle\int_{-\infty}^{\infty} dv_r \frac{\partial f_0(v)/\partial v}{(v_r - ip/k)} & \gamma > 0 \\[2em] \displaystyle\int_{-\infty}^{\infty} dv_r \frac{\partial f_0(v)/\partial v}{(v_r - ip/k)} + 2\pi i \left. \frac{\partial f_0(v)}{\partial v} \right|_{v=ip/k} & \gamma < 0 \end{cases} \tag{10.28}$$

The limiting case for $\gamma = 0$ is given by the Plemelj formula (see Appendix B.7). The contribution of the negative pole is therefore

$$-2\pi i \frac{\omega_{pe}^2}{n_0 k^2} \left. \frac{\partial f_0(v)}{\partial v} \right|_{v=ip/k} \tag{10.29}$$

The main contribution to $\in (k, p)$ comes from the integral along the real axis, v_r, outside the pole. It can be obtained by taking v as real and expanding the denominator in powers of ikv/p. Using the second version in Eq. (10.27) for $\in (k, p)$. Expanding $(p + ikv)^{-2}$ up to second order, one obtains

$$\in (k, p) = 1 + \frac{\omega_{pe}^2}{n_0 p^2} \int_{-\infty}^{\infty} dv_r f_0(v_r) \left(1 - \frac{2ikv_r}{p} - \frac{3k^2 v_r^2}{p^2} \right) - 2\pi i \frac{\omega_{pe}^2}{n_0 k^2} \left. \frac{\partial f_0(v)}{\partial v} \right|_{v=ip/k} \tag{10.30}$$

The real and imaginary parts of this equation can be used to find $\gamma(k, \omega)$ and $\omega(k)$. For large γ this is quite involved, but for weakly damped disturbances one can treat γ as a small correction to p. The integral of the first term in Eq. (10.30) is the unperturbed density, n_0. The integral over the second term vanishes because the plasma is at rest, $\langle v \rangle = 0$. The integral over the third term is $n_0 k_B T_e / 2m_e$. Keeping only these three terms and dropping the small contribution of the pole, the real part $\in_r (\omega, k)$ of the dispersion relation $\in$ takes the form

$$\in_r (\omega, k) \equiv 1 - \frac{\omega_{pe}^2}{\omega^2} - \frac{3\omega_{pe}^2}{\omega^2} \frac{k^2 v_{the}^2}{\omega^2} = 0 \tag{10.31}$$

This is the dispersion relation of Langmuir waves given in Eq. (9.32) for the one-dimensional case, $\gamma_e = 3$, if we replace one of the ω^2 in the denominator of the third

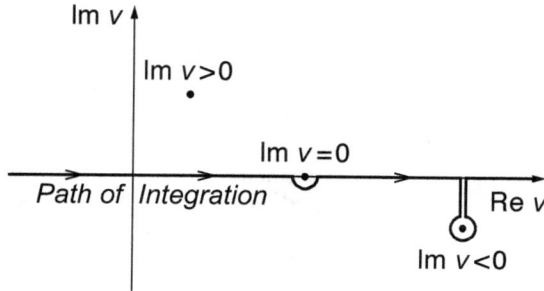

Figure 10.2: Integration contours for three positions of the pole. In this case the integration is again right-handed along the real axis. For inclusion of poles on the real axis and below it must be deformed accordingly. The pole on the real axis contributes half the residuum, while that below contributes a full residuum. For the positive pole the path needs to be not deformed.

term by ω_{pe}^2. This term turns out to be the thermal correction to the Langmuir waves and the dispersion relation, including the damping term, becomes

$$\omega_l = \pm \omega_{pe}\left(1 + \frac{3}{2}k^2\lambda_D^2\right) + i\gamma_l(k) \tag{10.32}$$

The collisionless damping decrement can now be determined by inserting $p = i\omega_l$ into the imaginary correction of Eq. (10.30) and calculating the derivative of the Maxwellian. Under the assumption that the damping is weak, $\gamma_l \ll \omega_l$, it is reasonable to expect that the imaginary part of $\in(k, \omega)$ is also small.

10.2.2 Damping Rate

Under this condition it is possible to develop a simple prescription to determine the damping rate from the dispersion relation. Let us split

$$\in(k, \omega, \gamma) = \in_r(k, \omega, \gamma) + i \in_i(k, \omega, \gamma) \tag{10.33}$$

into real and imaginary parts, $\in_r(k, \omega, \gamma)$, $\in_i(k, \omega, \gamma)$. Since $\gamma_l \ll \omega_l$, one can expand with respect to the real frequency at vanishing damping rate and obtain

$$\in(k, \omega, \gamma) = \in_r(k, \omega, 0) + i\gamma \frac{\partial \in_r(k, \omega, \gamma)}{\partial \omega}\bigg|_{\gamma=0} + i\in_i(k, \omega, 0) = 0 \tag{10.34}$$

Setting the real and imaginary parts of this expression separately equal to zero gives

$$\in_r(k, \omega, 0) \quad = \quad 0 \tag{10.35}$$

$$\gamma(k, \omega) \quad = \quad -\frac{\in_i(k, \omega, 0)}{\partial \in_r(k, \omega, \gamma)/\partial \omega|_{\gamma=0}} \tag{10.36}$$

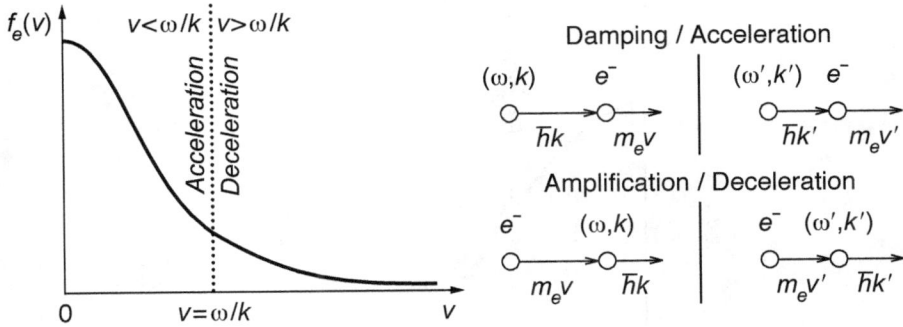

Figure 10.3: The mechanism of Landau damping. *Left:* The wave resonance is at location $v = \omega/k$. All electrons left to it are slow, those at the right are faster than the wave. Accordingly, in interaction, the faster electrons will give up energy to the wave while the slower will absorb energy from the wave. *Right:* This is shown in analogy of photon-electron collisions where a wave of frequency ω and wave number k has energy $\hbar\omega$ and momentum $\hbar k$.

The first of these equations is the usual dispersion relation depending only on real quantities. But the second equation is a very useful expression for calculating the damping rate of any weakly damped wave. In the following we will make extensive use of this expression. Applying it to the Langmuir wave dispersion relation (10.30) yields

$$\gamma_l(k) = -\left(\frac{\pi}{8}\right)^{1/2} \frac{\omega_{pe}}{k^3 \lambda_D^3} \exp\left(-\frac{1}{2k^2\lambda_D^2} - \frac{3}{2}\right) \tag{10.37}$$

as an expression for the Landau damping of high-frequency Langmuir waves in a collisionless unmagnetized plasma. This damping is not caused by particle collisions but is entirely due to particle decorrelation effects.

10.2.3 Physics of Landau-Damping

The appearance of collisionless dissipation in the dispersion of electron plasma waves is somewhat disturbing, since dissipation implies a preferred direction in time, while the Vlasov equation is reversible in time, as can be seen by substituting $t \to -t$ and $v \to -v$ into Eq. (10.1). However, Landau damping only affects a very small part of the distribution function. Only particles with velocities close to the phase velocity of the wave, $v_{ph} = \omega/k$ become resonant and are redistributed in phase space during their interaction with the wave. Hence, the directivity does not affect most of the distribution function and has thus no effect on the time symmetry of the Vlasov equation.

In order to understand the physics of Landau damping, let us consider a plasma wave propagating across a plasma in thermal equilibrium with a Maxwellian equilibrium distribution function, $f_0(v)$. The situation is depicted in the left-hand part of Fig. 10.3. Particles at position $v = v_{ph} = \omega/k$ are in resonance with the wave, since they are moving at the same speed as the wave in the plasma. Clearly, these particles

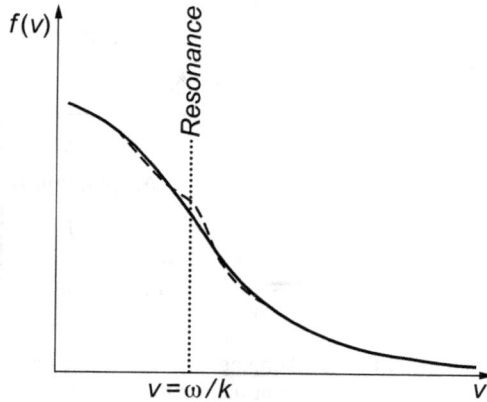

Figure 10.4: Attraction of particles by a resonance. The energy exchange retards the faster and accelerates the slower electrons thus 'attracting' them from both sides and generating a plateau on the distribution function around the resonance. This also causes some steeping of the distribution function on both sides outside the resonance. Resonant waves thus cause a local distortion on a thermal distribution function in velocity space which in principle implies a source of free thermal energy which may become available for other processes.

interact strongest with the wave electric field. Depending on the direction of this field the particles will either be accelerated or decelerated. On the other hand, any particles moving slightly faster or slower than the wave will experience a different kind of interaction.

We can investigate this kind of interaction by exploiting the simple analogy of a collision between two particles (right-hand side of Fig. 10.3), taking the wave as an uncharged particle of energy $\hbar\omega$ and momentum $\hbar k$. In a collision between the two particles (electron and wave) the one with the higher momentum will always speed up the lower momentum particle, whereby itself will loose energy.

This kind of interaction is of elastic nature and thus dissipationless. It exactly resembles the process of Landau damping as an elastic interaction between particles and waves with no preferred direction. Fast electrons will speed up the waves, while slow electrons are pushed by the wave and gain energy. But why then an effective damping of the wave? The reason is the asymmetry of the Maxwellian distribution function with respect to the plasma wave phase velocity. There are more low than high velocity particles, and, hence, the wave looses more momentum and energy in the interaction with low momentum particles than it gains back from interaction with higher momentum particles. Clearly, during this process the distribution function must necessarily become slightly distorted as shown in Fig. 10.4. The retarded and accelerated particles right and left of the resonance are attracted by the resonance and accumulate there.

We can make this argument a little more quantitative by estimating the change in energy which the electron distribution experiences during the interaction with a Langmuir wave. This change is given by the integral

$$\delta W_e = m_e \int_{-\infty}^{\infty} dv\, v \langle \Delta v \rangle f(v) \tag{10.38}$$

where $\langle \Delta v \rangle$ is the electron velocity change averaged over one Langmuir wavelength. It is convenient to transform to a system moving with phase velocity ω/k, choosing $v' = v - \omega/k$. The change in velocity, $\langle \Delta v \rangle$, is independent of such a transformation, and the above integral becomes

$$\delta W_e = m_e \int_{-\infty}^{\infty} dv' \left(v' + \frac{\omega}{k} \right) \langle \Delta v \rangle f \left(v' + \frac{\omega}{k} \right) \tag{10.39}$$

We now expand the distribution function around $v' = 0$ to obtain

$$\delta W_e = m_e \int_{-\infty}^{\infty} dv' \left(v' + \frac{\omega}{k} \right) \langle \Delta v \rangle \left[f \left(\frac{\omega}{k} \right) + v' \left. \frac{\partial f}{\partial v} \right|_{v'=\omega/k} \right] \tag{10.40}$$

It can be shown by using the electron equation of motion in an oscillating electric wave field of amplitude δE_0

$$m_e \frac{dv_e}{dt} = -e\delta E_0 \sin \left[kx(t) - \omega t \right] \tag{10.41}$$

that the average velocity variation, $\langle \Delta v \rangle$, is an odd function of v' (calculate $v_e(t), x(t)$ from the above equation of motion and substitute $x(t) = x_0 + v't$ back to determine Δv). Hence, of the four product terms in the above integral only the two terms containing the product $v' \langle \Delta v \rangle$ survive the integration from $-\infty$ to $+\infty$. After integrating Δv over one oscillation period, one finds

$$\langle \Delta v(t) \rangle = \frac{A}{k^2 v'^3} \left[\cos(kv't) - 1 + \frac{1}{2} (kv't) \sin(kv't) \right] \tag{10.42}$$

Between the constant of proportionality, A, and the wave energy density averaged over one wave oscillation, $W_w = \varepsilon_0 \,\delta E_0^2 / 2$, there is the relation

$$A = W_w \frac{\omega_{pe}^2}{n_0 m_e} \tag{10.43}$$

The change in particle energy is thus determined from the integral

$$\Delta W_e = m_e \int_{-\infty}^{\infty} dv' v' \langle \Delta v \rangle \left[f \left(\frac{\omega}{k} \right) + \frac{\omega}{k} \left. \frac{\partial f}{\partial v'} \right|_{v'=\omega/k} \right] \tag{10.44}$$

The first term in the brackets can be neglected because it adds only a small contribution which is independent of the shape of the distribution function

$$\Delta W_e \approx \frac{m_e \omega}{k} \left. \frac{\partial f}{\partial v'} \right|_{v'=\omega/k} \int_{-\infty}^{\infty} dv' v' \langle \Delta v \rangle \tag{10.45}$$

After integration and replacing $\omega \approx \omega_{pe}$, we get

$$\Delta W_e(t) = -A\pi t \frac{m_e \omega_{pe}}{k^2} \left. \frac{\partial f}{\partial v} \right|_{v=\omega_{pe}/k} \tag{10.46}$$

Hence, in the average over one wave oscillation period the electrons gain energy, if the derivative of the equilibrium distribution function in the vicinity of the resonance is negative. This energy is provided by the wave and leads to acceleration of the small number of resonant particles with velocities just below the wave phase speed. In equilibrium the energy transferred to the electrons per unit time equals the loss of wave energy

$$\frac{\Delta W_e(t)}{\Delta t} = -\frac{dW_w(t)}{dt} = 2\gamma_l W_w(0) \exp(-2\gamma_l t) \tag{10.47}$$

The second part of this equation results from the definition of the average wave energy, $W_w = \varepsilon_0 \delta E(t) \cdot \delta E^*(t)/2$, where after multiplication of the wave electric field with its conjugate complex part only twice the real part of the exponent survives. Inserting Eq. (10.46) for the gain in electron energy and Eq. (10.43) for the amplitude factor, A, one finds that the wave energy, W_w, appears on both sides of the equation and obtains another expression for the Landau damping

$$\gamma_l = \omega_{pe} \frac{\pi \omega_{pe}^2}{2n_0 k^2} \left. \frac{\partial f}{\partial v} \right|_{v=\omega_{pe}/k} \tag{10.48}$$

Inserting the Maxwellian distribution and remembering that it is normalized to the unperturbed density, n_0, one just recovers Eq. (10.37).

Equation (10.48) confirms that the derivative of the equilibrium distribution function in the vicinity of the resonance decides about the sign of the real part of p. Typically the derivative is negative, we have $\gamma_l < 0$ and the wave is damped. However, if the distribution function has a positive derivative at the resonance, we have $\gamma_l > 0$ and the wave extracts energy from the resonant particles and grows. Such *inverse Landau damping* implies instability and will be discussed in our companion book, *Advanced Space Plasma Physics*.

10.3   Unmagnetized Plasma Waves

Landau Damping does not only affect the Langmuir waves, but also the other wave modes that propagate in a warm unmagnetized plasma. In addition, a new wave mode appears in the kinetic treatment.

10.3.1 Ion-Acoustic Waves

So far we have suppressed the contribution of ion inertia. From the derivation of the dispersion relation, $\in(k, p) = 0$, it has become clear that the contributions of different species (electrons, ions, etc.) can be accounted for by adding a singular integral over the distribution function of the corresponding species of the same kind as in Eq. (10.27) to $\in(k, p)$. Hence, including the ion contribution requires solving the following dispersion relation

$$\in(k, p) = 1 + \frac{\omega_{pe}^2}{n_{0e}k^2} \int\limits_{-\infty}^{\infty} \frac{dv f_{0e}(v)}{(v - ip/k)^2} + \frac{\omega_{pi}^2}{n_{0i}k^2} \int\limits_{-\infty}^{\infty} \frac{dv f_{0i}(v)}{(v - ip/k)^2} = 0 \qquad (10.49)$$

At the high frequencies corresponding to electron plasma oscillations we can use the same expansion for the two integrals as before and find for the real part

$$1 - \frac{\omega_{pe}^2 + \omega_{pi}^2}{\omega^2} - \frac{3k^2}{\omega^4} \left(\omega_{pe}^2 v_{\text{the}}^2 + \omega_{pi}^2 v_{\text{thi}}^2 \right) = 0 \qquad (10.50)$$

which leads to a slightly modified dispersion relation of Langmuir waves, corrected for the effect of ions

$$\boxed{\omega_l^2(k) = \omega_{pe}^2 \left(1 + \frac{m_e}{m_i} \right) \left[1 + \frac{3k^2 \lambda_D^2}{1 + m_e/m_i} \left(1 + \frac{m_e^2}{m_i^2} \frac{T_i}{T_e} \right) \right]} \qquad (10.51)$$

The difference between the simple Langmuir dispersion relation and this corrected version is small. The plasma frequency is corrected by a term of the order of the electron-to-ion mass ratio. The correction of the Debye length turns out even smaller and becomes important only for extremely high ion temperatures.

Similarly, the Landau damping now contains an ion contribution. Denoting the complete Landau damping as $\gamma_l'(k)$, one obtains

$$\boxed{\gamma_l'(k, \omega_l) = \gamma_l(k, \omega_l) \left[1 + \left(\frac{T_e}{T_i} \right)^{3/2} \exp\left(\frac{\omega_l^2}{2k^2 \lambda_D^2} \frac{T_e - T_i}{T_i} \right) \right]} \qquad (10.52)$$

Ion damping at high frequencies is small compared to electronic Landau damping. The important contribution of ions to wave propagation is met at frequencies well below the electron plasma frequency and for wave phase velocities intermediate between the electron and ion thermal velocities, under the assumption that the ion temperature is considerably less than the electron temperature (this condition might not be satisfied in many astrophysical and space plasmas)

$$\frac{k_B T_i}{m_i} \ll \frac{\omega^2}{k^2} \ll \frac{k_B T_e}{m_e} \qquad (10.53)$$

The expansion of the electron integral must, under these conditions, be taken in the low phase velocity limit, $\omega/k \ll v$, while the ion integral is expanded in the large velocity limit, $\omega/k \gg v$. Such a double expansion leads to the following expression for the real part of the dielectric function

$$\epsilon(k,\omega) = 1 + \frac{1}{k^2\lambda_D^2} - \frac{\omega_{pi}^2}{\omega^2}\left(1 + \frac{3k^2}{\omega^2}\frac{k_B T_i}{m_i}\right) \tag{10.54}$$

Of course, because of the same kind of expansion, the ion term in this formula is of the same kind as the corresponding electronic term in the Langmuir dispersion relation. However, the resulting waves turn out to be very different from ion Langmuir waves. This difference is due to the second term on the right-hand side of the above formula, which is the low-frequency electronic contribution. Solving and iterating for the frequency yields

$$\omega_{ia}^2 = \frac{\omega_{pi}^2}{1 + 1/k^2\lambda_D^2}\left[1 + \frac{3T_i}{T_e}\left(1 + k^2\lambda_D^2\right)\right] \tag{10.55}$$

Up to the correction factor in parentheses this is the familiar ion-acoustic dispersion relation. In the long-wavelength limit, $k^2\lambda_D^2 \ll 1$, it reproduces the sound-wave branch of the ion-acoustic wave mode

$$\omega = \pm k c_{ia}' \tag{10.56}$$

where $c_{ia}^2 = k_B T_e/m_i$ is the ion sound velocity and $c_{ia}' = c_{ia}(1 + 3T_i/T_e)^{1/2}$ is a modified ion-acoustic speed. At short wavelengths it becomes a modified *ion plasma wave* with a dispersion like the electron Langmuir wave

$$\omega^2 = \omega_{pi}^2(1 + 3k^2\lambda_{Di}^2) \tag{10.57}$$

Here the ion Debye length has taken the position of the electron Debye length in the Langmuir wave. If the ion temperature is low, this wave becomes a simple ion plasma oscillation. Figure 10.5 shows a schematic plot of the ion-acoustic wave dispersion.

Following the procedure of the previous sections to calculate the Landau damping of the ion-acoustic wave from the imaginary part of the dielectric function

$$\epsilon_i(k,\omega) = -\left(\frac{\pi}{2}\right)^{1/2}\frac{1}{k^3\lambda_D^3}\frac{\omega}{\omega_{pe}}\left[1 + \left(\frac{T_e}{T_i}\right)^{3/2}\exp\left(-\frac{T_e}{T_i}\frac{1}{2k^2\lambda_D^2}\frac{\omega^2}{\omega_{pe}^2}\right)\right] \tag{10.58}$$

Assuming that the damping is small, one obtains in the long-wavelength domain

$$\gamma_{ia} = -\left(\frac{\pi}{8}\right)^{1/2}\left[\left(\frac{m_e}{m_i}\right)^{1/2} + \left(\frac{T_e}{T_i}\right)^{3/2}\exp\left(-\frac{T_e}{2T_i} - \frac{3}{2}\right)\right] \tag{10.59}$$

for the damping of the ion-sound wave. The first term is the ion damping term. If the electron temperature is very large, $T_e \gg T_i$, the damping reduces to only this term and is very weak

$$\gamma_{ia} \approx \omega_{ia}(\pi/8)^{1/2}(m_e/m_i)^{1/2} \tag{10.60}$$

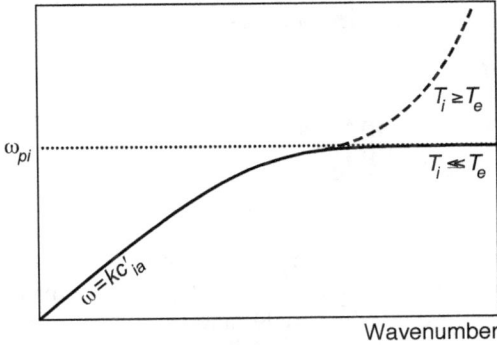

Figure 10.5: Ion-acoustic dispersion branches. For low ion temperatures one recovers the ion-acoustic dispersion relation known from simple fluid theory. However, for large ion temperatures the ion acoustic branch expands to higher than ion plasma frequencies. Such a situation is, for instance, given in the plasma sheet in Earth's magnetotail where $T_i > T_e$.

Because of this reason ion-acoustic waves at long wavelengths and in large electron temperature plasmas are practically undamped modes. Being one of the eigenmodes of an unmagnetized plasma, they play an important role in the dynamics of a collisionless plasma.

We can now calculate the ion-acoustic wave spectral energy density from Eq. (9.86), multiplying the response function by the frequency and differentiating. With $W_E = \varepsilon_0 |\delta E|^2 / 2$, we get an expression which becomes large at long wavelengths

$$W_{ia} = 2 \left(1 + \frac{1}{k^2 \lambda_D^2} \right) W_E \tag{10.61}$$

10.3.2 Electron-Acoustic Waves

Another interesting wave mode can be extracted from a dispersion relation similar to the ion-acoustic dispersion relation (10.49), if one assumes that the electron plasma consists of two independent components with different densities and temperatures. Designating the colder electron population by an index c, the hotter one by an index h, and splitting the electron distribution function into cold and hot distributions, $f_{0e}(v) = f_{0c}(v) + f_{0h}(v)$, the dielectric function (10.49) becomes

$$\in(k,p) = 1 + \frac{\omega_{pc}^2}{n_{0c}k^2} \int_{-\infty}^{\infty} \frac{dv f_{0c}(v)}{(v - ip/k)^2} + \frac{\omega_{ph}^2}{n_{0h}k^2} \int_{-\infty}^{\infty} \frac{dv f_{0h}(v)}{(v - ip/k)^2}$$

$$+ \frac{\omega_{pi}^2}{n_{0i}k^2} \int_{-\infty}^{\infty} \frac{dv f_{0i}(v)}{(v - ip/k)^2} \tag{10.62}$$

Due to the requirement of quasineutrality, the undisturbed densities satisfy the relation

$$n_{0c} + n_{0h} = n_{0i} = n_0 \tag{10.63}$$

The interesting range of phase velocities is

$$\frac{k_B T_i}{m_i} \sim \frac{k_B T_c}{m_e} \ll \frac{\omega^2}{k^2} \ll \frac{k_B T_h}{m_e} \tag{10.64}$$

This approximation permits to expand the hot electron integral in the small phase velocity limit, while the ion and cold electron integrals are expanded in the large phase velocity limit. This procedure which yields a result equivalent to Eq. (10.54)

$$\in(k, \omega) = 1 + \frac{\Theta}{k^2 \lambda_{Dh}^2} - \frac{\omega_{pc}^2}{\omega^2}\left(1 + \frac{3k^2}{\omega^2}\frac{k_B T_c}{m_e}\right) - \frac{\omega_{pi}^2}{\omega^2}\left(1 + \frac{3k^2}{\omega^2}\frac{k_B T_i}{m_i}\right) \tag{10.65}$$

where $\Theta \equiv 1 + n_{0c}T_h/n_{0h}T_c$. Multiplying this equation by ω^2 and iterating ω^2 on the right-hand side, one obtains the dispersion relation of electron-acoustic waves

$$\omega_{ea}^2 = \frac{\omega_{pc}^2\left(1 + \frac{m_e n_0}{m_i n_{0c}}\right)}{1 + \frac{1}{k^2 \lambda_{Dh}^2}}\left[1 + 3\left(k^2\lambda_{Dc}^2 + \frac{n_{0h}}{n_{0c}}\frac{T_c}{T_h}\right)\frac{1 + \frac{m_e n_0 T_i}{m_i n_{0c} T_c}}{\left(1 + \frac{m_e n_0}{m_i n_{0c}}\right)^2}\right] \tag{10.66}$$

This relation can be simplified by neglecting all electron-to-ion mass ratios, yielding

$$\boxed{\omega_{ea}^2 = \frac{\omega_{pc}^2}{1 + 1/k^2\lambda_{Dh}^2}\left(1 + 3k^2\lambda_{Dc}^2 + 3\frac{n_{0h}}{n_{0c}}\frac{T_c}{T_h}\right)} \tag{10.67}$$

as the dispersion relation of *electron-acoustic waves*. In this last form all ion contributions have been neglected, and the mode is purely electronic. Ion corrections become important only for high ion temperatures and low cold electron densities.

The term in front of the parentheses is of the same kind as the ion-acoustic wave dispersion relation. Hence, in the long-wavelength limit, $k^2\lambda_{Dh}^2 \ll 1$, the dispersion relation becomes that of an acoustic wave, with an acoustic wave speed of the order of the *electron-acoustic velocity*

$$c_{ea} = (n_{0c}/n_{0h})^{1/2}v_{thh} \tag{10.68}$$

This long-wavelength electron-acoustic wave satisfies

$$\omega = \pm k c_{ea} \tag{10.69}$$

In the short-wavelength limit, this wave becomes a modified Langmuir wave

$$\omega^2 = \omega_{pc}^2\left(1 + 3k^2\lambda_{Dc}^2 + 3\frac{n_{0h}}{n_{0c}}\frac{T_c}{T_h}\right) \tag{10.70}$$

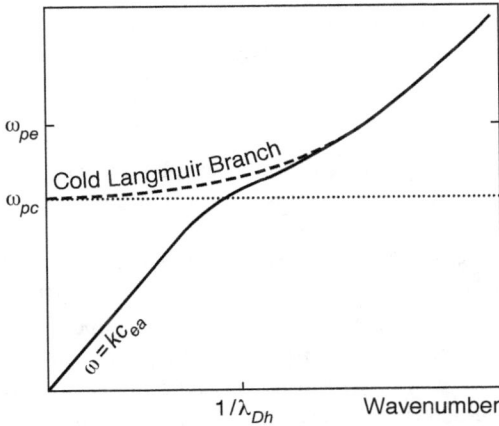

Figure 10.6: Electron-acoustic dispersion branches in a two-electron component plasma. At large wave numbers the dispersion of the electron-acoustic wave becomes the Langmuir wave branch. However, at long wavelengths one finds that the electron acoustic wave is the extension of the Langmuir branch to frequencies possibly far below the plasma frequency.

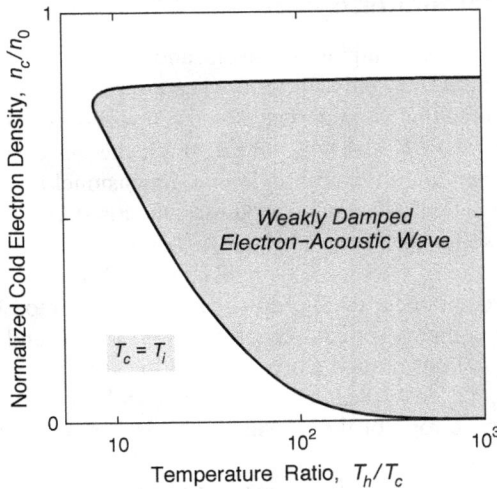

Figure 10.7: The weakly damped region of electron-acoustic waves where Landau damping is sufficiently small for electron acoustic waves to survive.

Hence, electron-acoustic waves at long wavelengths propagate like sound waves, while at short wavelengths they become cold electron Langmuir waves. Figure 10.6 gives an impression of the electron-acoustic wave dispersion.

These waves are heavily damped by electron Landau damping, as can be seen by rewriting the Landau damping of ion-acoustic waves for the electron-acoustic mode. In particular, the long-wavelength branch with linear dispersion is strongly damped.

Because of this strong damping, electron-acoustic waves are typically ignored. However, the intermediate branch corresponding to Eq. (10.67) and

$$\frac{n_h}{n_c}\frac{T_c}{T_h} < k^2\lambda_{Dc}^2 < 1 \tag{10.71}$$

is only weakly damped. Here the cold electrons play the role of the heavy ions in the ion-acoustic mode, while charge neutralization is provided by the mobile hot electrons. In this regime the frequency is relatively close to the cold plasma frequency.

Figure 10.7 shows, for $T_c = T_i$, where in parameter space the electron-acoustic mode is weakly damped. Typically, one must have $T_h/T_c > 10$ for weak damping, but at high cold electron densities no weakly damped regime exists. For shorter wavelengths the electron-acoustic waves become strongly damped again.

Calculating the wave spectral density by using Eq. (9.86), we find an expression similar to that for ion-acoustic waves, with W_E in Eq. (9.88),

$$\boxed{W_{ea}(k) = \left(1 + \frac{1}{k^2\lambda_{Dh}^2}\right)\left(2 - 3k^2\lambda_{Dc}^2 - 3\frac{n_{0h}}{n_{0c}}\frac{T_c}{T_h}\right)W_E} \tag{10.72}$$

10.3.3 Electromagnetic Waves

The calculations of Langmuir, ion-acoustic, and electron-acoustic waves have been performed in one dimension only. In an unmagnetized plasma this is no restriction because the only direction of importance is the direction of wave propagation, $\mathbf{k}$. For electrostatic waves, $\delta E(\omega, \mathbf{k}) = -i\mathbf{k}\,\delta\phi(\omega, \mathbf{k})$, the wave electric field is parallel to the wavenumber, and the situation is one-dimensional by itself. Hence, for the general case where the equilibrium distribution function depends on the full velocity vector, $f_0(\mathbf{v})$, it is sufficient to replace $k\partial f_0(v)/\partial v$ with $\mathbf{k}\cdot\partial f_0(\mathbf{v})/\partial\mathbf{v}$, and kv with $\mathbf{k}\cdot\mathbf{v}$. The integration over velocity is then taken over all three components, v_x, v_y, v_z. But since only the component parallel to k, say v_z, comes into play, the integrations over the two other components can trivially be performed, and the theory remains unchanged, with $f_0(v)$ understood as the distribution function integrated over v_x, v_y, $f_0(v_z) = \int dv_x dv_y f_0(\mathbf{v})$. This conclusion is of course violated if any anisotropy is introduced by, say, an anisotropy of the distribution function or by an external magnetic field.

Let us consider the case of electromagnetic waves propagating in an isotropic unmagnetized plasma with vanishing external electric and magnetic fields, $\mathbf{E}_0 = \mathbf{B}_0 = 0$. We must then use the full Vlasov equation, but the fields appearing are the magnetic field of the wave, $\delta\mathbf{B}$, and the transverse wave electric field, satisfying $\mathbf{k}\cdot\delta\mathbf{E} = 0$ or

$$\mathbf{k}\times\mathbf{k}\times\delta\mathbf{E} = -k^2\delta\mathbf{E} \tag{10.73}$$

Since only oscillating electromagnetic fields exist in this case, we immediately write down the linearized Vlasov equation for particles of charge q and mass m

$$\frac{\partial\delta f(\mathbf{v})}{\partial t} + \mathbf{v}\cdot\nabla\delta f(\mathbf{v}) + \frac{q}{m}\left(\delta\mathbf{E} + \mathbf{v}\times\delta\mathbf{B}\right)\cdot\frac{\partial f_0(\mathbf{v})}{\partial\mathbf{v}} = 0 \tag{10.74}$$

The isotropy of the plasma has been taken into account by letting the undisturbed equilibrium distribution function depend only on the modulus, v, of the velocity. Calculating $\partial f_0(v)/\partial \mathbf{v} = (\partial f_0/\partial v)(\partial v/\partial \mathbf{v}) = 3(\partial f_0/\partial v)\mathbf{v}$ and dot-multiplying with $\mathbf{v} \times \delta\mathbf{B}$, it becomes apparent that the last term in the parentheses vanishes, when dotted with the velocity derivative of the equilibrium distribution. Hence, the magnetic field contribution drops out of the Vlasov equation, leaving us with

$$\frac{\partial \delta f(\mathbf{v})}{\partial t} + \mathbf{v} \cdot \nabla \delta f(\mathbf{v}) + \frac{q}{m} \delta\mathbf{E} \cdot \frac{\partial f_0(v)}{\partial \mathbf{v}} = 0 \tag{10.75}$$

This equation can be understood as a linear inhomogeneous equation for $\delta f(\mathbf{v})$, which can be solved by superposition of plane waves or, in other words, by Fourier transformation both in space and time. The choice of this method, which is much simpler than the one applied in the previous electrostatic case, can be justified by arguing that we expect all physically reasonable distribution functions and fields to be analytic functions. In this case the dispersion relation will just depend on wavenumber, $\mathbf{k}$, and the complex variable $p = \gamma - i\omega$. Thus, interpreting the plane wave ansatz, $\exp[i(\mathbf{k} \cdot \mathbf{x} - \omega t) + \gamma t]$, as a complex Laplace factor, $\exp(pt)$, will yield the desired result.

With this philosophy in mind, we Fourier transform the linear equation, solve for the variation of the distribution function, and find

$$\delta f(\mathbf{v}, \mathbf{k}, \omega) = -\frac{q}{m} \frac{\partial f_0(v)/\partial \mathbf{v}}{\omega - \mathbf{k} \cdot \mathbf{v}} \cdot \delta\mathbf{E}(\mathbf{k}, \omega) \tag{10.76}$$

as a general expression for the disturbed distribution function in an isotropic plasma. It contains the velocity gradient of the equilibrium distribution and is proportional to the wave electric field.

In order to determine the dielectric function we need to know the plasma conductivity, which is calculated from the plasma current

$$\mathbf{j}(\mathbf{k}, \omega) = -\sum_{s=e,i} \frac{q_s}{m_s} \int d^3v\, \mathbf{v} \frac{\partial f_{s0}(v)/\partial \mathbf{v}}{\omega - \mathbf{k} \cdot \mathbf{v}} \cdot \delta\mathbf{E}(\mathbf{k}, \omega) \tag{10.77}$$

The factor in front of $\delta\mathbf{E}(\mathbf{k}, \omega)$ is the expression for the conductivity. When using it in the general dispersion relation derived in Chapter 9, we find the dispersion relation for electromagnetic waves in an isotropic plasma

$$(\omega^2 - k^2 c^2)\mathbf{I} = -\sum_{s=e,i} \frac{\omega_{ps}^2}{n_0} \int \frac{d^3v\, \omega}{\omega - \mathbf{k} \cdot \mathbf{v}} \mathbf{v} \frac{\partial f_{s0}(v)}{\partial \mathbf{v}} \tag{10.78}$$

From this expression we immediately find that for sufficiently high phase velocities, $\omega \gg \mathbf{k} \cdot \mathbf{v}$, the dispersion relation of the ordinary wave mode is recovered, since

$$(\omega^2 - k^2 c^2)\mathbf{I} = -\sum_{s=e,i} \frac{\omega_{ps}^2}{n_0} \int d^3v\, \mathbf{v} \frac{\partial f_{s0}(v)}{\partial \mathbf{v}} \tag{10.79}$$

and the integral just has the value $-n_0$ I. Hence, this expression finally yields

$$\boxed{\omega_{om}^2 = k^2 c^2 + \omega_{pe}^2} \tag{10.80}$$

where we have neglected the small ion plasma frequency correction. Such waves are undamped as is obvious from the disappearance of the resonance in the above expression. This mode, as we already know, is the only electromagnetic wave propagating in an isotropic unmagnetized plasma. Even a more precise calculation would show that it is practically undamped as long as relativistic particle effects are not included.

10.3.4   Plasma Dispersion Function

In the calculation of dispersion relations we have continuously encountered singular integrals of the kind

$$Z(\zeta) = \int_{-\infty}^{\infty} \frac{dx \, f_0(x)}{x - \zeta} \tag{10.81}$$

where $f_0(x)$ is some function related to the equilibrium distribution function which usually is an analytic function if its argument, x, is interpreted as the real part of a complex variable, z. The above integral is then taken along the real axis of the complex z plane. Integrals of this kind have been calculated in the previous sections. Clearly the value of the integral depends crucially on the choice of the distribution function. There is, however, a certain number of canonical distribution functions for equilibrium plasmas for which these integrals have been calculated. These functions are called *dispersion functions*. The best know dispersion function is the *plasma dispersion function* or *Fried-Conte function*. It is based on a Maxwellian equilibrium distribution. Therefore the plasma dispersion function is usually defined as

$$Z(\zeta) = \pi^{-1/2} \int_{-\infty}^{\infty} \frac{dx \, \exp(-x^2)}{x - \zeta} \tag{10.82}$$

The plasma dispersion function naturally plays an important role in most of the calculations of the linear properties of plasma waves propagating in an equilibrium background plasma. Its properties are listed in Appendix B.7.

In order to give an example of the application of the plasma dispersion function, we rewrite the dispersion relation of ion-acoustic waves in terms of the Z-function

$$\epsilon(\omega, k) = 1 - \frac{1}{k^2 \lambda_D^2} Z'(\zeta_e) - \frac{1}{k^2 \lambda_{Di}^2} Z'(\zeta_i) = 0 \tag{10.83}$$

The integrals containing the velocity derivative of the electron and ion Maxwellian equilibrium distribution functions have turned into the total derivatives of the plasma dispersion functions of the electron and ion arguments, $\zeta_e = \omega / k v_{the}$ and $\zeta_i = \omega / k v_{thi}$.

These arguments contain the frequency dependence. It is now simple to apply the small, $\zeta_e \ll 1$, and large, $\zeta_i \gg 1$, argument expansions given in Appendix B.7, to find the ion-acoustic dispersion relation and the associated Landau damping.

10.4 Magnetized Dispersion Relation

An external magnetic field introduces an anisotropy into the plasma. It affects the perpendicular particle motion while the parallel motion remains undisturbed. This must necessarily transform the scalar dielectric function into a tensor. In order to keep the mathematical difficulties at a minimum, this section does not make use of the Laplace method but returns to the easier though less precise procedure of Fourier transformation. We start from the linearized Vlasov equation of a magnetized plasma

$$\left(\frac{\partial}{\partial t} + \mathbf{v} \cdot \nabla + \frac{q}{m} \mathbf{v} \times \mathbf{B}_0 \cdot \frac{\partial}{\partial \mathbf{v}} \right) \delta f(\mathbf{v}, \mathbf{x}, t) = -\frac{q}{m} (\delta \mathbf{E} + \mathbf{v} \times \delta \mathbf{B}) \cdot \frac{\partial f_0(\mathbf{v})}{\partial \mathbf{v}} \quad (10.84)$$

where we have suppressed the species index, s. As before we have assumed that the undisturbed electric field vanishes, the ambient magnetic field and plasma densities are homogeneous. The equilibrium distribution function now depends on the velocity vector, in order to account for the anisotropy. The linearized current and charge densities are given by

$$\begin{aligned}
\delta \mathbf{j} &= \sum_s q_s \int d^3 v \, \mathbf{v} \, \delta f_s \\
\delta \rho_e &= \sum_s q_s \int d^3 v \, \delta f_s
\end{aligned} \quad (10.85)$$

Because δf depends on time and the six phase space coordinates, the left-hand side of Eq. (10.84) is the total time derivative of δf along a particle phase space orbit, while the right-hand side describes the change of the distribution function along this orbit under the action of the wave field. Hence, this equation can be rewritten as

$$\frac{d \, \delta f[\mathbf{v}(t), \mathbf{x}(t), t]}{dt} = -\frac{q}{m} \{ \delta \mathbf{E}[\mathbf{x}(t), t] + \mathbf{v}(t) \times \delta \mathbf{B}[\mathbf{x}(t), t] \} \cdot \frac{\partial f_0[\mathbf{v}(t)]}{\partial \mathbf{v}(t)} \quad (10.86)$$

Formally, one may calculate δf from this equation by integrating it over its entire history with respect to the time

$$\delta f[\mathbf{v}(t), \mathbf{x}(t), t] = -\frac{q}{m} \int_{-\infty}^{t} dt' \, \{ \delta \mathbf{E}[\mathbf{x}(t'), t'] + \mathbf{v}(t') \times \delta \mathbf{B}[\mathbf{x}(t'), t'] \} \cdot \frac{\partial f_0[\mathbf{v}(t')]}{\partial \mathbf{v}(t')} \quad (10.87)$$

But this procedure requires precise knowledge of the phase space orbit of all particles for all times $t' < t$, which is not available. But in a linearized theory, where the disturbance of the distribution function and the wave amplitudes remain small for all times, one can approximate the particle orbit by the orbit a particle would perform in a homogeneous and uniform external magnetic field, $\mathbf{B}_0 = B_0\hat{\mathbf{e}}_z$. This motion has been discussed in Chapter 2. It consists of a uniform motion along $\mathbf{B}_0$ and a gyration of frequency, ω_g, and gyroradius, $r_g = v_\perp/\omega_g$. Hence, the velocity components at any time can be represented by

$$\mathbf{v}(t'-t) = \{v_\perp \cos[\omega_g(t'-t) + \psi], v_\perp \sin[\omega_g(t'-t) + \psi], v_\|\} \tag{10.88}$$

where ψ is the initial phase angle. Correspondingly the position of the particle is given by the time integral of this expression

$$\mathbf{x}(t'-t) - \mathbf{x} = \omega_g^{-1}\{v_\perp \sin[\omega_g(t'-t) + \psi], -v_\perp \cos[\omega_g(t'-t) + \psi], v_\|(t'-t)\} \tag{10.89}$$

We now transform the time integration in Eq. (10.87) into an integration with respect to $\tau = t' - t$ in order to obtain

$$\delta f(\mathbf{v}) = -\frac{q}{m}\int_0^\infty d\tau\, [\delta\mathbf{E}(\tau) + \mathbf{v}(\tau) \times \delta\mathbf{B}(\tau)] \cdot \frac{\partial f_0[\mathbf{v}(\tau)]}{\partial \mathbf{v}(\tau)} \tag{10.90}$$

and introduce the plane wave ansatz, $\exp[-i\varphi(\tau)]$, for the electric and magnetic wave fields, with $\varphi(\tau) = \omega\tau + \mathbf{k}\cdot[\mathbf{x} - \mathbf{x}(\tau)]$. From Maxwell's equations we further deduce that these wave fields are related through Faraday's law, yielding $\mathbf{k} \times \delta\mathbf{E} = \omega\,\delta\mathbf{B}$, so that the magnetic wave field amplitude can be eliminated from Eq. (10.90). This expression then transforms into

$$\delta f(\mathbf{v}) = -\frac{q\delta\mathbf{E}(\mathbf{k}, \omega)}{m\omega} \cdot \int_0^\infty d\tau\, e^{-i\varphi(\tau)}\{\mathbf{k}\mathbf{v}(\tau) + \mathbf{l}[\omega - \mathbf{k}\cdot\mathbf{v}(\tau)]\} \cdot \frac{\partial f_0[\mathbf{v}(\tau)]}{\partial \mathbf{v}(\tau)} \tag{10.91}$$

This is the expression which has to be used in calculating the linear current in order to find the linear conductivity of the magnetized plasma. The expression for the current density contains an integral over velocity. This integral transforms into another integral over the new phase space volume element, $v_\perp dv_\perp dv_\| d\psi$, a transformation which completes our linear approach. What is left is to explicitly calculate the current integral

$$\delta\mathbf{j}(\mathbf{k}, \omega) = -\sum_s \frac{\varepsilon_0\omega_{ps}^2}{n_0\omega}\int_0^\infty\int_{-\infty}^\infty\int_0^{2\pi} v_\perp dv_\perp dv_\| d\psi$$
$$\delta\mathbf{E}(\mathbf{k}, \omega)\cdot\int_0^\infty d\tau\, e^{-i\varphi(\tau)}\{\mathbf{k}\mathbf{v}(\tau) + \mathbf{l}[\omega - \mathbf{k}\cdot\mathbf{v}(\tau)]\} \cdot \frac{\partial f_0[\mathbf{v}(\tau)]}{\partial \mathbf{v}(\tau)} \tag{10.92}$$

This calculation is performed in detail in Appendix C.6 for a gyrotropic equilibrium distribution function. With the help of Eq. (C.43) and splitting the wave vector into parallel and perpendicular components according to

$$\mathbf{k} = k_\perp\hat{\mathbf{e}}_\perp + k_\|\hat{\mathbf{e}}_\| \tag{10.93}$$

the magnetized dielectric function of the plasma takes the following form

$$\in(\omega, \mathbf{k}) = \left(1 - \sum_s \frac{\omega_{ps}^2}{\omega^2}\right)\mathbf{1} - \sum_s \sum_{l=-\infty}^{l=\infty} \frac{2\pi\omega_{ps}^2}{n_{0s}\omega^2}$$

$$\int_0^\infty \int_{-\infty}^\infty v_\perp dv_\perp dv_\parallel \left(k_\parallel \frac{\partial f_{0s}}{\partial v_\parallel} + \frac{l\omega_{gs}}{v_\perp} \frac{\partial f_{0s}}{\partial v_\perp}\right) \frac{\mathbf{S}_{ls}(v_\parallel, v_\perp)}{k_\parallel v_\parallel + l\omega_{gs} - \omega} \quad (10.94)$$

The tensor, $\mathbf{S}_{ls}$, appearing in the integrand is of the form

$$\mathbf{S}_{ls}(v_\parallel, v_\perp) = \begin{bmatrix} \dfrac{l^2\omega_{gs}^2}{k_\perp^2}J_l^2 & \dfrac{ilv_\perp\omega_{gs}}{k_\perp}J_lJ_l' & \dfrac{lv_\parallel\omega_{gs}}{k_\perp}J_l^2 \\[2ex] -\dfrac{ilv_\perp\omega_{gs}}{k_\perp}J_lJ_l' & v_\perp^2 J_l'^2 & -iv_\parallel v_\perp J_lJ_l' \\[2ex] \dfrac{lv_\parallel\omega_{gs}}{k_\perp}J_l^2 & iv_\parallel v_\perp J_lJ_l' & v_\parallel^2 J_l^2 \end{bmatrix} \quad (10.95)$$

and the Bessel functions, $J_l, J_l' = dJ_l/d\xi_s$, depend on the argument $\xi_s = k_\perp v_\perp/\omega_{gs}$.

Equation (10.94) is the most general expression for the linear dielectric function of a homogeneous nonrelativistic plasma immersed into a uniform magnetic field. It contains, when inserted into the general dispersion relation in Eq. (9.57), all electromagnetic and electrostatic wave eigenmodes, which can exist in such a plasma.

In the case of purely electrostatic (or longitudinal) modes with $\delta\mathbf{B} = 0$, the dielectric function simplifies considerably, since for such waves it is sufficient to consider the dielectric response function in Eq. (9.64). It is obtained by taking the dot-product of the dielectric tensor with the wave vector, $\mathbf{k}$, from both sides. Since $\mathbf{k}$ is in the (x, z) plane, one can show that after multiplication the term ω_{ps}^2/ω^2 in Eq. (10.94) is canceled by the $l = 0$ term of the sum. Furthermore, only two diagonal terms of the tensor $\mathbf{S}_{ls}$ survive, and their sum gives just J_l^2 (cf. Appendix B.7). Hence, the final result is

$$\in(\omega, \mathbf{k}) = 1 - \sum_s \sum_{l=-\infty}^{l=\infty} \frac{2\pi\omega_{ps}^2}{n_{0s}k^2} \int_0^\infty \int_{-\infty}^\infty v_\perp dv_\perp dv_\parallel$$

$$\left(k_\parallel \frac{\partial f_{0s}}{\partial v_\parallel} + \frac{l\omega_{gs}}{v_\perp} \frac{\partial f_{0s}}{\partial v_\perp}\right) \frac{J_l^2(\xi_s)}{k_\parallel v_\parallel + l\omega_{gs} - \omega} \quad (10.96)$$

Setting this function equal to zero, $\in(\omega, \mathbf{k}) = 0$, yields the dispersion relation for all the electrostatic waves propagating in a homogeneous uniformly magnetized plasma. In the following sections we will use it identify the dominant electrostatic modes.

10.4.1 Particle Resonance

As in the discussion of the Landau method for electrostatic electron waves the eigenmodes of a magnetized plasma are determined by the poles of the integrand of the dielectric tensor (10.94). These poles appear at the positions where

$$\omega - k_\parallel v_\parallel - l\omega_{gs} = 0 \tag{10.97}$$

which is the particle *resonance condition*. In the case of an unmagnetized plasma it reduces to the *Landau resonance*

$$\omega = k_\parallel v_\parallel \tag{10.98}$$

These two conditions are conditions on a specific group of particles in the plasma and are therefore particle resonances. They do not affect the whole plasma. This becomes obvious for instance from the Landau resonance, $l = 0$, which indicates that particles with velocities equal to the phase velocity of the wave are the only particles which contribute to the pole. These particles are in phase with the wave and the wave frequency seen by them is zero. Hence, these particles have a well defined parallel energy, $W_\parallel = m(\omega^2/2k_\parallel^2)$, the Landau resonant energy. In a magnetized plasma the selection of resonant particles becomes more complicated. Particles which move along the magnetic field case see the frequency of the wave Doppler-shifted to $\omega' = \omega - k_\parallel v_\parallel = l\omega_{gs}$. For $l = \pm 1$ this is just the gyrofrequency of species s; for $l \neq 1$ it is its l-th harmonic.

Hence, the group of particles whose parallel velocity just matches the parallel wave phase speed does not only see a constant parallel electric field of the wave, but if it is the right kind of particle also gyrates together with the perpendicular electric field component so that this component is also constant for them, or, for $|l| > 1$ sees a higher harmonic of its own gyration. Such particles interact strongly with the wave electric field because they become either accelerated or decelerated. It is these resonant particles who are responsible for the kinetic wave effects in magnetized and unmagnetized plasmas.

10.5   Electrostatic Plasma Waves

In this section we investigate the longitudinal eigenmodes of a warm magnetized plasma. These modes are the solutions of the dispersion relation

$$\in(\omega, \mathbf{k}) = 0 \tag{10.99}$$

where the response function in a magnetic field has been defined in Eq. (10.96). This function still holds for an arbitrary distribution function, but in the remainder of this chapter we will use an equilibrium Maxwellian

$$f_{0s}(v_\perp, v_\parallel) = \frac{n_{0s}}{\pi^{3/2} v_{\text{ths}\parallel} v_{\text{ths}\perp}^2} \exp\left(-\frac{v_\parallel^2}{v_{\text{ths}\parallel}^2} - \frac{v_\perp^2}{v_{\text{ths}\perp}^2} \right) \tag{10.100}$$

where for greater generality we permitted for anisotropy in the thermal velocities. The perpendicular velocity integral over the Bessel function can be substantially simplified. Making use of the Weber integrals given in Appendix B.7 and inserting f_{0s} from Eq. (10.100), the response function reduces to an integral over the parallel velocities

$$\epsilon(\omega,\mathbf{k}) = 1 + \sum_s \sum_{l=-\infty}^{l=\infty} \frac{2\omega_{ps}^2 \Lambda_l(\eta_s)}{\pi^{1/2} k^2 v_{ths\perp}^2} \int_{-\infty}^{\infty} \frac{dv_{\parallel}}{v_{ths\parallel}} \left(\frac{T_{s\perp}}{T_{s\parallel}} k_{\parallel} v_{\parallel} + l\omega_{gs} \right) \frac{\exp(-v_{\parallel}^2/v_{ths\parallel}^2)}{k_{\parallel} v_{\parallel} + l\omega_{gs} - \omega}$$

(10.101)

where we defined a new function

$$\Lambda_l(\eta_s) = I_l(\eta_s)\exp(-\eta_s)$$

(10.102)

and $I_l(\eta_s)$ is the modified Bessel function with the argument

$$\eta_s = \frac{k_\perp^2 v_{ths\perp}^2}{2\omega_{gs}^2} = \frac{k_\perp^2 T_{s\perp}}{m_s \omega_{gs}^2} = \frac{k_\perp^2 r_{gs}^2}{2}$$

(10.103)

The $v_{\parallel}$-integration can be performed with the help of the plasma dispersion function (Appendix B.7), yielding

$$\epsilon(\omega,\mathbf{k}) = 1 - \sum_s \sum_{l=-\infty}^{l=\infty} \frac{\omega_{ps}^2 \Lambda_l(\eta_s)}{k^2 v_{ths\perp}^2} \left\{ \frac{T_{s\perp}}{T_{s\parallel}} Z'(\zeta_s) - \frac{2l\omega_{gs}}{k_{\parallel} v_{ths\parallel}} Z(\zeta_s) \right\}$$

(10.104)

where $Z'(\zeta_s) = dZ/d\zeta_s$. The argument of the $Z(\zeta_s)$ function is $\zeta_s = (\omega - l\omega_{gs})/k_{\parallel} v_{ths\parallel}$. It depends on the species and on the running index of the sum, l. The zeros of this function are the electrostatic eigenmodes of the magnetized plasma.

10.5.1 Magnetized Langmuir and Ion-Acoustic Waves

Let us consider nearly parallel propagation first. At high frequencies the only wave is the Langmuir mode. Because for $k_\perp \to 0$ and $l \neq 0$ all $\Lambda_l \to 0$, the sum in Eq. (10.104) reduces to the term with index $l=0$. Neglecting the ion contribution, we obtain

$$\epsilon(\omega,\mathbf{k}) = 1 - \frac{\omega_{pe}^2}{k^2 v_{the\parallel}^2} \Lambda_0(\eta_e) Z'(\zeta_e) = 0$$

(10.105)

where $\zeta_e = \omega/k_{\parallel} v_{the\parallel}$, $\eta_e = k_\perp^2 v_{the\perp}^2/2\omega_{ge}^2$, and we allow for an anisotropy in the parallel and perpendicular thermal velocities. Now the plasma dispersion function is expanded in the large argument limit (see Appendix B.7), valid for Langmuir oscillations

$$1 - \frac{\omega_{pe}^2}{k^2 v_{the\parallel}^2} \frac{\Lambda_0}{\zeta_e^2} \left[1 + \frac{3}{\zeta_e^2} - \frac{2i\pi^{1/2}}{\zeta_e} \exp(-\zeta_e^2) \right] = 0$$

(10.106)

Solving for the real part of the frequency under the assumption of weak damping we find

$$\omega_l^2(k,\theta) = \omega_{pe}^2 \Lambda_0(\eta_e) \cos^2\theta \left(1 + 3k^2\lambda_{D\parallel}^2 \cos^2\theta\right) \tag{10.107}$$

Here $\lambda_{D\parallel}$ is the Debye length with respect to the parallel thermal velocity. Under the assumption that damping is weak, $\gamma_l \ll \omega_l$, one finds with the help of the general expression for the weak damping rate in Eq. (10.36)

$$\gamma_l(\omega_l,k,\theta) \approx -\left(\frac{\pi}{8}\right)^{1/2} \frac{\omega_l(k,\theta)\Lambda_0^{3/2}(\eta_e)}{k^3\lambda_{D\parallel}^3} \exp\left[-\frac{\Lambda_0(\eta_e)}{2k^2\lambda_{D\parallel}^2} - \frac{3}{2}\right] \tag{10.108}$$

for the Landau damping in a magnetized plasma. The requirement that the waves are weakly damped is thus restricted to the region where Λ_0 is sufficiently large. It depends on the angle of propagation. But because Λ_0 decreases monotonically for increasing argument, which implies that it decreases with increasing angle of propagation, oblique waves become ever stronger damped. In order to estimate up to what angle the damping is weak, we put $\Lambda_0(\eta_e) \approx 2k^2\lambda_{D\parallel}^2$. This yields

$$\exp(-\eta_e) \approx 2k^2\lambda_{D\parallel}^2 / I_0(\eta_e) \tag{10.109}$$

which can be rewritten as

$$\eta_e \approx \ln[I_0(\eta_e)/2k^2\lambda_{D\parallel}^2] \approx \ln(1/2k^2\lambda_{D\parallel}^2) \tag{10.110}$$

For not too large η_e we have $I_0(\eta_e) \approx 1$, which allows to write for the maximum weakly damped angle of propagation

$$\sin\theta < (kr_{ge})^{-1}\left|\ln(1/2k^2\lambda_{D\parallel}^2)\right|^{1/2} \tag{10.111}$$

For nearly parallel propagation η_e is small, and we can expand $\Lambda_0(\eta_e) \approx 1 - \eta_e$. This yields the magnetized Langmuir wave dispersion relation in compact form

$$\omega_l^2(k,\theta) = \omega_{pe}^2 \cos^2\theta \left[1 + 3k^2\lambda_{D\parallel}^2 \cos^2\theta \left(1 - \frac{T_\perp}{6T_\parallel} \frac{\omega_{pe}^2}{\omega_{ge}^2} \tan^2\theta\right)\right] \tag{10.112}$$

This dispersion relation is similar to that of ordinary Langmuir waves, but it shows that Langmuir waves in a magnetized plasma propagate essentially parallel to the magnetic field and depend only weakly on the ratio of plasma to gyrofrequency. The frequency of the wave decreases to zero with the wave vector turning to perpendicular propagation, but one must remember that the approximations made in the derivation of Eq. (10.112) break down for large angles of propagation. Nevertheless, in magnetized plasmas the Langmuir wave frequency at small k decreases to $\omega_l(\theta) < \omega_{pe}$ for oblique propagation.

Magnetized ion-acoustic waves can be treated along the same lines as Langmuir waves. The only difference is that instead of expanding the electron plasma dispersion function in the small argument limit, one expands the ion dispersion function in the large argument limit. As a final result one finds

$$\omega_{ia}^2 = \frac{\omega_{pi}^2 \Lambda_0(\eta_i) \cos^2\theta}{1 + \Lambda_0(\eta_e)/k^2\lambda_D^2} \left(1 + 3k^2\lambda_{Di}^2 \cos^2\theta\right) \tag{10.113}$$

Up to the Λ_0-correction this dispersion relation is similar to that of the unmagnetized ion-acoustic wave. This correction contains the effect of the magnetic field on wave propagation. But ion-acoustic waves in magnetized plasma do also propagate roughly parallel to the magnetic field, as is the case with the Langmuir wave.

10.5.2 Electron Bernstein Waves

We now turn to perpendicular propagation. If we put $k_\perp = 0$ in Eq. (10.96), the argument of the Bessel functions vanishes and only the term $l = 0$ in the sum survives. This implies that the influence of the magnetic field on the waves disappears and we recover the unmagnetized plasma case. Thus electrostatic waves in a plasma, which are susceptible to the presence of magnetic fields, propagate always oblique to the magnetic field. The most extreme case is to look for transverse propagation $k_\parallel = 0, k = k_\perp$. Inserting this into the response function, the dispersion relation can be written as

$$1 - \sum_s \frac{\omega_{ps}^2}{k_\perp^2} \frac{m_s^2}{k_B^2 T_s} \sum_{l=-\infty}^{\infty} \frac{l\omega_{gs}}{\omega - l\omega_{gs}} \int_0^\infty v_\perp dv_\perp J_l^2(\xi_s) \exp\left(-\frac{v_\perp^2}{v_{ths\perp}^2}\right) = 0 \tag{10.114}$$

or after performing the integration, again with the help of the Weber integral

$$1 - \sum_s \frac{\omega_{ps}^2}{\omega_{gs}} \sum_{l=-\infty}^{\infty} \frac{l I_l(\eta_s)}{\omega - l\omega_{gs}} \frac{\exp(-\eta_s)}{\eta_s} = 0 \tag{10.115}$$

This dispersion relation is free of any singular integrals. The only singularities are the resonances at the harmonics of the cyclotron frequencies of the various species of particles in the plasma. In particular, the $l = 0$ term is zero so that there is no wave below the first cyclotron harmonic. In the high-frequency range, $\omega \sim l\omega_{ge}$, the ion contribution to the dispersion relation can be neglected, and we find the dispersion relation of strictly perpendicular electrostatic electron-cyclotron waves or *Bernstein waves*

$$1 - \frac{\omega_{pe}^2}{\omega_{ge}^2} \sum_{l=-\infty}^{\infty} \frac{\omega_{ge} l \Lambda_l(\eta_e)}{\eta_e(\omega - l\omega_{ge})} = 0 \tag{10.116}$$

which shows that there is an infinite number of resonances or wave modes possible with frequencies $\omega \geq l\omega_{gs}$. Because of the symmetry of the modified Bessel functions with respect to their index, $I_{-l} = I_l$, Eq. (10.116) can be written as

$$1 - \frac{\omega_{pe}^2}{\omega_{ge}^2} \sum_{l=1}^{\infty} \frac{2l^2 \Lambda_l(\eta_e)}{\eta_e(v_{he}^2 - l^2)} = 0 \tag{10.117}$$

where $v_{he} = \omega/\omega_{ge}$. The different dispersion branches are concentrated near the harmonics. The behavior of the dispersion branches, $v_{he} = v_{he}(k_\perp)$, at large wavenumbers $k_\perp \to \infty$, is given by the asymptotic expansion of $\Lambda_l(\eta_e)$

$$\Lambda_l(\eta_e) \to (2\pi\eta_e)^{-1/2} \exp(-l^2\eta_e/2) \tag{10.118}$$

Substituting into Eq. (10.117), we find that with $\eta_e \to \infty$ the exponential vanishes and the dispersion relation can only be satisfied by $v_{he} = l$, which implies that all short wavelength waves are exact harmonics of the electron-cyclotron frequency, $\omega(k_\perp \to \infty) = l\omega_{ge}$. Moreover, the dispersion branches are bound to their harmonic bands, l, and cannot cross over into one of the adjacent bands, $l' = l \pm 1$. For small wavenumbers, $k_\perp \approx 0$, we have $\Lambda_l(\eta_e) \approx \eta_e^l/2l!$, and Eq. (10.117) turns into

$$1 - \frac{\omega_{pe}^2}{\omega_{ge}^2} \sum_{l=1}^{\infty} \frac{l\eta_e^{l-1}}{(l-1)!(v_{he}^2 - l^2)} = 0 \tag{10.119}$$

Two limiting cases are of interest. For $l \neq 1$ and $\eta_e \to 0$ this equation can be satisfied only if $v_{he} \approx l$. On the other hand, for $v_{he} \neq l$ only the term $l = 1$ survives and yields

$$\omega^2(k_\perp = 0) = \omega_{pe}^2 + \omega_{ge}^2 \tag{10.120}$$

This frequency is just the upper-hybrid frequency, ω_{uh}. The perpendicular electrostatic modes thus start at small wavenumber as a cyclotron harmonic oscillation with the exception of the band which contains the upper-hybrid frequency. Here the oscillation branch starts at ω_{uh}. For large perpendicular wavenumbers all dispersion branches approach their respective cyclotron harmonics. This behavior is sketched in Fig. 10.8.

To demonstrate this behavior explicitly for the lower harmonic bands, we neglect the ion contribution and write Eq. (10.116) up to the $l = 2$ term, using the small argument expansion for $\Lambda(\eta_e)$ together with the definition of η_e

$$\frac{\omega_{ge}^2}{\omega_{pe}^2} = \frac{1 - \eta_e}{v_{he}^2 - 1} + \frac{\eta_e(1 - \eta_e)}{v_{he}^2 - 4} \tag{10.121}$$

Let us now take $v_{he} = 1 + \delta_{e,1}$ with $\delta \ll 1$ as the resonance of the first term on the right-hand side. Then, we find to first order

$$\delta_{e,1} = \frac{3}{2(1 - \eta_e)}\left(1 + \frac{6}{1 - 2\eta_e}\frac{\omega_{ge}^2}{\omega_{pe}^2}\right) > 0 \tag{10.122}$$

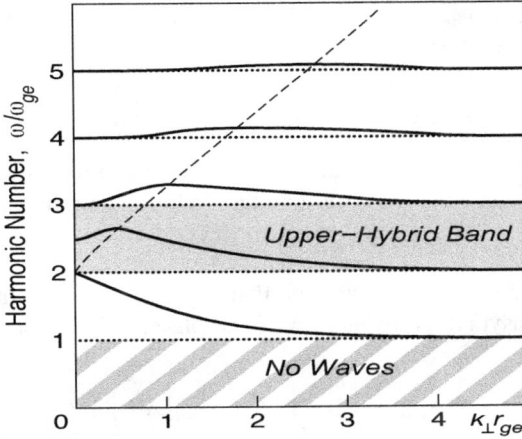

Figure 10.8: Dispersion branches of Bernstein modes. There are infinitely many cyclotron harmonics above ω_{ge} and no wave below ω_{ge}. The branches below and above the upper hybrid band have different character. Below they start at the upper harmonic and for larger perpendicular wavenumbers approach the lower harmonic. Above they start at the lower harmonic and go through a maximum located along the dashed straight line and fall back to the lower harmonic. The upper-hybrid band is unique in that the dispersion curve starts at the lower hybrid frequency for long wavelengths and vanishing wavenumber but otherwise shows the same behaviour as the higher harmonics.

a correction which is always positive, confirming that there is no dispersion branch below the first harmonic resonance. The next resonance, $v_{he} = 2 + \delta_{e,2}$, yields

$$\delta_{e,2} = -\frac{\eta_e}{4}\left(-\frac{3}{1-\eta_e}\frac{\omega_{ge}^2}{\omega_{pe}^2}\right) < 0 \qquad (10.123)$$

This dispersion branch starts at $v_{he} = 2$ and lies below the second harmonic. Only in the band containing the upper-hybrid frequency the dispersion branch starts at ω_{uh}. In the higher bands the branches all lie above the harmonic resonances.

Because the dispersion branches are all trapped into cyclotron harmonic bands, the general dispersion relation (10.117) of Bernstein waves can be written for each of the modes separately. Temporarily re-introducing ω instead of v_{he} and solving for the real frequency, ω, one finds as dispersion relation of each electron-cyclotron Bernstein mode

$$\omega_{ec}^2 = l^2 \omega_{ge}^2 \left[1 + \frac{2\omega_{pe}^2}{\omega_{ge}^2}\frac{\Lambda_l(\eta_e)}{\eta_e}\right] \qquad (10.124)$$

where the wavenumber dependence enters only through the Λ_l function. This dispersion relation holds for strictly perpendicular propagation. We can now look for a solution of the dispersion relation (10.104) at nearly perpendicular propagation. Knowing that only modes near $v_{he} = l$ play any role, the sum reduces to the $l = 0$ component plus one single l-component

$$\in(\omega, \mathbf{k}) = 1 + \frac{1}{k^2 \lambda_D^2} \left[1 - \Lambda_0(\eta_e) + \frac{v_{he} \Lambda_l(\eta_e)}{|k_\parallel r_{ge}|} Z\left(\frac{v_{he} - l}{|k_\parallel r_{ge}|} \right) \right] = 0 \qquad (10.125)$$

The index, $l = \pm 1, \pm 2, \ldots$, is arbitrary and refers to any of the harmonic bands. Let us ask for a weakly damped solution close to the cyclotron harmonic number $v_{he} = l$

$$v_{he}(\mathbf{k}) = l[1 + \delta_{e,l}(\mathbf{k})] + i\gamma_l(\mathbf{k}) \qquad (10.126)$$

with $\delta_{e,l}(\mathbf{k}) \ll 1$. Let us further assume that $|k_\parallel r_{ge}| \ll l\delta_{e,l}(\mathbf{k})$. In this case we can make use of the asymptotic expansion of the plasma dispersion function for large argument and find

$$\in(\omega, \mathbf{k}) \approx 1 + \frac{1}{k^2 \lambda_D^2} \left[1 - \Lambda_0(\eta_e) + \frac{l(1 + \delta_{e,l})\Lambda_l(\eta_e)}{|k_\parallel r_{ge}|} \left(i\pi^{1/2} e^{-\zeta_e^2} - \frac{k_\parallel r_{ge}}{l\delta_{e,l}} \right) \right] \qquad (10.127)$$

From here we obtain to first order in $\delta_{e,l}(\mathbf{k})$

$$\delta_{e,l}(\mathbf{k}) = \frac{\Lambda_l(\eta_e)}{k^2 \lambda_D^2 + 1 - \Lambda_0(\eta_e)} \qquad (10.128)$$

showing that the excursion of the dispersion curves for increasing harmonic number decreases both because of the inverse proportionality to l and because of the asymptotic behavior (10.118) of $\Lambda_l(\eta_e)$. The maximum of $\delta_{e,l}(\mathbf{k})$ decreases approximately as l^{-3}.

Investigating the behavior of the damping rate, $\gamma(\omega, \mathbf{k})$, we remind ourselves that γ is proportional to the imaginary part of the response function (10.127), which is proportional to $\exp(-l^2 \delta_{e,l}^2 / k_\parallel^2 r_{ge}^2)$. Hence, strictly perpendicular propagating Bernstein waves with $k_\parallel = 0$ have vanishing Landau damping and can propagate without any dissipation. However, damping becomes stronger with increasing angle of propagation so that Bernstein modes propagate nearly perpendicular to the magnetic field.

10.5.3   Upper-Hybrid Waves

The two most important branches of electron-cyclotron waves are the $l = 1$ harmonic branch, which is the electron-cyclotron resonance leading to the fundamental electron-cyclotron wave $\omega = l\omega_{ge}$, and the branch which contains the upper-hybrid frequency, ω_{uh}, because this is a natural plasma resonance frequency. The dispersion relation of upper-hybrid waves is found from the magnetized dielectric response function if we assume $\omega \approx \omega_{uh}$ and expand in the vicinity of perpendicular propagation, $k_\parallel \ll k_\perp$. Neglecting the minor contribution of the ions at such frequencies, we find that

$$1 - \frac{\omega_{pe}^2}{\omega^2} \frac{k_\parallel^2}{k_\perp^2} - \frac{\omega_{pe}^2}{\omega^2 - \omega_{ge}^2} \frac{k_\perp^2}{k^2} = 0, \qquad \omega \approx \omega_{uh} \qquad (10.129)$$

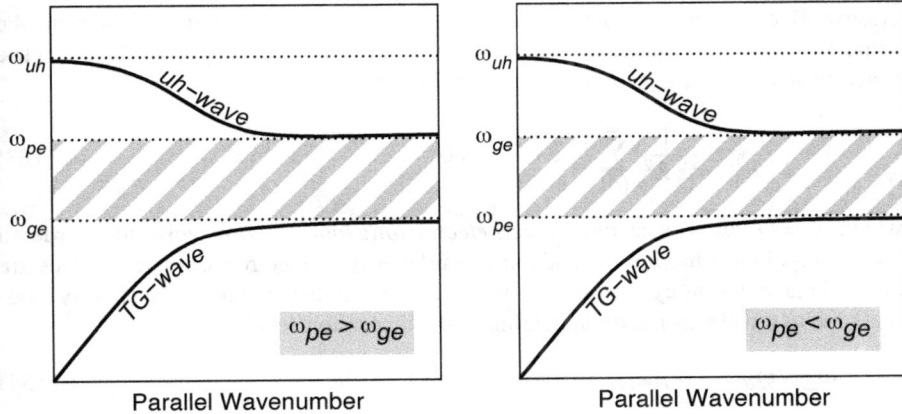

Figure 10.9: Dispersion branches of upper-hybrid and Trivelpiece-Gould waves.

In the limit of small gyrofrequency this equation describes oblique Langmuir waves (not including the thermal effects). The solution of this equation is

$$\omega^2 = \frac{1}{2}\omega_{uh}^2 \left[1 \pm \left(1 - \frac{4\omega_{pe}^2\omega_{ge}^2}{\omega_{uh}^4} \right)^{1/2} \right] \tag{10.130}$$

which for strictly perpendicular propagation becomes the upper-hybrid resonance. For oblique propagation the wave becomes strongly damped for

$$\cos^2\theta > \omega_{uh}^4/4\omega_{ge}^2\omega_{pe}^2 \tag{10.131}$$

Expanding the root in the above dispersion relation one obtains two different modes

$$\omega^2(\mathbf{k}) = \begin{cases} \omega_{uh}^2 - \dfrac{\omega_{pe}^2\omega_{ge}^2}{\omega_{uh}^2}\dfrac{k_\parallel^2}{k^2} \\[3mm] \dfrac{\omega_{pe}^2}{1+\omega_{pe}^2/\omega_{ge}^2}\dfrac{k_\parallel^2}{k_\perp^2} \end{cases} \tag{10.132}$$

The first of these modes is the *upper-hybrid wave*. It becomes an upper-hybrid oscillation for perpendicular propagation. The second is known as the *Trivelpiece-Gould mode* which is simply a modified oblique Langmuir mode. Their dispersion curves are shown schematically in Fig. 10.9 for $\omega_{pe} > \omega_{ge}$ and $\omega_{pe} < \omega_{ge}$.

10.5.4   Ion Bernstein Waves

In analogy to electron Bernstein waves, *ion Bernstein waves* can also propagate in a plasma perpendicular to the magnetic field. These waves are electrostatic ion-cyclotron resonances. Their dispersion relation is similar to the dispersion relation of electron Bernstein waves with the only difference that in addition to the ion terms the usual electron screening term $(k\lambda_D)^{-2}$ must be added to the real part of the dielectric function so that their general dispersion relation becomes

$$1 + \frac{1}{k_\perp^2 \lambda_D^2} - \frac{1}{k_\perp^2 \lambda_{Di}^2} \sum_{l=1}^{\infty} \frac{2l^2 \Lambda_l(\eta_i)}{v_{hi}^2 - l^2} = 0 \tag{10.133}$$

where $v_{hi} = \omega / \omega_{gi}$. Hence, one expects *electrostatic ion-cyclotron waves* to be ordered into ion-cyclotron harmonic bands in a similar way as electron Bernstein waves are ordered into electron-cyclotron harmonic bands. Their dispersion will be slightly modified due to the electron term appearing in the dispersion relation

$$\omega = l\omega_{gi}[1 + \delta_{i,l}(\mathbf{k})] \tag{10.134}$$

where

$$\delta_{i,l}(\mathbf{k}) = \frac{\Lambda_l(\eta_i)}{k^2 \lambda_{Di}^2 + 1 + (T_i/T_e) - \Lambda_0(\eta_i)} \tag{10.135}$$

There are, however, several differences in the behavior of ion-cyclotron waves and electron Bernstein modes. The first is that in the former the lower-hybrid frequency plays the role the upper-hybrid frequency plays in the latter. In the harmonic band containing the lower-hybrid frequency, ω_{lh}, putting $k_\perp = 0$ (which corresponds to putting $\eta_i = 0$) the dispersion relation (10.133) reduces to

$$\frac{\omega_{pe}^2}{\omega^2 - \omega_{ge}^2} + \frac{\omega_{pi}^2}{\omega^2 - \omega_{gi}^2} = 1 \tag{10.136}$$

which we are familiar with from earlier considerations and which has as solution for $\omega_{gi} \ll \omega \ll \omega_{ge}$, the lower-hybrid resonance frequency. Hence, in this band the dispersion curve starts at ω_{lh}. The second difference is the possibility that the ions may behave unmagnetized. This leads to waves propagating at the lower-hybrid frequency.

For long wavelength ion-cyclotron waves the dominant mode is the $l = 1$ mode. This becomes obvious from Eq. (10.135) in the small argument expansion, $\eta_i \ll 1$. Using the expansions of $\Lambda_l(\eta_i)$, this mode satisfies, with $v_{hi} = \omega / \omega_{gi} = 1 + \delta_{i,1}$

$$1 + \frac{1}{k_\parallel^2 \lambda_D^2} + \frac{k_\perp^2}{k_\parallel^2}\left(1 - \frac{\omega_{pe}^2/\omega_{ge}^2}{v_{hi}^2 - 1}\right) = \frac{\omega_{pi}^2}{\omega_{gi}^2}\frac{1}{v_{hi}^2} \tag{10.137}$$

The $\delta \ll 1$ solution of this equation defines the fundamental ion-cyclotron wave

$$\boxed{\omega_{ic} = \omega_{gi}\left(1 + \tfrac{1}{2}k_\perp^2 r_{gi}^2\right)} \tag{10.138}$$

This solution is good for nearly perpendicular propagation. If we calculate the energy contained in this particular wave mode, we find that it is $W_{ic} = W_E$. However, for the

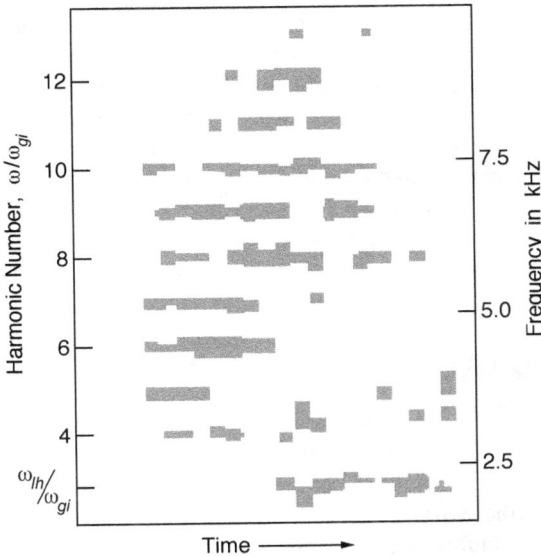

Figure 10.10: Spectrogram of ion-cyclotron harmonics showing the many steps of harmonic electrostatic ion-cyclotron emissions in an ionospheric observation.

more general mode, where the assumptions of very small parallel wavenumber have not been made, the spectral energy density becomes a function of $v_{hi} = l(1 + \delta_{i,l})$

$$W_{ic} = \frac{\omega_{pi}^2}{\omega_{gi}^2} \left(\frac{v_{hi}^2 W_{\perp E}}{(v_{hi}^2 - 1)^2} + \frac{W_{\parallel E}}{v_{hi}^2} \right) \tag{10.139}$$

and the electric field energy splits into parallel and perpendicular parts. Hence, depending on the ratio of ion plasma to ion-cyclotron frequencies the energy in the wave can be very different from the electric field energy.

An example of electrostatic ion-cyclotron waves is shown in Fig. 10.10. These waves have been excited during the injection of a heavy ion beam into the ionospheric plasma and show the harmonic structuring of the spectrum up to the 12th harmonic of the ion-cyclotron frequency. The lowest frequency in this figure is the lower-hybrid frequency, which is not at an exact harmonic, while ion plasma frequency is not seen in this case. This has been explained when we discussed Bernstein modes for electrons. A similar effect happens here in the case of ion-cyclotron waves where the lower-hybrid frequency appears in place of the upper hybrid frequency. Electrostatic ion-cycltron harmonics are the pendant to Bernstein modes and are therefore frequently also called ion-Bernstein modes. Still there is some difference. In electron-Bernstein modes the upper hybrid frequency is always above the electron cyclotron and plasma frequencies. This is not necessarily the case in ion Bernstein modes with the lower hybrid frequency which in dense plasma is the geometrical average of the electron and ion cyclotron frequencies while in dilute plasmas close to the ion plasma frequency. Hence in dense plasma it falls into a higher ion-cyclotron band and behaves similar to the upper hybrid frequency in Bernstein modes, as is the case in the above Fig. 10.10 which refers to the dense upper ionosphere. In dilute plasma, however, it may be quite below the ion cyclotron frequency and thus will not affect any of the ion-cyclotron harmonic bands. This is the case, for instance in the solar corona with its strong magnetic field as also in the lobes of the magnetosphere where the plasma density is very low. Any excitation of electrostatic ion-cyclotron harmonics there, for instance by ion beams escaping from the polar ionosphere during substorms, will not be affected by the lower-hybrid resonance frequency.

10.5.5   Lower-Hybrid Waves

When the ions are unmagnetized, we can make the approximation of large ion gyroradii, $k_\perp^2 r_{gi}^2 \gg 1$. In such a case the ions follow straight line orbits. It is clear that this approximation is particularly useful for high ion temperatures. Hence, it is complementary to the existence of ion-acoustic waves, which require low ion and high electron temperatures. We therefore use in addition $\eta_e \ll 1$ and the high phase velocity approximation, $\omega/k_\parallel \gg v_{the\parallel}$. The dispersion relation then becomes

$$\epsilon(\omega, \mathbf{k}) = \left(1 - \frac{\omega_{ge}^2}{\omega^2}\right)\frac{k_\parallel^2}{k_\perp^2} - \frac{1}{2k_\perp^2 \lambda_{Di}^2} Z'\left(\frac{\omega}{k v_{thi}}\right) = 0 \tag{10.140}$$

This dispersion relation simplifies for purely perpendicular propagation to

$$1 + \frac{\omega_{pe}^2}{\omega_{ge}^2} - \frac{\omega_{pi}^2}{\omega^2}\left(1 + \frac{3}{2}\frac{k_\perp^2 v_{thi}^2}{\omega^2}\right) = 0 \tag{10.141}$$

which has the approximate solution

$$\omega^2 = \omega_{lh}^2\left[1 + \frac{3}{2}\left(1 + \frac{\omega_{pe}^2}{\omega_{ge}^2}\right)k_\perp^2 \lambda_{Di}^2\right] \tag{10.142}$$

This dispersion relation describes perpendicular *lower-hybrid waves*. In the long wavelength domain, $k_\perp^2 \lambda_{Di}^2 \ll 1$, this wave becomes a lower-hybrid oscillation. Interestingly, the perpendicular lower-hybrid wave has dispersion similar to that of electron Langmuir waves in an unmagnetized plasma.

Keeping the finite though small parallel wavenumber term in the dielectric response function and expanding the plasma dispersion function for large argument produces the following general dispersion relation for lower-hybrid waves

$$\omega^2 = \omega_{lh}^2\left[1 + \frac{m_i}{m_e}\frac{k_\parallel^2}{k_\perp^2} + \frac{3k^2}{4k_\perp^2}\left(1 + \frac{\omega_{pe}^2}{\omega_{ge}^2}\right)k^2\lambda_{Di}^2\right] \tag{10.143}$$

We have assumed here that $k_\parallel \ll k_\perp$ in simplifying the right-hand side. This is justified insofar as the second term in the brackets suggests that the ratio of parallel to perpendicular wavenumbers must be of the order of $k_\parallel/k_\perp \approx (m_e/m_i)^{1/2}$ in order to affect the propagation of the waves. As we have mentioned these lower-hybrid waves propagate at a natural plasma resonance and are therefore of immense importance for the interaction between waves and particles. They can be used to heat a plasma and, in cases when the ion temperature is high, contribute to energy losses from the plasma. Some of these effects will be discussed in the following chapters.

Let us finally estimate the energy contained in the perpendicular lower-hybrid mode by using Eq. (10.141). If we multiply by frequency and take the derivative, we find that

$$
W_{lh} \approx \left(1 + \frac{\omega_{pe}^2}{\omega_{ge}^2} \right) W_E \tag{10.144}
$$

is larger than the electric field energy by a factor which in a dense plasma can be much larger than one. Lower-hybrid waves thus carry very much energy in dense plasmas, while in dilute plasmas their energy is just equal to the electric field energy.

10.6 Electromagnetic Plasma Waves

Electromagnetic wave propagation in a homogeneous plasma embedded into a uniform magnetic field must be treated on the basis of the general dispersion relation (9.52)

$$
D(\omega,\mathbf{k}) = 0 \tag{10.145}
$$

where $D(\omega,\mathbf{k})$ is the determinant of the full dispersion tensor (9.57)

$$
N^2 \left(\frac{\mathbf{kk}}{k^2} - \mathbf{I} \right) + \epsilon(\omega,\mathbf{k}) \tag{10.146}
$$

which contains the dielectric tensor (9.56). This tensor has been reduced to the general form in Eq. (10.94), valid for gyrotropic distribution functions. For a Maxwellian plasma the integrations over the parallel and perpendicular velocities can be performed with the help of the plasma dispersion function, Z, and its derivative, Z', using the Weber integrals, and taking advantage of the recursion relations between the Bessel functions. This calculation yields the dielectric tensor

$$
\epsilon(\omega,\mathbf{k}) = \mathbf{I} + \sum_s \begin{pmatrix} \epsilon_{s1} & \epsilon_{s2} & \epsilon_{s4} \\ -\epsilon_{s2} & \epsilon_{s1} - \epsilon_{s0} & -\epsilon_{s5} \\ \epsilon_{s4} & \epsilon_{s5} & \epsilon_{s3} \end{pmatrix} \tag{10.147}
$$

The components of the tensor inside the sum are given by

$$\epsilon_{s0} = \frac{2\omega_{ps}^2}{\omega k_\parallel v_{\text{ths}\parallel}} \sum_l \eta_s \Lambda_l'(\eta_s) \left[Z(\zeta_{s,l}) - \frac{k_\parallel v_{\text{ths}\parallel}}{2\omega} A_s Z'(\zeta_{s,l}) \right]$$

$$\epsilon_{s1} = \frac{\omega_{ps}^2}{\omega k_\parallel v_{\text{ths}\parallel}} \sum_l \frac{l^2 \Lambda_l(\eta_s)}{\eta_s} \left[Z(\zeta_{s,l}) - \frac{k_\parallel v_{\text{ths}\parallel}}{2\omega} A_s Z'(\zeta_{s,l}) \right]$$

$$\epsilon_{s2} = \frac{i \text{sgn}(q_s)\omega_{ps}^2}{\omega k_\parallel v_{\text{ths}\parallel}} \sum_l l \Lambda_l'(\eta_s) \left[Z(\zeta_{s,l}) - \frac{k_\parallel v_{\text{ths}\parallel}}{2\omega} A_s Z'(\zeta_{s,l}) \right]$$

$$\epsilon_{s3} = -\frac{\omega_{ps}^2}{k_\parallel^2 v_{\text{ths}\parallel}^2} \sum_l \left(1 - \frac{A_s}{A_s + 1} \frac{l\omega_{gs}}{\omega} \right) \left(1 + \frac{l\omega_{gs}}{\omega} \right) \Lambda_l(\eta_s) Z'(\zeta_{s,l})$$

$$\epsilon_{s4} = \frac{k_\perp}{2k_\parallel} \frac{\omega_{ps}^2}{\omega \omega_{gs}} \sum_l \left(A_s + 1 - \frac{l\omega_{gs}}{\omega} A_s \right) \frac{l\Lambda_l(\eta_s)}{\eta_s} Z'(\zeta_{s,l})$$

$$\epsilon_{s5} = -\frac{i \text{sgn}(q_s)}{k_\perp k_\parallel} \frac{\omega_{ps}^2}{2\omega \omega_{gs}} \sum_l \left(A_s + 1 - \frac{l\omega_{gs}}{\omega} A_s \right) \Lambda_l'(\eta_s) Z'(\zeta_{s,l})$$

(10.148)

where $\zeta_{s,l} = (\omega - l\omega_{gs})/k_\parallel v_{\text{ths}\parallel}$, the sum is taken from $l = -\infty$ to $l = +\infty$, and we defined the temperature anisotropy of species s as

$$\boxed{A_s = \frac{T_{s\perp}}{T_{s\parallel}} - 1}$$

(10.149)

In an isotropic plasma $A_s = 0$, and many of the above terms disappear.

10.6.1 Weak Damping

Under the assumption of weak wave damping in a magnetized plasma with damping rate $|\gamma(\omega,\mathbf{k})| \ll \omega(\mathbf{k})$, where ω is the real part of the complex frequency, we can find an expression for the damping rate, γ, from the general dispersion relation $D(\omega,\mathbf{k}) = 0$. The procedure is similar to that applied when calculating the weak damping rate of electrostatic waves, with the only difference that now the full dispersion relation replaces the plasma response function. Splitting into real and imaginary parts

$$D(\omega,\gamma,\mathbf{k}) = D_r(\omega,\gamma,\mathbf{k}) + iD_i(\omega,\gamma,\mathbf{k}) \tag{10.150}$$

the dispersion relation can be expanded around the real frequency as

$$D(\omega,\gamma,\mathbf{k}) = D_r(\omega,0,\mathbf{k}) + i\gamma \frac{\partial D_r(\omega,\gamma,\mathbf{k})}{\partial \omega}\bigg|_{\gamma=0} + iD_i(\omega,0,\mathbf{k}) = 0 \tag{10.151}$$

and one finds for weakly damped waves

$$D_r(\omega,0,\mathbf{k}) = 0 \tag{10.152}$$

$$\gamma(\omega,\mathbf{k}) = -\frac{D_i(\omega,0,\mathbf{k})}{\partial D_r(\omega,\gamma,\mathbf{k})/\partial \omega|_{\gamma=0}} \tag{10.153}$$

10.6.2   Electromagnetic Modes

High-frequency electromagnetic waves above the electron plasma or electron gyro-frequency, whichever is higher, are only weakly affected by the presence of the plasma. Those waves behave approximately like free space radiation to which we will return in the special case of *auroral kilometric radiation* when treating linear plasma instabilities. In dense cold plasmas, $\omega_{pe}^2 > \omega_{ge}^2$, no electromagnetic wave propagation is allowed between the electron plasma and the electron gyrofrequency. This conclusion is not changed for dense hot plasmas. In dilute cold plasmas, $\omega_{pe}^2 < \omega_{ge}^2$, electromagnetic waves can propagate in different modes at all frequencies, a conclusion which holds as well for hot plasmas, but the lower frequency branches are confined in both cases to the plasma. It is these lower frequency branches which we are interested in, because the specific plasma properties enter only at these lower frequencies. The three electromagnetic wave modes of interest are the whistler mode, the electromagnetic ion-cyclotron or ion whistler mode, and the Alfvén mode. These waves have already been met in the fluid plasma section. Here we show how they emerge from the more precise kinetic treatment.

10.6.3   Whistlers

Electromagnetic waves are transverse waves, satisfying the condition $\mathbf{k} \cdot \delta\mathbf{E} = 0$. Restricting, for simplicity, to parallel propagation, $k_\perp = 0$, and transforming to right- and left-hand circular polarizations

$$\delta E_{R,L} = \delta E_x \mp i\delta E_y \tag{10.154}$$

the dispersion relation for the two parallel propagating electromagnetic modes separates, as in the fluid plasma model treated in Chapter 9

$$\begin{pmatrix} N_\parallel^2 - 1 - \sum_s [\epsilon_{s1} + i\epsilon_{s2}] \\ N_\parallel^2 - 1 - \sum_s [\epsilon_{s1} - i\epsilon_{s2}] \end{pmatrix} \begin{pmatrix} \delta E_R \\ \delta E_L \end{pmatrix} = \begin{pmatrix} 0 \\ 0 \end{pmatrix} \tag{10.155}$$

All other components of the dielectric tensor vanish. For instance, the component ϵ_{s3} drops out, because it does not contribute to the electromagnetic wave for parallel propagation. Also, $\epsilon_{s0} = 0$ because $k_\perp = 0$, so that $\epsilon_{s1} = \epsilon_{xx} = \epsilon_{yy}$, and circular polarization makes the other components vanish. Clearly, the above equation splits into two separate dispersion relations for the R- and L-modes

$$N_{\parallel R,L}^2 = 1 + \sum_s (\epsilon_{s1} \pm i\epsilon_{s2}) \tag{10.156}$$

Although all particle components contribute to each of the R- and L-modes, they are independent of each other. In the low-frequency domain the R-mode will become the whistler mode, while the L-mode is the electromagnetic ion-cyclotron wave.

Because of the condition of vanishing perpendicular wavenumber, the argument of the $\Lambda_l(\eta_s)$ function vanishes, $\eta_s = 0$. Since $\Lambda_l(0) = I_l(0)$ and $I_{-l} = I_l \propto (\eta_s/2)^l$, only the $l = \pm 1$ terms in the components of the dielectric tensor (10.147) contribute. This is easily seen from the definition of ϵ_{s1}. To prove this for ϵ_{s2}, one uses the identity

$$\Lambda_l' = -\Lambda_l + I_l' \exp(-\eta_s) \tag{10.157}$$

and the recurrence relation $I_l' = 2(I_{l-1} + I_{l+1})$. Hence, in the sum over l only the terms $l = \pm 1$ are non-zero. This yields for the two surviving components

$$\epsilon_{s1} = \frac{\omega_{ps}^2}{2k_\parallel v_{\text{ths}\parallel}\omega} \left\{ Z(\zeta_{s,1}) + Z(\zeta_{s,-1}) - \frac{k_\parallel v_{\text{ths}\parallel}}{2\omega} A_s \left[Z'(\zeta_{s,1}) + Z'(\zeta_{s,-1}) \right] \right\}$$

$$\epsilon_{s2} = \frac{2i\,\text{sgn}(q_s)\omega_{ps}^2}{k_\parallel v_{\text{ths}\parallel}\omega} \left\{ Z(\zeta_{s,1}) - Z(\zeta_{s,-1}) - \frac{k_\parallel v_{\text{ths}\parallel}}{2\omega} A_s \left[Z'(\zeta_{s,1}) - Z'(\zeta_{s,-1}) \right] \right\} \tag{10.158}$$

Using these expressions in Eq. (10.156), the dispersion relation can be written as

$$N_{\parallel \text{R,L}}^2 = 1 + \sum_s \frac{\omega}{2k_\parallel v_{\text{ths}\parallel}} \frac{\omega_{ps}^2}{\omega^2} \left\{ [1 \mp \text{sgn}(q_s)]\tilde{Z}_{s,1} + [1 \pm \text{sgn}(q_s)]\,\tilde{Z}_{s,-1} \right\} \tag{10.159}$$

where

$$\tilde{Z}_{s,\pm 1} = Z(\zeta_{s,\pm 1}) - \frac{k_\parallel v_{\text{ths}\parallel}}{\omega} A_s Z'(\zeta_{s,\pm 1}) \tag{10.160}$$

Let us consider the R-mode. Its dispersion relation from Eq. (10.159) reads

$$\begin{aligned}
N_{\parallel \text{R}}^2 = 1 \;+\;& \frac{\omega_{pe}^2}{k_\parallel v_{\text{the}\parallel}\omega} \left[Z(\zeta_{e,1}) - \frac{k_\parallel v_{\text{the}\parallel}}{2\omega} A_e Z'(\zeta_{e,1}) \right] \\
+\;& \frac{\omega_{pi}^2}{k_\parallel v_{\text{thi}\parallel}\omega} \left[Z(\zeta_{i,-1}) - \frac{k_\parallel v_{\text{thi}\parallel}}{2\omega} A_i Z'(\zeta_{i,-1}) \right]
\end{aligned} \tag{10.161}$$

This equation is still fairly general. The further discussion depends on the approximations introduced. Clearly, the ions have no resonance at the electron cyclotron frequency, because in the whistler range, $\omega \le \omega_{ge}$

$$\zeta_{i,-1} = (\omega + \omega_{gi})/k_\parallel v_{\text{thi}\parallel} \approx \omega/k_\parallel v_{\text{thi}\parallel} \tag{10.162}$$

is usually much larger than one. This allows to use the large argument expansion for the ion terms in Eq. (10.161), which implies that ion damping can be neglected, because it is proportional to $\exp(-\zeta_{i,1}^2)$. The main damping comes from electrons near the cyclotron resonance frequency. Hence, entirely neglecting the contribution of the ions and assuming that the anisotropy is negligible and that we can use the large argument expansion for the electrons, we obtain

$$N_{\parallel \text{R}}^2 = 1 + \frac{\omega_{pe}^2}{\omega(\omega_{ge} - \omega)} + i \frac{\pi^{1/2}\omega}{k_\parallel v_{\text{the}}} \frac{\omega_{pe}^2}{\omega^2} \exp\left[-\left(\frac{\omega_{ge} - \omega}{k_\parallel v_{\text{the}}} \right)^2 \right] \tag{10.163}$$

This is the kinetic dispersion relation for parallel propagating whistler waves in a plasma. The first two terms are identical to the fluid dispersion relation (9.127) of whistler waves. But the presence of the imaginary part of the dispersion relation demonstrates that, similar to Landau damping of Langmuir waves, whistlers are also damped due to thermal effects. This damping is particularly strong near the electron-cyclotron frequency, where the argument of the exponential function in Eq. (10.163) vanishes. Far away from $\omega = \omega_{ge}$ the damping vanishes, and whistlers are only weakly damped. Under this assumption we can neglect the imaginary part and obtain the parallel whistler dispersion relation

$$
\boxed{\frac{k_\parallel^2 c^2}{\omega_{ge}^2} = \frac{\omega^2}{\omega_{ge}^2}\left[1 + \frac{\omega_{pe}^2}{\omega(\omega_{ge} - \omega)}\right]}
\tag{10.164}
$$

This equation shows the electron-cyclotron resonance of the fluid waves. But one should remember that it has been derived under the two restricting conditions of large $\zeta_{e,-1} \gg 1$, which at $\omega = \omega_{ge}$ is clearly violated, and the assumption of weak damping, which also does not hold at resonance. Here whistlers become strongly cyclotron damped, feeding their energy into the plasma electrons. Thus electromagnetic wave propagation in the whistler mode ceases at the electron cyclotron frequency.

We can easily estimate the cyclotron damping rate of whistlers sufficiently far away from the resonance, where the approximation of small damping is satisfied so that the dispersion relation (10.164) gives a valid expression for the whistler mode frequency. Assuming that $\gamma \ll \omega$ and using Eq. (10.153), we obtain

$$
\gamma \approx -\frac{\pi^{1/2}\omega}{k_\parallel \lambda_D}\frac{\omega_{pe}(\omega_{ge} - \omega)}{\omega_{pe}^2 - 2\omega(\omega_{ge} - \omega)}\exp\left[-\left(\frac{\omega_{ge} - \omega}{k_\parallel v_{\text{the}}}\right)^2\right]
\tag{10.165}
$$

for the weak damping rate of whistlers sufficiently far below the electron-cyclotron frequency. The above expression shows that whistlers are damped waves, but sufficiently far below the cyclotron frequency this damping is not very strong, because cyclotron resonance does not extend far from ω_{ge}. Moreover, for parallel whistler wavelengths much longer than the electron gyroradius, this damping becomes very small by the exponential factor. Such long-wavelength whistlers propagate practically without any cyclotron damping along the magnetospheric magnetic field lines.

10.6.4 Electromagnetic Ion-Cyclotron Waves

The ion equivalent to whistlers are electromagnetic ion-cyclotron waves. These waves have frequencies far below the electron-cyclotron frequency, $\omega \ll \omega_{ge}$. Hence

$$N_{\parallel L}^2 = 1 + \frac{\omega_{pe}^2}{\omega^2}\frac{\omega}{k_\parallel v_{\text{the}}}\left[Z(\zeta_{e,-1}) - \frac{k_\parallel v_{\text{the}}}{\omega}A_e Z'(\zeta_{e,-1})\right]$$

$$+ \frac{\omega_{pi}^2}{\omega^2}\frac{\omega}{k_\parallel v_{\text{thi}}}\left[Z(\zeta_{i,1}) - \frac{k_\parallel v_{\text{thi}}}{\omega}A_i Z'(\zeta_{i,1})\right] \tag{10.166}$$

We neglect the anisotropies and expand the dispersion functions in the large argument limit. For the electrons this expansion is obvious because of the large electron-cyclotron frequency, $\zeta_{e,-1} = (\omega + \omega_{ge})/k_\parallel v_{\text{the}} \approx \omega_{ge}/k_\parallel v_{\text{the}}$. For the ion component it requires that $\omega_{gi} - \omega \gg k_\parallel v_{\text{thi}}$. Hence, the parallel wavelength must be much larger than the ion gyroradius. Under these conditions the dispersion relation can be written as

$$N_{\parallel L}^2 = 1 - \frac{\omega_{pe}^2}{\omega \omega_{ge}} + \frac{\omega_{gi}^2}{\omega(\omega_{gi} - \omega)} + i\frac{\pi^{1/2}\omega_{pi}^2}{\omega k_\parallel v_{\text{thi}}}\exp\left[-\left(\frac{\omega_{gi} - \omega}{k_\parallel v_{\text{thi}}}\right)^2\right] \tag{10.167}$$

where we have neglected the exponentially small electron damping. The frequency in the electron term can be replaced by the ion gyrofrequency. Moreover, for weak damping the real part of the dispersion relation yields the electromagnetic ion-cyclotron wave

$$\boxed{\frac{k_\parallel^2 c^2}{\omega_{gi}^2} = \frac{\omega^2}{\omega_{gi}^2}\left[1 - \frac{\omega_{pi}^2}{\omega_{gi}^2}\left(1 - \frac{\omega_{gi}^2}{\omega(\omega_{gi} - \omega)}\right)\right]} \tag{10.168}$$

with a resonance at the ion-cyclotron frequency. Since we have neglected the electron term, electromagnetic ion-cyclotron waves are restricted to frequencies below the ion-cyclotron frequency, ω_{gi}. Sometimes the ion-cyclotron dispersion relation is given in the simplified form

$$\frac{k_\parallel^2 v_A^2}{\omega_{gi}^2} = \frac{\omega^2}{\omega_{gi}(\omega_{gi} - \omega)} \tag{10.169}$$

Clearly, for frequencies much below the ion-cyclotron frequency this dispersion relation goes over into the dispersion relation of Alfvén waves. Damping of electromagnetic ion-cyclotron waves is weak. It is given by the same expression as for whistlers, if the electron parameters are replaced by those of the ions.

10.6.5   Kinetic Alfvén Waves

As has been shown in Chapter 9, at extremely low frequencies both whistlers and ion-cyclotron waves become Alfvén waves. In this frequency domain the Alfvén wave has both polarizations, right-hand and left-hand, and can, depending on the amplitudes of the two circularly polarized components, assume any polarization.

These arguments hold for very long wavelengths. The dispersion relation of Alfvén waves in this domain can be obtained simply in the limit of very low frequencies, $\omega \ll \omega_{gi}$. There is, however, a range of wavelengths when the dispersion of the Alfvén wave changes. This happens when the wavelengths become either comparable to the ion gyroradius, $r_{gi} = v_{\text{thi}}/\omega_{gi}$, in a hot electron plasma, or to the electron inertial length, c/ω_{pe}, in a cold electron plasma. In these two ranges kinetic effects must be taken into account. The dispersion relations can be obtained from the above general dispersion relations for electron- and ion-cyclotron waves, but it is more instructive and also simpler to derive them separately by using a combined fluid-kinetic approach with the electrons taken to be fluid-like and the ions treated kinetic.

We allow for oblique propagation. Since Alfvén waves are electromagnetic waves, the wave electric potential is zero and the electric field is determined by the wave magnetic potential, $\delta \mathbf{E} = -\partial \mathbf{A}/\partial t$. Gyrotropy implies that the magnetic potential has two components. This is equivalent to introducing two different electric potential components, $\phi_\perp, \phi_\parallel$, and to represent the electric field as

$$\delta \mathbf{E} = (\delta E_\perp, \delta E_\parallel) = -(\nabla_\perp \phi_\perp, \nabla_\parallel \phi_\parallel) \tag{10.170}$$

Because no space charges are involved, Poisson's equation yields

$$\nabla_\perp^2 \phi_\perp + \nabla_\parallel^2 \phi_\parallel = \frac{e}{\varepsilon_0}(\delta n_i - \delta n_e) = 0 \tag{10.171}$$

and from the only non-vanishing component of Ampère's law one has

$$\nabla_\parallel \nabla_\perp^2 (\phi_\perp - \phi_\parallel) = \mu_0 \frac{\partial (j_{i\parallel} + j_{e\parallel})}{\partial t} \tag{10.172}$$

where the parallel currents are calculated from the ion and electron Vlasov equations.

Let us consider two extreme cases. When the electron temperature is so high that the electron thermal velocity is much larger than the Alfvén velocity, the electrons almost immediately respond to the electric potential of the wave and electron inertia can be neglected. The electrons can be treated as a mass-less fluid of temperature T_e, with their density given by Boltzmann's law. At such low frequencies the ions, which behave as kinetic particles, contribute only through the zero harmonic number, $l = 0$, in the ion term of the dispersion relation. We then find for the ion density as the zero-order moment of the Maxwellian ion distribution, keeping only the term $l = 0$

$$\frac{e\delta n_i}{\varepsilon_0} = -\frac{\omega_{pi}^2}{v_{\text{thi}}^2}[1 - \Lambda_0(\eta_i)]\phi_\perp + \frac{\omega_{pi}^2}{\omega^2}k_\parallel^2(1 - i\gamma_i)\Lambda_0(\eta_i)\phi_\parallel \tag{10.173}$$

and from Boltzmann's equation for the electron density

$$\frac{e\delta n_e}{\varepsilon_0} = \frac{\omega_{pe}^2}{v_{\text{the}}^2}(1 + \gamma_e)\phi_\parallel \tag{10.174}$$

where the Landau damping terms in the low-frequency range are given by

$$\gamma_{li} = 2\pi^{1/2}\beta_i^{-3/2}\exp(-\beta_i^{-1})$$

$$\gamma_{le} = \pi^{1/2}\beta_i^{-1/2}\left(\frac{m_e T_i}{m_i T_e}\right)^{1/2} \tag{10.175}$$

with $\beta_i = 2v_{\text{thi}}^2/v_A^2$. Electron Landau damping of the Alfvén waves is very small, unless the ion temperature is extremely high. In the following we neglect electron Landau damping. Similarly, we obtain for the parallel wave current densities

$$\mu_0 j_{e\parallel} = -\frac{\omega_{pe}^2}{v_{\text{the}}^2 c^2}\frac{\omega}{k_\parallel}\phi_\parallel$$

$$\mu_0 j_{i\parallel} = \frac{\omega_{pe}^2}{c^2}\frac{k_\parallel}{\omega}\Lambda_0(\eta_i)(1-i\gamma_{li})\phi_\parallel \tag{10.176}$$

Substituting these expressions into Poisson's and Ampère's laws and ignoring damping yields

$$\left[\Lambda_0(\eta_i)-\frac{\omega^2}{k_\parallel^2 c_{ia}^2}\right]\left[1-\frac{\omega^2}{k_\parallel^2 v_A^2}\frac{1-\Lambda_0(\eta_i)}{\eta_i}\right] = \frac{\omega^2}{k_\parallel^2 v_{\text{thi}}^2}[1-\Lambda_0(\eta_i)] \tag{10.177}$$

This dispersion relation shows how the Alfvén wave couples to the ion-acoustic wave for perpendicular wavelengths comparable to the ion gyroradius. For a sufficiently warm plasma the Λ_0-term in the first bracket can be ignored

$$\frac{\omega^2}{k_\parallel^2 v_A^2} = \frac{\eta_i}{1-\Lambda_0(\eta_i)} + \frac{T_e}{T_i}\eta_i \tag{10.178}$$

Clearly, this relation describes Alfvén waves. When its right-hand side becomes one, for $\eta_i \to 0$, the wave is the usual torsional Alfvén wave. In the more general case of $\eta_i \neq 0$, the Alfvén wave propagates obliquely to the magnetic field. This wave is known as the *kinetic Alfvén wave*. For not too small ion temperatures the argument of Λ_0 can be assumed small, $\eta_i \ll 1$. The above dispersion relation then simplifies further to

$$\omega_{ka}^2 = k_\parallel^2 v_A^2\left[1+k_\perp^2 r_{gi}^2\left(\frac{3}{4}+\frac{T_e}{T_i}\right)\right] \tag{10.179}$$

The kinetic Alfvén wave propagates across the magnetic field. Usually its parallel wavelength is much longer than the ion gyroradius. Therefore we have $k_\parallel \ll k_\perp \propto r_{gi}$, and the perpendicular wavenumber is much larger than the parallel wavenumber. This wave is Landau damped by the ions and, in addition, carries a non-vanishing parallel current. Moreover, it has a non-vanishing parallel electric field component.

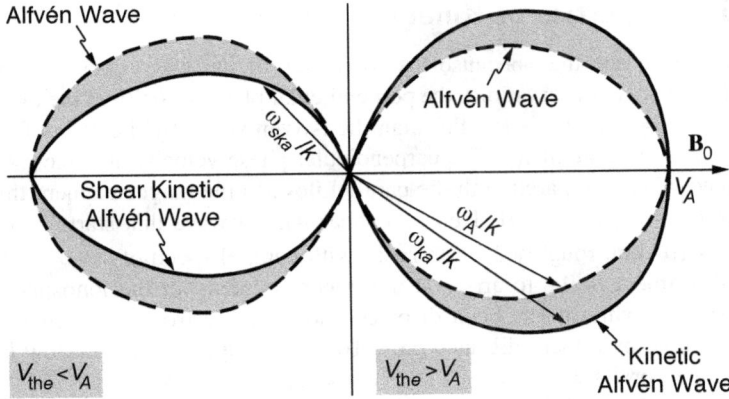

Figure 10.11: Angular dependence of the phase velocities of the kinetic Alfvén waves.

When considering the phase velocity of the kinetic Alfvén wave, we find that the parallel phase velocity is increased due to the factor in brackets in Eq. (10.179). This factor is of the order of two. Hence, kinetic Alfvén waves propagate with a parallel phase velocity which is about 50% higher than the normal Alfvén speed. Similarly, the group velocity of the kinetic Alfvén wave is enhanced

$$\left(\frac{\partial \omega_{ka}}{\partial k}\right)^2 \approx 2v_A^2 \left(1 + \tfrac{1}{8}\cos^4\theta\right) \approx 2v_A^2 \tag{10.180}$$

If the plasma is cool, such that $v_{the} < v_A$, electron inertia must be taken into account and the Boltzmann approximation for the electrons gets invalid. Instead one can use the fluid electron parallel equation, since their perpendicular motion still does not contribute at these low frequencies. The electrons merely perform a drift across the magnetic field in the perpendicular wave electric field. In this case the relevant length is not the ion gyroradius but the electron inertial length, c/ω_{pe}, and the dispersion relation reads

$$\boxed{\omega_{ska}^2 = k_\parallel^2 v_A^2 \frac{1 + k_\perp^2 r_{gi}^2}{1 + k_\perp^2 c^2/\omega_{pe}^2}} \tag{10.181}$$

The kinetic Alfvén wave described by this dispersion relation still contains the thermal effect of the ions, but additionally takes into account the inertia of the electron background plasma, which makes the wave 'heavier'. It effectively increases the mass which enters the Alfvén speed and slows the wave down. If we neglect the ion thermal effect, the wave is usually called *shear kinetic Alfvén wave*.

Figure 10.11 shows the angular dependence of the two kinetic Alfvén wave phase velocities for $k_\perp r_{gi} \approx 1$ and $k_\perp c/\omega_{pe} \approx 1$. In the first case the phase velocity is larger than the Alfvén velocity, in the second it is smaller.

10.6.6   Properties of Kinetic Alfvén Waves

It is easy to see that, because the frequency of the wave does not change (as required by energy conservation), the perpendicular phase velocity of the inertial shear wave $\omega_{ka}/k_\perp \sim v_A k_\parallel/k_\perp$ is smaller than the Alfvén speed by the ratio of parallel to perpendicular wave numbers. The perpendicular phase velocity and energy transport are reduced when compared with the parallel flow. In the magnetosphere this ratio is of the order of $k_\parallel/k_\perp \sim 0.01$. Hence, a shear Alfvén wave that starts at an altitude of $h_{magn} \sim 10^4$ km, roughly 2 Earth radii, with parallel speed of say $v_\parallel \sim 10^3$ km/s, will need a time $\tau \sim 9$ s to arrive at the upper boundary of the ionosphere at, say, $h_{ion} \sim 10^3$ km, while it has been displaced horizontally from its original field line within this time by a mere distance $\Delta x \sim 10$ km, the approximate narrow latitudinal width of an auroral arc.

The shear kinetic Alfvén wave in Eq. (10.181) is important insofar as in a cold plasma its parallel current is transported by the dense electron background and therefore can be rather strong. It is widely believed that these field-aligned Alfvénic currents transport the field-aligned energy along the magnetic field during all disturbances of the magnetosphere, causing the various auroral effects and producing the particle and wave phenomena observed during substorms.

Since the Earth's plasma sheet is a hot plasma region, kinetic Alfvén waves in the plasma sheet propagate in the non-inertial kinetic mode and will not carry bulk currents. The transition to the inertial shear kinetic mode will happen at that position in space where the electron thermal velocity drops below the local Alfvén speed. This becomes possible closer to the Earth, because here the plasma is cooler while the magnetic field strengthens. However, because of the high plasmaspheric density it can actually occur only at auroral latitudes, where the plasma density stays comparably low. The auroral lower magnetosphere is thus the only prospective candidate for the existence of inertial shear kinetic Alfvén waves. Writing the above condition as

$$\frac{v_{\text{the}}^2}{v_A^2} = \frac{m_i}{m_e}\beta \tag{10.182}$$

where β is the plasma beta, and assuming that the ions are sufficiently cold so that the plasma energy is carried by the electrons, one finds that inertial shear kinetic Alfvén waves with a dispersion given by Eq. (10.181) can propagate only if the condition

$$\beta < \frac{m_e}{m_i} \tag{10.183}$$

is satisfied. Otherwise, the kinetic Alfvén wave is non-inertial and satisfies the dispersion relation (10.179). Figure 10.12 shows the schematic boundary between the regions $\beta < m_e/m_i$ and $\beta > m_e/m_i$ in the auroral magnetosphere as function of the distance from the Earth's surface. For local electron temperatures below this curve the kinetic Alfvén waves are inertia-dominated, above they are non-inertial. We have included also a typical range of auroral region electron temperatures. Close to the Earth the electron temperatures fall into the inertial range while at distances larger than a few R_E the electron temperature rises above the regime of the shear kinetic Alfvén wave.

This behaviour is caused by the steep changes in plasma density and temperature when passing from the inner to the outer magnetosphere. In the inner magnetosphere the magnetic field energy density wins over the thermal plasma energy density even though the density increases considerably inside the plasmasphere. However, this plasma is fairly cold, its temperature being only of the order of at most few $\sim$eV while in the auroral region the density drops to very low values in particular during aurora when field-aligned electric fields are present over longer times and spatial locations.

Thus the inner magnetosphere favours inertial shear Alfvén waves to propagate parallel and perpendicular to the magnetic field, while the region outside the dashed line in Fig. 10.11 is where ordinary kinetic Alfvén waves propagate. These are fast while the inertial kinetic Alfvén waves are slow waves, whose phase velocity is reduced by the electron inertial effect.

If a signal in the outer magnetosphere is propagated into the inner magnetosphere, it is first transported by kinetic Alfvén waves along the field at large, in fact at super-alfvénic speed until reaching the inner magnetosphere and changing to become shear inertial Alfvén waves at low, in fact sub-alfvénic speed and accumulating large electric field amplitudes preferentially along the magnetic field which are important in particle acceleration. They also accumulate perpendicular electric fields which cause other additional effects like shear flow in the upper ionosphere. The latter is the reason for calling these kinetic Alfvén waves shear waves.

This is the way, at least most probably, by which low frequency (because Alfvén waves are low frequency oscillations, not free space electromagnetic radiation) signals are transported from processes in the outer magnetosphere into the inner magnetosphere, preferentially into the auroral zone during substorms and magnetic storms. At the same time this is also the way how signals can, without major compressions, be transported at some smaller distance across the magnetic field and at the same time become amplified. Possibly the kinetic Alfvén waves are the engine which generates substantial parallel electric fields in the upper ionosphere for particle acceleration and, at the same time causes perpendicular motion of the observed structures by imposing a large amplitude transverse electric field component.

There is an interesting consequence of the existence of a quite precisely defined spatial location of the transition from kinetic to shear Alfvén waves in Earth's magnetosphere. Any signal that approaches this transition at super-alfvénic speed becomes strongly decelerated to sub-alfvénic velocity at this location. The deceleration is not really sudden but possibly fast enough to generate a temporarily existing barrier or shock wave at this location. We will discuss shock waves later in Chapter 13, so here we just note that sudden transitions in signal speed from super- to sub-alfvénic are typical for shock generation. In the auroral magnetosphere where this could be expected such a shock would be weak in the sense that its magnetic and electric amplitudes would be moderate. The shock itself results from the fact that the retardation of the wave is felt first by the longer wavelengths of the signal. The shorter wavelength overrun them and cause the wave to steepen because of accumulation of ever shorter wavelength waves of the wave spectrum. This causes a magnetic step in the wave field which implies an electric potential drop across the step and thus causes local particle acceleration and other effects of considerable interest.

Kinetic Alfvén waves have a number of other applications as well. In deriving Eq. (10.177) we have seen that they are accompanied by density perturbations in both particle species δn and correspondingly also field-aligned current perturbations $\delta j_\parallel$ both determined basically through the parallel potential $\phi_\parallel$ and hence through a field aligned electric field $E_\parallel$. This field can accelerate preferentially electrons because of their high mobility. It is therefore believed that kinetic Alfvén waves propagating along the magnetic field for instance in the auroral magnetosphere (as discussed in the previous paragraphs) is responsible for part of the auroral electron acceleration. However, kinetic Alfvén waves, as we have seen, propagate also perpendicular, and then, if their perpendicular wavelength becomes of the order of the ion gyroradius or ion inertial length, then density and parallel current perturbations are

$$\frac{\delta n}{n} \approx \frac{m_i}{eB^2} \nabla_\perp E_\perp, \qquad j_\parallel = \frac{1}{\mu_0 v_A} \nabla_\perp E_\perp \qquad (10.184)$$

This current is carried by electrons in resonance with the wave for $\beta > m_e/m_i$. These populate only the high energy superthermal tail of the electron distribution. In the case $\beta < m_e/m_i$ which is the cold plasma case, there are no resonant electrons, and the current is transported by the bulk electron distribution which thus drifts with respect to the ions. This is an unstable situation along the magnetic field which leads to the excitation of current driven electrostatic waves and thus to dissipation of the current which for strong instability like the Buneman instability (see Chapter 11) may cause anomalous collision frequencies v_{an}, a most interesting case. The bulk slowly variable distribution function of the current electrons then evolves according to the one-dimensional Vlasov equation including a diffusive term on its right side

$$\left[\partial_t + v_\parallel \nabla_\parallel - \frac{e}{m_e} \partial_{v_\parallel}\right] f_e(v_\parallel, z) \approx -v_{an} f_e/v_\parallel, z) \qquad (10.185)$$

If one now calculates the velocity moment of this equation, the current $\delta j_\parallel = n \delta v_\parallel$ is obtained

$$\delta j_\parallel \approx \frac{ne^2(v_{an} - i\omega)}{k_\parallel^2 T_e} Z\left(\frac{\omega + iv_{an}}{k_\parallel v_e}\right) \delta E_\parallel \qquad (10.186)$$

where $Z(\zeta)$ is the plasma dispersion function given in the Appendix C. This current is to be combined with Maxwell's equations to obtain the dispersion relation including the anomalous collisional damping of the waves (which ultimately leads to the acceleration of electrons):

$$\frac{\omega^2}{k_\parallel^2 v_A^2} \approx \left[1 + k_\perp r_{gi}^2\left(\frac{3}{4} + \frac{T_e}{T_i}\frac{\omega}{k_\parallel v_e}z^{-1}Z^{-1}(z)\right)\right], \qquad z = \frac{\omega + iv_{an}}{k_\parallel v_e} \qquad (10.187)$$

which for $v_{an} \ll \omega$ just reduces to the full dispersion relation of kinetic Alfvén waves. When collisions become as large as the frequency of the wave ω, the wave becomes damped according to the dispersion relation

$$\omega^2 = \frac{k_\parallel^2 v_A^2}{1 + k_\perp^2 \lambda_e^2 (1 + i v_{an}/\omega)} \qquad (10.188)$$

with $\lambda_e = c/\omega_{pe}$ the electron inertial length scale. This dispersion relation describes damped kinetic Alfvén waves and thus dumping of wave energy into the electrons. This is believed to be important in auroral physics.

Kinetic Alfvén waves are, however, important in many other fields of space plasma physics as for instance in turbulence in the solar wind or in the magnetosheath. Under those conditions the plasma is not cold but in general hot, which is in contrast to the exceptional situation in planetary auroral regions like that of Earth or the other magnetized planets and even in astrophysical objects like neutron stars. Under hot plasma conditions the waves never become inertial but remain in the genuine kinetic range where they propagate at super-alfvénic speed and may easily steepen to generate small-scale shocklets which become important in turbulence. In fact, since the perpendicular wave length of kinetic Alfvén waves is of the order of the ion gyroradius or ion inertial scale, it becomes clear that whenever energy in a plasma is available on this scale then it will be deposited into kinetic Alfvén waves which transport it away, being a loss mechanism of electromagnetic energy. This may be the case in turbulence as well as in reconnection where the ion inertial scale is a typical scale as ions become non-magnetic there and may feed their magnetic energy into the waves. For this reason reconnection regions may act as sources of kinetic Alfvén waves. On the other hand, when those waves are present and enter a region with wavelength fitting the local ion inertial length, then particles may fall into resonance and extract wave energy.

10.7 Concluding Remarks

In the present chapter we discussed, from a kinetic theory viewpoint, the main wave modes in a homogeneous magnetized plasma under the condition that these waves are stable. The most important among these modes are the natural plasma resonances at the lower-hybrid, upper-hybrid and cyclotron frequencies, which provide the channels in $(\omega, \mathbf{k})$ space along which energy can be transported, once it has been injected into a certain frequency and wavenumber range. For instance, a right-hand polarized electromagnetic wave at a frequency below the electron-cyclotron frequency with a wavelength longer than the electron gyroradius, will necessarily propagate in the whistler mode. On the other hand, it is impossible to inject electromagnetic waves into a dense plasma between the electron cyclotron and the upper-hybrid frequency.

In the last two chapters we dealt with homogeneous plasmas only. Hence, the applicability of the present theory is restricted to cases when the wavelengths are very short in comparison with the characteristics scale of any inhomogeneity. Nevertheless,

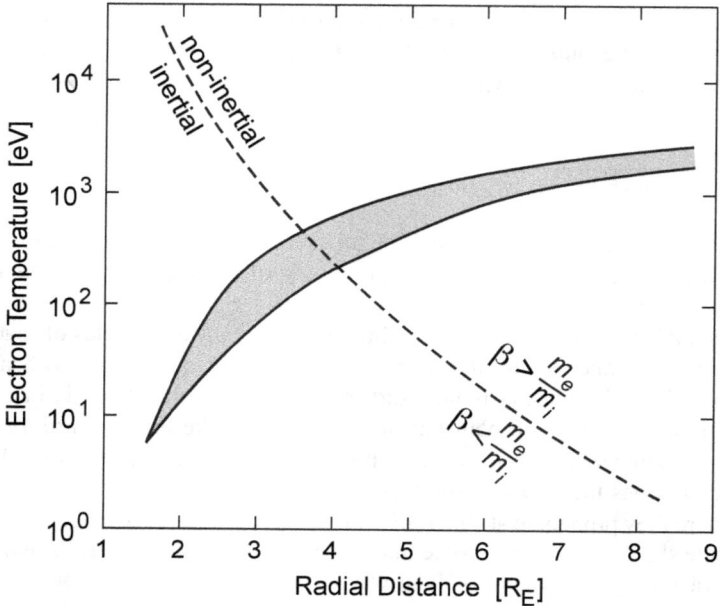

Figure 10.12: Regimes of kinetic and shear kinetic Alfvén waves in the equatorial magnetosphere as deduced from average magnetospheric plasma parameters. It is clearly seen that the nature of Alfvén waves in the magnetosphere depends heavily on the plasma β, i.e. on the ratio of thermal to magnetic energy densities in the plasma.

the range of applicability of the homogeneous theory is rather wide. When the inhomogeneity of the plasma is weak, the wave propagation can be treated as in optics as propagation of rays in a medium where the refraction index and phase velocities change slowly in space. This method is called *ray tracing*. On the other hand, when the changes are very steep, one can use the refraction and transmission theory at discontinuities, where the waves must satisfy the jump conditions derived in Chapter 8. In cases where the scale of the inhomogeneity becomes comparable to the wavelength, inhomogeneous theory must be applied, where one has to use eikonal methods and eigenvalue solutions of the Vlasov-Poisson system. The inhomogeneity does not only introduce a new kind of waves, called *drift waves*, but also changes the properties of the linear waves and allows *linear conversion* of waves from one mode to another. Drift waves have not been treated in this volume. In principle any plasma inhomogeneity introduces two sources of free energy available to drive new forms of waves. In this chapter we have not asked for the role of such free energy. We have considered possible modes and their propagation or damping. In inhomogeneous plasmas one may thus, if neglecting the possibility of new wave excitation, think of modification of wave propagation, wave number and frequency on the scale of the wave length. This causes the waves to becomes more strongly deformed than we have accounted for. In one case, however, inhomogeneity has been considered here as well. This is the case of diamagnetic drifts in a gradient region where the lower hybrid wave propagates. Also the resonance cone of the whistler mode indicates the effect of inhomogeneity when the whistler cannot propagate further. Inhomogeneity does however play an important role in instability theory.

We finally mention that the plasma dispersion function used in this chapter is based on a Maxwellian distribution function. One can define other plasma dispersion functions by using other distribution functions, as has been done for the kappa distribution (see Section 6.3). However, the kappa distribution is not a stable distribution function. For theoretical purposes it is more convenient to approximate unstable distributions as superposition of Maxwellians.

Further Reading

A classic general treatment of plasma waves, but with no emphasis on space physics, is the monograph [8]. More modern presentations are found in [9, 10]. A similar but more elaborate and educationally more suitable monograph including all the early knowledge on ionospheric, magnetospheric and solar coronal plasma wave propagation can be found in the early monograph [1]. The relativistic dielectric tensor applying to radiation belt particles is given in [3, 5]. A discussion of ion Bernstein modes can be found in [3] and [6]. Whistler waves are discussed in [7]. A brief collection of our knowledge about most aspects of Alfvén waves is contained in [2]. Different aspects of kinetic Alfvén waves are treated in [2, 4].

References

[1] V. L. Ginzburg, *The Propagation of Electromagnetic Waves in Plasmas*, 2nd ed. of Engl. translation (Pergamon Press, Oxford, 1970).

[2] A. Hasegawa and C. Uberoi, *The Alfvén Wave* (U.S. Dept. of Energy, Springfield, 1982).

[3] S. Ichimaru, *Basic Principles of Plasma Physics* (W. A. Benjamin, Reading, 1973).

[4] R. L. Lysak and W. Lotko, *J. Geophys. Res.* **101**, 5085–5094 (1996).

[5] D. C. Montgomery and D. A. Tidman, *Plasma Kinetic Theory* (McGraw-Hill, New York, 1964).

[6] D. R. Nicolson, *Introduction to Plasma Theory* (John Wiley and Sons, New York, 1983).

[7] S. Sazhin, *Whistler-mode Waves in a Hot Plasma* (Cambridge University Press, Cambridge, 1993).

[8] T. H. Stix, *Waves in Plasma* (American Institute of Physics, New York, 1992).

[9] D. G. Swanson, *Plasma Waves* (Academic Press, Boston, 1989).

[10] A. D. M. Walker, *Plasma Waves in the Magnetosphere* (Springer Verlag, Heidelberg, 1993).

Problems

Problem 10.1 *Show that the solution of the one-dimensional Poisson equation for the electric field in terms of the electron and ion distributions in the one-dimensional case has been correctly given in Eq. (10.2). Linearise the Vlasov and Poisson equations and find the linearised set of equations which completely describe the one-dimensional case. Derive the expression Eq. (10.20) for the dielectric function.*

Problem 10.2 *Prove that the form of the dielectric function given in Eq. (10.30) can be found by expansion.*

Problem 10.3 *Show what the contribution of the pole on the real axis in* Fig. 10.2 *is. Why is it half of that of the pole outside the real axis?*

Problem 10.4 *Explain the quantum physics picture of Landau damping.*

Problem 10.5 *Find the dielectric function for ion-acoustic waves. Solve it and discuss the ion-acoustic dispersion branch.*

Problem 10.6 *Prove that the form of the dielectric function given in Eq. (10.30) can be found by expansion.*

Problem 10.7 *Find the expression for the linear perturbation of the distribution function δf in the plasma for electromagnetic waves as function of the linear perturbation of the electric field δE from the Vlasov-Maxwell system of equations when no external magnetic field is present.*

Problem 10.8 *Find the same expression for the linear perturbation of the distribution function δf in the magnetised plasma. Show that it assumes the form of* Eq. (10.91).

Problem 10.9 *Derive the dispersion relation of lower-hybrid waves.*

Problem 10.10 *Confirm the dielectric tensor form for electromagnetic waves in a uniform magnetic field as given in* Eq. (10.147).

Problem 10.11 *Derive the whistler dispersion relation for parallel propagation. What form does it assume for frequencies far below the electron gyro-frequency but far above the ion-gyrofrequency? Discuss the properties of the wave. What does the damping of whistler waves mean? Give a physical explanation.*

Problem 10.12 *Compare the whistler case with the ion-cyclotron case. What differences and what similarities do you detect?*

Problem 10.13 *Why is it possible to use two potentials in the derivation of the dispersion relation for kinetic Alfvén waves? Explain what the transverse potential is in reality.*

Problem 10.14 *Find the expression for the parallel electric field of a kinetic Alfvén wave. Explain why the phase velocity of the wave has a transverse component. Show that also the group velocity is oblique. Discuss the consequences for energy transport.*

— 11 —

Plasma Instability

The last two chapters dealt with the propagation of waves in plasmas treating the plasma in various models. It was found that plasmas allow for a large number of waves to propagate. Whether a wave can or cannot pass across a plasma and in which range of frequencies and wavenumbers could be inferred from the appropriate linear dispersion relation $D(\omega, \mathbf{k}) = 0$ which is a function of wave frequency ω, wavenumber $\mathbf{k}$ and the plasma parameters entering the conductivity and dielectric tensor.

In order to find the propagating electrostatic and electromagnetic wave modes we have, however, so far considered just the real solutions of the dispersion relation even though we have already seen in Section 9.2 that the general dispersion relation Eq. (9.57) is a complex equation which contains the complex dielectric tensor $\in(\omega, \mathbf{k})$ which has been defined in (9.56) through the complex conductivity tensor $\sigma(\omega, \mathbf{k})$ of the plasma. Complex equations of this kind have, quite generally, complex solutions consisting of real and imaginary parts. Hence, a general solution of the dispersion relation Eq. (9.57) will always be of the form $\omega = \omega_r(\mathbf{k}) + i\gamma(\omega_r, \mathbf{k})$. Here ω_r is the real part of the wave frequency, the part which we have been familiar with up till now and which is a function of wave number and the plasma parameters, and $\gamma(\omega_r, \mathbf{k})$ is the imaginary part of the frequency which, so far, we have neglected. It is also a function of the real frequency ω_r, the wave number $\mathbf{k}$, and the plasma parameters.

The imaginary part of the frequency is an important quantity. In fact, large parts of our companion volume *Advanced Space Plasma Physics* are devoted to its investigation in great detail. In the present chapter we will as a preparatory introduction just sketch, without going into any detail, what its main implications are. The reason for including such a brief treatise is also the use of it we will make in the final chapter of this volume which deals with the physics of shock waves. Summarising its essence we may simply say, that it is the imaginary part of the frequency which decides about the existence or non-existence of a wave in plasma. Either the wave is supported by the plasma, grows and can propagate, or the plasma absorbs and damps the wave on a time scale of the inverse imaginary part of the frequency $\gamma^{-1}(\omega_r, \mathbf{k})$, the absorption time. In the former case the wave will gain energy and its amplitude will grow and survive, in the latter case the wave will die out when plasma eats up the wave energy and distributes it among the different plasma constituents: particles and electromagnetic fields. Thus, the imaginary part tells whether the wave gains or looses energy when passing across the plasma. Why this is so, will become clear in the next section.

11.1   Wave Growth Rate

In deriving the general wave dispersion relation Eq. (9.57) we have made use of the representation of a wave disturbance which enters the plasma from, say, outside as a superposition of plane waves. By Fourier's theorem this is always possible to assume for any arbitrary form of the disturbance. In the linear theory this superposition is exact even throughout the entire evolution of the wave because the linear equations hold for each of the Fourier components. We therefore can restrict ourselves to the consideration of only one of the plane waves which compose the arbitrarily shaped disturbance. Assigning it the frequency $\omega = \omega_r + \gamma$ and wave number $\mathbf{k}$ we have

$$\mathbf{E}(\mathbf{x},t) = \mathbf{E}(\omega,\mathbf{k})\exp\left[i\left(\mathbf{k}\cdot\mathbf{x} - \omega_r t\right) + \gamma t\right] + c.c. \tag{11.1}$$

where $\mathbf{E}(\omega,\mathbf{k})$ is the vector Fourier amplitude (in this case given for the electric field only) and the exponential function is the plane wave function the first part of which describes the plane propagating wave. Both depend on frequency and wave number. The appearance of the real exponential indicates that the wave can either be damped away exponentially in a typical decay time γ^{-1}, or the wave grows exponentially. We thus see that, indeed, the imaginary part γ of the frequency decides about the temporal fate of the wave. It thus becomes of vital interest to find a way of determining γ. Clearly, the dispersion relation, writing it as

$$D(\omega,\mathbf{k}) = D_r(\omega,\mathbf{k}) + iD_i(\omega,\mathbf{k}) = 0 \tag{11.2}$$

is just one algebraic relation from which one can find a relation only between ω and $\mathbf{k}$. For complex ω it can, in principle, be split into two equations, one for the real part $D_r = 0$ and another one for the imaginary part $D_i = 0$. From these two equations one could in principle determine ω_r and γ as functions of $\mathbf{k}$. However, inspecting the involved mathematical expressions for the dispersion relation as given in the two previous chapters, it turns out to be practically impossible to analytically provide such solutions, and one needs either to go straight away to a numerical treatment or to seek for simpler analytical methods.

   As always, one such method can be based on the assumption that the imaginary part of the frequency is small, $|\gamma| \ll \omega_r$. This assumption implies weak growth (or weak damping) of the wave which makes sense if one wants to speak about a wave which oscillates more than just a fraction of its period before it either disappears or explodes. The former case implies that at a time γ^{-1} after, say, injection of a wave into the plasma no wave exists anymore, which from the wave point of view is of interest only when dealing with absorption and wave heating, while the latter case is of considerable interest when asking for the natural sources, generation and amplification of waves by the plasma. It means that the plasma becomes *exponentially unstable* feeding energy into a number of plane waves until their amplitudes increase to values so large that a large fraction of energy of the plasma is absorbed into the waves, in which case the above assumption of linearity breaks down.

   There are also non-oscillatory instabilities which can be excited in a plasma (see below). In the following we will instead of dealing with these 'explosive' instabilities mostly consider the other limit when the wave has plenty of time to perform many oscillation before either being damped or undergo instability. It is the case of weak instability which interests us most. It can happen only if there is free energy available which the plasma can gradually spend in order to feed a wave until it grows and propagates away. Mechanisms of this kind produce waves from the thermal fluctuation background, the constantly present background noise in any temperate system.

11.1.1 Weak Instability

Under the above formulated condition of weak growth or damping which can be written as $|\gamma|/\omega_r \ll 1$ it is quite easy to derive a general formula for the growth (or damping) rate γ. When one may assure that the imaginary part $D_i(\omega, \mathbf{k})$ of the dispersion function $D(\omega, \mathbf{k})$ is solely due to the property of the frequency being a complex function, one can safely assume that also $|D_i| \ll D_r$. In this case the two terms in the dispersion relation can be expanded with respect to the ratio γ/ω taken at $\omega = \omega_r$ which then yields that to first order

$$D(\omega, \mathbf{k}) \approx D_r(\omega_r, \mathbf{k}) + (\omega - \omega_r) \frac{\partial D_r(\omega, \mathbf{k})}{\partial \omega}\bigg|_{\gamma=0} + iD_i(\omega_r, \mathbf{k}) = 0 \qquad (11.3)$$

where, of course, by definition $\omega - \omega_r = i\gamma$. Since the real and imaginary parts of this expression must vanish separately one immediately obtains the dispersion relation in the form we used it so far in the previous chapters, viz. $D_r(\omega_r, \mathbf{k}) = 0$, where we now can drop the index r since all quantities remain real. Its solution yields the real dispersion relation which defines the possible modes as functions of wavenumber and plasma parameters thus identifying their ranges of existence, phase and group velocities. On the other hand, $D_i(\omega_r, \mathbf{k})$ is now a well defined function obtained by first setting the frequency in the dispersion relation real and then splitting the dispersion relation into real and imaginary parts. With this simple prescription the imaginary part of the above equation yields the wanted general expression for the growth or damping rate

$$\gamma(\omega, \mathbf{k}) = -\frac{D_i(\omega_r, \mathbf{k})}{\partial D_r(\omega_r, \mathbf{k})/\partial \omega_r} \qquad (11.4)$$

This formula is particularly useful for inferring about the kinetic instabilities described by the kinetic dispersion relations of the previous chapter. These are high frequency instabilities of the waves we have discussed before. In the following we just list a few of them for later use without going into any detail. The derivations can be found in out companion volume.

11.1.2 Non-oscillatory Instability

When the growth rate is of the same order as or even larger than the frequency, then the wave will not have time to oscillate before either disappearing — the case of violent heating — or exploding. In the latter case the above assumption of weak growth is invalid. The plasma feeds energy into the wave on a scale faster than the oscillation period in order to re-arrange itself into a new state or equilibrium. Such cases in other branches of physics are typical for phase transitions from one state to another one. They are always accompanied by violent fluctuations of large correlation lengths affecting the entire plasma and thus cannot be considered simply as single growing wave modes. Also in space plasmas one knows about such cases which we will briefly describe later at some place. Since space plasmas are very extended, however, even in the cases of such non-oscillatory 'explosive' instabilities one encounters waves of relatively narrow spectrum and not necessarily violent turbulence. Examples of this will be mentioned below. Interestingly, such processes take place mainly at wave phenomena on the slow time scales of the evolution of fluids into turbulence. The kinetic wave limit is instead the domain of weak instability. The limit of non-oscillatory instability allows to approximate the frequency by the growth rate alone assuming that $|\gamma| \gg \omega_r$ in the dispersion relation, writing $\omega \approx i\gamma$ when solving the dispersion relation for the non-oscillatory growth rate $\gamma(\mathbf{k})$ of the wave. The dispersion relation in the non-oscillatory case becomes

$$D(\omega = i\gamma, \mathbf{k}) = 0 \tag{11.5}$$

where one has simply replaced the wave frequency with its imaginary part. In many cases this equation is much easier to solve than the general dispersion relation. One, however, must take care of the condition implied that a complex solution has to satisfy that $\omega_r/|\gamma| \ll 1$ as only in this approximation the above equation remains to be valid. This is mostly satisfied for the non-oscillatory fluid instabilities which can evolve in a plasma.

11.2   High-frequency Kinetic Instabilities

In this section we briefly list a number of 'weakly growing' instabilities in the kinetic theory approach. These are in the first place instabilities of high frequency waves like Langmuir and ion-acoustic waves. For then it is easiest to identify their sources of free energy, as it can be shown that they arise from the resonance of small groups of particles with waves which have phase velocity comparable to the particle velocity. Resonance in this sense means that the denominator of some complex integral vanishes and causes the dispersion relation to be dominated by the pole of a complex integral. We have already encountered this effect when having dealt with Landau damping of Langmuir waves in Section 10.2. Their we considered a thermal plasma which naturally led us to the derivation of a damping rate of the Langmuir waves in plasma. We did not ask for the source of the Langmuir waves, however. We simply assumed that they were given. This assumption is naive when considering a natural plasma as those in space because nobody would be there who could provide us with a given Langmuir or other wave. The wave would have to be produced by some

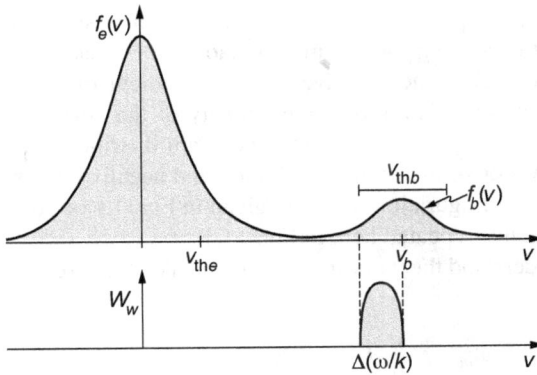

Figure 11.1: The plasma-beam configuration for resonant excitation of the Langmuir wave instability. It requires a fast beam of velocity $v_b \gg v_{the}$ which can be in resonance with a Langmuir wave of phase velocity $\omega_l \lesssim v_b$. In the positive gradient region of the beam distribution $f_b(v)$ Langmuir waves can be excited. The beam distribution has thermal spread $v_{thb} \ll v_b$. The lower panel shows the schematically expected Langmuir wave energy spectrum. Its spread in phase velocity corresponds approximately to the spread of the positive gradient part of the beam distribution.

mechanism. This mechanism is found not in the damping but in the resonant instability of the wave.

11.2.1 Resonant Excitation of Langmuir Waves

Langmuir waves are purely electrostatic plasma oscillations which at long wavelengths propagate at comparably slow phase velocity across the plasma. Their complex dispersion relation has been given in Eq. (10.32) with imaginary part

$$\gamma_l = \omega_l \frac{\pi \omega_{pe}^2}{2 n_0 k^2} \frac{\partial f_{0e}(v)}{\partial v}\bigg|_{v=\omega/k} \tag{11.6}$$

In the thermal plasma with the Boltzmann-Maxwellian electron distribution function $f_{0e}(v)$ of exponential decay with electron velocity v the derivative on the right-hand side taken at the Langmuir wave phase velocity $\omega/k = v$ is negative, and the Langmuir wave is damped by the plasma. The mechanism how this damping comes about has been explained in detail. The above expression suggest, however, that one will find $\gamma > 0$ if only the plasma exhibits a distribution function which at the location of the wave phase velocity obeys a positive gradient in velocity v.

Such a distribution can indeed exist in a plasma if only a beam of fast electrons traverses the plasma, for instance along the ambient magnetic field. In this case the distribution function consists of two parts, the thermal distribution of the immobile background electrons $f_{0e}(v)$ which have some thermal velocity spread v_{the} and the distribution $f_b(v)$ of a beam which crosses the plasma at beam velocity v_b and may have its own thermal spread $v_{thb} \ll v_{the}$. If the beam is sufficiently fast with $v_b > v_{the}$ substantially exceeding the thermal spread of the background distribution, one can to first order neglect the negative velocity gradient of the background plasma distribution with its few electrons at the velocity of the beam. In this case the only relevant velocity space gradient is that of the beam distribution, which is positive at the lower velocity side of the beam distribution and negative on its high velocity side. A sketch of such a configuration has been given in Fig. 4.1 of our companion volume and is for convenience repeated here in Fig. 11.1.

In order to understand the working of the resonance let us refer again to Eq. (10.27) in its first form

$$D(k, \omega) = 1 - \frac{\omega_{pe}^2}{n_0 k} \int_{-\infty}^{\infty} dv \frac{\partial f_0(v)/\partial v}{kv - \omega} \tag{11.7}$$

where we replaced $p = -i\omega$ with $\omega = \omega_r + i\gamma$ the complex frequency. From this expression it becomes immediately clear that the integral has a pole at velocity $v = \omega/k$, the phase velocity of a potentially present wave, implying resonance of the particles in the distribution function with the respective resonant velocity with the wave. As explained in the thermal background distribution this resonance leads just to damping while in the presence of the beam it causes amplification of Langmuir waves in the domain shown in Fig. 11.1. Clearly, waves on the thermal background and also on the high-speed side of the beam become damped and will disappear if they have been present for some reason in the region of the beam. Calculating the contribution of the pole at resonance then yields for

$$D(k, \omega) = D_r - 2\pi i \frac{\omega_{pe}^2}{n_0 k^2} \frac{\partial f_b(v)}{\partial v}\bigg|_{v=\omega/k} \tag{11.8}$$

The real part of the dispersion relation is obtained from expanding the integral outside the pole yielding

$$D_r(\omega, k) = 1 - \frac{\omega_{pe}^2}{\omega^2} - \frac{3\omega_{pe}^2}{\omega^2} \frac{k^2 v_{the}^2}{\omega^2} \tag{11.9}$$

It does not contain a contribution of the beam in this version since only the background distribution is involved. One can now replace here ω with the approximation ω_r for small γ and use the formula for weak growth Eq. (11.4) to find the above given growth rate (11.6) for the Langmuir waves where the distribution function $f_{0e}(v)$ is to be replaced by the beam distribution $f_b(v)$ which is assumed to be known and being proportional to the beam density. Hence the growth rate becomes proportional to the ratio n_b/n_0 of beam to plasma density.

What interests us here is much less the formalism of calculating a growth rate and the mathematical details which have been explicated in our companion volume, rather it is the lesson we learn about the natural origin of waves in a plasma. In the case of

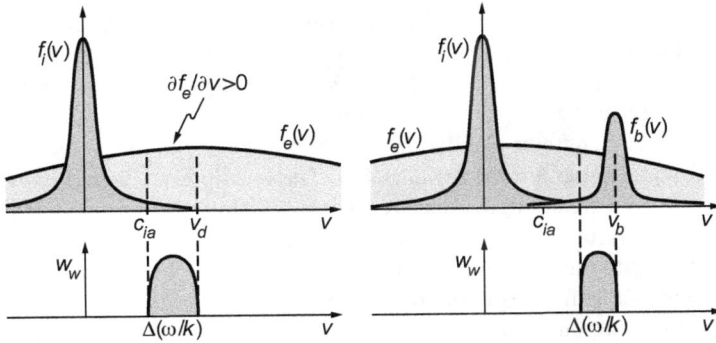

Figure 11.2: *Left*: The ion plasma-electron current configuration for resonant excitation of ion-acoustic waves. It requires a cold ion plasma at rest and a hot electron background plasma drifting at velocity v_d with thermal velocities $v_{the} \gg v_{thi}$ which can be in resonance with ion-acoustic waves of phase velocity $c_{ia} < \omega_l \lesssim v_d$ in the positive gradient region of the electron distribution $f_e(v)$. *Right*: A similar set up with background ions and warm electrons and a fast ion beam is also capable of exciting ion-acoustic waves in a narrow resonant range in velocity space. *Bottom*: The lower panels show the schematically expected ion-acoustic wave energy spectra for both cases. Their spread in phase velocities corresponds approximately to the spread of the positive gradient part of the distributions.

Langmuir waves it suffices to let a dilute electron beam traverse the plasma at sufficiently large velocity, say, along the magnetic field for keeping it one-dimensional. This beam provides the free energy to feed a previously present very weak seed spectrum of electric oscillations with the right frequency and wavelengths and to let it grow with the above calculated linear growth rate. Clearly, this growth cannot be kept ad infinitum because it is linear only in frequency while entering the exponential function and thus causes an exponential increase of the wave power and amplitude which soon will cause it to leave the linear range. We do not consider the nonlinear effects caused afterwards, at this place even though they in the long term are the most important. However, the recognition of instability as the mechanism which is responsible for the existence of a certain band limited spectrum of waves is extremely valuable. It does not just apply to the generation of Langmuir waves; it is a general and in fact the main mechanism of generation of waves and all effects in which waves are involved.

In the following we refer to a few other examples of weakly growing and also of non-oscillatory growing fluid-like waves which we will make use of in the next chapter.

11.2.2 Resonant Excitation of Ion-acoustic Waves

The next important wave in the context of application to some space problems in the Earth-plasma environment is the ion-acoustic wave. We have discussed its dispersion relation in depth already pointing out that it is a broad-band wave of fundamental importance for plasmas as it not only includes electrons but also the ions.

In this section we ask for the mechanism which can excite ion-acoustic waves, i.e. we ask for their free energy source. Naturally and similar to Langmuir waves there is not one unique energy source which can drive ion-acoustic waves unstable. In the case of Langmuir waves it was any electron velocity distribution which, in the phase velocity range of Langmuir oscillations, could provide a positive velocity space gradient for driving Langmuir waves unstable. The example we were choosing was a fast magnetic-field weak aligned electron beam which we will refer to later on. One can imagine many other distributions doing the same service. We restrict however paradigmatically to the simplest case here.

We choose the simple example of a field-aligned electric current resembling the beam case for Langmuir waves to show how ion-acoustic waves can be generated. But again, many other reasons could be called for serving as free energy sources of ion-acoustic waves like, for instance electron heat fluxes along the magnetic field or another fast ion beam crossing the plasma. Figure 11.2 (which is a copy of Fig. 4.3 in our companion volume) shows the velocity space set-up which is the basic paradigm for exciting ion-acoustic waves.

This figure is similar to that of the Langmuir wave set-up. The difference is that now a cold ion component $f_i(v)$ is included as shown on the left. The ion component has substantially smaller temperature (velocity spread) than the total background electron distribution $v_{thi} \ll v_{the}$ and, in addition is shifted against the centre of the electron distribution into the centre of mass thus being of rest. The difference in the central velocities of electrons and ions accounts for a current which is flowing in the plasma. This current can be a field-aligned current as known from the magnetosphere and in many other places in the Earth's environment. Thus, the region in the electron distribution between the ion distribution and the maximum of $f_e(v)$ exhibits a positive velocity space gradient, and as we may suspect from the investigation of the Langmuir wave dispersion relation, will cause the growth of some plasma waves. Clearly this growth will be restricted to the velocity range between the maximum of $f_e(v)$ which is found at a velocity v_d and a lower speed indicated by c_{ia} below that ion-Landau damping due to the negative velocity space gradient on the ion distribution will damp the waves away.

The model thus consist of a displaced total electron background distribution by the amount of $v = v_d$ in velocity to the right and a cold Maxwellian ion distribution under quasi-neutral conditions as

$$f_{0i} = \frac{n_0}{v_{thi}\sqrt{\pi}} \exp\left(-\frac{v^2}{v_{thi}^2}\right), \qquad f_{0e} = \frac{n_0}{v_{the}\sqrt{\pi}} \exp\left[-\frac{(v-v_d)^2}{v_{the}^2}\right] \qquad (11.10)$$

Since now the temperatures of both components must be explicitly taken into account, the calculation is a little more involved. Introducing the plasma dispersion function $Z(\zeta)$ it can, however, be simplified yielding the total dispersion relation

$$D(\omega,k) = 1 + \sum_{s=i,e} \frac{2}{k^2\lambda_D^2}\left[1 + \zeta_s Z(\zeta_s)\right] \quad \text{with} \begin{cases} \zeta_i = \omega/kv_{thi} \\ \zeta_e = (\omega - kv)/kv_{the} \end{cases} \qquad (11.11)$$

From Fig. 11.2 we read that the relevant conditions to be applied on the phase velocities of the ion-acoustic waves are $|\omega/k| \gg v_{thi}$ and $|\omega/k - v_d| \ll v_{the}$. Now, neglecting entirely the ion-Landau damping, which substantially simplifies the calculation, splitting the plasma dispersion function (see Appendix B) into real and imaginary parts and applying again the weak-growth formula, we find for the simplified growth rate

$$\gamma_{ia} = \left(\frac{\pi m_e}{8 m_i}\right)^{\frac{1}{2}} \frac{k c_{ia}}{\sqrt{(1+k^2 \lambda_D^2)^3}} \left(\frac{v_d}{c_{ia}} - 1\right) \tag{11.12}$$

It includes the ion-acoustic velocity c_{ia} and shows immediately that ion-acoustic wave growth is obtained for electron-current drift velocities $v_d > c_{ia}$ only. The current speed must thus be sufficiently large to excite ion-acoustic waves. This, however, is very often the case though some other instabilities compete as for instance the ion-cyclotron wave instability or the lower-hybrid drift wave instability both of which we do not discuss in the brief account of instabilities given in this volume.

11.2.3 Excitation of Weibel Instability

The Weibel instability is an electromagnetic instability which arises in a non-magnetic plasma which because of some reason develops a temperature anisotropy. This can be the case because of some preferential heating in one direction. In this case the electron pressure tensor of the wave can be written with for instance the streaming velocity **v** as preferential direction

$$\mathsf{P}_e = n k_B \left[T_{e\perp} \mathsf{I} + (T_{e\parallel} - T_{e\perp}) \mathbf{v}\mathbf{v}/v^2\right] \tag{11.13}$$

Assuming that with the two different temperatures the electron distribution becomes a bi-Maxwellian we obtain the linear electromagnetic dispersion relation

$$D(\omega, \mathbf{k}) = (n^2 - \epsilon_\perp)^2 \, \epsilon_\parallel = 0 \tag{11.14}$$

where $\epsilon_\perp, \epsilon_\parallel$ are the transverse and longitudinal response functions of which only the transverse one is of interest in electromagnetic modes:

$$\epsilon_\perp = 1 + \frac{\omega_{pe}^2}{\omega^2} \{1 - (A+1)[1 + \zeta Z(\zeta)]\} - \frac{\omega_{pi}^2}{\omega^2} \tag{11.15}$$

where $Z(\zeta)$ is the plasma dispersion function, $\zeta = \omega/k_\perp v_{the\perp}$ and $A = T_{e\parallel}/T_{e\perp} - 1 > 0$ is the thermal anisotropy. $\mathbf{k} = (k_\perp \sin\theta, 0, k_\perp \cos\theta)$ is the wave vector with two components only.

The above dispersion relation splits into the longitudinal and transverse relation of which we consider that for the transverse mode. It can be solved at very low frequencies with $\omega = i\gamma$ yielding for the growth rate

$$\gamma_W = \sqrt{\frac{2}{\pi}} \frac{k_\perp v_{\perp the}}{k_0 \lambda_e} \left(1 - \frac{k_\perp^2}{k_0^2}\right) (A+1)(k_0 \lambda_e)^3 \tag{11.16}$$

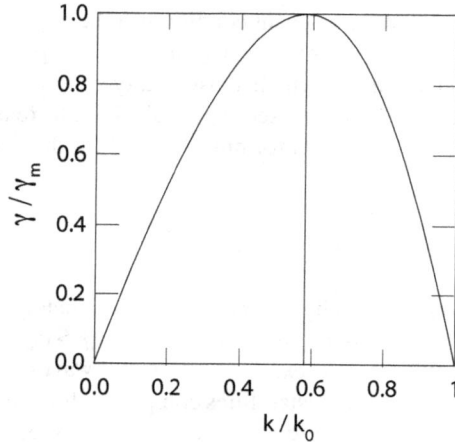

Figure 11.3: Growth rate of the Weibel instability under thermally anisotropic conditions. The growth rate maximises at perpendicular wavenumbers $k_\perp = k_0/\sqrt{3}$ and disappears for larger wavenumbers than $k_0\lambda_e$.

This growth rate is positive as long as $k_\perp < k_0$ where $k_0\lambda_e = \sqrt{A}$ is defined through the thermal anisotropy A and the electron inertial length $\lambda_e = c/\omega_{pe}$. It thus grows for wavelengths longer than λ_e. Its maximum it reaches, on the other hand at $k_{m\perp} = k_0/\sqrt{3}$ close to the electron inertial length. This behaviour is shown in Fig. 11.3.

11.3   Fluid Instabilities

Fluid instabilities are in general non-resonant though not necessarily non-oscillatory. The reason is simply that in a fluid you do not distinguish groups of particles. Therefore grouped resonances cannot be defined except for the entire fluid of one species, for instance. In this section we sketch only a few of them preparing for later reference. In most cases fluid instabilities do also involve the magnetic field because at the corresponding low frequencies the magnetic field in the ideal plasma is frozen-in into the flow and any change in the flow also causes a change in the magnetic field. However, there are also electrostatic fluid instabilities of importance. One of them is the two-stream (or Buneman) instability caused by a strong current in plasma and resembling the ion-acoustic mode. It becomes most important in many cases in current carrying plasmas and is thus worth of being mentioned here.

11.3.1   Electrostatic Two-stream (Buneman) Instability

When working in the fluid model with an electron and an ion fluid then one does not need any distribution function. The assumption that a current flow in the plasma (say, along the magnetic field) implies that the electron fluid is in motion with respect to the ions at velocity v_d as before, and we can again work in the ion frame because of the large inertia of the ions which do not participate in the current. Moreover, for simplicity, no temperature effects will be taken into account for sufficiently fast electrons $v_d \gg v_{the}$ and cold ions anyway. Then the dispersion relation becomes simply

$$D(\omega, k) = 1 - \frac{\omega_{pi}^2}{\omega^2} - \frac{\omega_{pe}^2}{(\omega - kv_d)^2} = 0 \tag{11.17}$$

as was shown in Section 9.6.1 if taking the non-magnetised field-aligned component ϵ_3 of the ϵ-tensor and Galilei transforming (Doppler shifting) the frequency in the electron term. This becomes a fourth-order equation for the frequency. The high-frequency wave $\omega = \omega_{pe} + kv_d$ decouples, and the dispersion relation reduces to a third-order equation. The low frequency case $\omega_{pi} \ll \omega \ll \omega_{pe}$ has wavenumber $k \approx \omega_{pe}/v_d$. This permits to write the dispersion relation as

$$\omega^3 \approx -(m_e/2m_i)\omega_{pe}^3 \tag{11.18}$$

which has one real negative frequency stable solution of no interest and two conjugate complex solutions which can be found when putting $\omega = \omega_r + i\gamma$ and separating into real and imaginary parts

$$\omega_r(\omega_r^2 - 3\gamma^2) = -m_e\omega_{pe}^3/2m_i \qquad \text{and} \qquad \gamma^2 = 3\omega^2 \tag{11.19}$$

which yields and oscillating wave of frequency $\omega_r = \omega_B$ and growth rate $\gamma = \gamma_B$ with

$$\omega_B = \omega_{pe}\left(\frac{m_e}{16m_i}\right)^{\frac{1}{3}} \qquad \text{and} \qquad \gamma_B = (3)^{\frac{1}{3}}\omega_B \tag{11.20}$$

The growth rate of this unstable wave exceeds its frequency a little showing that the instability is quite violent while still oscillating with non-vanishing frequency which is located between the electron and ion plasma frequencies. The Buneman mode is thus identified as a high-frequency oscillating but violently unstable mode which evolves under the condition that the electron-current drift velocity along the magnetic field exceeds the electron thermal velocity. In strong currents it will play an important role.

11.3.2 Electromagnetic Beam (Filamentation) Instability

Before proceeding we briefly refer to another interesting instability which is capable of generating very-low frequency magnetic fields in non-magnetised plasma. This instability is also called *filamentation instability* implying that a current flowing in a plasma filaments into current ropes which are surrounded by vortices of magnetic fields. This is a fluid version of the *Weibel instability*. Its electromagnetic dispersion relation in an unmagnetised plasma has been given earlier and can be written as

$$D_{xx}D_{zz} - |D_{xz}|^2 = 0 \tag{11.21}$$

where the D_{ij} are the components of the dispersion tensor $D(\omega,\mathbf{k}) = k^2 c^2 \mathbf{l} - \in(\omega,\mathbf{k})$ with the plasma dielectric tensor $\in(\omega.\mathbf{k})$. Assuming that the current of velocity v_b is flowing along x one has $D_{xz} = 0$ and the dispersion relation simplifies to

$$D_{zz} = n^2 - 1 + \sum_s \chi_{zz}^{(s)} \qquad \text{with} \qquad n^2 = k^2 c^2/\omega^2 \tag{11.22}$$

and the function $\chi_{zz}^{(s)}$ the susceptibility of species s which becomes in a symmetric electron/electron beam plasma configuration with immobile ions

$$\chi_{zz}^{(e)} = \frac{k^2 v_b^2}{\omega^2} \frac{\omega_{eb}^2}{\omega^2} + \frac{\omega_{pe}^2}{\omega^2} \left(1 + \frac{m_e}{m_i}\right)^{-1} \tag{11.23}$$

where again ω_{eb} is the beam-plasma frequency assuming that the beam configuration is symmetric, i.e. we would have two counter-streaming beams each of density $n_b/2$, which is the most simple case. Since the background plasma is at rest, the wave becomes non-oscillating (otherwise the frequency must be Doppler shifted by the amount $\mathbf{k} \cdot \mathbf{v}_0$ of background velocity v_0). Setting $\omega = i\gamma$ one solves for the growth rate and finds

$$\gamma_{ecfi} = k_{be} v_b \left\{ 1 + \frac{\omega_{be}^2}{k^2 c^2} \left[1 + \frac{2n_e}{n_b} \left(1 + \frac{m_e}{m_i}\right)^{-1} \right] \right\}^{-\frac{1}{2}} \tag{11.24}$$

that the growth rate of the electron current filamentation instability is proportional to the product of current beam velocity and electron-beam inertial wave number $k_{be} = \omega_{be}/c = \lambda_{eb}^{-1}$. This yields short wavelength magnetic field at about zero frequency and thus is an important instability which is capable of generating magnetic fields in a plasma which initially was free of any magnetic fields.

11.3.3 Firehose Instability

In the following we list three fluid instabilities which are of interest in the context of the space application section in this volume. The first of them is the so-called firehose instability discovered almost seventy years ago a the simplest case of an anisotropic pressure instability in magnetised plasma being responsible for the generation of the first ever known plasma wave, the Alfvén wave. Alfvén waves were proposed by Hannes Alfvén in 1942 in a less than one column note to the Nature journal which won him the Nobel prize 1970.

The firehose instability can be understood very simply as the dominance of the centrifugal force $F_{cf} = n_0(m_i v_{thi\parallel}^2 + m_e v_{the\parallel}^2)/R$ over the perpendicular pressure force including the self-restoring magnetic pressure on the magnetic field in a plasma with thermal velocity parallel to the magnetic field larger than perpendicular. In this case the parallel pressure $p_\parallel > p_\perp$ exceeds the perpendicular pressure. The condition of instability can be written

$$p_\parallel > p_\perp + \frac{B_0^2}{\mu_0} \qquad \text{or in terms of the plasma } \beta \qquad \beta_\parallel > 2 + \beta_\perp \tag{11.25}$$

where $\beta = 2\mu_0 n_0 k_B T / B_0^2$ is the plasma beta. The derivation of the growth rate starts from the single fluid equations of Section 7.3 with the double adiabatic pressure laws given in Eqs. (7.27) and (7.30):

$$\frac{d}{dt}\left(\frac{p_\perp}{nB}\right) = 0 \quad \text{and} \quad \frac{d}{dt}\left(\frac{p_\| p_\perp^2}{n^5}\right) = \frac{d}{dt}\left(\frac{p_\| B^2}{n^3}\right) = 0 \tag{11.26}$$

Linearization and use of the plane wave ansatz yields a dispersion relation which has a non-interesting real double-root solution $\omega = \pm k_\| \sqrt{3 p_\| / n_0 m_i}$ which is the ordinary parallel non-magnetic fluid-sound wave, and another solution

$$\omega^2 = \tfrac{1}{2} k_\|^2 v_A^2 \left(\beta_\perp - \beta_\| + 2\right) \tag{11.27}$$

which yields the non-oscillatory growth rate of the firehose mode under the above condition of parallel pressure dominance

$$\gamma_{fh} = \frac{k_\| v_A}{\sqrt{2}} \sqrt{\beta_\| - \beta_\perp - 2} \tag{11.28}$$

where v_A is the Alfvén velocity, and the wave propagates parallel to the magnetic field, being a zero frequency oscillation of the magnetic field line which is a parallel propagating Alfvén wave. In fact, since the instability is non-oscillatory this wave has no time to evolve into a real oscillation before growing. It thus grows almost explosively under the above conditions. However, the model is highly idealised and neglects the differences between electrons and ions as well as any kinetic effects. If these would be included then the wave became travelling with real frequency

$$\omega_{fh} = \gamma_{fh} k_\| / k_c \tag{11.29}$$

where, as has been shown in Section 3.4 of our companion volume, k_c is some cut-off wave number which depends on the plasma parameters. Similar behaviour is found in almost all fluid-like non-oscillatory instabilities which, when kinetic effects are taken into account, in most cases become oscillatory with frequencies small compared to the linear growth rates. In addition all those fluid-like non-oscillatory waves are vulnerable to plasma convection because of their frozen-in nature. In addition, they tend to saturate quickly because of their low group velocities and large growth rates.

Hence, the mechanism of anisotropic pressure is one possibility of exciting low frequency Alfvén waves. Other mechanisms are, if remembering that Alfvén waves are the lowest wave number branches of ion- and electron whistlers, to excite lowest frequency waves of these kinds on kinetic ways for instance by shooting an ion beam across the plasma along the magnetic field. In this case ion whistlers are excited by an ion-ion beam instability which for low frequencies are Alfvén waves.

11.3.4 Mirror Instability

Another instability which can arise under unstable pressure conditions but now in the case opposite to that of the firehose mode is the mirror instability. Though it

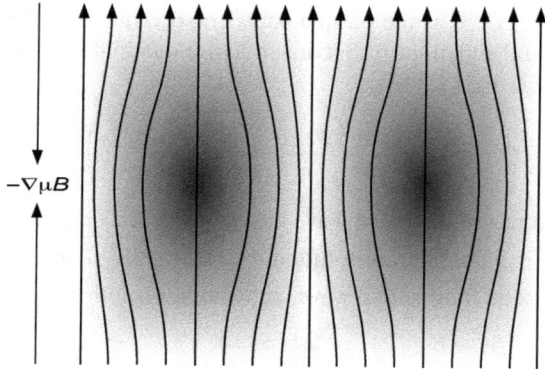

Figure 11.4: Two mirror bubbles adjacent to each other evolving in the mirror instability. The shading suggests the concentration of plasma inside the bubbles whose pressure acts to inflate the magnetic field in the perpendicular direction. The figure is somewhat unrealistic because under natural conditions the bubbles will be displaced against each other along the magnetic field such as to create a structure in the field and plasma where for energetic reasons the belly region of one bubble will lie adjacent to the neck region of the adjacent bubble.

is essentially a fluid mode, the derivation of its dispersion relation is best done in an kinetic theory approach which yields for the growth rate the expression

$$\gamma_{mirr} = \sqrt{\frac{2}{\pi} \frac{\beta_{i\parallel}}{\beta_{i\perp}}} \left[\sum_s \beta_{s\perp} \left(\frac{\beta_{s\perp}}{\beta_{s\parallel}} - 1 \right) - 1 \right] k_\parallel v_{thi} \qquad (11.30)$$

This expression is positive for positive large bracket which thus yields the condition for instability. The resulting mirror mode is again not a propagating wave, rather it is a structure in the plasma which is convected across the plasma thus locally by Doppler shift of the frequency generates an out-of phase fluctuation between the pressure, density and the magnetic field fluctuations.

Such fluctuations are frequently observed in the solar wind and in the magnetosheath and occasionally also in the plasma-sheet and magnetotail. The mechanism of how they are generated is not yet fully understood. However, under certain conditions when the perpendicular pressure in a place becomes large enough, the magnetic field is locally inflated thereby sucking in plasma from the environment which even further inflates the magnetic field thus causing a magnetic bubble or magnetic hole shown in Fig. 11.4 as it is sometimes called with high pressure inside and high magnetic pressure outside in equilibrium. Usually many such holes evolve which balance themselves creating a certain structure of adjacent holes and magnetic walls in the plasma. On the surfaces of the holes diamagnetic currents are flowing which however cause repulsing forces on neighbouring holes. Observation of such current sheets thus does not imply antiparallel fields as could erroneously be concluded. Instead, one is confronted with a large number of tangential discontinuities which structure the flow on the meso-scale while being convected with the flow.

Mirror bubbles also represent magnetic mirrors in which the trapped particles oscillate along the magnetic field. However one may expect that the bubbles are elongated along the magnetic field direction with the parallel lengths $\lambda_\parallel \sim 2\pi/k_\parallel$ of the bubbles being much longer than their perpendicular radius $\rho_\perp \sim \lambda_\perp \sim 2\pi/k_\perp$. Kinetic theory moreover shows that the perpendicular scale $\lambda_\perp \sim r_{gi}$ is mainly of the order of the thermal ion gyro-radius while the parallel scale is much longer. We should, however mention, that the restriction of the mirror mode to ions is barely justified. The above dispersion relation, obtained from kinetic theory including electrons also contains electron terms similar to ions. The electrons may as well exhibit a pressure anisotropy $A_e = \beta_{e\perp}/\beta_{e\parallel} - 1$ contributing to the ion growth rate. This contribution has two effects, it slightly modulates the ion growth rate, which in linear theory is at the best marginally important. The more interesting effect is that, independent of the ion mode, an electron anisotropy also drives an independent separate electron mirror mode which is however on scales the order of the electron inertial length $\lambda_e = c/\omega_{pe}$. This electron-mirror mode may occur anywhere inside as well as outside the ion mirror bubbles where it generally would appear as independent excursions on the magnetic profile of much shorter wavelength and smaller amplitude, as has sometimes been observed.

From a physical perspective the mirror instability is somewhat problematic. It is basically a zero frequency structure of the plasma which in principle propagates only if one also accounts for subtle inhomogeneities in plasma like large-scale density gradients, finite gyro-radius effects and other higher order kinetic contributions. Such effects cause so-called drift mirror modes. Neglecting all that makes a chain of mirror bubbles just stationary and transported by the general flow. This is what happens when observing mirror modes in the magnetosheath between the Earth's bow shock and the magnetopause, the very region where they have been found first. However, as an instability such structures should not evolve at all as to some extent they seem to violate the basic physical, i.e. thermodynamic conditions of stationary states. Appendix A will revoke this problem which we here just mention only.

11.3.5 Kelvin-Helmholtz Instability

As our last example of a plasma instability w consider a fluid instability which is caused by the shear flow at the magnetopause. We have noted it already when discussing the flow around the magnetopause in connection with the shape of the magnetopause in Section 8.6 where we showed its effect on the duskside magnetopause in Fig. 8.15.

The two different plasma configurations leading to the Kelvin-Helmholtz are shown in Fig. 11.5. Their essence is that they imply the presence of shear flows in two adjacent plasmas or plasma and another non-plasma medium. The first case is realised at the magnetopause in particular on its dusk side where the plasma of the ring current flows in sunward direction while the solar wind outside the magnetopause flows in anti-sunward direction. The plasma velocity then shears across the magnetopause over the width of the transition region. The reconstruction technique which we have

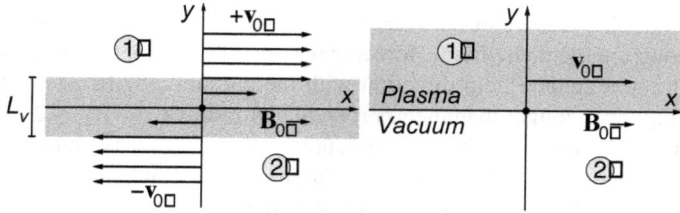

Figure 11.5: Two configurations in which the Kelvin-Helmholtz instability can occur. The first case is most probably realised at the dusk side of the magnetopause and causes instability when the ring current and magnetospheric convection become amplified during magnetic storms. The second case may be realised at the boundary between the ionosphere and the atmosphere for strong horizontal winds.

describe for the magnetopause indicates that this transition region is by no means of negligible width. It is therefore not surprising that the Kelvin-Helmholtz instability evolves at the boundary of the magnetopause, and is also not surprising that the evolution conditions on the dawn and dusk sides are quite different.

In order to describe the evolution of the Kelvin-Helmholtz instability one needs to include the inhomogeneity of the flow which is present in the flow shear tangential to the magnetopause. In addition one needs to include the pressure change across the magnetopause because of the condition of pressure balance. This all is a little tedious to do because by the different velocities on the two sides of the transition also the wave frequency is Doppler shifted. After some algebra by using the plasma fluid equations (see Section 3.3. of our companion volume) one arrives at the following dispersion relation for a single linear plane wave of frequency ω and wavenumber $\mathbf{k}$

$$n_{02}\left[\omega^2 - (\mathbf{k} \cdot \mathbf{v}_{A2})^2\right] + n_{02}\left[(\omega - (\mathbf{k} \cdot \mathbf{v}_0)^2 - (\mathbf{k} \cdot \mathbf{v}_{A1})^2\right] = 0 \qquad (11.31)$$

where we simply assumed that in region 1 the plasma streams at velocity $\mathbf{v}_0$ while being at rest in region 2 outside the transition, thus having transformed to region 2. Moreover, in the above expressions the Alfvén velocities in the two regions enter.

Solving for the frequency yields a solution which is unstable when the following condition on the flow in region 1 is satisfied

$$(\mathbf{k} \cdot \mathbf{v}_0)^2 > \frac{n_{01} + n_{02}}{n_{01}n_{02}}\left[n_{01}(\mathbf{k} \cdot \mathbf{v}_{A1})^2 + n_{02}(\mathbf{k} \cdot \mathbf{v}_{A2})^2\right] \qquad (11.32)$$

which is the case when the flow in the direction of the wave vector is sufficiently large with respect to the sum of the two Alfvén velocities in the adjacent streaming media. This is most probably the case when the flows are counter-streaming as one expects for the dusk of the magnetopause and particularly during magnetic storms when both magnetospheric convection and ring current become strongly amplified and, moreover, the magnetosphere is compressed such that the transition layer becomes comparably narrow. In this case one finds experimentally that the dusk side magnetopause is eroded, develops a wave structure of large amplitude and leads to strong mixing of plasma of magnetospheric and magnetosheath origins.

11.4 Thermal Fluctuations

As a final section in this brief overview and farther leading chapter we turn to an account of the broad field of thermal fluctuations.

Plasmas are hot places what concerns the energies of the particles which are substantially higher than 1 eV simply because plasmas, in particular when being collisionless, arise from ionisation and thus at least their electron component will have energy higher than the lowest ionisation energy. When becoming more dilute they may, of course, cool to lower energies but if sufficiently long time is available as normally realised in astrophysical of space physics conditions will recombine with an occasional ion they meet. One thus expects that plasma always are temperate. Temperate systems, however, exhibit a thermal ground state in which the spontaneous emission of fluctuations and the reabsorption balance. This is the microscopic essence of thermodynamics equilibrium which is thus only in existence when thermal fluctuations are allowed to be present.

In plasma thermal fluctuation also play another most important role. As we have seen in this chapter instability excites waves by feeding energy into them. However, the wave must already be present in the plasma because otherwise the instability will run into the nothing as nothing would be there to grow. Thermal fluctuations fill this gap by eliminating the nothing and always providing a substantial though very low level of a fluctuating background at all frequencies and wavenumber respectively wavelengths. The instability then just chooses from this level by feeding the available free energy into the range where the growth rate is largest. If then is sufficient time available, this range will become strongest amplified, and the waves contained in it will rise above thermal fluctuation noise background level to become visible as real propagating waves which are capable of transporting energy and information, something what thermal fluctuation noise does not do because the thermal noise does not carry information other than its temperature.

The calculation of thermal noise levels is, however, a formidable task as it requires knowledge about the microscopic state of the interactions in thermal equilibrium. Each of the multitude of waves in plasma has its own thermal fluctuation level because it obeys its own dynamics which is different from the dynamics of other waves. Nevertheless there is a well defined though complicated theory of how the thermal level of any plasma mode can be determined. Unfortunately, however, this theory can only be given its formulation when referring to the basic quantum mechanics of a many body system in interaction with the electromagnetic field. This prevents an easy representation even though the theory requires nothing else than the knowledge of the linear dispersion relation and the linear absorption coefficient.

The reason for the sufficiency of a knowledge of the linear properties of the plasma is that the thermal level — in spite of being a nonlinear equilibrium state between emission of waves and absorption of waves — implies only very small amplitudes of the fields or, in other words, small deviations from equilibrium only. Such small deviation, we have seen, can be described sufficiently accurately by linear theory. Their smallness permits to expand the highly nonlinear kinetic equations with respect to these small fluctuations (even though at this level of thermal fluctuations there is no large number against which the smallness could be measured; in fact justification for such an approach can be given only by reference to quantum theory which is the appropriate theory of treating thermal fluctuations). The expanded equations are just the full set of linear equations we have treated in determining the wave modes and instabilities. It is this no surprise when all the linear modes reappear in fluctuation theory as the basic modes. The recipe of calculating the thermal equilibrium fluctuation level is thus provided by balancing wave emission and wave absorption and determining the *average* correlation function of the fluctuating fields in equilibrium far away from the parameter range of instability.

11.4.1   General Expressions for the Fields

If we acknowledge that not more to know is needed than just the linear properties of the plasma in the equilibrium state when it exhibits small-amplitude fluctuations, then it suffices to refer to the linear dispersion tensor of the plasma which we derived in full generality already in Section 9.2 writing it here in component form instead of the dyadic form preferred so far

$$D_{ij}(\omega,\mathbf{k}) = n^2\left(\frac{k_i k_j}{k^2} - \delta_{ij}\right) + \epsilon_{ij}(\omega,\mathbf{k}) \qquad \text{with} \qquad n^2 = \frac{k^2 c^2}{\omega^2} \qquad (11.33)$$

In vacuum one would replace the dielectric tensor ϵ_{ij} contained in this expression by the unit tensor δ_{ij} thus obtaining the vacuum dispersion tensor D_{ij}^0. With these definitions the average spectral fluctuation of the current density can be written as

$$\langle j_i j_j\rangle(\omega,k) = i\omega\varepsilon_0 k_B T \sum_{\ell,m} D_{i\ell}^0 [D_{\ell m}^{-1} - D_{\ell m}^{-1*}] D_{mj}^0 \qquad (11.34)$$

where $j_i(\omega,\mathbf{k})$ is the Fourier component of the current density $\mathbf{j}(\mathbf{x},t)$, and the average in the above current correlation function is taken over time and space. The asterisk in the correlation function indicates the complex conjugate. In addition, for small damping the imaginary part of the dispersion tensor is small. Then the current correlation can be brought into the form

$$\langle j_i j_j\rangle(\omega,k) = \frac{\varepsilon}{2}\omega k_B T \sum_{\ell,m} D_{i\ell}^0 \tilde{D}_{\ell m} D_{mj}^0 \, \delta\left[D(\omega,\mathbf{k})\right] \qquad (11.35)$$

Here D is the dispersion relation, i.e. the determinant of the dispersion tensor D_{ij}, and $\tilde{D}_{\ell m}$ is the cofactor of D_{ij} in the determinant D, i.e. in the expression $\sum_m \tilde{D}_{im} D_{mi} = \delta_{ij} D$. The δ-function on D in the last formula tells that the spectral density is to be taken along the branches of the waves which are defined by the dispersion relation $D(\omega,\mathbf{k}) = 0$ which allows to calculate the fluctuation spectrum of the current density for each of the wave modes separately. Remembering in addition that the charge and current densities are related through the continuity equation and the electric field together with the current obeys Ohm's law for the waves, one can also express the density fluctuations $\langle(\delta n)^2\rangle(\omega,\mathbf{k})$ and electric field fluctuations $\langle\delta E_i \delta E_j\rangle(\omega,\mathbf{k})$ through those of the current. The same holds for the magnetic fluctuations by Faraday's law.

11.4.2 Fluctuations in Isotropic Plasma

These expressions become simpler in an isotropic plasma where the dispersion tensor reduces to the transverse ϵ_T and longitudinal ϵ_L response functions, yielding

$$\langle(\delta n)^2\rangle(\omega,\mathbf{k}) = \frac{2\varepsilon_0 k_B T}{\omega} \frac{\text{Im}(\epsilon_L)}{|\epsilon_L|^2}$$

$$\langle\delta E_i \delta E_j\rangle(\omega,\mathbf{k}) = \frac{2k_B T}{\varepsilon_0 \omega}\left[\frac{k_i k_j}{k^2}\frac{\text{Im}(\epsilon_L)}{|\epsilon_L|^2} + \left(\delta_{ij} - \frac{k_i k_j}{k^2}\right)\frac{\text{Im}(\epsilon_T)}{|\epsilon_T - n^2|^2}\right]$$

$$\langle\delta B_i \delta B_j\rangle(\omega,\mathbf{k}) = \frac{2\mu_0 k_B T}{\omega}\left(\delta_{ij} - \frac{k_i k_j}{k^2}\right)\frac{n^2 \text{Im}(\epsilon_T)}{|\epsilon_T - n^2|^2} \qquad (11.36)$$

In the case of electrostatic waves like Langmuir, electron-acoustic and ion-acoustic waves one has $\epsilon_T = 0$, and the expression for the electric fluctuation simplifies substantially while naturally the magnetic fluctuations disappear. Nonetheless, the explicit calculation of the fluctuation spectrum still remains a tedious problem. For the plasma being Maxwellian and isothermal with electrons and ions having the same temperature $T_e = T_i$ one can evaluate the above expression for Langmuir wave electric fluctuations which yields for the energy distribution with respect to wavenumber

$$\int \frac{d\omega}{2\pi}\langle|E|^2(\omega,\mathbf{k})\rangle = \frac{2}{\varepsilon_0}\frac{k_B T_e}{1+k^2\lambda_D^2} \qquad (11.37)$$

The Langmuir instability of a beam passing through the plasma finds this energy distribution as function of wavenumber and amplifies it in the resonant range. Integrating over the full wavenumber space one obtains the total energy density in the thermal Langmuir fluctuations

$$\frac{\varepsilon_0\langle|E|^2\rangle}{2} = \int \frac{d^3k}{8\pi^3}\frac{k_B T_e}{1+k^2\lambda_D^2} \approx \frac{k_B T_e}{\lambda_D^3} \qquad (11.38)$$

Similar expressions can be derived for ion-acoustic waves or electron acoustic waves. However, in the presence of a magnetic field the simplification introduced by the response function is not possible anymore, and one has to return to the full dispersion tensor of the plasma.

As for an example of magnetic fluctuations in the isotropic plasma we chose the thermal Weibel instability. In contrast to Langmuir and ion-acoustic waves one now needs just the transverse response function. Writing it explicitly for the case of an anisotropic Maxwellian, which in thermal equilibrium is the appropriate distribution function,

$$\epsilon_T = 1 + \frac{\omega_{pe}^2}{\omega^2}\left\{1 - \frac{T_\parallel}{T_\perp}\left[1 - \Phi(z) + i\sqrt{\pi}z e^{-z^2}\right]\right\} - \frac{\omega - pi^2}{\omega^2} \qquad (11.39)$$

where $z = \omega/\sqrt{2}k v_{the\perp}$, $n^2 = \epsilon_\perp$ is the refractive index of transverse magnetic fluctuations, and the real function $\Phi(z) \approx 2z^2$ for $z \ll 1$, and going to the limit of zero frequency $\omega \to 0$ corresponding to the Weibel case, one finds for the thermal Weibel magnetic fluctuation spectrum

$$\langle |B|_w^2(\mathbf{k}, \omega = 0)\rangle = \frac{\mu_0}{\omega_{pe}} \sqrt{\frac{\pi k_B T_\perp}{m_e c^2}} \frac{m_e c^2 (A+1)^2 k \lambda_e}{(A+2)[k^2 \lambda_e^2 - A - m_e/m_i]^2} \qquad (11.40)$$

The Weibel instability chooses this level in order to amplify the Weibel-unstable magnetic field wave in the range discussed earlier. One may note that the spectral energy density of the Weibel fluctuations has a k-dependence of $\sim k^3 \lambda_e^3$. Assuming that the anisotropy vanishes, we may put $A = 0$ in this expression for the stable Weibel case.

11.4.3 Summary

In this brief chapter we have introduced the concept of instability in plasma which is fundamental to all the dynamic plasma processes. These processes, like for instance the formation of shocks like the bow shock, the evolution of the structure of the magnetosheath, the structure of the magnetopause, and the various processes in the magnetosphere, they all start with the evolution of waves of one or the other kind which grow by instability being fed by the free energy which is injected into the magnetosphere mostly from the solar wind side and which needs to be transported to such places where it can be absorbed and dissipated. Such processes thus always need waves and these waves are both excited by instability and grow due to instability.

Instability theory is a linear theory. It provides linear growth rates and cannot inform about the fate of the wave and the energy when the unstably excited wave grows and reaches its nonlinear state. In many cases the propagating wave readily leaves the region of access to free energy remaining at finite amplitude. The wave then does not need to dissipate the energy before entering the region of dissipation. Waves of low speeds may stay for long in the unstable region growing along their path. These waves are convectively unstable because they grow underway. When they reach nonlinear amplitudes in this region they start then reacting on the plasma and may reconfigure the state of the plasma, the particle distribution and even affect the energy source. These cases are as well of large interest. In this chapter we have not studied any of them. We have just informed about the possible unstable state of waves. For the treatment of non-linear processes the reader may consult our companion volume.

Finally we have given a very brief account of the importance of thermal fluctuations as the background level from which any linear instability takes its start. Thermal fluctuations are unavoidable in a plasma which is either in thermal equilibrium when all its quantities fluctuate, or is in non-equilibrium when an instability arises and picks out some part of the fluctuation spectrum to amplify it. In this case the fluctuations spectrum becomes distorted.

The expressions for the fluctuation spectra in the fields turn out to be quite involved. However one does not need to invoke any nonlinear processes to account for the thermal level. The linear approximation suffices. This is a consequence of the small field amplitudes as we have explained already. Interestingly, any observation will always be polluted by the thermal noise. The thermal noise itself organises along the different modes in plasma. Thus, a measurement of sufficient sensitivity will always map all the wave branches which a plasma is capable of in thermal fluctuations. Because of the weakness of the magnetic fluctuations which are smaller by a relativistic factor than the electric fluctuations, one will mostly encounter the electric fluctuation fields, however. The only exception seem to be the Weibel fields which we have given some attention at the end of this chapter. It seems that they, however, are not extraordinarily important in space plasmas, possible with the exception of the physics of collisionless shocks. In astrophysical plasmas, though, they may have been at the centre of the production of magnetic fields in the early universe or in gamma ray bursts. Whether they play a role or not in the generation of seed magnetic fields for the generation of the geomagnetic field in the Earth's core has not yet been investigated.

Further Reading

Classic general treatments of plasma instabilities, but with no emphasis on space physics sometimes including also inhomogeneous processes are contained in the monographs [4, 7, 8, 10]. Space aspects, in particular for the ionosphere, magnetosphere and solar corona are included in the early extended and excellent monograph [2]. Detailed treatises on instabilities are found in [4] and [3, 11], in the former in a somewhat unfamiliar presentation, in the latter two in view of space applications and including non-linear evolution. A classical about complete collection of linear instability theory in plasmas with no emphasis on space was given long ago in [6] still not being completely out of date. Thermal fluctuation theory is discussed rarely in plasma physics in spite of its importance for practical problems as we have argued when an instability starts growing from scratch. Two of the rare references are [1] and [9].

References

[1] A. I. Akhiezer, I. A. Akhiezer, R. V. Polovin, A. G. Sitenko and K. N. Stepanov, *Plasma Electrodynamics, Vol. 2. Nonlinear Theory and Fluctuations* (Pergamon Press, Oxford, 1975).

[2] V. L. Ginzburg, *The Propagation of Electromagnetic Waves in Plasmas*, 2nd ed. of Engl. translation (Pergamon Press, Oxford, 1970).

[3] A. Hasegawa, *Plasma Physics Instabilities and Nonlinear Effects* (Springer-Verlag, Berlin-Heidelberg-New York, 1975).

[4] S. Ichimaru, *Basic Principles of Plasma Physics* (W. A. Benjamin, Reading, 1973).

[5] D. B. Melrose, *Instabilities in Space and Laboratory Plasmas* (Cambridge University Press, Cambridge, 1989).

[6] A. B. Mikhailovskii, *Theory of Plasma Instabilities, Volumes 1 & 2* (Consultants Bureau, New York, 1974).

[7] D. R. Nicholson, *Introduction to Plasma Theory* (John Wiley & Sons, New York, 1983).

[8] T. H. Stix, *Waves in Plasma* (American Institute of Physics, New York, 1992).

[9] A. G. Sitenko, *Electromagnetic Fluctuations in Plasma* (Academic Press, New York, 1967).

[10] D. G. Swanson, *Plasma Waves* (Academic Press, Boston, 1989).

[11] R. A. Treumann and W. Baumjohann, *Advanced Space Plasma Physics* (Imperial College Press, London, 1997).

Problems

Problem 11.1 *Write the general expression for an arbitrarily shaped travelling wave packet $A(x,t)$ in the superposition representation. The amplitude of the wave should be $A(k, \omega)$. If you assume that this is the electric field which is related to the current through Ohm's law with conductivity $\sigma(x,t)$, what is the expression for the representation when taking into account causality, i.e. that only past states and not future states can affect the current state of the field $E(x,t)$?*

Problem 11.2 *Give a strict mathematical proof for the weak growth expression of the growth rate.*

Problem 11.3 *What is the justification for setting $\omega = i\gamma$ in the non-oscillatory instability? If one would do it exactly, expanding the dispersion relation, what does one find?*

Problem 11.4 *Give a derivation of the Langmuir dispersion relation and growth rate in the presence of a Maxwellian beam including Landau damping by the background distribution. Under what condition do Langmuir waves grow? Show that one needs $v_b \gtrsim 3v_{the}$.*

Problem 11.5 *Derive the growth rate for ion-acoustic waves in the model given in the text but include the ion term and discuss the restrictions it implies on ion-acoustic growth.*

Problem 11.6 *Show that the Weibel instability obeys the dielectric function $\in_\perp$ and that $\in_\parallel$ plays no role. Derive the growth rate of the anisotropic Weibel mode.*

Problem 11.7 *Confirm the Buneman frequency and growth rate.*

Problem 11.8 *Find the firehose mode growth rate and instability condition.*

Problem 11.9 *Give a physical discussion of the Kelvin-Helmholtz instability growth rate at the example of the magnetopause boundary flow.*

Problem 11.10 *Thermal fluctuations are always given as correlation functions like the one given here without derivation for the current density. If you use the representation of the current as in Problem 11.1, show that the form of the correlation function $\langle j_i j_j \rangle$ we have found is correct. You will have to use the field representations and the dispersion tensor for this purpose.*

— 12 —

Collisionless Reconnection

The last two chapters of this book deal with two particular cases of application of the plasma physics developed in the previous chapters. The first of these applications is to the important problem of reconnection. We have become familiar with it already in Sections 5.2, 5.6 and 8.6 in the context of substorms, the general magnetospheric convection, and the physics of the magnetopause. In this chapter we describe the current knowledge about the process of reconnection in some detail even though research in this field has not yet settled.

Reconnection was proposed about seventy years ago as a process which may change the magnetic topology, release magnetic energy that would previously be stored in magnetic fields in an energetically 'excited state', i.e. in an energetically unfavourable or complicated magnetic topology, and accelerate particles. The simplest of such an 'unfavourable' magnetic topology is a topology of adjacent antiparallel magnetic fields. It implies the presence of a separating current sheet. Such a state has low entropy and is thus vulnerable to instability in the course of which the released energy heats the plasma and the entropy grows. Equivalently, since by Maxwell's equations magnetic field topologies are related to current flows, one may also speak about storage of energy in current configurations.

The first models of reconnection included finite electrical conductivities and were thus resistive models. We have, however, in the former chapters become familiar with the near-Earth plasma of being practically ideal and of infinite conductivity. In such a case the frozen-in concept does not permit the magnetic field to escape from the enslavement to the plasma which is contained in the hydromagnetic theorem. Clearly, reconnection requires violation of either, the frozen-in condition or the hydromagnetic theorem.

Reconnection in near-Earth space when it occurs is necessarily collisionless. This requires such a violation and thus poses a severe problem to the theory. It is the reason why reconnection today more than half a century after its introduction into plasma and space physics is still an unresolved and active field of research. Many interesting result have been found and enormous effort has been invested into theory, observation and experiment.

The observations have by now unambiguously confirmed that reconnection is by no means just a naive theoretical concept drawing antiparallel field lines, merging, cutting and reconnecting them. It has been confirmed many times from remote and also in situ. Nevertheless, though we know that reconnection is present, though we know that it is an important and probably one of the most important actors behind the

365

scene in the behaviour of near-Earth plasma, we still have not disclosed its mysteries and do not know in detail how it works in the plasma-physical reality of space.

12.1   Reconnection Models

Reconnection was originally proposed on the simple assumption of the interaction of magnetised fluids containing anti-parallel magnetic fields. These models intended to infer about the energy source of solar eruptions based on the knowledge that solar activity reflects itself in the occurrence of large numbers of magnetised regions on the solar surface known as rising and decaying sunspots. The physical theory behind these original reconnection models was plasma magneto-hydrodynamics. In the frame of the interface of interaction between the two approaching plasmas the only existing electric fields were the convection electric fields $\mathbf{E} = -\mathbf{v} \times \mathbf{B}$ to both sides of the interface. Since both the velocity and magnetic field changed direction when crossing the interface, $\mathbf{E}$ was tangential to the interface, of same sign and thus continuous. Any parallel electric field component was assumed vanishing except inside the interaction layer where $E_\parallel \neq 0$ was assumed as a result of the interaction of the antiparallel magnetic field lines and being necessary for field line merging.

12.1.1   Sweet-Parker-Petschek-Sonnerup

It is easily seen from the non-ideal Ohm's law (neglecting all terms related to the particle mass or temperature differences) that this might indeed be the case, when antiparallel magnetic fields approach each other and the magnetic flux tubes become squeezed between them to form a thin transition layer (see, e.g., Section 5.2). Write Ohm's law in the most simple form

$$\mathbf{E} + \mathbf{v} \times \mathbf{B} = \eta \mathbf{j} = \frac{\eta}{\mu_0} \nabla \times \mathbf{B} \tag{12.1}$$

with η a uniform resistivity which is assumed to be small on the time scale of the flow $\tau = L/v$. In other words (cf. Section 7.5.1) $\eta \ll \mu_0 vL = \mu_0 L^2/\tau$. As long as this relation holds, which is assumed to be the case outside the transition layer of thickness d, the right-hand side of Ohm's law can be neglected and the plasma is frozen-in. However, when the thickness of the transition layer is sufficiently decreased to $L = d$, one may claim that the current term cannot be neglected anymore and the frozen-in state breaks down inside the transition layer. Then $E_\parallel \approx \eta j \neq 0$, and one supposes that the antiparallel magnetic field lines can merge. Inserting for $j \approx B/\mu_0 d$ yields readily for the parallel electric field in the current layer and thus from Faraday's law (cf. Eq. (B.51)) for the rate of annihilation $d\Phi/dt$ of magnetic flux Φ per unit of length in the invariant direction along the sheet perpendicular to the current

$$E_\parallel \approx \frac{\eta}{\mu_0} \frac{B}{d} \approx \frac{d\Phi}{dt} \approx -V_{rec} B \tag{12.2}$$

This is a finite value even if the resistivity η is small everywhere. The right-hand side has been written here as the product of a 'reconnection velocity' V_{rec} and the magnetic field B suggesting that the generated parallel electric field is caused due to the inward transport of magnetic field by this velocity into the current layer. This reasoning has been taken for long time as justification for performing resistive fluid simulations of reconnection. However, reconnection is restricted because all the plasma must necessarily flow through the current layer where its velocity is turned around by 90° to both sides, and the plasma is squeezed out by the magnetic pressure into the reconnection jets mentioned earlier. The maximum velocity the plasma can assume in this incompressible flow is the Alfvén speed v_A. Thus, from mass flux conservation (remember that reconnection is a process of flow and thus slow with respect to c such that radiation or particle annihilation does not occur and mass is strictly conserved) one finds for the velocity at which magnetic field is provided for reconnection

$$V_{rec} \approx \frac{v_A d}{\ell} \approx \frac{\eta}{\mu_0 d} \tag{12.3}$$

where ℓ is the length of the current layer along the magnetic field, i.e. the lengths of the reconnection region. This may be solved for the thickness d of the reconnection region (current layer) and inserted into the reconnection velocity. Introducing the Lundqvist number $S = \mu_0 \ell v_A / \eta$ (which is the magnetic Reynolds number for the Alfén velocity) one obtains finally

$$d \approx S^{-\frac{1}{2}} \ell \quad \text{and} \quad V_{rec} \approx S^{-\frac{1}{2}} v_A \tag{12.4}$$

Thus the ratios of current-layer thickness to current-layer length and reconnection-speed to Alfvén velocity scale both with the inverse root of the Lundqvist number. However, since the latter is large these ratios are small indicating that the reconnection rate in this fluid model of Sweet and Parker is fairly inefficient and the current layer long extended in order to digest the large amount of plasma that must pass the current layer.

The last expression for the reconnection velocity can be used to estimate the fraction of field lines which undergo merging in this model when the plasma fluid approaches with inflow velocity v_{in}. This number is given by the ratio $E_\parallel / E_\perp$ of the parallel electric field generated in reconnection to the convection electric field $E_\perp = -v_{in} B$ of the inflow. With the above expressions one estimates this fraction to

$$E_\parallel / E_\perp \approx S^{-\frac{1}{2}} v_A / v_{in} \approx \left(M_A \sqrt{S} \right)^{-1} \tag{12.5}$$

where $M_A = v_{in}/v_A$ is the Alfvénic Mach number of the plasma inflow. Hence, there is some competition between the Mach number which, for instance in the solar wind, is large and the Lundqvist number which is also large. However, because of the small resistivity S is usually a huge number and thus the reconnected fraction of field lines will be small in this model. Historically, it was felt disappointing that reconnection in this celebrated model worked so slowly. Observation suggest that reconnection

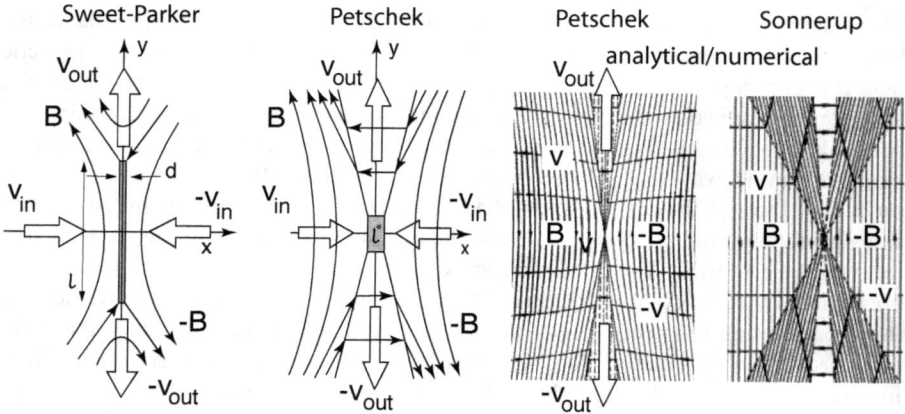

Figure 12.1: The two basic fluid models of reconnection. *Left*: Sweet-Parker's long thin current sheet model with all plasma flow across the sheet. *Centre*: Petschek's small (unspecified) diffusion region/slow shock model. The slow shocks are the four separatirces emanating from the corners of the diffusion region of lengths ℓ^* in this figure. *Right*: The semianalytical solutions of Petschek's and Sonnerup's extension of Petschek's model which includes another discontinuity.

does indeed proceed at a very fast rate, much faster than predicted above. In order to improve on this a modification of the Sweet-Parker model was proposed by Petschek who suggested to replace the unpleasant long length ℓ of the current sheet by a much shorter length ℓ^*. This model was improved by Sonnerup adding more boundary conditions to adjust for the flow and by Vasyliunas correcting a numerical error. The resulting very short reconnection region now assumed the wanted size of an X-line. Since now not all the plasma can pass across the current layer, four slow shocks had to be attached to the corners of the reconnection site as physical separatrices between inflow and outflow regions. Plasma could now pass across the separatrix shocks and join the outflow. Accounting for the boundary conditions a shortest length ℓ^* was estimated which made the expression for V_{rec} practically independent of S yielding an expression

$$V_{rec}^* \approx \frac{\pi v_A}{8 \ln S} \qquad \text{with} \qquad \ell^* \approx \frac{\ell}{S} \tag{12.6}$$

such that the length of the current layer became at least as short as the current layer width d. Comparing ℓ^* with d as given before indeed suggests a contradiction. These models are shown in Fig. 12.1.

This Petschek expression was for long time used as the valid reconnection rate though numerical simulations did not reproduce it. Ultimately it was found that the slow shock model in magnetohydrodynamics with a constant resistivity contradicts the Ohm's law to second order implying that slow shocks would become unstable and the short current layer would quickly expand to become an elongated Sweet-Parker layer.

This leaves open the possibility that for a localised resistivity a Petschek-Sonnerup-like reconnection mechanism might evolve. The problem is then shifted to the question of how a localised resistivity could be generated in collisionless plasma during reconnection. This is a delicate problem which cannot be described in a simple way. It may, nevertheless, happen locally in the reconnection process being restricted to the X-line environment only. As we have already noted in the previous chapter, unstably generated plasma waves may scatter particles out of their orbits thus faking real collisions. Processes like these may indeed generate an equivalent to normal collisions called anomalous collisions which then implies that an anomalous resistivity η_{an} is caused in the plasma. This replaces the η used in the above formulae and if large can substantially reduce the Lundqvist number $R_{Lan} \ll R_L$ to values much smaller than R_L in the Sweet-Parker model. The resulting reconnection rate would substantially increase in this case, and the required length of the current layer would shrink as well. However, since the anomalous resistivity is generated merely in the narrow small diffusion region it is not possible to simply use it in the homogenous resistive Parker model which does not restrict it to the reconnection site. Reconnection, in this anomalous case, becomes inhomogeneous. It has therefore been investigated mostly by numerical simulation methods.

12.1.2 The Ohm's Law Terms

Resistivity whether binary or anomalous is only one way of breaking the frozen-in state of the magnetic field in plasma. We have shown in Chapter 7 that magneto-hydrodynamic theory of a plasma is very different from magneto-hydrodynamics of fluids. The difference is expressed in the various forms of Ohm's law. In the previous section we have used it in a form which holds for fluids and plasmas in as far as it only includes isotropic resistivity. Plasmas may not obey isotropic resistivity at all. The resistivity, in particular when anomalous, will be anisotropic with different values along and perpendicular to the magnetic field. In a less simplified version Ohm's law in single fluid plasma theory has been given in Eq. (7.53) which we repeat at his place for further discussion:

$$\mathbf{E} + \mathbf{v} \times \mathbf{B} = \eta\mathbf{j} + \frac{1}{ne}\mathbf{j} \times \mathbf{B} - \frac{1}{ne}\nabla \cdot \mathbf{P}_e + \frac{m_e}{ne^2}\frac{\partial \mathbf{j}}{\partial t} \tag{12.7}$$

The right-hand side of this equation contains three terms in addition to the first resistive term. Still, in this version Ohm's law is not yet complete not even in simple single fluid approximation. It lacks a number of terms which include the correlation between fluid velocity and current and fluid turbulence. And it lacks a possible term which may result from the slow-time variation of the low frequency spectra of waves which pass across the plasma or have been excited in the plasma. This term exerts an average force on the plasma known under the name *ponderomotive force* which we will not discuss in this volume. Its proper definition can be found in our companion volume.

The Hall-term

For the purposes of this brief introduction into the reconnection problem we stay with the above form of Ohm's law which shows that resistivity is just one of the available ways of breaking the frozen-in condition and possibly lead to field-line merging and reconnection. One may remember what we have said in Chapter 7 about the origin of these terms: that they are the consequence of the plasma of consisting of two separate components not only of different sign of charge but also of different masses and therefore completely different dynamics. Hence, all the additional terms in Ohm's law we are going to discuss in relation to reconnection are due to the lighter component, viz. the plasma electrons.

Neglecting resistivity whether scalar or anisotropic, the first next term on the right in Eq. (12.7) is proportional to the cross product of current density and magnetic field. Remembering that in the derivation of Ohm's law the current — up to a minor contribution of the ions which is proportional to the mass ratio m_e/m_i — is given by the electron current $\mathbf{j} \approx -en\mathbf{v}_e$, we recognise that this term is very close to the electron Hall effect with $\mathbf{v}_e \approx \mathbf{E} \times \mathbf{B}/B^2$ the electron-electric cross-field drift. This Hall term is perpendicular to the current and magnetic field. In other words, it is non-dissipative as it does not produce Joule heat which is seen when scalar multiplying by the electric field. Thus in homogeneous plasma with constant density n this term will not be involved directly in breaking the frozen-in condition. If, however, the plasma is inhomogeneous — as is the case in a Harris current layer as shown in Section 7.4, — then the Hall current term will contribute to the electric field. In order to see whether any of those terms is capable of contributing to reconnection is does not suffice to consider just Ohm's law and infer about the contribution to the electric field. The electric field must, in addition, be an inductive field. Otherwise $\nabla \times \mathbf{E} = 0$ and it does not contribute to the general induction equation (7.99) which we reproduce here

$$\frac{\partial \mathbf{B}}{\partial t} = \nabla \times \left(\mathbf{v} \times \mathbf{B} - \frac{1}{ne}\mathbf{j} \times \mathbf{B} + \frac{1}{ne}\nabla \cdot \mathbf{P}_e - \frac{m_e}{ne^2}\frac{\partial \mathbf{j}}{\partial t} - \eta\mathbf{j} \right) \tag{12.8}$$

In considering here only the Hall term we replace the current by Ampère's law where the Hall term becomes just a sum of the gradient of the magnetic pressures $\nabla(B^2/2\mu_0)$ and tensions $\nabla \cdot \mathbf{B}\mathbf{B}/\mu_0$ as usual. Taking the curl the magnetic pressure in homogeneous plasma disappears. Hence the Hall term in this equation reduces to

$$-\nabla \left(\frac{1}{en}\mathbf{j} \times \mathbf{B} \right) = \nabla \left(\frac{1}{en} \right) \times \nabla \cdot \left(\frac{B^2}{2\mu_0}\mathbf{I} - \frac{\mathbf{B}\mathbf{B}}{\mu_0} \right) - \frac{1}{en}\nabla \times \nabla \cdot \frac{\mathbf{B}\mathbf{B}}{\mu_0} \tag{12.9}$$

and we see that in homogeneous plasma where the density is constant the first term disappears. The last term can be written as

$$-\frac{1}{en}\nabla \times \left(\nabla_{\parallel}\frac{B^2}{2\mu_0} - \frac{B^2}{2\mu_0}\frac{\mathbf{n}}{R_c} \right) \tag{12.10}$$

where $\mathbf{n}$ is the outer normal of the magnetic field and R_c the curvature radius. Hence, if the field lines are stretched and curvature can be neglected and in addition the magnetic

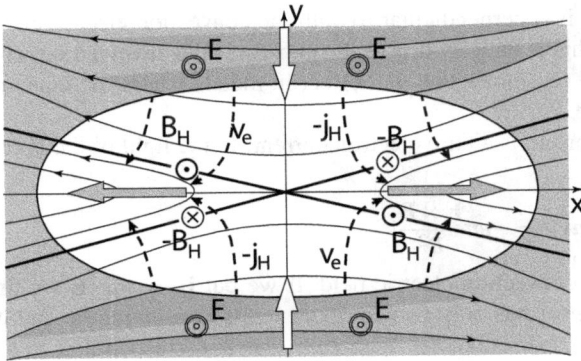

Figure 12.2: The geometry of the Hall effect in the vicinity of the neutral X-line in collisionless reconnection. The white region is the domain of distance $r \lesssim \lambda_i$ from the X-line accounting for a potential asymmetry between the inflow and outflow directions which elongates the region into the outflow direction. In this region the ion motion is irregular. Electrons are still magnetised and feel the convection electric field **E** performing the inward E×B-drift as indicated by the dashed lines which are in the plane perpendicular to the magnetic and electric fields. The Hall current $\mathbf{j}_H$ flows exactly opposite to the electrons as indicted in two places. It forms four half-loops and generates the quadrupolar structure of the Hall magnetic field $\mathbf{B}_H$. The open white arrows show the inflow at low speed and the outflow at high speed from the reconnection site. The convection electric field points out of the plane. Ions are accelerated in this field out of the plane.

field is about constant along the field lines in the vicinity of the reconnection site then this term disappears entirely. When the plasma is inhomogeneous the first term in the Hall term contributes whenever the gradient of the density is not parallel (or anti-parallel) to the magnetic pressure gradient or the normal on the magnetic flux tube.

In order to infer about the Hall effect it is, however, more appropriate to leave the one-fluid theory and consider the electron and ion components separately. Clearly, if the current sheet near the reconnection site becomes thin, as for instance during compression prior to the onset of a substorm, then the electron and ion motions decouple almost completely on the inertial scale of the ions. In order to see this we consider the ratio of ion gyro-radius $r_{gi} = v_{i\perp}/\omega_{gi}$ to ion inertial scale $\lambda_i = c/\omega_{pi}$. It is easy to show that $r_{gi}^2/\lambda_i^2 = \beta_{i\perp}$ is just the perpendicular plasma beta of the ions. Thus, when $\beta_i > 1$ the ion-inertia will dominate the motion of the ions, and the ions decouple from the magnetic field, becoming effectively non-magnetic. This happens at thin current sheets like the Harris layer where the magnetic field weakens. The ion motion n the scale of the ion inertial length becomes about irregular, while the electrons on the same scale still remain magnetised. They perform their gyration and cross-electric field drift, and are convected together with the magnetic field toward the current layer. Moving against the ions with the convection, their velocity is the $E \times B$ velocity in the convection electric field across the magnetic field, i.e. perpendicular to the magnetic field. Thus the electrons constitute a current, and it is easy to see that this current is a

Hall current as it is perpendicular to both the convection electric and magnetic fields respectively. This is entirely due to the narrow scales involved on which the ions are motionless and resembles the Hall effect in metals where the ions are fixed into the lattice structure.

The Hall current is a pure electron current. In the light of the above explained we can write it as

$$\mathbf{j}_H = -en\mathbf{v}_E = -en\frac{\mathbf{E} \times \mathbf{B}}{B^2} \tag{12.11}$$

where $\mathbf{E}$ is the convection electric field. If we put $\mathbf{j}_H = \sigma_H \cdot \mathbf{E}$ we find that the Hall conductivity tensor has only two components, is antisymmetric and has components only outside the main diagonal with $\sigma_H = en/B$ corresponding to a Hall resistivity $\eta_H = B/en$, independent of any collision rate. We also see that this Hall resistance is an inductive blind resistance and plays no role in the generation of heat.

The importance of the Hall current is more in it being an indication of the presence of collisionless reconnection because it can evolve only in the absence of collisions. Collisions, even anomalous would destroy the frozen-in state of the electrons by scattering them out of their paths and would inhibit the formation of Hall currents. In addition, Hall currents in the vicinity of a reconnection site X-line generate a Hall magnetic field in the direction of the current flow perpendicular to the symmetry plane of the external magnetic field but parallel to the current and the convection electric field $\mathbf{E}$. This Hall field has quadrupolar structure and, though being weaker than the external magnetic field, introduces a guide field into the reconnection, opening the possibility of accelerating the electrons in the direction anti-parallel to the electric field along the guide field. The role of guide fields will be elucidated later below. This is shown in Fig. 12.2.

The Pressure-term

The next term is the electron pressure term. This might be a complicated term because in general the electron pressure in plasma is not isotropic. However, under strictly collisionless conditions as those in space plasma near the Earth the pressure tensor could be anisotropic but should, in the frame of the magnetic field be locally diagonal assuming the canonical form

$$\mathsf{P}_e = p_\perp \mathsf{I} + (p_\parallel - p_\perp)\frac{\mathbf{BB}}{B^2} \tag{12.12}$$

where I is the unit tensor, and the components $p_\perp, p_\parallel$ of P_e are scalar quantities referring to the directions perpendicular and parallel to the magnetic field. It is often claimed that terms not diagonal of P_e would be responsible for reconnection. Such terms in general pressure tensors are always related to viscosity and are thus dissipative. They can be caused, for instance, by finite Larmor radius effects which cause ion-viscosity in which case the ion-pressure tensor P_i would contain such terms. However, in the electron pressure tensor such terms do not appear under strictly collisionless conditions as long as the electrons are magnetised. Electron finite Larmor radius

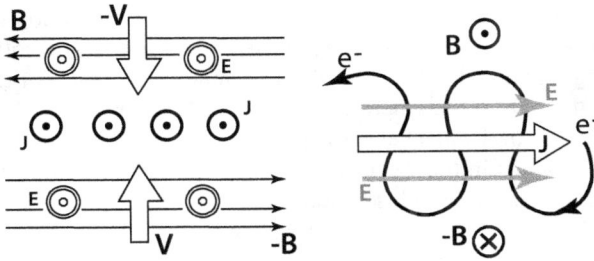

Figure 12.3: Meandering orbit of an electron in the centre of the Harris neutral sheet current. *Left*: Harris neutral sheet current **J** geometry with magnetic field **B**, electric convection and sheet current field **E**. *Right*: The meandering orbit of the electron with its double turning gyration parts leading to a drift directed oppositely to the electric field thus enforcing the current. This kind of meandering orbit may imply finite Larmor radius effects and thus electron viscosity which causes non-diagonal elements in the electron pressure tensor $\mathbf{P}_e$ which are, however, restricted to the inner part of the Harris current layer.

effects are most probably not important because of the smallness of the electron gyroradius.

Before discussing this point we return to the induction equation. Inserting the pressure tensor for the pressure tensor divergence term yields

$$-\left(\nabla\frac{1}{en} - \frac{1}{en}\nabla\right) \times \left[\nabla p_\perp + \nabla_\parallel (p_\parallel - p_\perp) - (p_\parallel - p_\perp)\left(\nabla_\parallel \ln B + \frac{\mathbf{n}}{R_c}\right)\right] \quad (12.13)$$

This shows that in a homogenous plasma only the very last curvature term contributes because all other terms can be written as gradients of which the curl vanishes. For example, the second last term becomes $\nabla_\parallel (\ln B)^{(p_\parallel - p_\perp)}$. In inhomogeneous plasma the curvature term survives and all terms which are not parallel to the density gradient. Under adiabatic conditions these are only the terms in the last parenthesis. Hence it is essentially the curvature of the magnetic field which contributes. However, there is some subtlety in the argument. The frame from which one should consider the reconnection process is not the local frame of the magnetic field but the (x, y)-frame of the X-line. In that frame the pressure tensor has indeed all nine components and is far from being diagonal with the out-of diagonal terms being composed of the pressure components and the geometric magnetic contributions. These terms are indeed off-diagonal and thus might, in this frame of reference, geometrically contribute to the magnetic induction equation in a very non-transparent way acting as fake viscosities. In this case one is confronted with the example of the importance of geometry in a physical problem even in classical non-general relativistic physics. Simulations which have been performed under these propositions assuming the presence of non-diagonal terms in the electron pressure tensor in the fluid picture and seem to confirm the occurrence of reconnection.

There might also be some deeper justification for the assumption the presence of real non-diagonal non-geometric terms in the electron pressure tensor. These would be caused by similar effects as ion viscosity is generated in the ion pressure tensor by finite Larmor-radius effects. For this, however, one needs to resolve the electron inertial scale region around the neutral sheet current in the centre of the Harris layer because this is the only place where electron finite Larmor radius effects may come into play on the scale of the local electron gyro-radius $r_{ge} > \lambda_e$ which here exceeds the electron-inertial lengths in the weak central magnetic field. The electron gyro-orbits in the anti-parallel magnetic fields become meandering orbits (as shown schematically in Fig. 12.3) if not scattered by plasma wave effects or disturbed otherwise. Such meandering orbits give rise to electron viscosity and in the fluid picture generate non-diagonal pressure terms. One should, however, be careful in application as these effects are restricted just to the centre of the neutral layer which in a fluid picture is not resolved. Therefore, even though suggestive, the electron viscosity picture can on the fluid scale be applied only if the fluid simulation resolves the electron inertial scale.

Electron-inertial Current-term

Finally the last term is caused by fluctuations in current density. It is a pure electron term caused by electron inertia as can be recognised from its proportionality to the electron mass m_e. If we introduce the electron plasma frequency $\omega_{pe}^2 = ne^2/\varepsilon_0 m_e$ we find that this term corresponds to an electron-inertial electric field $\mathbf{E}_{inert} \approx \eta_{inert}\mathbf{j}$ where the electron-inertial conductivity is given by

$$\sigma_{inert} = \eta_{inert}^{-1} \approx \varepsilon_0 \omega_{pe}^2 \tau_j \tag{12.14}$$

and τ_j is the typical time of variation of the current. This inertial resistivity can be compared with the plasma resistivity. In a plasma of $n = 1$ cm^{-3} we thus have $\sigma_{inert} \approx 0.04\tau_j$ or $\eta_{inert} \approx 30\tau_j^{-1}$. This may become important at fast time variation of the current in particular when the plasma is assumed to be ideally conducting and thus $\eta = 0$. Thus the electron-inertial resistivity is one possibility of breaking the frozen-in condition of the plasma and to contribute to field-line merging whenever the neutral sheet current exhibits time variations of frequency the order of, say, $v_j \equiv \tau_j^{-1} \gtrsim 0.1$ Hz.

When using this electron-inertial resistivity in the Lundqvist number we may re-express it in terms of the ion inertial length $\lambda_i = c/\omega_{pi}$ as $S_{inert} = \omega_{ge}\ell/v_j\lambda_i$ which can be used in the expressions for the Sweet-Parker reconnection rates. We find that expressed in terms of the ion inertial length and the frequency of current variation

$$\frac{d_{inert}}{\lambda_i} \approx \left(\frac{\ell}{\lambda_i}\frac{v_j}{\omega_{ge}}\right)^{\frac{1}{2}} \quad \text{and} \quad \left.\frac{E_\parallel}{E_\perp}\right|_{inert} \approx \frac{1}{M_A}\left(\frac{\lambda_i}{\ell}\frac{v_j}{\omega_{ge}}\right)^{\frac{1}{2}} \tag{12.15}$$

This shows that turbulence in the current layer of reconnection of sufficiently high frequency can contribute to the merging by breaking the frozen-in condition through electron inertia. In this case, for instance, for a thin current layer of the order of $d_{inert} \sim \lambda_i$ the length of the current layer $\ell_{inert} \sim \lambda_i(\omega_{ge}/v_j)$ in terms of the ion inertial length is just the ratio of the electron gyro-frequency to the frequency of the turbulent current oscillations, and the parallel reconnection electric field generated in the layer becomes the order of

$$(E_{\parallel}/E_{\perp})_{inert} \sim M_A^{-1}(v_{\mathbf{j}}/\omega_{ge}) \tag{12.16}$$

Since in reconnection $M_A \lesssim 1$ is not a large number, the value of the parallel electric field generated is mainly determined by the ratio of the frequency of the current fluctuation to the electron gyro-frequency. The above formulae of electron-inertial reconnection thus favour a reconnection state of fluctuating currents with electromagnetic variations of frequency $v_{\mathbf{j}} < \omega_{ge}$. Such fluctuations are in the lower-hybrid frequency or whistler range $\omega_{gi} < v_{\mathbf{j}} \ll \omega_{ge}$. They favour thin current sheets $d \sim \lambda_i$ of extension $\ell \sim \sqrt{m_i/m_e}\,\lambda_i$ yielding a substantially large reconnection rate

$$(E_{\parallel}/E_{\perp})_{inert} \sim M_A^{-1}\sqrt{m_e/m_i} \tag{12.17}$$

One should note that in this expression the smallness of the Mach number M_A competes with the smallness of the root of the mass ratio. Thus the reconnection rate is by no means small; at the contrary, it is orders of magnitude larger than the Sweet-Parker rate. In the ordinary Sweet-Parker model the Lundqvist number is of the order of $S = 10^{10} - 10^{20}$ yielding a reconnection rate of $\sim 10^{-10} - 10^{-5}$ which contrasts with the above rate which for $M_A = 1$ becomes $\lesssim 2\%$, depending on the inflow Mach number which, as one should also note, is sub-magnetosonic in almost all reconnection problems. Estimates on the assumption of large Mach numbers which sometimes are published are incorrect because at large Mach numbers shock waves evolve outside the current sheet which reduce the Mach number to values below unity.

The numbers obtained here are not unreasonable. In addition, though, they suggest that a Sweet-Parker mechanism which is based on electron-inertial effects cannot be stationary. Reconnection in the electron-inertial case will always be a temporarily variable process which switches on and off on time-scales $\tau_{rec} v_{\mathbf{j}} > 1$ which should be larger than the fluctuation time-scale of the reconnection currents.

Turbulence

We have noted several times that the complete generalised Ohm's law in single-fluid plasma theory contains additional terms which include the correlations between spatial current and flow velocity distributions. These terms have not been discussed here. They are of the form

$$-\frac{m_e}{e^2 n}\nabla \cdot [\mathbf{vj} + \mathbf{jv} - e(n_i - n_e)\mathbf{vv}] - \frac{1}{en}\mathbf{F}_{pmf} \tag{12.18}$$

with the last term disappearing for quasineutrality and $\mathbf{F}_{pmf}$ a possible ponderomotive force. The current-velocity product terms imply the inclusion of turbulence theory. It is, however, not unreasonable to assume that such correlations may drive reconnection possibly on a substantially faster scale than the resistive and pressure terms we have considered in the previous subsections. In the literature some possible effects of magneto-hyrodynamic turbulence on reconnection have been studied numerically in view of astrophysical application only. There it has been suggested that the turbulence will substantially speed up reconnection by creating many turbulent current filaments and causing kind of a 'glowing' reconnection state. Such attempts are, however, not self-consistent but put the turbulence in as an initial condition.

Moreover, we have also not taken into account any ponderomotive forces of the average wave field fluctuations $\langle \delta E_i \delta E_j \rangle, \langle \delta B_i \delta B_j \rangle$ if these quantities vary with location or time. At the low frequencies in a weak magnetic field which is rather vulnerable to large-amplitude magnetic fluctuations such forces might become large. Ponderomotive forces have so far experienced very little attention yet.

The previous discussion suggests that reconnection does not proceed on large global scales. It is a process which takes place on the smallest plasma scales, the scales of the ion inertial length and the electron scale, and insight which does not preclude its global application and the cause of violent global scale effects in the near-Earth plasma.

12.2   Small-scale Physics of Reconnection

The above fluid theories are incapable of any proper description of the reconnection process. The most they provide are some global pictures of magnetic field and plasma flow lines in the vicinity of the reconnection and the current, and some global and uncertain estimates of reconnection rates and extensions of the current layer which can be expected in the models. Moreover these estimates differ by orders of magnitudes. This inhibits any deeper understanding of reconnection as a process and not just taking it as a means for drawing relaxed magnetic geometries or waving hands about the acceleration of plasma and release of magnetic energy. Since it has become clear that reconnection is by no means a simple process — as was initially believed — but a complicated micro-physical transition from one plasma state into another one, it has also become clear that there will be no way of providing a simple analytical solution. Petschek's model has turned out to be incorrect to second order while Parker's model is just global and not satisfactory.

Other analytical calculations introduced sequences of different boundaries in the form of staples of discontinuities in order to adjust for the transitions while nevertheless being idealisations the realisation of which is questionable and does not lead to a more profound understanding. So only way to progress is numerical simulations. Such simulations have been performed since the last quarter of the past century first in solving the magneto-hydrodynamic and later the two-fluid plasma equations numerically. The realisation that the Hall effect might play a role has also led to the attempt to describe reconnection in a pure electron fluid plasma by so-called electron-magneto-hydrodynamics which includes the Hall effect. Still all these fluid approaches do not cover the micro-physics and contribute only insufficiently to the global physics.

The above discussion has already suggested that collisionless reconnection proceeds on kinetic, more precisely on electron scales. Reconnection is an electron effect being much faster than on fluid scales. The coupling to the ions occurs probably not before the ions have become magnetised which happens on scales of the order of the ion-inertial length λ_i. Hall magnetic and guide fields may become also important in this respect. Coupling to global scales poses another problem.

The best way of treating reconnection on these scales has been found in full particle-PIC codes (standing for Particle-In-Cell) where one assumes an initial magnetic geometry including a current plus a particle distribution and solves the microscopic equations of the particle dynamics in the self-consistently generated field separately for each of the particles. Such simulations are very useful even though they have their drawbacks in the large numbers of particles which have to be treated as well as in the satisfactory treatment of initial and boundary conditions. In the following we briefly describe some off the interesting results obtained with these methods.

12.2.1 Absence of Guide Field

The overwhelming majority of all numerical investigations of reconnection starts from the symmetric two-dimensional initial state imposing a stable Harris current sheet as described earlier with the antiparallel fields becoming homogeneous at large distances from the Harris sheet to both of its sides. Such models are void of any guide fields.

Since Harris sheets are stable solutions of the microscopic plasma equations the evolution of reconnection has to be initiated artificially in one or the other way. The methods differ between
– introducing a small artificial resistivity in the centre of the Harris layer,
– inserting an initial 'seed-X-point' into the current layer,
– imposing a very small amplitude magnetic fluctuation spectrum, or
– compressing the layer locally until it becomes very thin at one location, usually the centre of the simulation box, and undergoes instability.
– In a few cases the system has also been tailored in such a way that reconnection starts from mock 'thermal fluctuations', i.e. randomly imposed numerical fluctuations.

Comparing with reality so all these cases may be realised in one or the other place. As boundary conditions one assumes either periodic conditions which imply considering two antiparallel current sheets, i.e. doubling the box, or open boundary conditions in which case plasma can freely escape from the box. Depending on the different settings the results may differ as well.

No Hall Field: Boltzmannian Electrons — The 'Hybrid' Case

Particle simulations have for long been performed by considering the ions as the main drivers of reconnection and simply assuming that the electrons form a thermal fluid background with Boltzmannian electrons which react instantaneously to any electric potential fields $\phi(\mathbf{x}, t)$ according to the Boltzmann law for the electron density

$$n_e(\mathbf{x}, t) = n_0 \exp\left(\frac{e\phi_\parallel(\mathbf{x}, t)}{k_B T_{e\parallel}} + \frac{e\phi_\perp(\mathbf{x}, t)}{k_B T_{e\perp}}\right) \tag{12.19}$$

This form allows for purely scalar $\phi_\parallel = \phi_\perp$ electric potential fields and for vector potentials and also takes into account a possible anisotropy of the electron temperature. However, no electron dynamic effects can be resolved in such codes which are known as *hybrid codes*. As we have already mentioned, though, the reconnection process includes ions but is not an ion effect; it requires the electrons to be separated from the magnetic field lines when the ions have already become nearly insensitive to the magnetic field in the Hall region on scales $\sim \lambda_i$ away from the neutral line frequently — though mistakenly because there is no diffusion — called the *ion diffusion* region. Any code must therefore resolve the electron inertial scale region $\sim \lambda_e$ around the X-line, the *electron inertial* region, if it intends to understand the physical mechanism of reconnection.

No Hall Field: Electron-Positron Pair-Plasma

Such codes have become available only recently. Hybrid-simulations with Boltzmann electrons nicely reproduce the X-line configuration of the reconnection site and several other effects like plasma ejection and so on. They cannot reproduce the Hall effect in reconnection, however. They are thus realistic only in the case when a pair plasma is considered consisting of equal numbers of electrons and positron. Because them $m_e = m_i$ the electron and positron dynamics are identical up to the sign of gyration and other drifts, and thus no Hall current can be produced. An example of an electron-positron plasma simulation is shown in Fig. 12.4. The upper pat of this figure shows the magnetic flux tubes which map the magnetic geometry. Several neutral lines have developed while the main neutral line has become broad and of very weak magnetic field. The electric field is mainly concentrated around the magnetic islands. On the right the acceleration of one of the particles in the simulation is shown when it after meandering encounters the electric field of reconnection, is accelerated, increases its gyro-radius and is just entering the next strong electric field to become further accelerated.

Electron-Proton Plasma: Hall Fields

More interesting than this equal-mass pair-plasma case — which may be applied to astrophysical problems — is the case of an ion-electron plasma. These simulations show the effect of the Hall current but are more difficult to perform because of the large mass difference. In order to resolve the electron scale and see the Hall effect the simulation time must be long and the box extended. Moreover, since reconnection is by nature at least a two-dimensional process, one-dimensional simulations as for instance useful in shock simulations make no sense here.

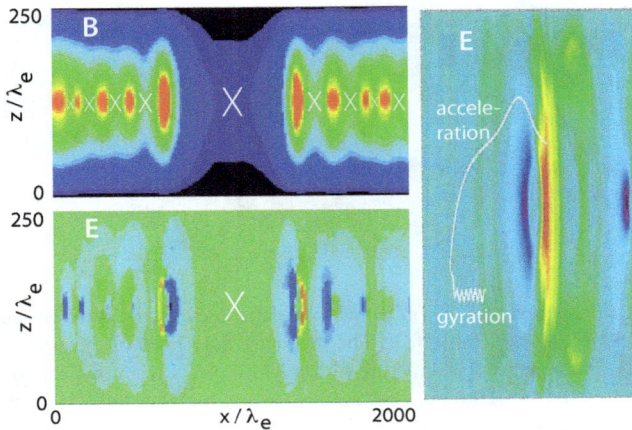

Figure 12.4: Non-Hall reconnection simulations in three dimensions, high resolution and large domain. *Top*: The magnetic field in colour representation showing the magnetic flux contours the boundaries of which are the magnetic topology. One recognises the X-line topology with nearly zero flux at the centre and high closed loop flux in several separated points. In the simulation which has been initiated in the centre, rapidly many neutral lines evolve which all have their own dynamics. *Bottom*: The electric field in the same simulation evolving around the magnetic islands. Red is positive fields, blue negative fields. Strong electric fields evolve around the first magnetic islands at the main X-line. Here the main particle acceleration takes place. Plasma jetting in each neutral line is to both sides. In this three-dimensional simulation the neutral lines have finite extension in the not shown third dimension into the plane. *Right*: Acceleration of one selected electron in (an enlarged view of) the electric field of the first right magnetic island. Originally the electron performs the meandering gyration motion (white line small oscillation) until experiencing the reconnection electric field and becoming accelerated as seen by the large orbit in z and x directions. The electron is just entering the next acceleration phase in the strong electric field on the back of the magnetic island.

Whether Hall fields directly affect the process of reconnection is not known. Hall currents are strictly perpendicular to the primary magnetic and electric fields. They, moreover, do not produce any Joule heating. From that point of view they should be a side product of reconnection. However, plasmas are highly dynamical systems with the electrons and ions reacting differently to the presence of fields. Because of this reason Hall electron currents and Hall magnetic fields in the 'ion diffusion region' around the reconnection site exert some effect on the behaviour of the plasma.

The first of these effects is the mere presence of the magnetic Hall field and its quadrupolar structure. Superimposed on the initial field the Hall field causes the total magnetic field to be twisted with respect to the simple initial two-dimensional geometry. This twisting is seen already in electron magneto-hydrodynamic and in more detail in electron-proton numerical PIC simulations of reconnection as shown in Fig. 12.5. The reconnection process thus becomes magnetically three-dimensional. Though this twist is felt only outside the centre of the neutral sheet, i.e. outside the electron inertial ('electron diffusion') domain and therefore does not directly influence the electron

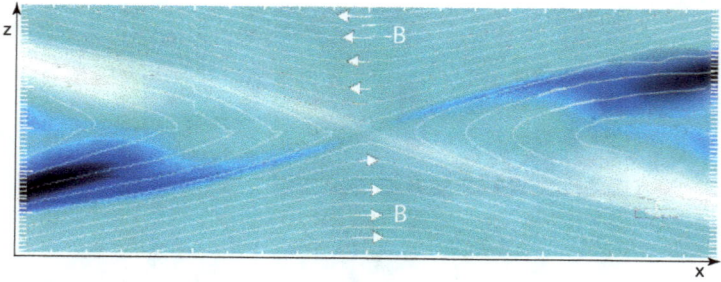

Figure 12.5: Reconnection simulation in two dimensions including different electron and ion dynamics and exhibiting the Hall effect. The magnetic field lines of the background field are shown in white. The quadrupolar structure of the Hall field is made visible by the colouring. White is outward pointing Hall field, dark is inward pointing Hall field. It is interesting to see that the Hall field in this two-dimensional simulation evolves only inside the separatrix which indicates a strongly elongated 'ion diffusion' region in the direction of the outflow. Note that this structure differs from the theoretical expectations where the extension of the inertial region was assumed to be much less dependent on the location of the separatrix. The twist in the magnetic field caused by the Hall field extends along the separatrices and is caused by magnetic tension.

dynamics there, it is not yet known in which way the merging on the electron scale proceeds.

$$\tau \approx \lambda_i/v_{eH} \approx \lambda_i B/E_{conv} \qquad (12.20)$$

During this time the velocity of the electrons along the Hall field increases as $v_{eH\parallel} = e\tau E_{conv}/m_e \approx e\lambda_i B/m_e$ which yields a large parallel energy gain of the electrons

$$\frac{1}{2}m_e v_{eH\parallel}^2 = ev_{eH\parallel}E_{conv}\tau \approx \lambda_i^2 \frac{e^2 B^2}{m_e} = m_i c^2 \frac{\omega_{ge}^2}{\omega_{pe}^2} \qquad (12.21)$$

Aside of the magnetic twist causing magnetic three-dimensionality but otherwise being just a geometric effect, the main effects of the presence of the Hall field are more subtle and more important. The presence of the Hall field implies that it functions as a guide field along the external convection electric field even if no guide field is imposed. This can be seen from Fig. 12.3 and has the consequence that (1) the magnetised electrons in the ion inertial region do not only E×B drift and producing the Hall current, they also see the convection electric field being along the Hall field and become accelerated in the direction opposite to $\mathbf{E}_{conv}$ a process in which the electrons gain energy in the direction along the neutral line. The energy gain can be estimated from the time the Hall electrons need to cross the ion inertial region of radius λ_i to arrive at the centre of the neutral sheet. This time is given by Eq. (12.20).

Since the rest energy of a proton is $m_i c^2 \approx 1$ GeV, the electrons would be accelerated to the square of the fraction of gyro-to-plasma frequency of this energy. If this fraction is $\omega_{ge}/\omega_{pe} \sim 0.1$ the presence of the Hall- magnetic field implied a parallel energy of electrons of ~ 10 MeV. One thus expects very high energy electrons to be present in collisionless reconnection with Hall fields. High-energy electrons have indeed been observed in reconnection, however at one order of magnitude less energy. The resolution of this discrepancy is probably that the reconnection site is three-dimensional with the neutral line extending into the third dimension only for a relatively short distance Δy not giving the electrons enough time to reach high energies. Limiting the energy to a fraction α, this distance can be estimated from the convection electric potential drop

$$e\Delta\phi_{conv} = eE_{conv}\Delta y = \frac{\alpha}{2}m_e v_{eH\parallel}^2 \tag{12.22}$$

when using the above estimate for the maximum gained electron energy when experiencing E_{conv} on the length Δy and setting it equal to the fraction α suggested by observation. This yields for Δy measured in ion inertial lengths λ_i the fraction

$$\frac{\Delta y}{\lambda_i} \approx \alpha \frac{m_i}{m_e}\left(\frac{v_A}{c}\right) \approx 2 \times 10^3 \alpha \left(\frac{v_A}{c}\right) \tag{12.23}$$

The ratio of Alfvén to light velocity is usually a small number in reconnection, in the magnetospheric tail being of the order of $\sim 10^{-3}$ thus compensating for the mass ratio. Hence the extension of the reconnection site into the third dimension should also be of the order of λ_i.

$$\tau \approx \lambda_i/v_{eH} \approx \lambda_i B/E_{conv} \tag{12.24}$$

During this time the velocity of the electrons along the Hall field increases as $v_{eH\parallel} = e\tau E_{conv}/m_e \approx e\lambda_i B/m_e$ which yields a large parallel energy gain of the electrons

$$\frac{1}{2}m_e v_{eH\parallel}^2 = ev_{eH\parallel}E_{conv}\tau \approx \lambda_i^2 \frac{e^2 B^2}{m_e} = m_i c^2 \frac{\omega_{ge}^2}{\omega_{pe}^2} \tag{12.25}$$

One immediately sees that the relative extension of the X-line in the third dimension shrinks with increasing density and decreasing magnetic field. Since one cannot expect Δy to be substantially shorter than λ_i one obtains a bound on the fraction of electron acceleration

$$\alpha_{max} \lesssim 5 \times 10^{-4}\left(\frac{c}{v_A}\right) \tag{12.26}$$

This energy of the streaming electrons usually exceeds the thermal energy of the electrons. Since the unmagnetised ions are accelerated as well by the electric field and move in the opposite direction, we are facing the classical situation of the Buneman and modified-two stream instabilities when Buneman and lower-hybrid waves can be excited and cause non-linear plasma waves, electron holes, heating and possibly even anomalous dissipation. We will later go into more detail with respect to these small-scale structures. In this way the Hall current becomes an important factor in reconnection.

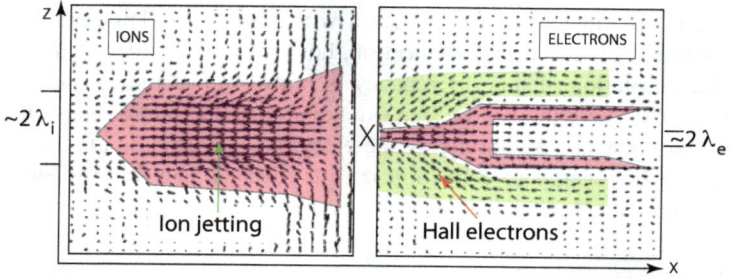

Figure 12.6: Particle velocities near X-line in collisionless reconnection with small mass ratio $m_i/m_e = 25$. *Left:* Left half (negative x) of space for ion velocity. Ions flow in from positive and negative z and become diverted to negative x by the jetting from the reconnection site in a region of width z proportional to the ion inertial length λ_i. *Right:* Right half (positive x) of space for electron velocity. Electron inflow continues across the ion inertial region until close to the centre of the current sheet. This inflow of electrons maps the electron Hall current flow. In the current sheet centre the electron velocity turns around by $180°$ to participate in the plasma jetting (here in positive x direction) in a region of width proportional to the electron inertial length λ_e.

This happens all in the ion inertial region only outside the inner part of the neutral current layer, and it remains unclear how the magnetic field lines from the two sides of the current can be brought into contact in order to reconnect and to form an X-line.

Figure 12.6 shows the differences in electron and ion jetting in a two-dimensional simulation caused in presence of reconnection close to the X-line. The left part of the figure is the left ion-half space, the right part the right electron-half space. Ions are seen deflected into the plasma jet to the left from the neutral line in a region of width of the order of λ_i. The right part shows the same jetting of electrons but being restricted to a region of the order of λ_e. In this simulation model the ratio $\lambda_i/\lambda_e = 5$. The most interesting observation is the undisturbed inward flow of the electrons on the scale of λ_i which maps the electron-Hall current thus confirming the completely different dynamics of the two particle components in collisionless reconnection. Figure 12.7 on its left gives an impression on the jetting velocities of both kinds of particles and on its right on the electron acceleration into the simulation plane thus enhancing the current by far over the ions. The ion plasma jetting corresponding the mass flow is barely half the Alfvén speed. It is the electrons which reach jetting speeds of up to roughly $2v_A$. The current speeds show deceleration of the ions at the reconnection site while the electron velocity is turned around such that the entire current is now carried by the high-speed electrons.

12.2.2   Guide Field Reconnection

Under natural conditions very rarely two magnetised plasmas approach each other strictly in an anti-parallel fashion. In most cases approaching magnetic fields are mutually inclined. Under such conditions a magnetic field is frequently present along the

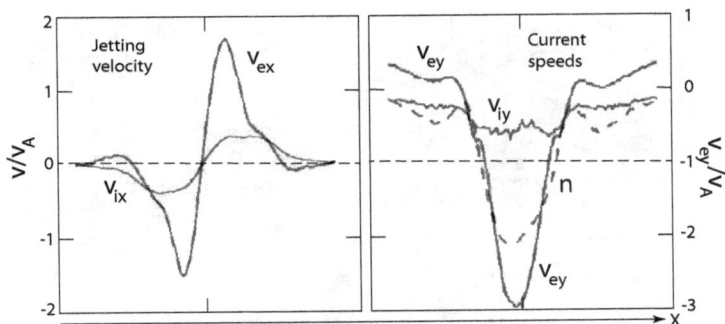

Figure 12.7: *Left*: Ion and electron jetting velocities in the two-dimensional reconnection. Jetting velocities of the electrons are three times as high as those of the ions. *Right*: The electron acceleration into the reconnection plane opposite to the current direction. The electron velocity (right ordinate) is more than twice the ion current speed.

current direction, called a guide field. In fact, the presence of guide fields will be the normal case. Sometimes the guide magnetic field has also a component across the current layer. Such magnetic guide fields play an important role in reconnection which for long time has been underestimated.

The first action of a guide field is to magnetise the electrons across the entire current layer. At first glance this seems to make reconnection more difficult or even to inhibit it entirely. Indeed, first simulations showed that reconnection with guide fields sets on slower and later thus having smaller growth rate than reconnection in strictly antiparallel fields. On the other hand, reconnection with magnetised electrons was also at the centre of some early and ingenious analytical theory which found that reconnection might under those conditions set on explosively. This theory has been described in our companion volume in more detail and will not be referred here to.

A homogeneous guide field B_y^g parallel or antiparallel to the Harris current is the most natural extension of the non-guide field case. It does not provide insurmountable difficulties. Figure 12.8 shows some of the graphical outputs of those simulations. Here we list a number of results with guide field included into two-dimensional full particle PIC-simulations.

– The first result of inclusion of a guide field is the expected reduction of the reconnection rate with increasing guide field strength if the guide field becomes stronger than the original anti-parallel field B_0. Weaker guide fields have no appreciable effect on the reconnection rate.

– Guide fields cause an asymmetry of the reconnection pattern by tilting the X-line geometry against the z axis, i.e. the normal to the Harris current layer. The X becomes an inclined X.

– Guide fields do, however, not completely destroy the Hall field. The quadrupolar structure of the Hall geometry becomes tilted and deformed because the guide field favours one foot of the Hall field.

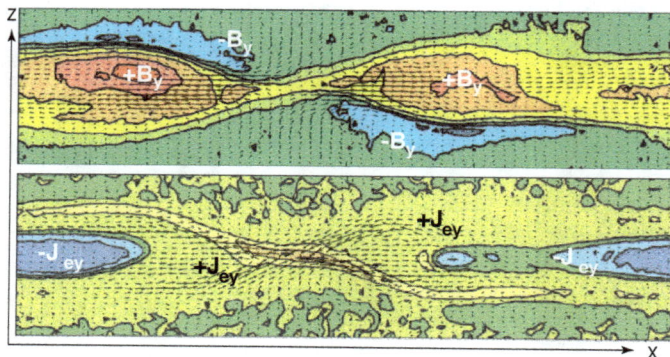

Figure 12.8: *Top*: Ion velocity (arrows) overlaid by the out-of-plane (along the Harris sheet current) component of the magnetic field. This field includes the original guide field B_y^g and the self-consistent Hall field showing the anisotropy in the field evolving in the reconnection with guide field. The ion flow pattern also shows the distortion of the symmetry of the flow and X-line. *Bottom*: The highly asymmetric electron flow pattern evolving in guide field reconnection overlaid by the out of plane electron currents J_{ey}. Yellow-red colours indicate positive, blue-green colours negative values. One may notice that the electron current is mainly in positive direction along the induction electric field and of highest concentration in the yellow-reddish diagonal. This is the indication that it has been accelerated along the inductive convection field which has been amplified by reconnection.

Plasma Waves in Reconnection

Reconnection itself is not a smooth process; at the contrary, it is necessarily accompanied by fluctuations in the fields. Historically this was realised half a century ago in the mid-sixtieth when reconnection was understood as the instability of a thin current sheet against the self-contraction of the current, the so-called *tearing mode* which in the current sheet creates a chain of magnetic X-lines and O-points, the latter being called magnetic islands (because they contain only closed magnetic field lines) or also plasmoids (because in their centres the magnetic field should vanish) and which are expelled to the sides from the reconnection X-line by the jetting process. Other instabilities which might be excited in a plane current sheet are fluid modes of the kind of *sausage* modes, *kink* modes, *ballooning* modes and *pinch* modes which may grow once the current has undergone filamentation by the tearing mode after reconnection. These modes have been treated in our companion volume but are not of primary interest here.

The first important wave related to reconnection is the lower-hybrid drift wave. This lower-hybrid wave has frequency $\omega_{lh} \approx \omega_{ge}\sqrt{m_e/m_i} \approx \omega_{ge}/43$ in a proton electron plasma of weak field. It causes ion heating and scatters electrons along the magnetic field producing field-aligned electron currents. It also contributes to anomalous collision frequency and anomalous resistivity but this effect is less important than the generation of fast electron current and beams from the Hall region. The accelerated electrons and electron currents flow along the magnetic field in highest concentration

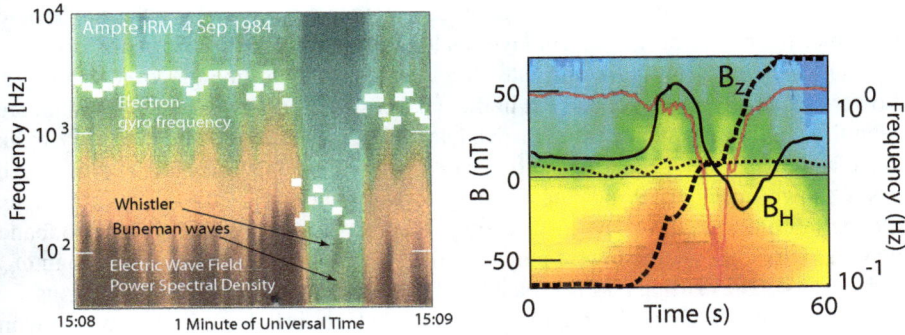

Figure 12.9: Close spacecraft approach to a reconnection site at the magnetopause showing low wave activity. *Left*: An early observation by Ampte. The white line is the electron gyro-frequency as obtained from the magnetic field. In the extremely low magnetic field region — the steep drop in gyro-frequency — the magnetic field has briefly changed direction. In the lowest magnetic field, two types of waves are observed, a comparably intense and broadband emission below the electron gyrofrequency in the Buneman wave band, and just before a weak drifting emission below the electron cyclotron frequency in the whistler band. *Right*: A Polar spacecraft high resolution observation at very low frequencies below those of the left panel. The red line is the ion-cyclotron frequency. In the low magnetic field region lower-hybrid drift waves are detected which probably have leaked in from the sides.

close to the separatrices. They have been found in simulations to generate Buneman waves which evolve into electron holes. This process has been described by us in the previous chapter on instabilities. Their role is to heat the electrons and to break the current flow.

The lower-hybrid wave is excited by the lower-hybrid drift instability in the density gradient of the Harris current sheet in the region where the gradient is steepest. Example of rare spacecraft approaches to a possible reconnection site as seen in the measured electric power spectral density (colour coded) are shown in Fig. 12.9.

The lower-hybrid drift instability has frequency $\omega \approx \omega_{lh}$ but requires the presence of a magnetic field because it is excited by the diamagnetic electron current (see Section 7.4) across the stationary (unmagnetised) ions which means the presence of a density gradient $\nabla_\perp p_{e\perp}$. It can, however, also be excited by the electron-Hall current then being independent of the density gradient. It does not evolve in the centre of the Harris sheet where the gradient and Hall currents vanish and the magnetic field is zero and is quenched when the plasma-$\beta > 1$. The latter is the case in the centre of the neutral layer but in the presence of a sufficiently strong guide field which helps holding β down. The lacking density gradient and Hall current do however not allow growth there even in the presence of a guide field. On the other hand, it propagates mainly perpendicular to the magnetic field with wave vector $\mathbf{k}_{lh} = (k_\perp, k_\parallel), k_\parallel \ll k_\perp$. Its parallel phase speed is thus large and its perpendicular phase speed is finite of the order of the diamagnetic drift or Hall current velocity. Because of its small $k_\parallel$ is accelerates electrons along the magnetic field.

Waves excited in the density gradient can move across and along the guide field and enter the centre of the current layer as long as the magnetic field is not too weak (i.e. $\omega < \omega_{gi}$ locally) and contribute to anomalous effects like electron scattering. Lower-frequency waves are excited in the weaker gradient close to the X-line and are seen first; higher-frequency waves arrive later. The observations show only very weak wave activity in the X-point in the few cases when it was approached. The two such cases shown in Fig. 12.9 indicate the drop in wave activity at the reconnection site. On the left one sees the higher frequency part above ~ 30 Hz where some broadband wave activity exists in the lowest magnetic field region as indicated by the drop in ω_{ge} (white dots). The drifting part is probably whistlers while the broadband emissions are Buneman waves which indicate the presence of electron holes and thus two-stream instability. On the right in this figure the lowest frequency part $\omega < 1$ Hz is shown for another approach to the X-line. Wave activity is seen below and in the deepest magnetic field drop exceeding the ion gyro-frequency (red line). Interestingly, this activity is largest where the Hall fields are strongest and thus is probably caused by the Hall-current driven lower-hybrid instability with the waves leaking into the reconnection site by the above propagation mechanism. Their intensity there is in the nonlinear regime but insufficient to produce enough resistance for driving reconnection.

Another important instability is related to the electrons which are accelerated in the electric induction field. These electrons form fast beams and currents along the current layer, as we have seen above. The related electron currents flow along the guide field and perpendicular to the initial magnetic field and excite the modified two-stream instability with strong wave field again around the lower-hybrid frequency. This time, however, the waves may become strong enough, in particular in the presence of a magnetic guide field, to generate electron holes in the same way as the Buneman holes along the magnetic field. Once this happens, the current breaks locally off, a process reminiscent of anomalous resistivity and thus important for magnetic diffusion and merging.

However, even more important is that in the presence of the guide field the fast electron current is concentrated in the centre of the current layer were it is strong enough to drive the zero-frequency Weibel-or the Weibel-filamentation mode. Either of these modes create a magnetic field in the very centre of the current sheet, i.e. in the regions which the initial anti-parallel magnetic fields cannot penetrate. The Weibel field is either parallel or anti-parallel to the initial field non-guide field, of short scale and provides the seed-X-points which are needed in order to ignite reconnection. This process may happen as well in non-guide field reconnection where it starts in the Hall field region, and in guide field reconnection where it occurs in the centre of the current layer.

12.2.3 Driven Reconnection

Initiating reconnection is a problem. We have discussed this at the beginning. One obvious possibility is to force a continuous inflow of plasma into the current sheet from its sides until reconnection sets on. This is realised in nature and can be done by imposing an external electric field at the boundaries of the simulation box.

So far the conditions have been assumed two-dimensional. When the guide field is inclined against the current layer it has not only a tangential but also a continuous — for $\nabla \cdot \mathbf{B} = 0$ — normal component B_z to the Harris current. In this case the stabilising effect of the field becomes dramatic and the electrons in the enter of the Harris layer cannot decouple from this field. Reconnection in this case becomes necessarily three-dimensional not only on the scale of the 'ion-diffusion' region but also locally on the electron scale. In this three-dimensional case one can use an initial two-dimensional generalised Harris sheet model which is derived from a vector potential with the only (initial) component

$$\frac{A_{0y}(x,z)}{wB_0} = \ln\left\{\frac{\cosh\left[zF(x)/w\right]}{F(x)}\right\}, \qquad \frac{B_{0x}(x,z)}{B_0} = -F(x)\tanh\left[\frac{zF(x)}{w}\right] \qquad (12.27)$$

where $F(x)$ is a slowly varying arbitrary function. This vector potential yields the generalised Harris field $B_{0x}(x,z)$ and density

$$n(x,z) = n_{bg} + n_0 F^2(x)\mathrm{sech}^2\left[zF(x)/w\right] \qquad (12.28)$$

with n_{bg} a uniform background density, and the width of the sheet is given by $w/F(x)$ being variable in direction x along the sheet. The free function $F(x)$ can be chosen as $F(x) = \exp(\varepsilon x/w)$ with $\varepsilon \ll 1$ which yields for the assumed weak normal magnetic guide field $B_{0z}(x,0) = \varepsilon B_0$.

In order to initiate reconnection one imposes the convection electric field in the direction of the Harris current of the above model

$$E_y(x,t) = E_{0y}f(t)\mathrm{sech}^2(x/\Delta x_E) \qquad (12.29)$$

with $f(t)$ a possible free time function and Δx_E the scale of the forcing convection electric field along x. Thus the forcing field is maximum at $x = 0$ but may vary with time according to $f(t)$. This field creates an E×B plasma inflow with peak speed $E_{0y}/B_0F(0)$. This kind of simulation requires open boundary conditions for letting the plasma freely flow out in the x-direction.

Simulations performed with such a model were done by switching on the driving field $E_y(x,t)$ for one ion gyro-period and keeping it afterwards stationary. Such simulations yield a relatively long waiting time since the signal has to move in from the boundary into the current to be felt by it. During this time the current layer is compressed and thins, forming an electron layer in its centre and causing reduction of the B_z component. This reduction is essentially an inductive effect in the current layer. Then, however, reconnection starts faster and, as expected, yields a higher reconnection rate than compared with the non-driven case. Four points are of interest:

– First, the reconnection rate seems to be insensitive to the mass ratio m_i/m_e.
– Second, the forcing causes a strong time effect in the normal magnetic field component B_z suppressing the stabilising effect of B_z on the electrons. This is a time dependent process in which

– the normal magnetic field is quasi-periodically reversed and re-establishes itself.
– This causes pulsed reconnection with the release of large-B_z fronts (large normal magnetic field components) which propagate away to both sides of the X-line. Such fronts resemble the dipolarization fronts known from observations of substorms in the Earth's magnetotail current layer where an initially present normal magnetic field is quite natural.

The important conclusion learned from this kind of simulation is that the stabilising effect of a finite B_z is overcome by forcing reconnection from the outside by plasma inflow over a sufficiently long time through imposing a convection electric field.

12.3   Magnetopause Reconnection

The two final sections of this chapter present two cases where reconnection plays the central role in magnetospheric physics: at the magnetopause and in the geomagnetic tail plasma sheet.

Reconnection at the magnetopause is a particularly complex case. It takes place in a very narrow layer of the order of an ion gyroradius or less, the magnetopause current sheet which is generated by the pressure balance between the solar wind/magnetosheath plasma and the geomagnetic field. The plasmas to both sides of the magnetopause have grossly different properties. The magnetosheath plasma is dense, in particular close to the magnetopause stagnation point, the nose of the magnetopause, and also in the cusp regions. Densities of the order of $n \sim 30$ cm^{-3} are the rule. The temperatures of ions and electrons are not identical. Electrons have temperatures of about $k_B T_e \sim 30 - 50$ eV, with ion temperatures being larger by a factor of $3 - 5$. Magnetic field strength are of the order of $20 - 30$ nT, and the field can have any direction partly following the Parker spiral but also affected by the bow shock and the magnetosheath turbulence. Normal inflow velocities are not zero. Therefore, from this point of view reconnection is forced here.

The plasma of the magnetosphere on the inside of the magnetopause is rather dilute with particle densities of $n \sim 0.1 - 1$ cm^{-3}. Temperatures are high, in the keV range for ions and electrons though not equal. Flow velocities are low of the order of ~ 10s of km/s determined by inner magnetospheric convection substantially less than in the magnetosheath. The magnetic field is mainly directed northward and of magnitude of ~ 100 nT.

These different environmental properties suggest that reconnection at the magnetopause is (1) highly asymmetric, (2) generally includes guide fields and (3) is forced, being driven by the continuous low-Mach number inflow from the magnetosheath. Close to the stagnation point the Mach number is small $M_A \lesssim 1$ while at larger distances the flow is mainly tangential. It is, however, probably free of a normal magnetic component though even that is not certain because, as we have seen in Section 8.6, a fluid model of the magnetopause could consist of an intermediate shock which is in pressure balance but nevertheless contains a normal magnetic field component.

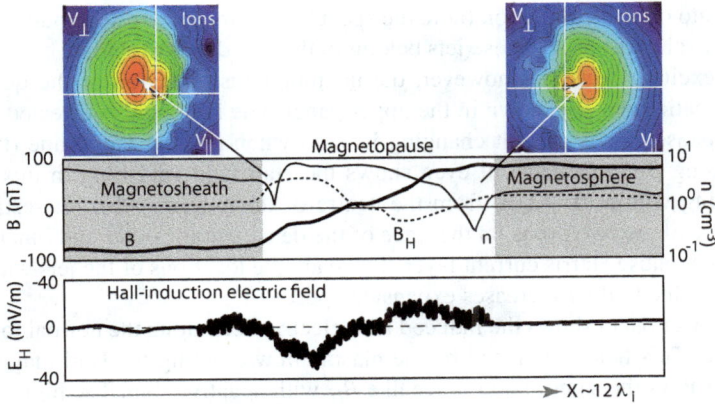

Figure 12.10: Schematic of a magnetopause spacecraft passage by the Polar spacecraft of a reconnection site at the dayside close to the stagnation point in plasma density and fields. For orientation: the total passage time was 75 s. The total length (shown by the arrow at bottom) was roughly 12 λ_i. The data (all schematised) have been transformed to the de Hoffmann-Teller frame, i.e. the frame comoving with the plasma tangential to the magnetopause. *Central panel*: Density n, main antiparallel field component B and Hall-magnetic field B_H. *Bottom panel*: The induced Hall electric field E_H. *Top*: Velocity space of ions schematically showing the two oppositely directed accelerated ion jets (for illustration taken from other similar snapshot measurements by the Themis spacecraft) indicated by white arrows.

12.3.1 Observations

Reconnection at the magnetopause was first confirmed by the observation of symmetric plasma jetting from the magnetopause X-line. This has been stated already in Section 8.6. It took roughly another thirty years until precise plasma and field measurements became available for resolving a reconnection site crossing at the magnetopause. Relevant data from such an observation are schematically shown in Fig. 12.10 during a passage of length roughly $\Delta x \sim 12\lambda_i$. This magnetopause passage was on the dayside not far away from the magnetopause stagnation point (nose). For cleaning purposes the data have been transformed into the $(E \times B)_{tang}$ frame of the magnetosheath plasma flow along the magnetopause. This procedure eliminates the main convection electric field component tangential to the magnetopause. This electric field component, by Maxwell's equations, is continuous across the magnetopause.

The two panels show the plasma density dropping from the high magnetosheath value near $n \approx 70$ cm^{-3} to the (still high) density value $n \approx 15$ cm^{-3} adjacent to the magnetopause from the inside of the magnetosphere indicating the generation of a dense inner boundary layer during reconnection. The high density in the central part of the panel coincides in location with the transition of the main magnetic field component B from negative magnetosheath values near -80 nT to positive values of same magnitude in the magnetosphere. This transition in B and n is typical for a Harris layer. On the top of the figure two observations of ion jets (from another similar crossing) emitted into opposite direction (note the specular symmetry) are schematically shown in the ion velocity space. These jets belong to the low density regions.

The exciting facts are, however, the unambiguous detection of the quadrupolar Hall magnetic field B_H shown in the upper panel. The Hall field is directed along the magnetopause Harris current changing its sign when passing the X-line (the region of vanishing B-component). It even shows the Hall field vanishing in this location, maximising within $(1-2)\lambda_i$ from the centre of the current. Also interesting is the occurrence of the two drops on the trace of the density at the outer and inner edges of the magnetopause Harris current layer. These are the locations of the jets which dilute the plasma due to their increases expansion speeds.

The lower panel shows the induced Hall electric field along the normal to the magnetopause. This field is induced by the plasma flow crossing the Hall magnetic field and thus shows the same spatial trace like B_H with negative sign. The high variability of this field is noticeable as it implies the presence of electric field fluctuations of very low frequency in the 'ion diffusion' region. During the crossing the electric field also shows variations in the other two components perpendicular and parallel to the main antiparallel magnetic field. These fields are of bursty nature thus being localised in moving structures which belong to large amplitude waves and electron holes and are accompanied by electron acceleration. Examples of such wave measurements have been shown in Fig. 12.9.

The measurement of the plasma parameters, velocity distributions of ions and electrons, fields and wave fields near an X-line at the magnetopause is made difficult by the small size of the region of interests where reconnection takes place. The total size of this region is of the order of the ion inertial length λ_i. In the magnetosheath plasma of $n \sim 30$ cm^{-3} the ion plasma frequency is $\omega_{pi} \sim 7$ kHz yielding $\lambda_i = c/\omega_{pi} \sim 40$ km. Reconnection takes place on the electron inertial scale $\lambda_e \sim 1$ km. A spacecraft is slow enough only on the apogee of its orbit to measure for sufficiently long in order to map the fields, plasma parameters and wave fields. Monitoring the entire velocity space of electrons or ions is hardly possible during the short time of crossing which in Fig. 12.9 was in both cases just of the order of 30 s for the entire 'ion-diffusion' region and barely 10 s for the magnetic field drop in the inner part of the reconnection site. In order to have a good chance to get long enough observations the spacecraft must thus be designed to skim the magnetopause for long time. Such spacecraft have been Ampte, Equator-S, Cluster, Polar and Themis.

Two of the globally most surprising findings were the confirmation that (1) the neutral line along the equatorial magnetopause might extend over a very long distance from the magnetopause nose to the dawn side of the magnetopause, a distance of the order of ~ 10 R$_E$, (2) reconnection in the cusp region probably takes place for long time of the order of many hours during northward periods of the solar wind/magnetosheath magnetic field as indicated by the continuous Image-spacecraft optical Lyman-alpha observation of proton auroral activity at the auroral footpoint of the cusp which could be correlated with Polar observations of plasma jetting at the northern cusp-magnetopause boundary. Optical Lyman-alpha is emitted from excited electrons that have been captured by precipitating cusp protons which could be

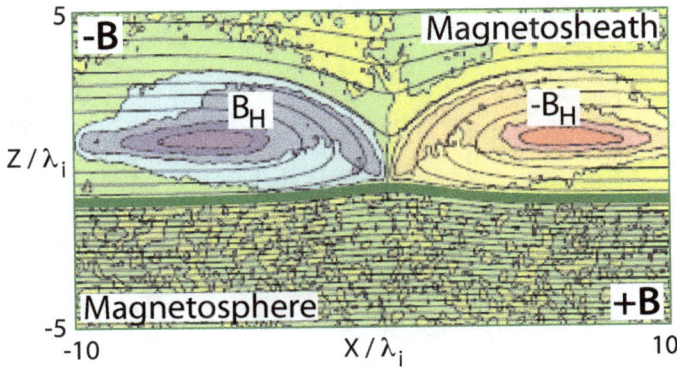

Figure 12.11: A non-forced non-guide field simulation of asymmetric magnetopause reconnection with main magnetic field lines **B** overlaid on the colour-coded Hall field B_H. The heavy solid field line is the nominal magnetopause, i.e. the transition from disturbed to purely northward field $+B_x$. The last sheath field line is the indented line outside the Hall spots. Notably the Hall field occurs exclusively on the magnetosheath side. The corresponding reconnected field lines are already connected to the Earth. The plasma on them forms the boundary layer.

measured over extended periods by Image. Both these observations indicate that reconnection at the magnetopause can be in a 'quasi-steady' state for long time. Of course. quasi-steady implies just that reconnection is continuously ignited as it is intrinsically a non-stationary process. The same caveat holds for the enormous extension of the neutral line along the magnetopause surface. It means that along this line reconnection continuously reproduces itself on scales of the order of the ion inertial lengths with the line where this occurs being the preferential site for reconnection under about stationary external driving conditions.

12.3.2   Comparison with Simulations

The difficulty of observing a reconnection site at the magnetopause and the prevailing uncertainty about what mechanism is responsible for reconnection in addition to the complexity of the basic equations which are believed to govern the process forces one to consider full particle PIC simulations as the only way out of the dilemma.

Summarising the difficulties any theory of reconnection encounters we note that (1) reconnection requires the violation of the frozen-in theorem. (2) This violation must occur on the electron scale. (3) It occurs in narrow current sheets. (4) These imply inhomogeneity in density, temperature, current density, velocity, pressure and magnetic field as also anisotropy and curvature effects. (4) It also implies plasma motion and thus electric convection fields and pressure-gradient caused electric fields. (5) Localisation of the gradients implies shear motions and field-aligned electric field components. (6) These cause field-aligned currents and particle acceleration, in the first place electron acceleration. (7) Reconnection includes plasma jetting. It includes (8) localisation and (9) a large number of free sources for wave excitation. (10) These

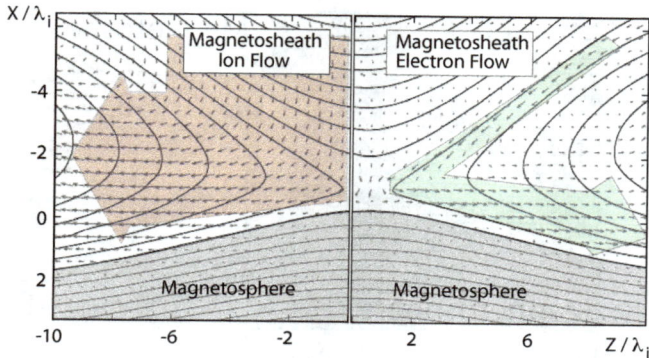

Figure 12.12: Ion and electron flow in forced non-guide field simulation of asymmetric magnetopause reconnection overlaid on the magnetic field. Shown are the left half-space for ions, the right half-space for electrons. Plasma is forced to flow in from above. At the X-line the magnetopause is the least compressed. The main particle flows are indicated by the big coloured arrows. Ions flow in and out over the entire region turning around at right angle. Electrons flow in and out mainly along the separatrices.

waves may grow to large amplitudes and cause additional nonlinear effects like heating, energy transport to different places etc. (11) The different ion and electron dynamics cause Hall currents and Hall magnetic fields which twist the field geometry and provide guide fields. (12) Guide field are also naturally generated when the original mutually impacting fields are not strictly anti-parallel. (13) Guide field effects include the possibility of parallel particle acceleration, the suppression of high-β, pressure anisotropies and so on. (14) All these effects suggest that reconnection is probably three-dimensional and in addition non-stationary. (15) Stationarity might occur only in the average. (16) Frequently, as in the case of the Earth, reconnection also must take place in a magnetic configuration of non-vanishing normal magnetic field component. (17) Moreover, no convincing mechanism is known yet which lets start reconnection from scratch, i.e. from the thermal fluctuation level. The reason is that magnetic field are fairly rigid against bending for which one needs high-β conditions. On the other hand, in field which are very weak such that they do not play any dynamical role, reconnection is not of vital interest. This case does, however, not apply to the geophysically interesting applications at the magnetopause and in the magnetospheric tail. It would still be possible that it is the 'tearing mode' which generates many 'reconnection islands' which may then mutually interact due to their different dynamics, may overlap and favour one or more of these islands along the current sheet to produce an extended reconnection X-line. Finally (18) reconnection needs probably driving or forcing which in the Earth's case at the magnetopause is given by the persistent inflow of plasma from the solar wind into the magnetosheath and towards the magnetopause.

    In addition to these general difficulties, reconnection at the magnetopause provides the already noted extra difficulty of being asymmetric by nature. Any simulation which intends to be applied to magnetopause reconnection must thus include asymmetry with weak magnetosheath magnetic fields and strong magnetospheric magnetic fields. Figure 12.11 shows the schematic results of such a simulation still not taking into account guide fields, inhomogeneity and forcing and being strictly two-dimensional. The asymmetry of the simulation is seen in the different densities of magnetic field lines to both sides of the magnetopause boundary. There are a few interesting conclusions which can be drawn from this picture of reconnection in its quasi-final state. The first conclusion is that a boundary between reconnected and un-reconnected fields remains intact after reconnection. It can be defined as the next-to the inner separatrix lying flux tube on the strong field side. Second, the reconnected fields are all located outside of this boundary which de facto is an inner magnetopause. Third, however, on the low field side one finds the expected and well-expressed Hall-magnetic field structure which now, due to the compression of the inner separatrix, is just dipolar and not anymore quadrupolar. Fourth, this region of this Hall field is already connected to a limited number of field lines from the high-field side as is seen from the broad X-line region which is roughly $\Delta z \sim 3\lambda_i$ of magnetosheath ion inertial scales wide. Fifth, in application to the magnetosphere, these field line have their feet in the dayside auroral region. Inflow of particles along the field will cause dayside aurora.

Similar simulations including one-sided driving by the magnetosheath flow are at variance with a few of the above results. Driving, as expected, increases the reconnection rate by some factor of the order of a few. Another result which could have been expected in driving reconnection by forcing inflow is that now the 'ion diffusion' region is compressed on the magnetosheath side while the Hall region breaks somewhat in into the magnetosphere to the strong field side re-establishing the quadrupolar structure of the Hall field and making the entire reconnection more similar to the symmetric case. The last unaffected 'magnetopause' field line is pushed farther in. Thus, the reconnection boundary layer which in the un-driven case was broad and entirely on the magnetosheath side with dipolar Hall field has now become narrower while being extended into the magnetosphere with quadrupolar Hall field and stronger Hall-aligned electric field. The flow fields are shown in Fig. 12.12 exhibiting the asymmetry of flow in ions and electrons.

Finally, we comment on the effect of guide fields on magnetopause reconnection. These have recently been included into reconnections of the above kind. There are two kinds of guide fields, those along the current layer and those transverse to it. As expected the assumption of a guide field along the current layer, which is the case of interest for the magnetopause, distorts the quadrupolar symmetry of the Hall fields with respect to north and south for reconnection at the magnetopause nose. The Hall currents become strongly amplified on one side and depleted on the other depending on the direction of the guide field. In addition the electron flow component along the magnetic field causes another contribution to the magnetic field.

Similar effects are seen in the electric fields and electron fluxes which all become asymmetric. In addition, however, the strong electron fluxes readily undergo instability and form electron vortices. This indicates that in the presence of guide fields reconnection at the magnetopause can only in the time average be in a steady state. Otherwise it is time dependent on the scale of generation and disappearance of the electron vortices. Such electron vortices imply that single magnetic flux tubes of the size of the electron vortices behave independent. Spatially localised flux tubes of this kind have long been observed and have been named *flux transfer events* or FTEs. They have always been consider of being the signature of time-dependence of reconnection in addition to spatial limitation and three-dimensionality. Their observation in two-dimensional simulation of forced-guide field reconnection might be a hint on their nature. It is hence clear that strong effects are caused on the evolution of reconnection at the magnetopause in the presence of guide fields when combined with forcing by driving the plasma inflow.

12.4 The Reconnecting Magnetotail Current Sheet

In contrast to magnetopause reconnection the magnetotail presents a much cleaner picture of reconnection. It is to good approximation symmetric with respect to the normal direction of the neutral sheet Harris current layer if one neglects the bending of the plasma sheet caused by the inclination of Earth's axis and by the impact of the solar wind on global scales much larger than the micro- or meso-scales of reconnection. Reconnection here is, in contrast to the magnetopause, mostly weakly driven only. On the other hand, it is subject to variations along the plasma sheet, is not free of a normal magnetic field component, possibly contains guide fields and is embedded into a broad plasma sheet which sometimes thins in an irregular and not well known way. Another peculiarity of the tail is that it is generally highly vulnerable to changes in the solar wind and internal changes in the magnetosphere which affect the form and stability of the neutral sheet current. This implies that it is highly time variable. Finally, the magnetic field on the side of Earth is frozen into the body of Earth while on the side of the tail the field is about open, quite weak and thus much more vulnerable than at the magnetopause where solar wind compression increases its strength and partially stabilises it. This fixing field lines on one side and leaving them open on the other causes an anisotropy in tail-reconnection which is unlike the quasi-symmetry at the dayside magnetopause. It inhibits plasmoids from penetrating deep into the inner magnetosphere while on the tailward side any plasmoid generated in the reconnection process can easily form and become expelled from the magnetotail into the solar wind causing loss of mass from the magnetosphere and compensates for the addition of mass from the solar wind on the dayside or in high latitudes.

These properties of tail reconnection, in particular its asymmetry with respect to the Earth-anti-Earth direction and its high time-variability make it much more difficult than at the magnetopause to identify reconnection in situ and to monitor its properties. Usually the X-line in the magnetotail current sheet is not stationary but moves along the current layer readily shifting away from the spacecraft location. Though the mere number of observations in Earth's magnetotail is larger than observation of reconnection at the magnetopause and though most of them relate in one or the other way to reconnection in the tail current layer and plasma sheet, direct evidence for crossing an X-line and the amount of data collected is not overwhelmingly large.

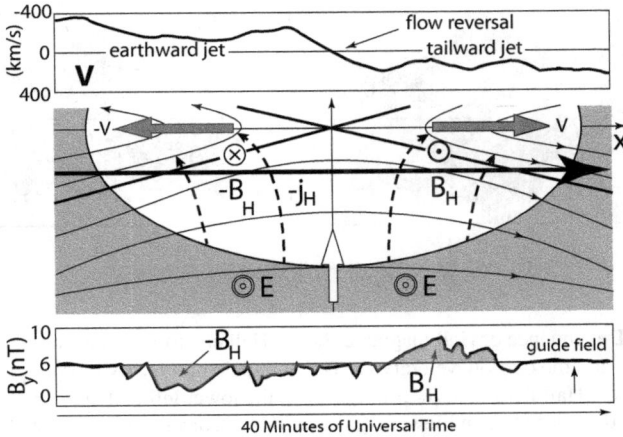

Figure 12.13: Wind spacecraft observations of reconnection in the tail current sheet combined with the general Hall reconnection picture. The long dark arrow below the X-line is the spacecraft path which is along x. *Top panel*: The observed flow velocity indicating plasma jetting to both sides of the reconnection site. The spacecraft has crossed part of the outflow regions with flow reversing from earthward to tailward. *Bottom panel*: The out of plane magnetic field component showing the dipolar Hall field structure in the lower part of the current sheet. The Hall field oscillates around a finite value of $B_y \approx 6$ nT which is a strong indication of the presence of a stationary guide field in tail reconnection.

12.4.1 Observations

The quiet magnetotail provides little chance of observing reconnection in the tail. Reconnection under quiet condition proceeds in the distant tail at a closest distance of, say, $x \sim 30$ R_E. A spacecraft at such distance has a long orbital period, and passing by chance across the distant X-line has a low probability. On the other hand, the tail current sheet is quite unstable against reconnection which is reflected in the relatively high occurrence rate of substorms. Under substorm conditions, as we have mentioned in Section 5.6, according to theory a new secondary X-line forms in the neutral sheet current layer at distances between 10 $R_E < x < 20$ R_E which are well accessible to spacecraft. Hence the information on tail reconnection is provided solely from substorm observations. These are, however, highly non-stationary. Nevertheless, there have been some key observations in the near past from which it could be concluded that reconnection does indeed take place in the near-Earth current sheet, being the main responsible for the occurrence of substorms and in addition unambiguously showing that substorm reconnection in Earth's magnetotail is collisionless. We have already noted that for this to show one needs to demonstrate that the reconnection distinguishes between electrons and ions in such a way that Hall currents and Hall fields are created. After some preliminary hints on this being actually the case, it could be demonstrated ultimately ten years ago by two spacecraft, Geotail and Wind. Neither of

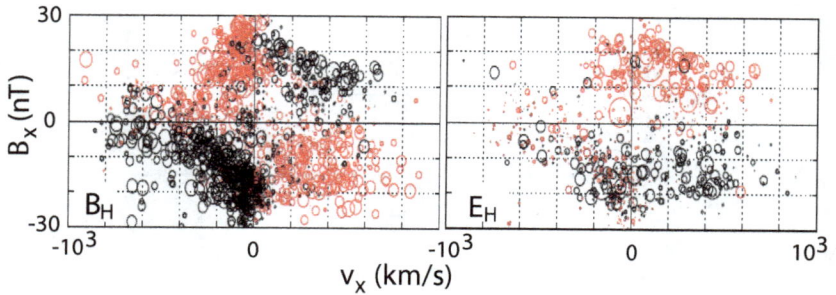

Figure 12.14: Dependence of Hall magnetic B_y and Hall electric E_z fields on the primary reconnection field B_x and the outflow velocity v_x. The magnetic plot nicely reproduces the quadrupolar structure of the Hall field with $B_y > 0$ (black) in the lower left and upper right quadrants and $B_y < 0$ (red) in the upper left and lower right quadrants. Correspondingly the Hall electric field is positive in the lower and negative in the upper half plane. For large outflow velocities the Hall region is at small initial fields and closer to the centre of the current layer at $B_x = 0$.

these spacecraft crossed the neutral line; they, however, as shown in Fig. 12.13 passed two feet of the quadrupolar Hall magnetic field and, in addition observed the opposite jetting of plasma during such a passage. Figure 12.14 then shows the Hall magnetic and electric field observations of a large number of similar passages by the Cluster spacecraft in dependence on the initial anti-parallel tail magnetic field amplitude and the observed earthward or anti-earthward jetting of the reconnected plasma.

This figure nicely shows that, statistically, the quadrupolar geometry of the Hall magnetic field is also well reproduced. It turns out that for large initial antiparallel fields $\pm B_x$ the jetting velocities $\pm v_x$ are small, while high jetting speeds correspond to small initial B_x. It thus seems that small fields have a higher acceleration rate in reconnection than high fields. The Hall electric field in the tail turns statistically out to dominate over the convection electric field. This is a surprising result.

Another important result in the Wind and Geotail observations is that the geomagnetic tail seems to frequently contain quite a strong guide field. In the case of Wind this guide field was of the order of $B_y \sim 6$ nT. Its nature is not quite clear because the tail seems to be about symmetric on the small scale of the reconnection site.

Nevertheless the reality of the guide field cannot be denied. The most probable interpretation is that it results from a distortion of the tail symmetry by twisting or by some undulation of the current sheet on a larger scale. Otherwise it could also be attributed to the free penetration of the Parker spiral magnetic field from the solar wind into the magnetospheric tail around the ecliptic plane in the course of formation of the tail current layer even though no viable mechanism is known which could be made responsible for a free penetration of this kind.

Independent of the nature of the guide field its mere presence shows that the naive assumption of total symmetry of the tail current sheet and thus reconnection is probably not justified.

12.4.2 Magnetotail-relevant Simulations

The differences between tail and magnetopause reconnection have already been noted. Some of the properties of magnetopause reconnection can be transferred also to tail reconnection: the onset of reconnection when the current sheet becomes thin of the order of λ_i, generation of Hall fields and the case of forced reconnection. We will not enter into a discussion of the high variability of tail reconnection, the fast motion of the neutral line along the tail and dipolarisation. These questions belong into the wide and unresolved field of the substorm the discussion of which would fill at least an own chapter and, though very important for magnetospheric physics, is not a genuine space plasma physics problem. Rather it presents a whole complex of problems interwoven into the fabric of the substorm phenomenon.

The remaining main questions of tail reconnection are two-fold. (1) How can a magnetic field enter into the central current sheet in order to ignite reconnection? (2) What is the effect of the residual positive normal component B_z of the tail magnetic field inside the distance between Earth and the distant X-line? The first question has not been clarified by simulations which in all cases have been initiated artificially, in the collisionless case mostly by imposing the presence of a seed X-line in the centre of the current layer. Any other initiation corresponds to the assumption of some initial localised dissipation like anomalous resistance or ion viscosity. The second question is vital for tail reconnection. When — for instance at substorm onset — a new near-Earth X-line develops then it has to deal with reconnection at finite $B_z > 0$ simply because the near Earth magnetic field lines are the stretched geomagnetic field lines which close across the plasma sheet and neutral current sheet in the tail.

Answering the first question — neglecting for the moment any finite B_z — is possible when the presence of a guide field is assumed. In this case the current sheet electrons become accelerated by the cross tail electric field along the guide field. The PIC simulations have shown that this acceleration takes place already in the Hall region. The electrons readily reach a high speed in this process along the magnetic field, causing a temperature anisotropy $T_\parallel/T_\perp > 1$ along the guide and Hall fields. Under such conditions the Weibel instability takes over and generates a very low frequency field with components perpendicular to the original guide and Hall fields. This field is thus in the direction of the original antiparallel field. Since it has short wavelengths of the order of the electron inertial length λ_e it causes vortices in the centre of the tail current layer which act like a chain of seed-X- lines and can initiate reconnection. It is interesting that this kind of chain concentrates the tail current in current filaments with the magnetic field evolving into magnetic flux ropes. Thus the reconnected structure of the tail current sheet becomes very complicated.

Reconnection including a finite $B_z > 0$ has been the subject of vital analytical approaches in an attempt to get an idea how reconnection can proceed at all under such conditions. We have reviewed this theory in our companion volume. It partially stabilises the reconnection and tearing mode which needs to overcome the magnetisation of the electrons in the field B_z. Here, we briefly review recent simulations on this case.

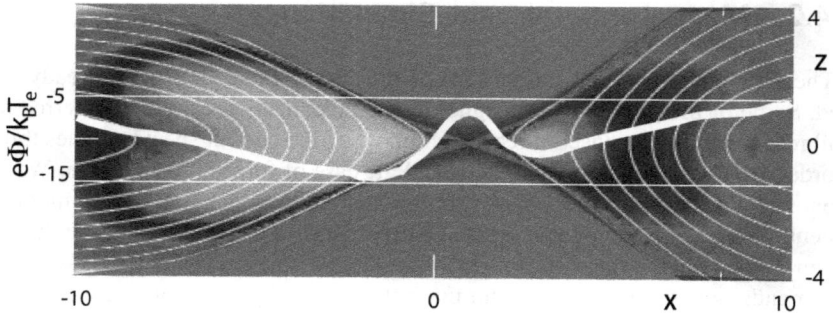

Figure 12.15: The normalised electric potential energy in the vicinity of the X-line in the simulation with finite $B_z > 0$. The potential energy is normalised to $k_B T_e$ and is negative. Due to the asymmetry introduced by the finite B_z the potential is asymmetric in x. The overlaid curve is the measured value of the potential along x.

The full particle two-dimensional PIC simulations relevant for the tail case are performed with the Harris-sheet model introduced in Eqs. (12.27)–(12.29). This simulation thus includes forcing of tail reconnection by a switched-on external electric field, this time symmetric to both sides of the current layer. After switching on during a brief period the field is held constant. This causes a continuous thinning of the current layer until reconnection initiates. In addition an initial $B_z > 0$ is organically included into the simulation from beginning through Eq. (12.27) in modelling the residual northward dipole field component. An important point is that the mass ratio in these simulation is assumed about realistic $m_i/m_e = 1600$. In addition open boundary conditions are imposed in x. Both particles and fields can freely cross the box in x direction with the lost particles replaced by the inflow assuming an inflowing Maxwellian. The main results of such simulations have already been mentioned in the Subsection on Driven Reconnection.

Figure 12.15 shows the normalised electric potential energy $e\Phi/k_B T_e$ in grey scale after reconnection has evolved. The potential is restricted to the close vicinity of the X-line and is non-symmetric, being larger ion the left which is the earthward direction. The potential is negative. The heavy white line is the potential along x through the centre. One recognises the two negative though asymmetric excursions to the left and to the right of the X-line with the asymmetry being caused by the presence of $B_z > 0$.

Further interesting results of this kind of simulation are the occurrence of a strong electron temperature anisotropy $1 < T_{e\parallel}/T_{e\perp} < 8$ found on the inflow side in a narrow region close to the X-line with exclusion of the X-line itself. Obviously electrons are heated in this small region by a strong parallel electric field. This field has a well expressed quadrupolar structure each of its feet extending $< \lambda_i$ in the x direction measured from the X-line and being restricted in z to a fraction of λ_i only. Already earlier we noted that these simulations show that driven reconnection including a finite B_z cannot be stationary. They form electron layers which are ejected into the direction of

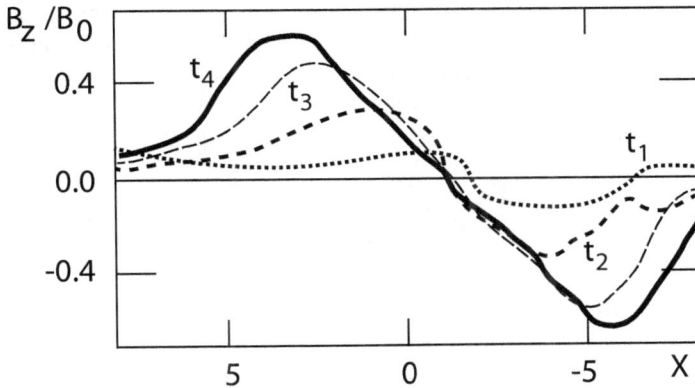

Figure 12.16: Evolution of 'dipolarisation fronts' in the simulation with finite $B_z > 0$. The evolution of the field B_z is shown at four consecutive times. These fronts move to the left in the direction of Earth.

$-x$ forming progressing regions containing high B_z similar to 'dipolarisation fronts'. This is of considerable interest for tail reconnection and the theory of substorms.

Figure 12.16 shows the evolution of the fronts in B_z for four consecutive times. The normal component has been normalised to the initial external antiparallel field. Its initial value was small as still seen at time t_1. At time t_4 it has grown to more than 50% of the initial field by maximum amplitude.

12.5 Summary

Summarising our knowledge of reconnection today we should mention that reconnection theory and observation are still in fast evolution. In the present chapter we are far from having given an exhaustive account of our present knowledge on reconnection. The future will be taking into account all the effect we have listed here in addition to three-dimensionality and the use of very large simulation boxes with many more particles whose orbits will be calculated. Both simulations and observations have, however, convinced us that reconnection in magnetised plasmas is a real effect. It takes place at a violently fast speed under completely collisionless conditions being much faster there than in the presence of resistivity which the first models of reconnection assumed for sure. But at the same time it is much more complex than originally believed. It not only serves for reconnecting magnetic fields, it changes topology and accelerates plasma, it connects initially unconnected regions, causes mixing of plasma of different kinds, and it leads to many other effects like the acceleration of electrons, Hall currents, twisting of magnetic fields, excites various kinds of waves, causes electron holes and field-aligned currents and becomes involved into aurora on the magnetised planets and into flares on the sun. Its small-scale-physical effects however, are not restricted to local regions. Reconnection has violent global effects like driving convection and causing substorms. Simulations have not yet proceeded to the proper inclusion of the micro-scale physics of reconnection into global simulations. This is a difficult task which is the great goal for future investigation of reconnection in order to apply it to the magnetosphere.

Further Reading

The first excellent and comprehensive review of the old reconnection theory is found in [6]. It reviews the four basic theories that were available at that time: the resistive Sweet-Parker-model of elongated current sheets, the resistive Petschek model of narrow current sheets, the Sonnerup model and the Yeh-Axford model showing that these models were consistent with fluid theoretical assumptions and arguments and with observations as available at that time. It did, however not account for neither the Hall effects nor the inconsistencies of all these models which came about by more precise observations and numerical particle simulations. A quarter century later the monograph [3] published the most complete collection of MHD simulations on these models as a handbook for applications but not going further than [6]. An excellent brief and unique view on reconnection as a problem where electrons and ion behave differently though not in the ordinary and common treatment of fluid theory but as an instability boundary value problem of collisionless tearing modes is contained in [4]. Though this model contains many very interesting ideas and insight, it has until recently never been considered further in the literature; it has been completely replaced by numerical simulations. The Hall effect has been introduced into reconnection first in [5] as an important step ahead. An account of (mainly) fluid simulations of reconnection including the Hall effect and distinction between electrons and ions though being more oriented on application in the laboratory and fusion research is contained in [1]. A recent review of reconnection rather from the laboratory than theoretical point of view and emphasising mainly the various fluid approaches with focus on anomalous processes while just sketching the simulational approach to collisionless reconnection is contained in [7]. The space aspect is only mentioned in passing in a few places in this review which is, what concerns the theory, not quite modern emphasising mainly the fluid aspects of reconnection but providing an extended bibliography.

To the bibliography below we added a number of more recent papers on reconnection which the reader might consult if he wants to inform himself about particular problems and applications beyond what is contained in the main bibliography. The list of these papers is by far not complete; it is however contemporary, being at the front of the intense research in reconnection. Most of the papers refer to numerical simulations. Some, however, are mainly observational. We have been very selective from our view and do not cover the whole literature.

References

[1] D. Biskamp, *Magnetic Reconnection in Plasmas* (Cambridge University Press, Cambridge, UK, 2000).

[2] F. S. Mozer and P. L. Pritchett, *Magnetic field reconnection: A first-principles perspective*, Phys. Today June issue 34–39 (2010).

[3] E. R. Priest and T. Forbes, *Magnetic Reconnection - MHD Theory and Applications* (Cambridge University Press, Cambridge, UK, 2000).

[4] R. Z. Sagdeev, *The 1976 Oppenheimer Lectures: Critical problems in plasma astrophysics. II. Singular layers and reconnection*, Rev. Modern Physics **51**, 11–20 (1979). (1989).

[5] B. U. Ö. Sonnerup, in *Solar System Plasma Physics*, Vol. III (L. J. Lanzerotti, C. Kennel, and E. Parker, eds., North-Holland Publ., New York, 1979) pp. 45–108.

[6] V. M. Vasyliunas, *Theoretical models of magnetic field line merging, 1.*, Rev. Geophys. Space Physics **13**, 303–336 (1975).

[7] M. Yamada, R. M. Kulsrud and H. Ji, *Magnetic Reconnection*, Rev. Modern Physics **82**, 603–664 (2010).

Additional selected more specialised references

[8] M. André *et al.*, *Ann. Geophys.* **19**, 1471–1481 (2001).

[9] Y. Asano *et al.*, *J. Geophys. Res.* **109**, A02212 (2004).

[10] Y. Asano *et al.*, *J. Geophys. Res.* **113**, A01207 (2008).

[11] S. D. Bale, F. S. Mozer, and T. Phan *Geophys. Res. Lett.* **29**, 2180 (2002).

[12] H. Che, J. F. Drake, and M. Swisdak, *Nature* **474**, 184–187 (2011).

[13] W. Daughton *et al.*, *Nature Phys.* **7**, 539–542 (2011).

[14] J. F. Drake *et al.*, *Science* **299**, 873–877 (2003).

[15] J. F. Drake, M. A. Shay, and M. Swisdak, *Phys. Plasmas* **15**, 042306 (2008).

[16] J. P. Eastwood *et al.*, *J. Geophys. Res.* **115**, A08215 (2010).

[17] H. U. Frey, T. D. Phan, S. A. Fuselier, and S. B. Mende, *Nature* **426**, 533–537 (2003).

[18] M. Fujimoto *et al.*, *Geophys. Res. Lett.* **24**, 2893–2896 (1997).

[19] M. Fujimoto, I. Shinohara, and H. Kojima, *Space Sci. Rev.* **24**, 10.1007/s11214-011-9807-7 (2011).

[20] F. S. Mozer, S. D. Bale, and T. D. Phan *Phys. Rev. Lett.* **89**, 015002 (2002).

[21] T. Nagai *et al.*, *J. Geophys. Res.* **106**, 25 929–25 949 (2001).

[22] R. Nakamura *et al.*, *J. Geophys. Res.* **111**, A11206 (2006).

[23] R. Nakamura, W. Baumjohann, A. Runov, and Y. Asano, *Space Sci. Rev.* **122**, 29–38 (2006).

[24] R. Nakamura *et al.*, *Adv. Space Res.* **36**, 1444–1447 (2005).

[25] R. Nakamura *et al.*, *J. Geophys. Res.* **113**, A07S16 (2008).

[26] M. Øieroset *et al.*, *Nature* **412**, 414–417 (2001).

[27] G. Paschmann *et al.*, *Nature* **282**, 243–246 (1979).

[28] G. Paschmann *et al.*, *Ann. Geophys.* **23**, 1481–1487 (2005).

[29] A. A. Petrukovich, W. Baumjohann, R. Nakamura, and H. Rème, *J. Geophys. Res.* **114**, A09203 (2009).

[30] T. D. Phan *et al.*, *Nature* **404**, 848–850 (2000).

[31] T. D. Phan *et al.*, *Geophys. Res. Lett.* **30**, 16-1–16-4 (2003).

[32] P. L. Pritchett, *J. Geophys. Res.* **106**, 3783–3789 (2001).

[33] P. L. Pritchett, *J. Geophys. Res.* **113**, A06210 (2008).

[34] P. L. Pritchett and F. S. Mozer, *J. Geophys. Res.* **114**, A11210 (2009).

[35] A. Retinò *et al.*, *J. Geophys. Res.* **113**, A12215 (2008).

[36] A. Runov *et al.*, *Geophys. Res. Lett* **36**, L14106 (2009).

[37] A. Runov *et al.*, *J. Geophys. Res.* **116**, 5216 (2011).

[38] A. Runov *et al.*, *Geophys. Res. Lett* **30**, 1036 (2003).

[39] M. Scholer *et al.*, *Phys. Plasmas* **10**, 3521–3527 (2003).

[40] V. Sergeev *et al.*, *Geophys. Res. Lett.* **36**, L21105 (2009).

[41] M. A. Shay *et al.*, *Geophys. Res. Lett.* **26**, 2163–2166 (1999).

[42] M. I. Sitnov, P. N. Guzdar, and M. Swisdak, *Geophys. Res. Lett.* **13**, 1712–1715 (2003).

[43] M. I. Sitnov, M. Swisdak, and A. V. Divin, *J. Geophys. Res.* **114**, A04202 (2009).

[44] B. U. Ö. Sonnerup *et al.*, *J. Geophys. Res.* **86**, 10049–10067 (1981).

[45] T. Terasawa, *Geophys. Res. Lett.* **10**, 475–478 (1983).

[46] R. A. Treumann and W. Baumjohann, *Front. Physics* **1**, 31 (2013).

[47] M. Yamada *et al.*, *Rev. Modern Physics* **82**, 603–664 (2010).

— 13 —

Collisionless Shocks

This last chapter of the present volume is devoted to one of the most important applications of kinetic wave theory in space plasma physics: the physics of collisionless shocks. In previous chapters we have several times talked already about shocks in plasmas, there however only from the viewpoint of fluid theory which cannot give any clue of the interior, generation or general behaviour of shocks and all the processes it may trigger in the environmental plasmas. This deficiency we will resolve — at least in part — in the present chapter where we are going to apply some of the knowledge we have accumulated by now in the three last chapters on the complicated physics of waves in a collisionless plasma. One may, in this context, at the beginning stumble across the term *collisionless* shock which seems to be a contradiction in itself. If something is shocked then probably by collision, isn't it? True, this may hold in real life. But plasmas exhibit a strange behaviour as we have shown in many cases in as far as they behave as if collisions would act in the plasma though there are no collisions. The reason, we have elucidated, is that plasmas contain long-range interactions which are provided by the electromagnetic field acting on the charged particle component. These long-range actions provide correlations and force the particle into enslavement. Particles are not really free in plasmas; they have to follow the dictatorship of the fields, even when these fields have been generated by the particles themselves as a consequence of their own dynamics under the action of the Lorentz force.

Shocks in a collisionless plasma can evolve in two ways. One is the case we are already familiar with when a large obstacle stands in the flow which the flow is unable to push away. If the Mach number of the flow as seen in the frame of the obstacle is $M_{ms} > 1$ then the flow will evolve a shock wave in front of the obstacle. In principle this is also the case when a blast wave expands into the flow from an explosion like a solar flare or solar mass ejection, the impact of a large asteroid or comet on one of the planets or an atomic explosion. One then only has to change the frame in order to see the similarity. The driver of the blast is the extended obstacle whose relative speed to the flow is super-magnetosonic and the flow heads against the driver creating the blast shock in front of the driver.

We have also become familiar with the other possibility when speaking of instability and wave growth. This growth of the wave may be local only, and then the wave itself evolves into a large-amplitude localised entity with steep flanks which may exhibit shock behaviour. In the other case, when the wave is substantially faster or slower than the flow, it presents an obstacle to the flow. The last case happens for instance when a wave flows against a fast flow with speed large enough to exceed the flow speed even slightly and thus has large Mach number in the flow frame. In both

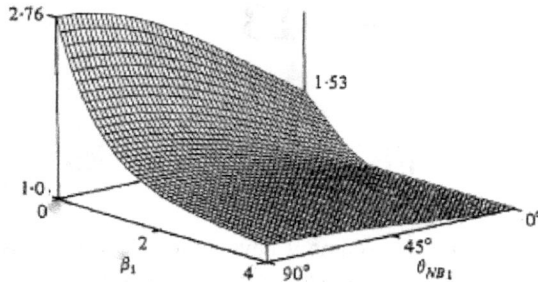

Figure 13.1: The dependence of the critical magneto-hydrodynamic Mach number for dissipative shocks on the upstream shock-normal angle Θ_{Bn} and on the upstream plasma β_1. The critical Mach number is a strong function of the latter and maximising Θ_{Bn} for perpendicular shocks $\Theta_{Bn} = 90°$ and cold upstream flow which can become quite small otherwise.

cases shocks evolve as well. In the latter case, however, those shocks will be relatively slow, just exceeding the Mach number bound above which shocks evolve. Such shocks are *subcritical* in contrast to blast wave shocks which are mostly *super-critical*. We may thus distinguish between these two classes. In the following we will briefly discuss subcritical shocks before switching to the more interesting super-critical shocks the discussion of which will occupy most of this chapter and of which the Earth's bow shock is the most distinguished and best investigated example in the entire Universe.

13.1 Subcritical Shocks

An important property of collisionless shocks is their criticality. Subcritical shocks have Mach numbers $M < M_c$ less than critical. Criticality refers to the capability of a shock wave to maintain its shock character solely by resistive (Joule) dissipation. Since collisionless shocks are free of binary interaction between the particles, criticality is defined with respect to some Joule dissipation. In this sense criticality is defined implicitly; a collisionless shock that does not generate any Joule dissipation is supercritical for all Mach numbers $M > 1$. Usually, a collisionless shock generates some weak dissipation, however. Therefore, it makes sense to define a largest critical Mach number M_c by assuming that the Mach number downstream of the shock should not be larger than unity implying that the downstream plasma cannot be further heated by any dissipation. Hence, from the magneto-hydrodynamic Rankin-Hugoniot jump relations putting the downstream Mach number $M = 1$ defines the critical upstream Mach number. The numerically determined *largest* critical Mach number is

$$M_c \approx 2.76 \tag{13.1}$$

The dependence of the critical Mach number on plasma β and Θ_{Bn} is shown in Fig. 13.1. Critical Mach numbers are small and decrease with increasing shock normal angle Θ_{Bn}. Some of the shocks of very low Mach number in the solar wind are

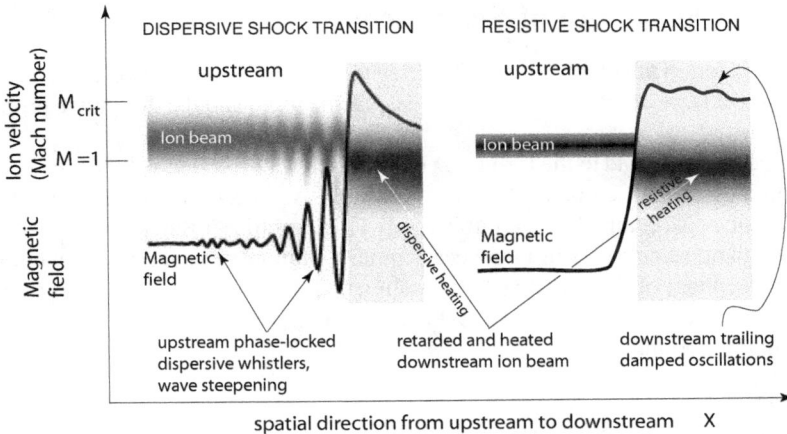

Figure 13.2: Two types of subcritical shocks in ion phase space and magnetic field profile, dispersively dominated (left) and dissipatively dominated (right). The dispersive shock has upstream whistlers attached, being a nonlinear whistler itself. The upstream ion flow is sub-critical. The flow retardation in the shock is dominated by dispersive effects. In resistive shock transition the shock forms a steep ramp lacking upstream oscillations. Heating and retarding is due to Joule heating. Downstream the magnetic field may evolve into trailing oscillations.

subcritical, but in general the shocks in the heliosphere and in particular Earth's bow shock are super-critical with their Mach numbers exceeding the critical Mach number. For instance the bow shock has canonical Mach number of $5 < M < 12$ with the most probable value around $M = 8$.

In subcritical shocks the cooperation of dissipation inside the shock front and dispersion of waves is sufficient for providing the necessary dissipation/dispersion of sustaining the shock transition from upstream to the downstream flow. For illustration, Fig. 13.2 shows two schematic subcritical shock profiles, one of them a dispersion dominated shock transition resulting from nonlinear steeping of waves (in this case whistlers), the other the extreme case of a purely resistive shock transition.

Dispersion alone cannot create a shock transition. It produces some kind of wave that is localised like the various kinds of solitary waves and BGK modes we will briefly note below. These are all structures that are connected with local electric potential wells. Depending on the polarity of these potentials they reflect one sort of particles out of the upstream low energy component while they trap and accelerate particles from the other component. Hence, in the complete absence of any kind of dissipation a subcritical 'shock' can exist only when it reflects and traps some particles. Then, however, it is a marginal case of shock, a non-dissipative structure which by definition is reversible. Such structures belong to the family of solitary waves. If this is not the case, the subcritical shock must be capable of generating some collisionless dissipation even when it is dominated by dispersion like in the first case shown.

13.1.1   Mechanism of Subcritical Shocks

The evolution of subcritical shocks is due to the competition between the nonlinear steeping of a large amplitude low frequency plasma wave and the dispersive properties of the plasma.

Dispersion can lead to the formation of localised waves called *solitons*. In a medium that contains a small amount of dissipation the waves become weakly damped, and these localised structures evolve into a ramp which mimics a shock. It is important to note that the dissipation is in many cases quite unimportant as long as the amplitude and the steepness of the wave packet remain small. However when both increase, the gradient scale enters the scale of local dissipative interactions, and dissipation starts becoming important.

Wave Steeping in Dispersion Dominated Shocks

There is no problem with the understanding of dispersion in plasma. Shock waves grow out of waves that are generated in the impact of the super-magnetosonic flow onto the obstacle. The obstacle reflects ions back upstream thus creating an ion-ion beam situation which is unstable with respect to low-frequency magnetosonic waves propagating upstream with velocity V away from the obstacle. Under collisionless conditions their velocity evolves with time according to

$$dV/dt = (\partial V/\partial t) + V \partial V/\partial x = 0 \tag{13.2}$$

During propagation, the main effect on the shape of a sinusoidal disturbance $V \sim \sin k(x - c_{ms}t)$ comes from the action of the nonlinear term. This term can be written as $Vk\cos k(x - c_{ms}t)$. Inserting for V this becomes $\sim \frac{1}{2}\sin 2k(x - c_{mc}t)$.

Harmonic sidebands of half wavelength and half the amplitude are generated which, by the same mechanism, also generate sidebands on their own now at quarter original wavelength and amplitude, and so on, with increasingly shorter wavelengths. The total amplitude is the superposition of all these sideband harmonics which propagate at the same magnetosonic velocity c_{ms}. They superimpose locally and add to the wave amplitude, causing the wave to steepen until the gradient lengths become so short that dissipation takes over. If this does not happen, the wave will turn over and break like water waves do when running against the beach. In the other case when dispersion lets the shorter wavelength waves escape from the transition, wave steeping is arrested at a maximum wave amplitude and minimum wavelength.

We refer to a one-dimensional cold ($\beta_1 \ll 1$) two fluid damped model of a stationary magnetosonic wave in pressure balance

$$\nabla_x B_z = \mu_0 env_y, \qquad \nabla_x E_y = 0, \qquad \nabla_x nv_x = 0 \tag{13.3}$$

$$\nabla_x \left(\frac{1}{2} m_i nv_x^2 + \frac{B_z^2}{2\mu_0} \right) = 0, \qquad m_e nv_x \nabla_x v_y = -enE_y + env_x B_z - v_{an} m_e nv_y$$

We now define $\xi = x/\lambda_e$. Transformation to the shock velocity frame V_{sh}, one obtains

$$\frac{d^2 b}{dx^2} = b - 1 + Ab(1 - b^2) - a\frac{db}{dx} \qquad (13.4)$$

with $b(x) \equiv B_z(x)/B_1$, the ratio of the magnetic field to the upstream magnetic field, $A \equiv B_1^2/2\mu_0 nm_i V_{sh}^2$, and $a \equiv v\lambda_e/V_{sh}$. This equation can be identified as the equation of motion of a hypothetical particle with coordinate b and time ξ, including frictional dissipation in the first derivative. In the absence of dissipation one defines the Sagdeev pseudo-potential $S(b)$

$$2S(b) = (b-1)^2 \left[A(b+1)^2 - 2 \right], \qquad b < b_{max} = (2V_{sh}/V_{A1}) - 1 \qquad (13.5)$$

Negative values of $S(b)$ constitute a potential trough for the hypothetical particle if the field amplitudes $b < b_{max}$ are smaller than a maximum value b_{max}. For a given $S(b)$ the hypothetical particle performs a stationary oscillation in this potential trough with amplitude equal to the distance between the walls. The shape of the trough is shown as the heavy line marked $s = 0$ on the right in Fig. 13.3 for the special case $M_A = 1.6$. The maximum possible amplitude is reached for $S(b) = 0$. The ratio $V_{sh}/V_{A1} = \mathcal{M}_{sh}$ is the shock-Mach number, i.e. the Mach number of the possible stationary solutions in this dissipationless case. Including the resistive damping, this amplitude decreases during the oscillation and the pseudo-particle will ultimately settle in the final state at the minimum of the pseudo-potential $S(b)$. The Sagdeev potential minimum is at the stationary downstream value of the normalised magnetic field

$$2b(S_{min}) \equiv 2b_2 = \left(1 + 8M_{sh}^2\right)^{\frac{1}{2}} - 1 \qquad (13.6)$$

It corresponds to the final stationary shock state of a subcritical fast magnetosonic shock. It is independent of the dissipation even though it has been reached only due to the action of the anomalous dissipation a.

A shock of this kind is a weak shock since the plasma pressure contribution has been neglected and the plasma has been assumed to be cold such that the heating of the plasma by the shock itself is also small and does not appear anywhere. The shock profile can be determined from the solution of the equation for b. It is found to be a spatial oscillation reaching maximum at the shock ramp and decreasing exponentially behind the shock with spatially damped amplitude $b(\xi) \sim \exp(-a\xi)\sin(\xi\sqrt{M_{sh} - 1})$. Because of the simplifying assumptions this solution holds only for small shock amplitudes. At large dissipation one obtains instead a non-oscillating shock ramp, the cases shown in Fig. 13.2.

13.1.2   Causes of Anomalous Dissipation

Dissipation sets on when the scale of steepness of the wave compares to the dissipation scale L_d. Then, in the evolution equation of the velocity the next higher order derivatives with respect to x cannot be neglected anymore, yielding

$$dV/dt = (\partial V/\partial t) + V\nabla_x V = \nabla_x D\nabla_x V - \beta\nabla_x^3 V + \cdots \qquad (13.7)$$

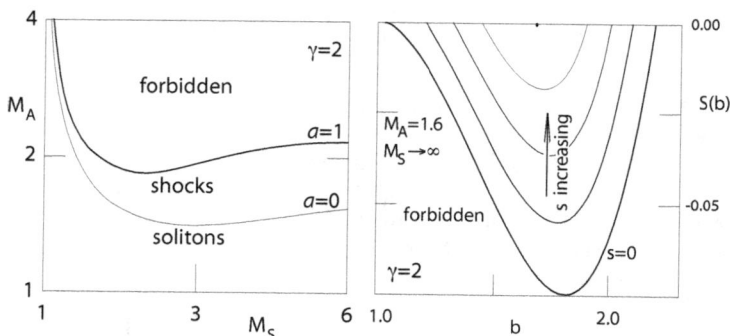

Figure 13.3: *Left*: The allowed regions in (M_A, M_s)-space for solitons and shocks depending on the dissipation coefficient a. Subcritical shocks exist only in the narrow domain between the two curves are for $0 < a < 1$. *Right*: The shape of the Sagdeev pseudo-potential $S(b)$ for $M_A = 1.6, M_s \rightarrow \infty$ as function of the entropy density s. For $s = 0$, corresponding to $a = 0$, the Sagdeev pseudo-potential has its largest excursion into the negative domain. For this case, solitons exist throughout the entire region inside the curve. With increasing s the domain shrinks, and no soliton solutions exist anymore, being replaced by shock solutions. A shock starts at one of those curves and wanders upward in the diagram until it reaches the maximum entropy point on the $S = 0$ axis (black dot).

now with non-vanishing right-hand side. The first of the new terms is second-order in the gradient ∇_x and corresponds to the anomalous diffusion. The second (third-order in ∇_x) term describes the higher-order dispersion. One may recognise this equation as the shock model equation (13.5) if taking the stationary case and integrating one time with respect to space x. It then leads to the definition of the Sagdeev pseudo-potential.

At the example of the above equation one can discuss the effects of dissipation and dispersion. If diffusion dominates, entropy is produced, and steeping ceases, a ramp evolves, and the wave turns into a shock. If dispersion dominates, the shorter wavelength waves run away and a stationary wave packet of finite size is produced which is not a shock, but can, under certain circumstances, also evolve into a shock if the wavelength of the packet shrinks further until it reaches the dissipation scale L_d. In subcritical shocks this effect ultimately takes place.

Hence the question arises, how the required anomalous dissipation is generated? We do not go into detail of this process as it is strongly nonlinear. The idea is that particles — mostly electrons — are scattered in the self-excited waves which becomes equivalent to a braking of their speed. This is just the effect of friction, resistance and dissipation because it is the electrons which carry the current. The anomalously produced collision frequency is given by the Sagdeev formula

$$\nu_a \simeq (W_{sat}/nk_B T_e)\omega_{pe} \tag{13.8}$$

where W_{sat} is the saturation amplitude of the waves, mostly ion-acoustic waves excited by the current, which replaces the thermal fluctuation level when the ion-acoustic waves become unstable in the shock.

The dissipation depends crucially on the particular wave spectrum W_k. There is a large number of possibilities of which kind of waves is excited by the electric current in the shock front. In subcritical shocks, identification of the waves is necessary, while in supercritical shocks it is not of primary importance for the shock dynamics. Candidates for unstable waves are the two-stream instability which becomes unstable at current drifts $V_\| > v_e$, i.e. at very high current speeds, i.e. preferentially in supercritical shocks. In subcritical shocks it is mostly the ion-acoustic instability which is unstable at thermal speeds $v_e > V_\| > c_{ia}$, when the current drift speed exceeds the ion acoustic velocity c_{ia}, the latter causing an anomalous collision frequency $v_a \simeq 0.01 (V_\|/c_{ia})(T_e/T_i)\omega_{pi}$, being favoured by high electron temperatures $T_e \gg T_i$.

In subcritical shocks two other waves may be more important: the lower-hybrid-drift instability and the modified-two-stream instability. The former arises when the scale of width of the shock front is comparable to the ion inertial length $\Delta_{sh} \sim \lambda_i$. The latter can be excited when the currents flow perpendicular to the magnetic field, as is the case for the electron current in the shock ramp. Both instabilities have much lower thresholds than the two-stream and ion-acoustic instabilities. They yield comparably large anomalous collision frequencies $v_a \sim \omega_{lh}$ of the order of the lower-hybrid frequency $\omega_{lh} \simeq \sqrt{m_e/m_i}\,\omega_{ce}$, being candidates for the required dissipation in subcritical shocks.

Cross-shock Potential and Dissipation Scale

Subcritical shocks do not reflect ions. All upstream ions pass the shock. However, the shock ramp, being of the order of the ion inertial length $\Delta_{sh} \sim \lambda_i$ allows for the ions to be non-magnetised while the electrons are tied to the magnetic field. As a consequence, the shock ramp contains a finite electric potential U which retards the ions, however is not large enough to stop their motion and to reflect them. Neglecting any upstream thermal spread of the ions which move at upstream bulk velocity V_1, the ion speed v_x across the shock is therefore given by

$$v_x(x) = [V_1^2 - 2eU(x)/m_i]^{\frac{1}{2}}, \qquad n(x) = n_1[V_1/v_x(x)] \qquad (13.9)$$

independent of how complicated the real motion of the ions would be in crossing the ramp. Pressure balance requires $m_i n v_x^2 + P + B^2/2\mu_0 = \text{const}$, $(P \equiv P_i + P_e)$. Substituting for v_x yields with $\bar{U} \equiv 2eU/m_i V_1^2$ and upstream flow ram pressure $P_{1\text{ram}} = \frac{1}{2} m_i n_1 V_1^2$

$$\frac{n(x)}{n_1} = \frac{1}{[1-\bar{U}(x)]^{\frac{1}{2}}} \left\{ 1 + \frac{3}{4}\frac{\beta_i}{M_A^2}\frac{\bar{U}(x)}{[1-\bar{U}(x)]^2} \right\}, \qquad \frac{P_i(x)}{P_{1\text{ram}}} = \frac{\beta_i/M_A^2}{[1-\bar{U}(x)]^{\frac{3}{2}}} \qquad (13.10)$$

as an implicit expression for the shock potential $\bar{U}(x)$ at position x in the shock ramp. At the top of the ramp the ions have velocity $V_R(x_R)/V_1 = (1-\bar{U}_{\text{tot}})^{\frac{1}{2}}$ where $\bar{U}_{\text{tot}}$ is the total normalised ramp potential drop. The corresponding velocity is less than the upstream speed but does not coincide with the downstream velocity V_2 determined from the Rankine-Hugoniot relations. The difference $V_R - V_2$ gives the downstream gyration speed and the downstream ion temperature $T_2 \sim m_i(V_R - V_2)^2/2$. The potential drop obeys the condition $\bar{U} < 1 - V_2^2/V_1^2$. Ion deceleration in the ramp is solely due to the shock potential.

The dissipation length of the shock transition is obtained from shock width Δ_{rmsh} and the induction equation

$$\partial \mathbf{B}/\partial t = \nabla \times \mathbf{V} \times \mathbf{B} - (\eta_{an}/\mu_0)\nabla^2 \mathbf{B} \tag{13.11}$$

Dissipation is active only during the convection time $\tau_{sh} = \Delta_{sh}/V_1$ yielding the dissipation time $\tau_d = L_d^2 \mu_0 \eta_{an}^{-1} < \tau_{sh}$ is shorter than the crossing time. Alternatively, the 'dissipation scale' $L_d < \Delta_{sh}$ is shorter than the shock width. Dimensionally, we obtain for the dissipation scale

$$L_d^2 \lesssim (\eta_{an}/\mu_0)(\Delta_{sh}/V_1) \tag{13.12}$$

which when inserting for the anomalous resistivity yields a condition on the anomalous collision frequency generated in the shock transition

$$\nu_{an} \gtrsim \alpha \left(\frac{V_1}{\lambda_e}\right)\left(\frac{\Delta_{sh}}{\lambda_e}\right) \qquad \text{or equivalently} \qquad \nu_{an}\tau_{sh} \gtrsim \alpha \left(\frac{\Delta_{sh}}{\lambda_e}\right)^2 \tag{13.13}$$

where $\alpha \lesssim 1$ is a numerical factor of proportionality. For anomalous dissipation to be sufficiently large to sustain the shock, the second version of this condition suggests that the ratio of transition-time to collision-time must thus be larger than a fraction α of the square of the shock width measured in electron inertial lengths.

As for an example let us assume that the shock width is $\Delta_{sh} = 1000$ km like in the bow shock. Then, for a plasma density of $n \sim 5 \times 10^6$ m^{-3} and a subcritical flow velocity of $V_1 = 100$ km/s the anomalous collision frequency should be larger than $\nu_{an} > 10^4\alpha$ Hz, which is of similar order as the electron plasma frequency $f_{pe} \sim 20$ kHz. Since such high anomalous collision frequencies are unrealistic, one must require $\alpha \sim 0.1$ corresponding to a substantially narrower dissipation scale $L_d \sim 0.3\Delta_{sh}$ or, correspondingly, narrower current sheets inside the shock transition.

13.2   Quasi-perpendicular Shocks

Subcritical shocks probably exist in multitude in the near-Earth space plasma. However, in most cases their Mach numbers are just marginally larger than $M = 1$ such that they just represent one component of the general turbulence as long as they are not of very large amplitude. On the other hand, super-critical shocks are relatively rare. When they occur, however, they are accompanied by large and sometimes violent effects. One of them, Earth's bow shock wave, is permanently present in near Earth space allowing for a monitoring and investigation of super-critical shock properties.

The first of these properties is that super-critical shocks do not form one class but decay into two families of shocks depending on the angle between the shock normal and the direction of the upstream magnetic field, the so-called *shock normal angle*

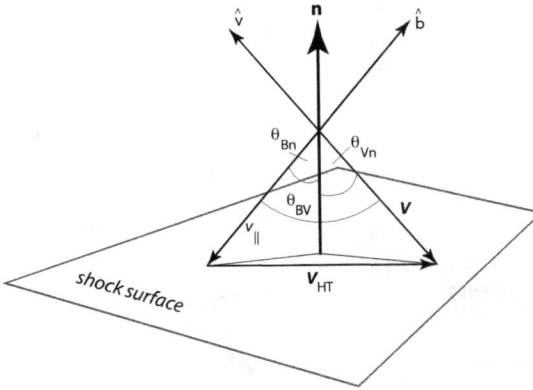

Figure 13.4: The shock coordinate system showing the shock normal **n**, velocity and magnetic field directions $\hat{v}, \hat{b}$, the three angles $\Theta_{Bn}, \theta_{Vn}, \theta_{BV}$ between $\hat{b}$ and **n**, velocity **V** and **n**, and velocity **V** and $\hat{b}$, respectively. The velocity $\mathbf{V}_{HT}$ in the shock plane is the de Hoffmann-Teller velocity.

Θ_{Bn}. Referring to this angle it is customary to distinguish between quasi-perpendicular and quasi-parallel shocks whether $\Theta_{Bn} > \frac{1}{4}\pi$ or $\Theta_{Bn} < \frac{1}{4}\pi$, respectively. Such a distinction looks artificial though reasonable. It turns out, however, that when considering the particle dynamics during an encounter with a super-critical shock this distinction is a natural condition on two families of shocks exhibiting completely different behaviour.

13.2.1 Dynamics of Particle Reflection

Figure 13.4 shows the coordinate frame used at the planar shock, which is assumed to be super-critical with shock normal **n**, magnetic and velocity unit vectors $\hat{b}, \hat{v}$, respectively. Shown are the angles $\Theta_{Bn}, \theta_{Vn}, \theta_{BV}$. The velocity vector $\mathbf{V}_{HT}$ is the de Hoffmann-Teller velocity which lies in the shock plane and is defined in such a way that in the coordinate system moving along the shock plane with velocity $\mathbf{V}_{HT}$ the plasma flow is along the magnetic field. $\mathbf{V} - \mathbf{V}_{HT} = -v_{\parallel}\hat{b}$. The guiding centres of the particles in this frame move all along the magnetic field.

For the latter reason it is convenient to consider the motion of particles in the de Hoffmann-Teller frame. Hence, the velocity vector has the two components

$$v_{\parallel} = V(\cos\theta_{Vn}/\cos\Theta_{Bn}) \qquad (13.14)$$

$$\mathbf{V}_{HT} = V\left[-\hat{v} + \hat{b}\cos\theta_{Vn}/\cos\Theta_{Bn}\right] \equiv (\mathbf{n} \times \mathbf{V} \times \mathbf{B})/\mathbf{n} \cdot \mathbf{B} \qquad (13.15)$$

Because B_n and the tangential electric field are both continuous, the de Hoffmann-Teller velocity is the same to both sides of the shock ramp. There is no induction electric field $\mathbf{E} = -\mathbf{n} \times \mathbf{V} \times \mathbf{B}$. The remaining problem is two-dimensional (see the coplanarity theorem which holds strictly in this purely kinematic case). The particle

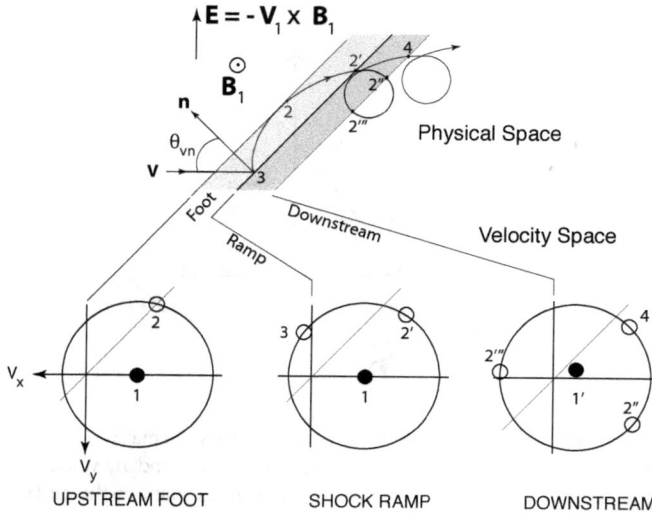

Figure 13.5: *Top*: Reflected ion orbits in the foot of a quasi-perpendicular shock in real space. The ion impact under an instantaneous angle θ_{vn}, is reflected from the infinitely thin shock, and performs a further partial gyration in the upstream field $\mathbf{B}_1$ where it is exposed to the upstream convection electric field $\mathbf{E} = -\mathbf{V}_1 \times \mathbf{B}_1$ in which it is accelerated as is seen from the non-circular section of its orbit in the shock foot. It hits the shock ramp a second time now at energy high enough to overcome the shock potential, passing the ramp and arriving in the compressed downstream magnetic field behind the shock where it performs gyrations of reduced gyro-radius. *Bottom*: The ion distribution function mapped into velocity space v_x, v_y for the indicated regions in real space, upstream in the foot, at the ramp, and downstream of the shock ramp. Upstream the distribution consists of the incoming dense plasma flow (population 1, dark circle at $v_y = 0$) and the reflected distribution 2 at large negative v_y. At the ramp in addition to the incoming flow 1 and the accelerated distribution 2' there is the newly reflected distribution 3. Behind the ramp in the downstream region the inflow is decelerated 1' and slightly deflected toward non-zero v_y, and the energized passing ions exhibit gyration motions in different instantaneous phases, two of them (2", 4) directed downstream, one of them (2''') directed upstream.

velocity is described by the motion along $\hat{b}$ plus the gyromotion in the plane perpendicular to $\hat{b}$:

$$\mathbf{v}'(t) = v'_{\|}\hat{b} + v_\perp[\hat{x}\cos(\omega_{ci} t + \phi_0) \mp \hat{y}\sin(\omega_{ci} t + \phi_0)] \tag{13.16}$$

The unit vectors $\hat{x}, \hat{y}$ point along the orthogonal coordinates in the gyration plane of the ion, the phase ϕ_0 accounts for the initial gyro-phase of the ion, and $\pm$ accounts for the direction of the upstream magnetic field being parallel (+) or antiparallel to $\hat{b}$. In specular reflection the velocity component along $\mathbf{n}$ is reversed, and (for cold ions) becomes $\mathbf{v}' = -v_{\|}\hat{b} + 2v_{\|}\cos\Theta_{Bn}\hat{n}$, which (with $\phi_0 = 0$) yields

$$v'_{\|}/V = [\cos\theta_{Vn}/\cos\Theta_{Bn}](2\cos^2\Theta_{Bn} - 1) \tag{13.17}$$

$$v_\perp/V = 2\sin\Theta_{Bn}\cos\theta_{Vn} \tag{13.18}$$

A reflected particle returns to the shock when the upstream component of the velocity $v_x = 0$, which for $\phi_0 = 0$ yields

$$\mathbf{x}'(t) = v'_\parallel t \hat{b} + (v_\perp / \omega_{ci})\{(\sin \omega_{ci}t)\hat{x} \pm (\cos \omega_{ci}t - 1)\hat{y}\} \tag{13.19}$$

After scalar multiplication of this expression with **n**, the ion displacement normal to the shock in upstream direction becomes

$$\mathbf{x}'_n(t^*) = v'_\parallel t^* \cos\Theta_{Bn} + (v_\perp / \omega_{ci}) \sin\Theta_{Bn} \sin \omega_{ci}t^* = 0 \tag{13.20}$$

It vanishes at time t^* when the ion re-encounters the shock with normal velocity $v_n(t^*) = v'_\parallel \cos\Theta_{Bn} + v_\perp \sin\Theta_{Bn} \cos \omega_{ci}t^*$. Setting this to zero, one obtains for the maximum displacement time

$$\omega_{ci}t_m = \cos^{-1}\left[(1 - 2\cos^2\Theta_{Bn}) 2\sin^2\Theta_{Bn}\right] \tag{13.21}$$

which must be inserted in $\mathbf{x}_n$ yielding for the distance a reflected ion with gyro-radius $r_{ci} = V/\omega_{ci}$ can reach in upstream direction

$$\Delta x_n = r_{ci}\cos\theta_{Vn}[\omega_{ci}t_m(2\cos^2\Theta_{Bn} - 1) + 2\sin^2\Theta_{Bn}\sin\omega_{ci}t_m] \tag{13.22}$$

This distance, for a perpendicular shock $\Theta_{Bn} = 90°$, is

$$\Delta x_n \simeq 0.7 r_{ci}\cos\theta_{Vn} \tag{13.23}$$

which is less than an ion gyro radius. Note that this distance depends strongly on the angle the velocity makes with the normal of the shock, and on the shock normal angle.

The important conclusion is drawn from consideration of the argument of $\cos^{-1}$ in the expression (13.21) for $\omega_{ci}t_m$ which exceeds unity for $\Theta_{Bn} \leq 45°$. Hence, there are no solutions for such angles.

Reflected ions return to the shock only when the magnetic field makes an angle with the shock normal larger $\Theta_{Bn} > 45°$. For less inclined shock normal angles the reflected ions escape along the magnetic field upstream of the shock and do not return. This sharp distinction between shock normal angles $\Theta_{Bn} < 45°$ and $\Theta_{Bn} > 45°$ thus provides the clear natural discrimination between quasi-perpendicular and quasi-parallel shocks we were looking for. Of course, this distinction holds just under the simplifying assumption of specular reflection made earlier.

13.2.2 Shock Foot and Reformation

The previous section showed that super-critical shocks necessarily reflect particles. This has been accounted for by inverting the normal velocity component of an incoming particle at the shock. The mechanism of reflection has not yet been clarified. It remains an active field of shock research. Clearly the cross shock electric field is involved. However the mechanism is more complicated because numerical simulations show that the incoming particles experience some violent retardation which a simple electric field can hardly explain. We do not go into detail of the processes involved at this place.

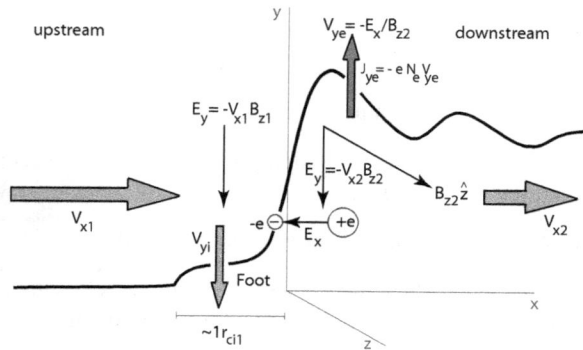

Figure 13.6: Geometry of an ideally perpendicular supercritical shock showing the field structure and sources of free energy. The compressive profile of the shock stands for the compressed profile of the magnetic field $|\mathbf{B}|$, density n, temperature T, and pressure nT. The inflow of velocity V_1 and outflow of velocity V_2 are in x direction, and the magnetic field is in z direction. Charge separation over an ion gyroradius r_{ci} in the shock ramp magnetic field generates a charge separation electric field E_x along the shock normal which reflects the low-energy ions back upstream. These ions become accelerated in the inflow convection-electric field E_y along the shock front. The magnetic field of the current carried by the accelerated ions causes the magnetic foot in front of the shock ramp. The shock electrons are accelerated antiparallel to E_x perpendicular to the magnetic field. The shock electrons perform an electric field drift in y-direction in the crossed E_x and compressed B_{z2} fields which lead to an electron current j_y along the shock. These currents are sources of free energy which drive various instabilities in different regions of the perpendicular shock.

Production of the Foot

Referring to the reflected particles at a strictly perpendicular one asks where they could go. At a perpendicular shock they are not permitted to go far upstream of the shock because they cannot leave the magnetic field. With their parallel velocity they can move along the magnetic field thus staying tangential to the shock ramp. The answer is illustrated in Fig. 13.5 in the plane perpendicular to the magnetic field. This figure has been deduced from real measurements at Earth's bow shock wave in a place where it was quasi-perpendicular. An ion of some initial velocity is reflected back from the strong shock magnetic field into the weak solar wind magnetic field. With its perpendicular velocity it just penetrates one ion gyroradius back into the solar wind. If there would be no solar wind velocity, the ion would perform one gyration in the solar wind field and return to the shock, being possibly reflected again.

The presence of the solar wind changes the case because in the shock frame the solar wind is flowing against the shock with its Mach number which corresponds to the solar wind speed $\mathbf{v}$ which has a component perpendicular to the magnetic field. In fact, the velocity is perpendicular to $\mathbf{B}$ producing the solar wind electric field $\mathbf{E} = -\mathbf{v} \times \mathbf{B}$. This field, in the strictly perpendicular case is thus directed strictly tangential to the shock but perpendicular to the magnetic field. This is the field, the reflected ion experiences when returning for its half gyro-orbit into the solar wind. Seeing the field,

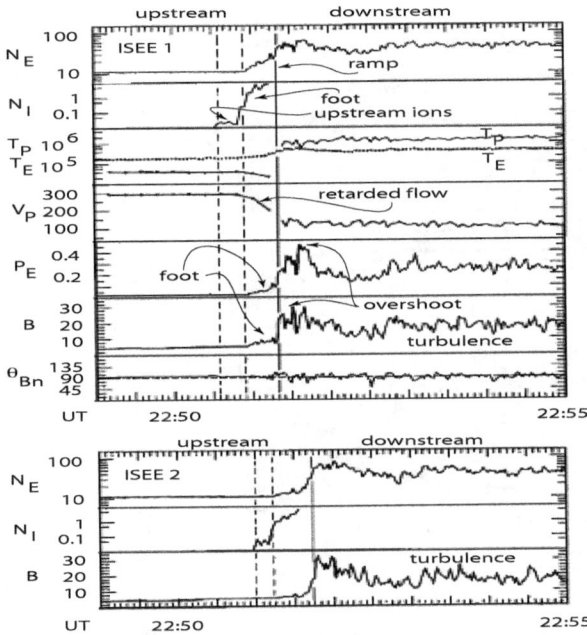

Figure 13.7: Time profiles of plasma and magnetic field parameters across a real quasi-perpendicular shock, Earth's bow shock that had been crossed by the ISEE 1 and 2 spacecraft on November 7, 1977 [Courtesy of The American Geophysical Union]. The measurement is typical for a quasi-perpendicular shock. N_E is the electron density, N_I the reflected ion density, both in cm^{-3}, T_p, T_E are for protons and electrons in K. V_P is the proton (plasma) bulk velocity in $km\,s^{-1}$, P_E electron pressure in $10^{-9}\,N\,m^{-2}$, B the magnitude of the magnetic field in nT, and Θ_{Bn}. The vertical lines mark the first appearance of reflected ion, the outer edge of the foot in the magnetic profile, and the ramp in the field magnitude, respectively.

however, it becomes accelerated along **E**, i.e. along the shock perpendicular to the magnetic field thereby gaining perpendicular energy from the solar wind and increasing the lengths of its gyropath along the shock. It may gain sufficient energy to overcome the shock reflection when entering the shock again and cross the shock as shown in the figure. In a quasi-perpendicular shock the electric field is not strictly tangential to the shock. Nevertheless the ion still experiences a tangential electric field component and is accelerated. Since this happens to all reflected ions, quasi-perpendicular shocks possess a hot reflected and accelerated ion layer adjacent to them in the upstream solar wind. The ions in this layer flow along the shock and constitute an ion current which has its proper magnetic field. It is this magnetic field of the tangential ion current which produces the foot on the quasi-perpendicular shock. Figure 13.6 is a picture of the shock structure for the case when it would be strictly perpendicular.

The example of a real observation of the changes of the plasma parameters and fields across Earth's quasi-perpendicular bow shock, which had been traversed many

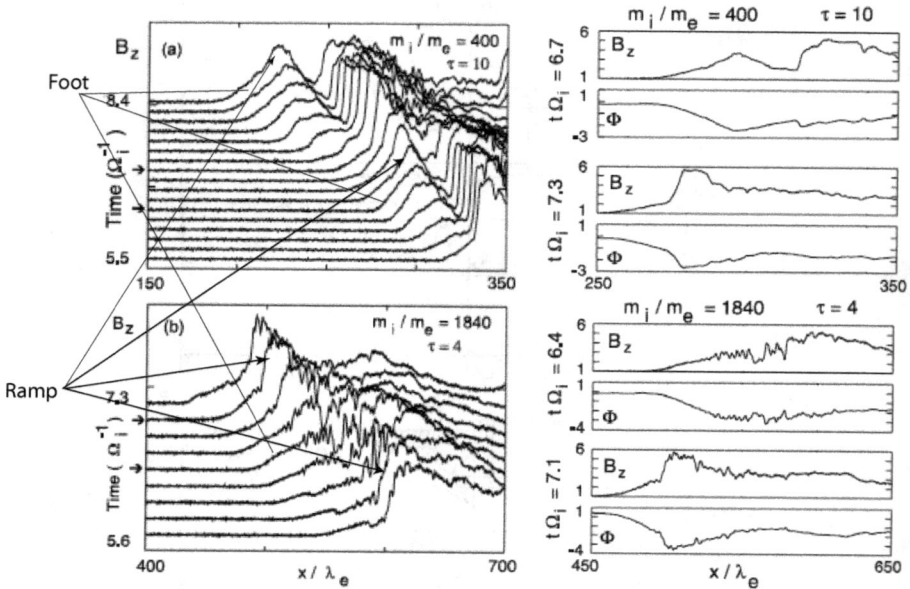

Figure 13.8: *Left*: 1D-PIC simulations of quasi-perpendicular $\Theta_{Bn} = 87°$ shock reformation (mass ratios $m_i/m_e = 400$ and 1840). Time is in ω_{ci}^{-1}, space in λ_e and small $\tau = \omega_{pe}^2/\omega_{ce}^2$. Higher mass ratios show strong time evolution. Reformation is due to the evolution of shock feet. *Right*: Spatial profiles at two time sections (see arrows on the left). The higher mass ratio shows structure in B_z and shock potential Φ. Potential drops appear in foot and ramp [courtesy American Geophysical Union].

times by the ISEE spacecraft, is given in Fig. 13.7. The foot of the shock is exhibited in the density of energetic ions N_I and magnetic field B for both spacecraft. The electrons also do react to the presence of the foot which is, however, a secondary effect due to instability of the foot current. We will briefly discuss it later when presenting numerical simulations.

Quasi-perpendicular Shock Reformation

An important question arises when asking whether the super-critical shock can remain in a stationary state when the foot forms. Imagine that ions will be reflected from the shock transition, pass into the solar wind, gyrate, become accelerated, increase their velocity along the shock ramp and cause a strong current flowing along the ramp. If there would be no flow-electric field upstream of the shock, the ions would just gyrate and return into the shock forming an ion layer and weak current simply because they have no faster gyration than the upstream solar wind ions.

However, when accelerated, they firstly extract energy from the solar wind which will be retarded in the foot region, they secondly amplify the foot current. The magnetic field of this foot current, which is a current sheet, is strongest at the outer edge of

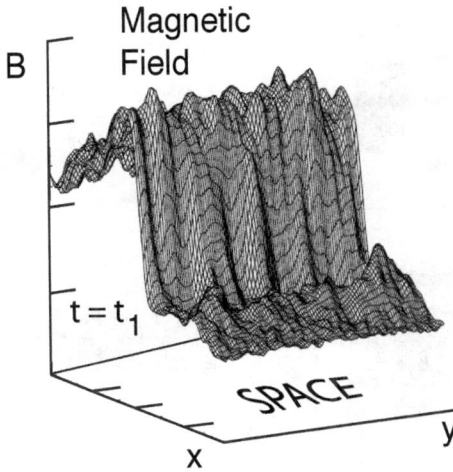

Figure 13.9: Ripples evolving on the shock ramp plane found in two-dimensional numerical simulations of quasi-perpendicular shocks. The ripples deform the shock along the shock plane during reformation. The figure shows the magnetic field amplitude at one particular simulation time as a function of the two space coordinates x, y across and along the shock. These simulations use a low mass ratio, but it is not expected that at realistic mass ratios no rippling occurs.

the foot where it grows during amplification. This implies that the solar wind running into the shock already encounters a magnetic ramp located at the upstream edge of the foot and starts compressing it. This again leads to its growth and to reflection of ions at its place which now form a new foot along the outer edge of the foot. At the same time an already retarded solar wind arrives at the shock which consequently weakens, loses its function and disappears while a new shock forms at the outer edge of its foot. This process is repeated about periodically. A supercritical quasi-perpendicular shock which builds up a foot consisting of reflected ions thus cannot be stable. It does continuously and reform itself as a consequence of foot formation, each time when reformation takes place jumping ahead by one reflected ion gyroradius.

Since, in the solar wind frame, the shock moves against the solar wind, it turns out that this motion is not continuous but consists of a sequence of periodic jumps of the shock front. In the stationary frame of the Earth one instead observes a motion of the shock back and forth with the period of reformation and a scale comparable to the reflected ion gyroradius. Moreover, since this happens locally at the shock, the shock surface must exhibit a rippled structure as seen in Fig. 13.9 identifying the shock as a non-stationary phenomenon. This reformation is shown in one-dimensional numerical full-particle simulations in Fig. 13.8. Corresponding two-dimensional simulations confirm the irregular structure of the real quasi-perpendicular shock surface which is far from being a plane as was assumed in the simple Rankine-Hugoniot fluid approach in plasma magneto-hydrodynamics of Chapter 8.

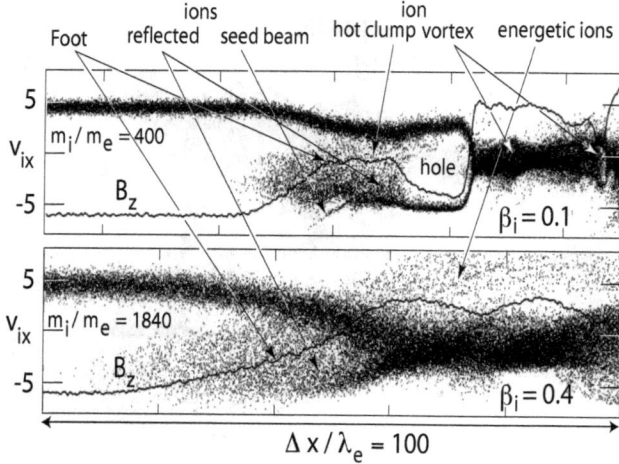

Figure 13.10: Ion phase space for $\beta_e = 0.2$ but different mass ratios and β_i [Scholer2003] at same Mach number. *Top*: $\beta_i = 0.1$ simulation. A low-field hot clump-vortex is formed which scatters reflected and upstream ions. Reformation cycles appear in the downstream distribution as ion vortices (holes). *Bottom*: Realistic mass ratio but $\beta_i = 0.4$. Hot foot ions smear out the gap between inflowing and reflected ion beams. Large numbers of diffuse energetic ions appear in this case [courtesy American Geophysical Union].

13.2.3   Particle Dynamics

Ion Dynamics in Quasi-perpendicular Shocks

Figure 13.8 gives a clear idea of the behaviour of ions in the reformation process. The corresponding ion phase space plots are shown in Fig. 13.10 for the two mass ratios $m_i/m_e = 400$ (top) and 1840 (bottom) respectively. Both plots show just the enlarged shock foot transition region over the same scale of $100c/\omega_{pe}$ for the same $\beta_e = 0.2$, which has been kept constant in both simulations, while the β_i has been changed. Only the normal component of the ion velocity is shown for the nearly perpendicular supercritical shock. In both plots the magnetic field B_z is drawn as a thin continuous line showing the magnetic shock profile over the spatial distance Δx.

In the low-mass-ratio low-β_i simulations the cold dense ion inflow at velocity $v_{ix} \sim$ 5 (in units of the upstream Alfvén velocity) is retarded to nearly $M \approx 1$ already when entering the foot. This retardation is due to its interaction with the intense but cold (narrow in velocity space) reflected ion beam which is seen as the narrow negative v_{xi}-velocity beam originating from the shock ramp. This reflected ion beam needs a certain distance to interact with the upstream plasma inflow. This distance is the growth-length of the beam-beam excited waves. At this location the reflected ions are scattered into a hot ion clump in addition to being turned around by gyration. Both effects cause a reduction in velocity v_x of the reflected ions which, being accelerated in the convection electric field, turn to flow in y direction, causing the magnetic bump that develops in this region of the foot. In the (v_x, x)-plane the reflected ions close with the upstream flow into an hot ion ring distribution (vortex) just in front of the ramp of which the hot ion clump that brakes the inflow is the upstream boundary.

Behind the ramp, which is the point of bifurcation of the ion distribution, i.e. the location where the reflection is at work, a broad hot ion distribution arises which at some locations shows rudimentary remains of ion vortices from former reformation cycles. Their magnetic signatures are the dips seen in the magnetic field.

The next reformation cycle can be expected to completely close the ion vortex in the foot and to transform the ramp from its current position to the position of the foot. The first sign of this process is already seen in the foot-ion distribution, which shows the birth of a faint new reflected ion beam at high negative speeds. This beam is not participating in the formation of the ring but serves as the seed of the newly reflected population. A similar behaviour is found in realistic mass-ratio simulations as long as β_i is small (this is obvious from the realistic mass-ratio magnetic field shown in Fig. 13.8). The shock in this case undergoes reformations also for realistic mass ratios. In other words, as long as the plasma is relatively cool or the magnetic field strong the real shocks found in nature should develop feet which at later times quasi-periodically become the shock ramp.

This changes when β_i increases, the plasma is either hot or the magnetic field weak, as is suggested by the lower panel in Fig. 13.10. There, a realistic mass ratio has been assumed, but $\beta_i = 0.4$. No reformation is observed, at least not for the simulation time. Instead, the shock develops a very long foot region that extends upstream about twice as far as for low-β_i. The high ion temperature smears out the reflected ion population over the entire gap region between upstream and reflected beam regions, and no current vortex respectively new magnetic ramp can develop. This implies that the foot remains smooth and does not evolve into a secondary ramp.

Reformation in two dimensions will thus be suppressed only when the thermal speed v_i of the ions is large enough to bridge the gap between the reflected and incoming ion beams, i.e. large enough to fill the hole in phase space between them by scattering the ions in their self-generated waves by ion-ion beam instability. Semi-empirically based on the simulations one can establish a condition for shock reformation as $v_i < \alpha V_{n1}$ when taking into account that the normal speed of incoming ions is specularly turned negative. Since this is never exactly the case, the coefficient will roughly be in the interval $1.5 < \alpha < 2$. This condition for reformation to occur can be written as $\beta_i < \alpha M_A^2$, where the Alfvénic Mach number is defined on V_{1n}. The larger the Mach number the less suppression of reformation takes place.

Electron Dynamics in Quasi-perpendicular Shocks

Observations *in situ* the quasi-perpendicular Earth's bow shock wave (plotted in Fig. 13.11) indicate that electrons are strongly heated when crossing the shock from upstream to downstream. Their reduced (field-parallel) distribution changes from Maxwellian to hot flat-top type with the intermediate distribution observed in the shock ramp showing signs of a shock-reflected electron beam. This heating and electron reflection is typical for supercritical shocks.

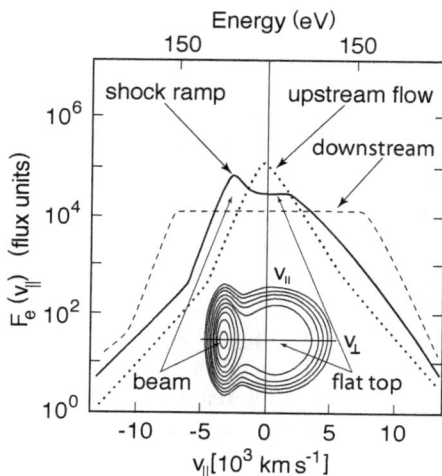

Figure 13.11: Successive reduced parallel electron distributions $F_e(v_\|)$ in the supercritical Earth's bow shock. The transition from the Maxwellian-plus-halo upstream flow distribution through the shock ramp distribution to the shock downstream distribution is monitored. The shock ramp distribution is intermediate in evolving from beam into a flat-top distribution. In its upstream directed part it contains a shock-reflected electron beam of velocity of a few 1000 km s^{-1} which is sufficiently fast to excite electron plasma waves.

In the foot region of a perpendicular highly supercritical shock the velocity differences between reflected ions and electrons from the upstream plasma inflow is large enough for the excitation of the Buneman two-stream instability and heating the electrons which contributes to shock formation. Full particle PIC simulations in strictly perpendicular shocks for small mass ratio of $m_i/m_e = 20$, $\beta_i = \beta_e = 0.15$, and Alfvénic Mach numbers $3.4 \le M_A \le 10.5$ discovered that the Buneman two-stream instability can indeed be at work.

Figure 13.12 shows the results in expanded view on the right in electron phase space. Development of electron holes generated by the Buneman two-stream instability is seen. The electrostatic field E_y in the bottom panel shows the bipolar electric field structure caused by the holes with zero mean. E_y assumes large values in the holes, the behaviour expected for solitons and BGK modes trapping and heating electrons and accelerating passing electrons to high speeds happening in the simulations near the shock: Three such holes are completely resolved, with decreasing amplitude closer to the shock ramp. They contain a small number of trapped electrons over a wide range of speeds which on the gross scale in the left panel fakes the high temperature of the electrons. In addition the electron velocity shows two accelerated populations, one with positive velocity about 2–3 times the initial electron speed, the other reflected component with velocity almost as large as the positive component but in the opposite direction suggesting that the electron current in the holes is almost compensated by the electron distribution.

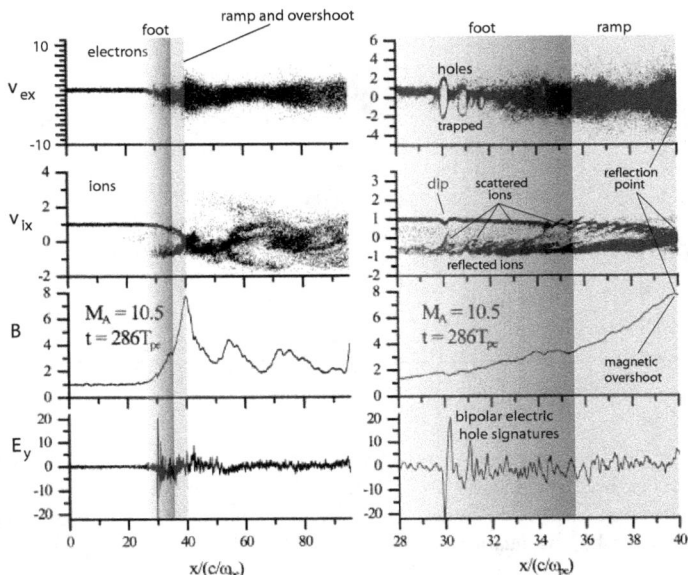

Figure 13.12: Electron dynamics in 1D-PIC simulations for $m_i/m_e = 20$, $M_A = 10.5$, $\Theta_{Bn} = 90°$ for quasi-perpendicular shocks. *Left*: Electron/ion phase spaces, magnetic field, electric field. Second panel: ion reflection and foot formation. Bottom: Foot-electron heating in large electric field amplitudes. *Right*: Expanded view of shaded regions. Electron heating related to hole formation. Three Buneman holes are formed with trapped electrons. Second panel: Ion retardation in interaction with holes due to retarding electric potential (lowest panel) in the overshoot. Ion distribution is highly structured due to interaction with many small-scale electron holes.

The magnetic signature confirms that the reflection of the main incoming ion beam takes place at the location of the magnetic overshoot and not in the shock ramp in the strictly perpendicular supercritical shock. The actual ramp region is narrow. Its width is only of the order of $\Delta \sim (1-2)c/\omega_{pe}$.

An extended electron tail is hereby generated on the electron distribution (Fig. 13.13 in a log-lin representation). On a log-log scale the tail has a power law slope $F(\varepsilon) \propto \varepsilon^{-\alpha}$, notably with power $\alpha \approx 1.7$, close to the marginally flattest power $\alpha = \frac{3}{2}$ below that an infinitely extended power law energy distribution cannot exist. The effect does not occur for small Mach numbers, too small for the Buneman two-stream instability to be excited. However, once excited, the heating increases strongly with M_A. Over the range $5 < M_A < 20$ the increase in electron temperature (electron energy in the tail of the distribution) is a factor of 40–50, which demonstrates the strong non-collisional but anomalous transfer of flow energy into electron energy via the two-stream instability.

Recently, the Polar satellite, when crossing the quasi-perpendicular Earth's Bow Shock, provided *in situ* measurements of very strong localised electric fields that

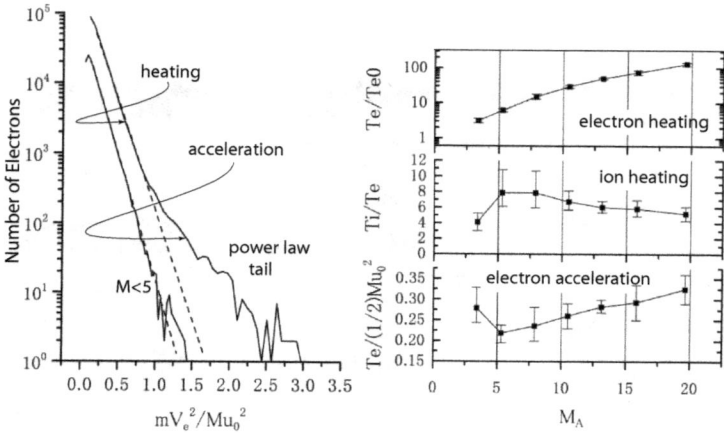

Figure 13.13: *Left*: Heating and energetic tail formation in shock-electron distribution from Buneman electron hole interaction at $M_A > 5$. The tail has power law shape $F\varepsilon \propto \varepsilon^{-\alpha}$, with $\alpha \approx$ 1.7. Corresponding to marginal flatness $\alpha = \frac{3}{2}$. *Right*: Evolution of average electron temperature ratio as function of M_A. All quantities are in computational units.

exist on scales $\lesssim \lambda_e = c/\omega_{pe}$ and reach values of $\lesssim 100$ mV m^{-1} parallel and $\lesssim$ 600 mV m^{-1} perpendicular to the magnetic field. These fields are related to the electron dynamics in the shock ramp. Presumably they play a substantial role in quasi-perpendicular shock dynamics.

13.2.4 Waves

Plasma Waves and Radiation in Quasi-perpendicular Shocks

Figure 13.14 on its left shows six minutes of electric wave field spectra detected during an Ampte spacecraft crossing of Earth's bow shock. The spacecraft approaches the shock from the magnetosheath downstream of the bow shock seeing the strongly turbulent magnetosheath plasma with the irregular broadband wave spectra followed by the shock crossing which is indicated by the strong intensification of the low and medium frequency waves. In the upstream plasma it records the general solar wind low frequency ion-acoustic turbulence at frequency $\lesssim 1$ kHz. Centred at $\simeq 31$ kHz the highly enhanced wave intensity shows the presence of intense electron plasma waves with frequency $\omega \simeq \omega_{pe}$. They are generated by the electron beam which has been accelerated along the electron foreshock boundary and reaches the spacecraft along the foreshock boundary magnetic field line that connects the spacecraft with the shock. These emissions are highly irregular in intensity, bandwidth and time. Some signature of these highly fluctuating waves is also seen near 18 kHz, and due to nonlinear effects occasionally down to 1 kHz as for instance at 2250:30 UT. The electron plasma waves disappear when approaching the shock at 0553 UT.

The sequence of spectra at the right taken at another crossing, this time of a travelling interplanetary reverse shock in the solar wind, shows the sharp plasma wave peak

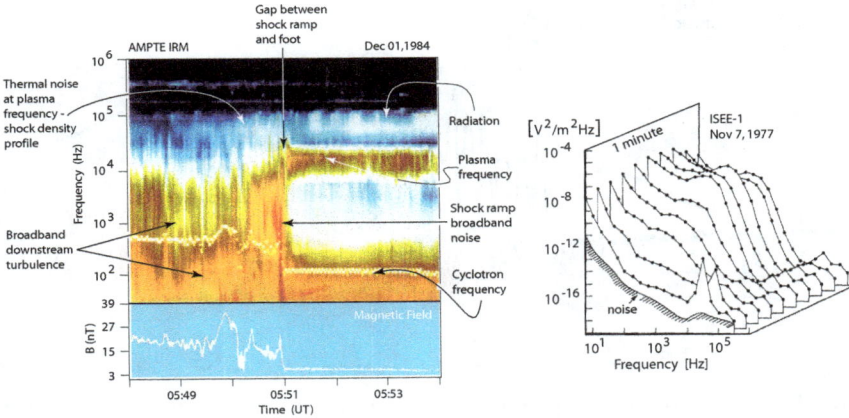

Figure 13.14: Electric wave spectra measured during spacecraft crossings of shocks in space. *Left:* An outbound bow shock transition by Ampte. The top panel is the dynamic spectrogram of power spectral energy density showing the intense low and medium frequency turbulence in the magnetosheath downstream of the shock, the violent shock transition with strong broadband wave amplification, and the solar wind low frequency ion-acoustic waves separated in frequency from the intense Langmuir waves at solar wind plasma frequency caused by the field-aligned electron beam emanating from along the electron foreshock boundary. The weak band at twice the plasma frequency is the electromagnetic radiation. The lower left panel shows the change in the magnetic field along the orbit of the spacecraft. *Right:* A sequence of shock electric spectra during another (this time inbound) crossing by Isee-1 of an interplanetary shock. Single spectra are shown exhibiting the dramatic increase of the low frequency wave power when the spacecraft approaches and crosses over the shock. Behind the shock the power remains high but lower than in the transition region. The bump around a few 100 Hz is the most interesting from the point of view of instability. These waves are excited by electron-ion instabilities discussed in the next section. The sharp peak at 60 kHz in the solar wind is the electron plasma frequency excited by the foreshock boundary electron beam.

near 60 kHz in the first few spectra and the gradual evolution of the broadband noise peaking around 0.5 kHz related to the Buneman instability and also to a Modified Two-Stream instability of a current perpendicular to the magnetic field similar to the Buneman instability. It also indicates the presence of localised electron holes in the shock as we discussed earlier.

Radiation from the shock can be recognised upstream at higher frequencies in the left panel during bow shock crossing at about twice the electron plasma frequency. It is emitted as second harmonic radiation $\omega \simeq 2\omega_{pe}$. It is produced by the foreshock-boundary electron beam-excited electron plasma waves in a narrow frequency band. The mechanisms of generating radiation in supercritical collisionless shocks are well known to be a three-wave 'collision' between three waves, two Langmuir waves and the escaping free space electromagnetic wave, symbolically written as L+L' → T, where L,L' are the two involved longitudinal waves and T is the radiated transverse wave.

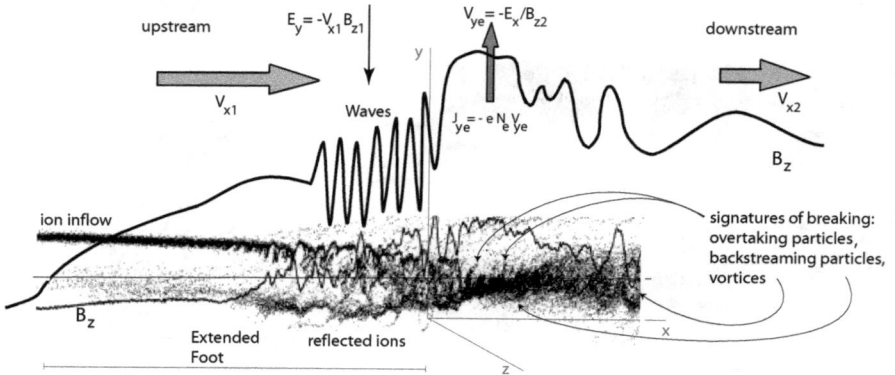

Figure 13.15: Schematic of the profile of a highly supercritical quasi-perpendicular shock with waves just before shock reformation and signatures of the beginning of wave breaking. The sketch has been completed with a copy of the ion phase space from simulations showing the structure of the ions in the ramp with the signatures of overtaking ions and backstreaming ions as well as ion vortices, all as an indication of the onset of breaking in the particle component.

Summary of Quasi-perpendicular Shocks

Figure 13.15 gives a graphical summary of the structure of quasi-perpendicular shocks.

1. The magnetic profile shows the extended shock foot which, close to the shock ramp, is quite irregular, being modulated by low frequency propagating waves. The ramp is the transition to a smoother shock overshoot profile and from there to the downstream region.

2. The ion phase space exhibits both the incoming and reflected ion beams. In the foot, the former is already retarded, showing the signatures of the slowed down ions which have been scattered in the wave fields inside the foot which have been excited by the reflected ions and the foot ion current through instabilities of low frequency mostly electrostatic plasma waves.

3. When the incoming ion beam starts interacting with the reflected beam and the foot current-generated waves, a sequence of phase-space structures is formed which, closer to the shock ramp, evolve into large holes in ion phase space being practically void of ions, incoming or reflected.

4. The shock ramp is the location of the ultimate ion reflection and heating of the bulk distribution. Signatures of onset of wave breaking are seen in the overtaking particles. However breaking seems not to take place which is probably due to the environment of a shock of being deep, i.e. not bounded in any spatial direction.

5. The entire transition region, including the foot and downstream regions, contains a diffuse hot energetic ion component that has been accelerated in the shock.

6. The plasma state just downstream of the shock is nothing else but the collection of the old shock ramps which have been left over from former shock-foot reformation cycles and, relative to the shock frame, move in the direction downstream of the shock.

7. The simulations show that, in more than one dimension, the shock front is far from being a plane surface. It exhibits a strong variability in time and space. Some of this variability can be explained as surface waves propagating along the shock front. These surface waves are mostly due to the quasi-periodic shock foot-reformation process described. Otherwise they might also be driven or additionally amplified by the reflected foot-ion current-flow along the quasi-perpendicular shock surface in similarity to a Kelvin-Helmholtz instability.

13.3   Quasi-parallel Shocks

The surprisingly sharp division between quasi-parallel and quasi-parallel shocks inferred from the investigation of specular reflection has its not less surprising pendant in real space. It has, in fact, been observed as a permanent property of Earth's bow shock which by the nature of its geometry is a curved shock.

Figure 13.16 summarises the observations in front of the concavely bent bow shock — as seen from Earth — for the special case of an interplanetary magnetic inclination angle of 45° against the Earth-Sun line in the ecliptic. In this figure the bow shock is the hyperbolic curve in the lower part of the left panel and on the right in the right panel. At the 45° magnetic field inclination angle in the left panel the field lines are tangential to the bow shock at its evening flank. Θ_{Bn} exceeds 45° up to about the nose of the bow shock. In this region the shock is quasi-perpendicular. Reflected particles form a foot, shown as the narrow shaded region in front of the bow shock which ends at the bow shock nose. The solar wind velocity V splits into parallel and perpendicular parts $V_\parallel, V_\perp$ as shown. Because of the direction of $V_\parallel$ in this region no particles escape along the magnetic field upstream of the quasi-perpendicular shock.

13.3.1   The Foreshock

To the right of the bow shock nose on the left in Fig. 13.16 one has $\Theta_{Bn} > 45°$. Here the reflected particles do escape upstream along B filling an extended region in front of the pre-noon side of the bow shock, called the foreshock.

The boundary of the foreshock is thus a sharp line in space which is, however, more inclined against the solar wind magnetic field because during their upstream motion of escape the particles are convected by the solar wind velocity together with their magnetic field line toward the bow shock. Since the electrons have higher velocity than the ions — we do not ask here, how they can escape from the bow shock and where they receive their high upstream directed velocities — the inclination angle of the electron foreshock is less than that of the ion foreshock occupying a larger fraction

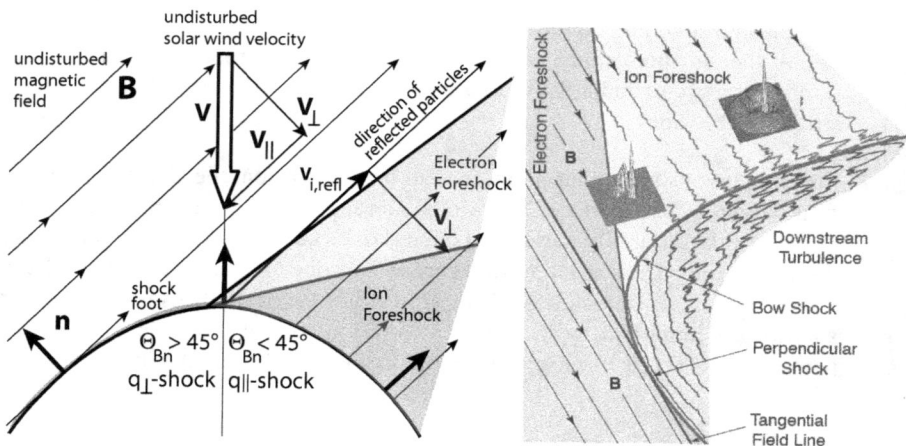

Figure 13.16: Two schematic views of the relation between a curved shock and its foreshock in dependence on the direction of the upstream magnetic field **B**, shock-normal **n**, and shock-normal angle Θ_{Bn}. *Left*: The special case when the magnetic field is inclined at 45° with respect to the symmetry axis of the shock. In this case the upper half of the shock becomes quasi-parallel ($\Theta_{Bn} < 45°$), the lower half is quasi-perpendicular ($\Theta_{Bn} > 45°$). The velocity of reflected particles is along the magnetic field. However, seeing the flow the field-line to which they are attached displaces with perpendicular velocity. This velocity shifts the foreshock boundary toward the shock as shown for electrons (light shading) and ions (darker shading). The ion foreshock is closer to the shock because of the lower velocity of the ions than the electrons. For the electrons the displacement of the electron foreshock boundary is felt only at large distances from the shock. *Right*: The different velocity space ion distribution function (inserts) along the ion foreshock boundary and deep inside the ion foreshock. Also shown is the different character of the magnetic fields in the quasi-perpendicular and quasi-parallel shocks. Note also the quiet state of **B** in the electron foreshock region.

of the upstream foreshock space than the ions. This double structure of the foreshock is what has been constantly observed at the bow shock and is a typical property of quasi-parallel super-critical shocks.

The foreshock of a supercritical shock is the most important structural property of quasi-parallel shocks. It forms a fan in front of the quasi-parallel shock which extends as far upstream as the fastest particles manage to travel along the magnetic field during the solar wind convection time.

These fastest particles occupy the flux tube at the edge of the foreshock. Upstream propagating electrons are first encountered at the edge of the electron foreshock, and upstream propagating ions are located at the edge of the ion foreshock. Here, at these spatially separated edges the escaping particles form upward streaming beams of particles which in the solar wind frame are more energetic than the corresponding solar wind particles. Roughly spoken, in the solar wind frame these particles propagate at twice the solar wind velocity because their downstream velocities have been inverted into upstream. Together with the solar wind they form a beam-beam plasma system, in which the solar wind is the hot background and the upstream beam is cold because it consists solely of the fast particles along the edges of the foreshock. Thus, for the electrons one has the classical configuration of a hot solar wind electron background distribution traversed by a colder fast electron beam, the configuration in which one expects strong excitation of Langmuir waves by the mechanism noted at the end of the last section on quasi-perpendicular shocks. Similarly for the ions, the reflected ion beam at the ion-foreshock edge forms a fast ion beam interacting head-on with the solar wind ion stream.

The slower upstream particles, electrons and ions, feel the downstream convection stronger than the faster particles. As a consequence the solar wind convection flow sweeps them readily away from their own foreshock boundary into the centre of the foreshock. The foreshock is therefore filled with a broad population of particles of lesser energy and slower speed than those particles in the foreshock-edge beams. Moreover, these lower energy particles interact with the spectrum of waves present in the foreshock and become scattered in phase space. For these reasons they possess a broadened distribution function than the solar wind and the reflected beam. This is shown on the right in Fig. 13.16 where two measured ion-velocity space distributions are shown as insets in the ion foreshock. The first inset belongs to the ion-foreshock ion beam distribution. The square plane is the (v_x, v_y)- velocity plane. The distributions are plotted a number of particles in this plane. The sharp large peak in the centre of the plane is the solar wind ion distribution (drawn in the solar wind frame) which has very little spread due to the very low temperature (velocity spread) of the solar wind ions. Displaced from it a smaller peak of slightly larger velocity spread is moving at upstream speeds (displaced into upward direction, i.e. solar direction from the solar wind peak). This is the warm reflected ion beam along the ion-foreshock edge. Its spread is larger than that of the solar wind in this case because the foreshock-edge cannot be resolved sharply in the observations; it fluctuates highly in space. For this reason one needs to average over a certain space interval in order to obtain a reliable measurement. In addition, however, the ion beam is unstable with respect to several kinds of low-frequency plasma waves which scatter the ions in the beam causing a substantial spread of the beam in velocity.

The second inset is an example of an ion distribution much deeper in the ion foreshock. Still the undisturbed solar wind flow sticks out. But at the same time, the reflected ions which have been swept into the foreshock form a nearly uniform ring around the solar wind peak which does not yet merge into the solar wind in order to form a smooth tail on the solar wind distribution. Rather it remains to be separated from the solar wind showing that the reflected ions have been scattered nearly uniformly in velocity. This scattering results from both the adding of the lower velocity part of the reflected ion distribution being swept towards the shock and the vulnerability of the reflected ions against excitation and scattering of waves. The foreshock contains two ion components. One is the ions of the foreshock boundary which become convected toward the shock. However, at the same time the shock still reflects ions upstream which try to propagate into the solar wind. This however is not easily

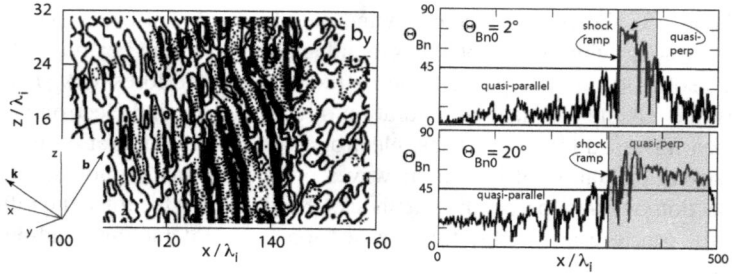

Figure 13.17: Two-dimensional simulations of the evolution of upstream waves. *Left*: Upstream waves in a super-critical quasi-parallel shock in the (y,z)-plane. Contour plot of the two (normalised to upstream magnetic field) components of magnetic fluctuations. Plane magnetic wave fronts inclined against the shock in direction x evolve for wavelengths $\sim 10\,\lambda_i$ in z and shorter in x. Near the shock the fronts turn parallel to the shock producing a non-coplanar magnetic component $|b_y|$ of same order as $|b_z|$. The shock is not stable exhibiting structure in z produced by the reflected upstream particles and waves. *Right*: Evolution of the shock normal angle Θ_{Bn} with distance from the shock in two-dimensional hybrid simulations for two initial quasi-parallel shock-normal angles $\Theta_{Bn0} = 2°$ and $\Theta_{Bn0} = 20°$, respectively. The horizontal line at $45°$ is the division between quasi-perpendicular and quasi-parallel shock normal angles. In both cases Θ_{Bn} evolves from quasi-parallel direction into quasi-perpendicular direction. The shaded areas identify the locally quasi-perpendicular domains of the quasi-parallel shock as consequence of the upstream waves.

possible because of the turbulent structure of the foreshock magnetic fields. These ions thus penetrate upstream into the foreshock in a diffusive wave in which the foreshock waves are actively involved.

The interaction with the waves indicates that the motion of the reflected ions is a combination of convection and diffusive scattering into upward direction in the waves generated close to the shock. The spatial decay into the upstream direction of the diffuse ion density $n_i(W,z) \sim \exp[-z/L(W)]$ differs for particles of different energy W. The e-folding distance $L(W) \sim W$ increases linearly with energy, suggesting that low energy reflected particles are practically confined to the shock and do not diffuse upstream. The higher the ion energy, the deeper the ions can overcome the convection and diffusively penetrate back into the upstream flow. In contrast to the ion beam thus, whose source is at the quasi-perpendicular/quasi-parallel shock boundary, the source of the *diffuse* ions is located at the quasi-parallel shock.

The e-folding distance for the diffuse ions is given by $L(W) = \kappa_\parallel(W)/V_1$, with spatial diffusion coefficient $\kappa_\parallel(W) = \frac{1}{3}v\ell_\parallel(W)$, where $\ell_\parallel$ is the diffusion length, and v the particle velocity. From the balance between convective inflow and upstream diffusion, one writes

$$\ell_\parallel(E) = 3L(W)\sqrt{W_1/W} \sim \sqrt{W\,W_1} \tag{13.24}$$

with upstream flow energy W_1. In the solar wind the flow energy is a few keV, and a 20-keV diffuse ion will have a typical parallel diffusion length (or diffusive mean free path) of $\ell_\parallel \sim (1-2)\,R_E$, a rather short upstream distance from the bow shock only,

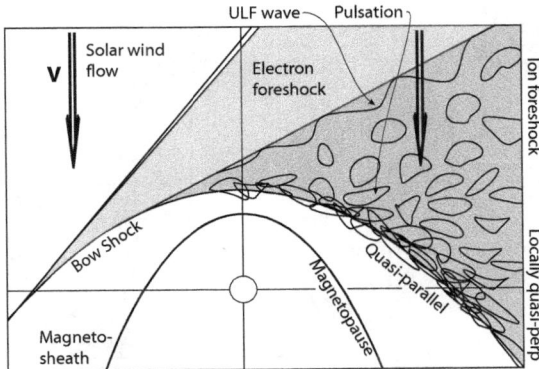

Figure 13.18: Quasi-parallel supercritical shocks reform by accumulation of upstream magnetic pulsations convected into the shock from the ion foreshock, thereby causing an irregular shock structure and contributing to the turning of the magnetic field into a direction to the shock normal that is more perpendicular, i.e. the magnetic field is more parallel to the shock surface with the shock surface itself becoming very irregular. It is shown schematically that the pulsations grow out of the small amplitude upstream waves which are generated in the broad ULF-wave-unstable region in greater proximity to the ion-foreshock boundary. When the ULF waves evolve to large amplitude and form localised structures and pulsation these convected toward the shock, grow, steepen, overlap, accumulate and lead to the build up of the irregular quasi-parallel shock structure which overlaps into the downstream direction. Downstream turbulence behind quasi-parallel supercritical shocks is thus large due to pulsations that emerge from the shock and move downstream.

orders of magnitude shorter than the collisional mean free path of an ion. Hence, the diffusion estimate suggests that strong wave-particle interactions in the shock transition are responsible for the scattering and acceleration of the diffuse particle component. The corresponding upstream-ion collision frequency $\nu_{c,ui} \simeq v/\ell_\parallel$ for the 20 keV-upstream ions in the solar wind yields $\nu_{c,ui} \sim 0.2$ Hz, being comparable to the ion cyclotron frequency $\omega_{gi}/2\pi = (0.1 - 0.3)$ Hz in the $B \simeq 8$ nT upstream to $B \simeq 30$ nT shock ramp magnetic field in observations.

13.3.2 Quasi-parallel Shock Reformation

In quasi-parallel supercritical shocks there is not such a stringent distinction between the region upstream of the shock and the shock itself like in quasi-perpendicular shocks. Strictly speaking the foreshock and the shock itself cannot be considered separately. This is due to the presence of the reflected and diffuse particle components in the foreshock. These are the sources of a large number of waves which are excited in the foreshock by an ion-ion instability between the solar wind flow and the isotropic through inhomogeneous diffuse upstream ion component. The excited waves are, however, slow in comparison to the super-Alfvénic solar wind. This implies that they cannot propagate upstream from the location of their excitation. On the contrary, through

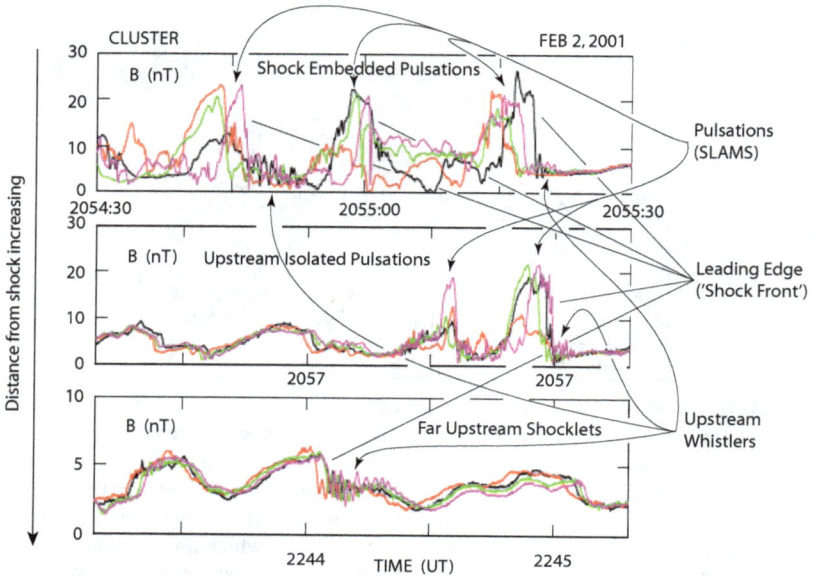

Figure 13.19: Magnetic field of pulsations in the quasi-parallel foreshock for the 4 Cluster (colour coded) spacecrafts. *Top*: Accumulated pulsations in the shock transition being of small scale and large amplitude. *Centre*: Isolated pulsation at larger distance seen almost simultaneously at the 4 spacecrafts being of larger size but lower amplitude. *Bottom*: Far upstream shocklet and ultra-low-frequency waves with attached upstream whistlers.

flowing against the solar wind they become swept downstream against the shock and grow. The interaction of these large-amplitude electromagnetic ion waves — called foreshock magnetic pulsation — with the shock is one of the main issues in quasi-parallel shock physics.

The Role of Upstream Pulsations

The shock radiates energy away towards upstream in the reflected diffuse particle component which exhibits a steep upstream ion-density gradient. On their paths when approaching the shock the low-frequency electromagnetic ion-plasma waves that have been excited in the foreshock by the electromagnetic ion-ion instability encounter the diffuse ion gradient and experience the exponentially increasing number of diffuse particles the shorter the distance between the waves and the shock becomes. Interacting with the ever denser diffuse ions the electromagnetic waves become strongly amplified until reaching large amplitudes of the same order as the background solar wind magnetic field in the foreshock.

At this stage they undergo nonlinear deformation and steeping, forming short wavelengths magnetic foreshock pulsations of larger tangential than normal wavelengths. Their **k**-vectors turn away from the background of solar wind magnetic field

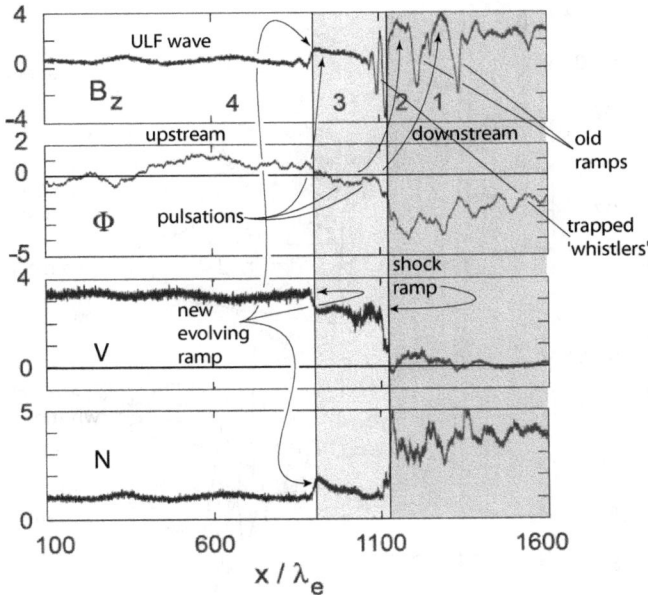

Figure 13.20: 1D full particle PIC simulations of quasi-parallel shock evolution. *From top:* Magnetic field B_z, electric potential Φ, flow velocity V, density N (in simulation units). Numbers indicate 3 pulsations. Pulsation 1 was the old shock. Pulsation 2 is the actual shock (coinciding with drop in V to zero and steep increase in density and potential). Pulsation 3 is just evolving. Number 4 will become a pulsation. The actual shock ramp has some trapped whistlers.

when growing nonlinear and the components of **k** assume comparable values parallel to **B** and parallel to the shock normal **n**.

Close to the shock, where the backstreaming ion density is highest, the waves have short wavelengths, and **k** is almost parallel to **n** implying that the magnetic field of the waves has turned away from the quasi-parallel direction and has become about quasi-perpendicular. This has a very strong effect on the evolution of quasi-parallel shocks which we are going to explain below.

Before proceeding we show high Mach number quasi-parallel shock simulations in Fig. 13.17 in which these effects are seen. In these simulations reflected ions were artificially injected with same Mach number as the incoming flow but with much higher temperature $v_{thi} \approx 14\, v_A$, forming an initially spatially uniform ion component.

The upstream waves have two components, b_z, b_y, of low amplitude at large distance from the shock but reach very large amplitudes simultaneously in both components during shock approach while bending and assuming structure in z-direction that is different from the regular elongated shape at large distance. This deformation of wave front implies that the shock has structure on the surface in both directions x and z and is not anymore planar. The shock becomes locally curved on the scale of the shock-tangential wavelength. The waves deform the shock and are of same

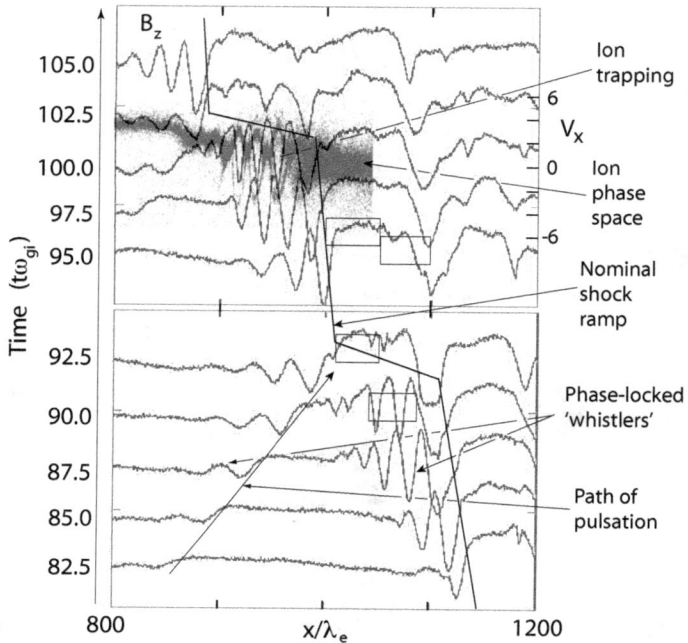

Figure 13.21: Magnetic field B_z in simulation frame for subsequent simulation times (shifted upward by $\Delta t \omega_{gi} = 2.5$). The shock moves to the left into the upstream direction. Reformation results from exchange with incoming pulsation. The magnetic field trace at time $t\omega_{gi} = 100$ has been overlaid on the ion phase space at this time. Heavy steps show location of nominal shock ramp moving upstream until new pulsation arrives and it suddenly jumps forward by roughly $100\lambda_e$. A pulsation arrives at the shock at $t\omega_{gi} = 92.5$ to take over the role of the shock. In the minima of the 'whistler' field fluctuations ($t\omega_{gi} = 100$) ions are trapped, forming vortices in phase space centred around local minima of the electric potential Φ (not shown). The little boxes indicate where particle phase space distributions have been determined.

amplitude as the shock ramp. They become gradually indistinguishable from the shock. The shock is, so to say, the first and largest-amplitude upstream magnetic pulsation. In addition the quasi-parallel shock-magnetic field is not coplanar, because the waves have contributed a substantial component b_y that points out of the coplanarity plane.

Reformation

Quasi-parallel shock reformation and much of its physics is predominantly due to the presence of the large-amplitude and spatially distinct upstream waves. The important conclusion is that large-amplitude upstream pulsations are the physical generators of the shock.

The presence of large amplitude magnetic pulsations lets the quasi-parallel shock locally change its character. It becomes variable in time and position along the shock

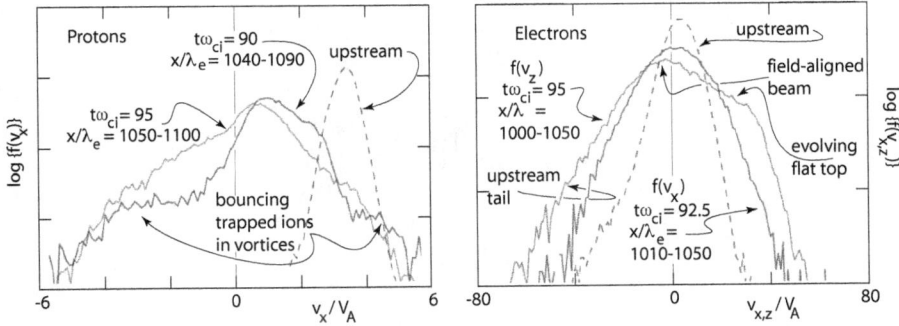

Figure 13.22: Velocity distributions for different phases of shock evolution taken in the boxes of Fig. 13.21. *Left*: Proton distributions at $t\omega_{ci} = 90, t\omega_{gi} = 95$. The first period is in the arriving pulsation. The upstream distribution is heated with upstream tail due to trapped ions. At $t\omega_{gi} = 95$ the trapped hot ion distribution dampens the waves. The incident plasma is slowed down. *Right*: Electron distributions $f_e(v_x)$ at $t\omega_{gi} = 92.5$ and $f_e(v_z)$ at $t\omega_{gi} = 95$ in the new shock built up from a fresh pulsation and in the well-developed shock but along the main magnetic field, respectively. The parallel distribution is non-symmetric, heated, has an upstream tail, forming a flat top and a remaining upstream-beam-like part similar to that measured.

surface and — on the smaller scale close to the shock transition — is 'less quasi-parallel' (or 'more perpendicular'), i.e. the shock-normal angle Θ_{Bn} has increased on the scale of the upstream waves because of the presence of the out-of coplanarity-plane component that is introduced by the upstream waves.

This can be realised from the right part of Fig. 13.17 where the evolution of the local shock normal angle Θ_{Bn} is plotted for two simulations of different initial quasi-parallel shock normal angle in the course of the approach of a spectrum of upstream large amplitude waves which have been generated self-consistently. When approaching close to the shock ramp, Θ_{Bn} rapidly increases from quasiparallel to quasi-perpendicular in both cases until it is found deep in the domain of quasi-perpendicular shocks. The evolution of Θ_{Bn} is not smooth.

Transition to local quasi-perpendicularity occurs for the initially nearly parallel case right at the nominal shock ramp on a short scale, while for the initially more inclined quasi-parallel case it occurs already at an upstream distance of about $\lesssim 100\lambda_i$ from the shock. The latter case is sufficiently far away for the upstream flow ions to feel the change in the shock normal. This transition, being on the ion scale, implies that in the region close to the shock the upstream-flow ions occasionally experience the shock as being quasi-perpendicular and become reflected. This is the way in which the quasi-parallel shock regenerates its upstream reflected ion component.

The behaviour of the shock normal angle gives a rather clear identification of the location of the shock transition in the quasi-parallel case, as indicated on the right in Fig. 13.17 by shading. Three distinctions can be noticed:

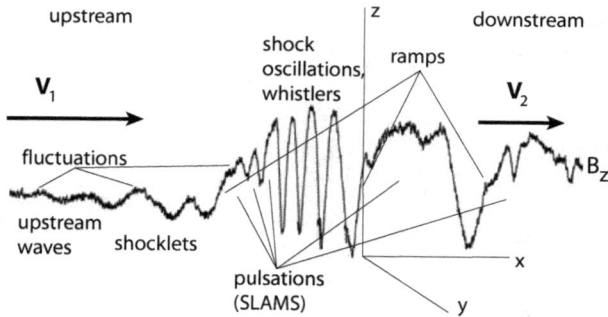

Figure 13.23: Schematic one-dimensional profile taken along the nominal instantaneous shock normal to a supercritical quasi-parallel shock as seen in the magnetic field component B_z. This is the analogue to the quasi-perpendicular shock profile. It shows the main features in the vicinity of the quasi-parallel shock transition: the large amplitude upstream waves with the turbulent fluctuations on top of the waves, the formation of shocklets, i.e. steep flank formation on the waves exhibiting small-scale fluctuations on top of the wave, which act already like small shocks, very-large amplitude pulsations (magnetic pulsations or) which turn out to be the building blocks of the shock, multiple shock-ramps at the leading edges of the pulsations belonging to diverse ramp-like steep transitions from upstream to downstream lacking a clear localisation of the shock transition and any attached phase-locked whistlers. (Note that the entire figure is, in fact, the shock transition, as on this scale no clear decision can be made where the shock ramp is located.) Not shown here are the out of plane oscillations of the magnetic field that accompany the waves. The particle phase space is not shown.

– at larger initial shock-normal angles the transition to quasi-perpendicular angles occurs earlier, i.e. farther upstream than for nearly parallel shocks. This is due to the strong effect of the large amplitude upstream waves;

– at larger initial shock normal angles the quasi-perpendicular shock transition is considerably broader than for nearly parallel shocks, i.e. it extends farther downstream before the main quasi-parallel direction of the magnetic field in the downstream region takes over again and dominates the direction of the magnetic field;

– at an initial shock-normal angle of $20°$, this region is roughly $\sim 150\lambda_i$ wide, implying that the magnetic field direction behind a quasi-parallel shock remains to be quasi-perpendicular over quite a long downstream distance measured from the shock ramp. For the nearly parallel shock this volume is only about $\sim 50\lambda_i$ wide.

Three representative examples of such upstream waves that support the above theoretical inference, have been plotted in Fig. 13.19. The shock region consists of many embedded magnetic pulsations of very large amplitudes, having steep flanks and irregular shapes. They exhibit higher frequency oscillations in the whistler mode sitting on their feet or shoulders. Pulsations in the shock are short-scale; the 4 different Cluster spacecrafts — at spacecraft separation < 1000 km — do not observe a coherent picture of a particular pulsation. The magnetic field directions differ from spacecraft to spacecraft and from pulsation to pulsation, and even for one pulsation at its front and trailing edges, and the magnetic normal directions change on the short spatial scales across a single pulsation.

Full particle PIC simulations of quasi-parallel shock reformation shown in Fig. 13.20 strongly support this picture. On the left the reformation of the shock due to exchange of pulsations is shown. On the right this process is followed in time. The nominal shock front progresses stepwise into upstream direction when an old magnetic pulsation is replaced by one that is newly arriving.

In the phase space, the magnetic oscillations are well correlated with ion holes with the field maxima coinciding with the vertexes and the minima corresponding to the centres of the holes, i.e. to the locally largest spread in the ion distribution or, in other words, the locally highest "ion temperatures". The evolution of the corresponding velocity distributions for ions and electrons in this reformation process is given in Fig. 13.22. Heating of ions and electrons takes place in different phases, and there is an indication of acceleration of an electron beam.

Quasi-parallel shocks form and reform in the course of an interplay between the shock-produced upstream diffuse energetic-ion component and the magnetic pulsations which grow out of the wave-excited upstream waves and are amplified in the interaction with the diffuse ions during their approach of the shock. On the other hand and supporting the above formulated view, quasi-parallel supercritical (high Mach number) shocks do not reform in the absence of diffuse upstream ions and the waves that are excited by them. A graphical summary of the structure of quasi-parallel shocks is given finally in Fig. 13.23.

13.4 Summary

In this last chapter we have given an application of the physics of space plasma to one of the most interesting and most complex problems in space plasma physics, the formation of subcritical and super-critical collisionless shocks.

This problem has been chosen because it contains all of plasma physics: fluid physics which provides the global jump conditions, single particle theory which allows to distinguish between quasi-parallel and quasi-perpendicular collisionless shocks in the super-critical case, steeping of waves and the achievement of a stationary state in balance of steeping and dispersion in the case of subcritical solitary waves which can be described by Sagdeev's pseudo-potential, the insufficient role of dissipation in shock formation in particular in the case of super-critical shocks, and the problem of instability and wave generation.

Because in shocks entropy is produced continuously, collisionless super-critical shocks cannot be and are not in thermodynamic equilibrium. They are continuously forming and reforming. This process is one of most interesting in the entire physics of collisionless shocks, and we have focussed on it. The surprising result is that reformation is completely different for quasi-perpendicular and quasi-parallel shocks.

Both kinds of collisionless shocks reform with the help of particle reflection. This is necessary for reasons of entropy because the time to generate sufficient diffusion in order to prevent shock wave breaking in the collisionless super-critical plasma is too long. The shock has simply not enough time to digest the inflowing amount of energy which it can get rid of only by not permitting the excess energy to pass the shock transition. In other words, the shock rejects the excess energy inflow by reflecting its carriers, a large fraction of upstream ions, back upstream.

Quasi-perpendicular shocks use these reflected particles to form a shock foot and shock-foot current, the proper magnetic field of which grows in front of the shock ramp to large amplitudes until it becomes large enough to itself start reflecting particles. In this way it becomes the new shock front. The quasi-perpendicular shock reforms itself quasi-periodically in this way.

Quasi-parallel shocks on the other hand release a large fraction of particles back upstream forming a shock foreshock which contains high fluxes of those upstream particles. These upstream particles create a high level of upstream low frequency electromagnetic plasma waves via an ion-ion beam instability. Since these newly generated waves have low phase and group speeds they cannot travel far upstream but become convected into the shock ramp, grow during the convection time and evolve into large-amplitude magnetic foreshock pulsations.

It is these pulsations which reform the shock by becoming the shock along small tangential patches of the shock surface. In this process the old shock front in the patch is ejected downstream where it becomes the main component of the downstream large amplitude turbulence as for instance in the magnetosheath, while the shock is reformed by the pulsation that had newly arrived at this particular patch. For this reason quasi-parallel shocks are very variable and in fact turbulent. For this reason they can form under collisionless conditions when fluid shocks cannot exist. From this statement we learn in addition that any collisionless super-critical shock will always be magnetised as only then they can form and reform itself which is the condition for its existence. Probably, in the absence of an initial magnetic field, the shock will take care to generate its own field in some way. One way is through a violent Weibel instability.

Having presented these basic ideas, mostly by quoting numerical simulations and not by deductively deriving them from basic principles, we should note that we have not only suppressed their deduction but also kept silent about a large number of other effects which are related to collisionless shocks. Among such effects are the production of entropy in general and the production of dissipation which we did not discuss at all as this is a main and unresolved question of thermodynamics when dealing with collisionless shocks.

Other important effects include the generation of radiation by shocks. Astrophysical shocks and also some of the shocks in the heliosphere like Earth's and Jupiter's bow shocks as well as interplanetary shocks are strong banded radiators and are visible in radiation of plasma and electromagnetic free space waves. These processes are related to plasma instability and nonlinear interaction.

Finally one of the most interesting questions is whether shocks can accelerate ions and electrons to the high energies which are occasionally observed in space. Customarily shocks are made responsible for the generation of very high energy particles like cosmic rays and solar energetic particles. The bow shock, on the other hand, has not been found as a strong accelerator. The bow shock is capable of injecting particles and reflecting particles, but the energies obtained are very moderate. Why this is so remains unclear. Possibly the geometric size of the bow shock is too small and the size of the transition region, the magnetosheath, is too narrow and the level of turbulence in the magnetosheath is too low for the shock and its environment to become a strong particle accelerator.

In all these mentioned processes waves are involved, and we will have to await much larger scale simulations in three dimensions until possibly being in the position to answer some of the related questions.

Further Reading

The first and classic general treatment of the physics of collisionless shocks going far beyond earlier hydrodynamic theories, but with no emphasis on space physics, is the comprehensive and still valid review article by Sagdeev [5] of 1966. This article is full of important ideas and deep insight into the shock problem. A first and highly mathematical treatise of plasma instabilities related to the formation of solitons, also including anomalous dissipation to obtain shocks is found in the short monograph [4] written five years later lacking, however, the depth of understanding as contained in [5]. Useful observational and theoretical accounts on various main and scattered aspects of collisionless shocks summarising the state of the arts in 1985 an be found in the two AGU conference proceedings [6, 8] which, however, are rather inhomogeneous and of highly variable importance. So are the two Cospar [3] and AIP [2] conference proceedings of 1995 and 2005, respectively. A recent more complete and systematic account of the current knowledge on non-relativistic collisionless shock physics including applications to the shocks in the heliosphere can be found in [1]. For application aspects to astrophysics one may consult [8].

References

[1] A. Balogh and R. A. Treumann, *Physics of Collisionless Shocks, The Space Plasma Shock Waves* (Springer Verlag, Heidelberg-New York, 2011).
[2] G. Li, G. P. Zank, and C. T. Russell (eds.), *The Physics of Collisionless Shocks*, AIP Conference Proceedings Vol. 781 (American Institute of Physics, Melville, New York, 2005).
[3] C. T. Russell (ed.), *Physics of Collisionless Shocks*, Adv. Space Phys. **15**, Number 8/9 (Cospar-Pergamon-Elsevier, Oxford, 1975).
[4] D. A. Tidman and N. A. Krall, *Shock Waves in Collisionless Plasmas* (Wiley-Interscience, New York, 1971).
[5] R. Z. Sagdeev, *Cooperative Phenomena and Shock Waves in Collisionless Plasmas* in: Rev. Plasma Phys. Vol. 4, ed. by M. A. Leontovich, (Consultants Bureau, New York, 1966), pp. 23-92 (1989).
[6] R. G. Stone and B. T. Tsurutani (eds.), *Collisionless Shocks in the Heliosphere, A Tutorial Review* (AGU Monograph 34, AGU, Washington, D.C., 1985).
[7] R. A. Treumann, *Astron. Astrophys. Rev.* **17**, 409–535 (2009).

[8] B. T. Tsurutani and R. G. Stone (eds.), *Collisionless Shocks in the Heliosphere, Reviews of Current Research* (AGU Monograph 35, AGU, Washington, D.C., 1985).

Additional selected more specialised references

[9] S. D. Bale *et al.*, *Astrophys. J.* **575**, L25–L28 (2002).

[10] S. D. Bale, F. S. Mozer, and T. S. Horbury, *Phys. Rev. Lett.* **91**, 265004 (2003).

[11] S. D. Bale and F. S. Mozer, *Phys. Rev. Lett.* **98**, 205001 (2007).

[12] D. Biskamp, *Nucl. Fusion* **13**, 719–740 (1973).

[13] D. Burgess *et al.*, *Space Sci. Rev.* **118**, 205–222 (2005).

[14] M. Hoshino and T. Terasawa, *J. Geophys. Res.* **90**, 57–64 (1985).

[15] C. Kennel, J. P. Edmiston, and T. Hada, in *Collisionless Shocks in the Heliosphere: A Tutorial Review* (R. G. Stone and B. T. Tsurutani, eds., AGU, Washington DC) pp. 1–36 (1985).

[16] A. Kis *et al.*, *Geophys. Res. Lett.* **31**, L20801 (2004).

[17] E. A. Lucek *et al.*, *Ann. Geophysicæ* **20**, 1699–1710 (2002).

[18] E. A. Lucek *et al.*, *J. Geophys. Res.* **109**, A06207 (2004).

[19] E. A. Lucek *et al.*, *J. Geophys. Res.* **113**, A07S02 (2008).

[20] S. Matsukiyo and M. Scholer, *J. Geophys. Res.* **111**, A06104 (2006a); S. Matsukiyo and M. Scholer, *Adv. Space Res.* **38**, 57–63 (2006b).

[21] M. Scholer, I. Shinohara, and S. Matsukiyo, *J. Geophys. Res.* **108**, 1014 (2003).

[22] S. J. Schwartz, M. F. Thomsen, and J. T. Gosling, *J. Geophys. Res.* **88**, 2039–2047 (1983).

[23] S. J. Schwartz and D. Burgess, *Geophys. Res. Lett.* **18**, 373–376 (1991).

[24] S. J. Schwartz, D. Burgess, and J. J. Moses, *Ann. Geophysicæ* **14**, 1134–1150 (1996).

[25] N. Sckopke *et al.*, *J. Geophys. Res.* **88**, 6121–6136 (1983).

[26] N. Shimada and M. Hoshino, *Astrophys. J.* **543**, L67–L71 (2000).

[27] N. Shimada and M. Hoshino, *J. Geophys. Res.* **110**, A02105 (2005).

[28] E. S. Weibel, *Phys. Rev. Lett.* **2**, 83–84 (1959).

[29] P. H. Yoon, *J. Geophys. Res.* **99**, 23481–23488 (1994).

— 14 —

Final Remarks

Space plasma physics started almost a century ago with the discovery of the ionosphere and a few years later with the realisation that the variations in the geomagnetic field detected another century earlier seemed to be related to something violent taking place in space. This was, not completely wrongly, suspected of having to do with something going on at the Sun. Almost at the same time energetic cosmic rays were discovered whose origin remained completely obscure for long time. These were high energy particles coming from somewhere in the sky. Guided by their existence it was attempted to explain the variations in the geomagnetic field by the pressure exerted by those energetic particles on the magnetic field of Earth causing its compression during the observed magnetic storms. It took another twenty years until, from the observation of comets with their tails pointing radially away from the Sun, it was concluded that the Sun apparently continually blows out a stream of plasma, the solar wind whose origin could not be illuminated by theory or observation but that stimulated a theory of its expansion and the first model of its interaction with the geomagnetic field. The advent of the space age brought the confirmation of all this and the stimulation of theoretical and experimental investigation of collisionless plasmas in parallel in two direction: the single particle approach and the fluid approach, both being discussed in the present book.

We have shown how both of these approaches are related in an ascending approach from single particle to fluid and then to the more detailed and more involved kinetic theories of collisionless plasmas. In a tribute to the origins of space physics we have ended the book with a comprehensive account of reconnection and collisionless shocks which are now widely believed to be responsible for the existence and acceleration of fast particles, possibly even including the acceleration of the solar wind by small scale though large amplitude shock waves in the solar corona and blowing the solar wind, the continuous plasma flow, out of the solar atmosphere into interplanetary space.

The theoretical part of the present book closes with the linear kinetic theory of waves in a plasma. In this edition we added a chapter on instabilities in order to round the discussion up without the necessity of opening our companion volume which deals with instabilities in a more complete manner also discussing the nonlinear effects occurring when unstable waves grow to large amplitudes. As applications we have added a chapter on collisionless reconnection and the chapter on collisionless shocks. Two more subjects, the theory of substorms, and a chapter on auroral physics might also have been in place. These, however, would have unreasonably blown up this volume. We therefore decided to stay just with reconnection and shocks, two of the most interesting and best investigated plasma-physical subjects in the complicated field of space plasmas. In the last decades both progressed substantially up to a state where it makes sense to summarise the achievements even in a textbook, intended for giving the reader a few examples of the validity and importance of collisionless space plasma physics. Nonetheless, these chapters break with the tutorial style of this volume. The reader may forgive us. Description of sophisticated applications in a necessarily short chapter would be impossible. We just provide a brief overview of the interesting physics involved such that the reader might smell the odour of the ongoing research. Of course, the physics of substorms and aurora would also have deserved a similar service; however these fields are much more geophysical than plasma physical. Hence making a cut between them and the other two might make sense.

In Chapter 12 we briefly explained how reconnection works explicating where the problems are buried and where progress is to be expected within the next decades. Such progress is unimaginable without the growing efficiency of numerical work and computational capability. In Chapter 13 we explained how shock waves grow out of waves which are excited in a plasma, a question related to the stability of a plasma, the spontaneous, induced, or forced generation of oscillations and waves and their growth to large amplitudes. We did not go into detail of how these waves react onto the plasma, however, genuinely a nonlinear and dissipative process. These processes are of general importance in space plasma physics, solar physics, and astrophysics. Some of them can be found in our companion text, *Advanced Space Plasma Physics*.

The present *Third Edition*, however, refers in *Appendix A* to four more recent developments, the precise theory of the *Kappa Distribution* which had so far without justification been introduced as a fit to observed particle distributions in space. We present its statistical mechanical theory which justifies its wide use and supports its further applications. It describes stationary states in a collisionless plasma far from equilibrium. We then turn to a discussion of magnetic *Mirror Modes* whose physics until now is barely understood. They are nonlinear plasma states in the diluted, high temperature, ideally conducting space plasma, where they evolve under certain conditions, resembling metallic superconductivity. We also added a section on *Plasma Turbulence*, a rapidly growing rather complex field which is heavily grounded in observations, spectral theory, the theory of complexity, and in numerical simulations based mainly on magnetized fluid-theory. Finally, we superficially describe a new engineering field in Space Plasma Physics called *Space Weather*, which has the character of a *Space Meteorology* (or in an even wider context *Space Climatology*) in an attempt of predicting any hazards civilization may become exposed to, from processes taking place in the space plasma environment of Earth. This scientific activity has recently achieved much attention which we felt should be acknowledged even in a non-engineering text as the present one.

— Appendix —

A — Additions to the Third Edition

In this Appendix A we take the opportunity to add a few sections which are intended to extend the general text in order to account for some more recent achievements in Space Plasma Physics. Thanks to the increasing availability of single and in particular multiple spacecraft flying near Earth and to other planets, their moons, comets, and even to selected asteroids, space plasma physics evolves with enormous speed. Covering the entire field and even the most spectacular new discoveries present an impossible task. However, the overwhelming majority of achievements is in observations, not theory. Most of them either fit into the known theoretical framework, which is covered by the content of this book and our companion monograph (which should of course be updated as well), or they are still awaiting to be explained and built into the accepted theoretical framework of space plasma physics which to a large degree is already completed, just awaiting some refinements. Because of the complexity of the fundamental equations, these however, require mainly numerical work which with higher spatial and temporal resolution as well as in higher dimensions. This lies outside the scope of the present book.

Since space and time inhibit an full account of the mass of knowledge which has accumulated during the past decades, we in the two appendices below just pick out to more detail two theoretical points which should from our restricted angle of view be added. In addition we add a third appendix which is devoted to a brief account of space weather, a newly installed more engineering space science which has grown in the recent decades because of its vital importance for civilisation and humanity. We stay with a few remarks only not to expand the book unreasonably and also not to confuse the reader. This book is a textbook on the basics of space plasma physics, not a monograph which wants to cover the whole field. This limits the selections to those problems only which with the help of the former chapters are accessible to the student.

A.1   Olbert's Kappa Distribution

In Chapter 6 we introduced the so-called kappa-distribution as paradigm of a naturally observed probability or particle distribution function which obeys a Gaussian core but to its flanks deviates from the usual bell-shaped probability distribution exhibiting pronounced wings which in most observations are not exponential but asymptotically power law. This power law behaviour has been confirmed in many space plasmas under various conditions. It has been found in the more energetic particle distributions measured in space, both ions and electrons, as also in wave power distributions. Examples are the famous cosmic ray particle energy spectra which span, with several excursions indicating different particles and also different processes, roughly twenty orders of magnitude in energy and momentum. Other examples are the energy flux spectra of well developed stationary turbulence which can be understood with reference to Kolmogorov's ingenious now eighty years old insight that the spectral energy flux in cascading turbulence should be stationary. Though this assumption very well reproduces the observations at least in some spectral ranges, its very mechanism is still far from being understood, which is reflected in the vivid observational and theoretical attempts to develop the theory of turbulence.

From the observational point of view the Kappa distribution simply serves as a fit function to the observations reproducing either the particle energy flux or spectra. It was invented as an attempt to fit particle spectra in the magnetosphere by Stan Olbert at MIT in 1965 who realized that a simple rational function including just one free exponent κ, a real rational number, provided a reasonable fit to a skewed Maxwell distribution. Two of his students were given this master function to apply it to ion and electron fluxes recorded by the Imp spacecraft in the Magnetosphere. The success was overwhelming even though the obtained fitted temperatures and the origin of the functional form of the Kappa distribution, as it was subsequently called, remained obscure for very long time.

That power law tails on Maxwell's distribution could be generated by wave particle interaction was of course known for quite long. All such distributions result from nonlinear interaction between waves and charged particles, a time dependent process which is rooted in diffusion theory in momentum space. The fundamental equation is the Fokker-Planck equation for the probability $p(x,t)$ of a particle

$$\frac{\partial p}{\partial t} = a\nabla_x(xp) + D\nabla_x^2 p \tag{A.1}$$

where a and D constants, a is the coefficient of the convective term, and D is the diffusion coefficient. For $a = 0$ one simply has the heat conduction (diffusion) equation whose solution is the time dependent decreasing Gaussian profile of a heat pulse

$$p(x,t) = (2\pi Dt)^{-1/2}\exp\left(-\frac{x^2}{2Dt}\right) \tag{A.2}$$

The more interesting case $a \neq 0$ yields a stationary Gaussian profile solution $\partial_t p = 0$ when the convective gain just compensates for the diffusive losses

$$p(x) = \left(\frac{a}{2\pi D}\right)^{1/2}\exp\left(-\frac{ax^2}{2D}\right) \tag{A.3}$$

which suggests that with non-stationary coefficients the Gaussian profile can become distorted. This is achieved with the inclusion of convective and diffusive terms in the Boltzmann equation for the distribution function $f(\mathbf{x},\mathbf{v},t)$, the genuine Fokker-Planck equation

$$\left[\partial_t + \mathbf{v}\cdot\nabla + \frac{q}{m}(\mathbf{E}+\mathbf{v}\times\mathbf{B})\cdot\nabla_\mathbf{v}\right]f = -\nabla_\mathbf{v}\cdot(f\langle\Delta\mathbf{v}\rangle) + \frac{1}{2}\nabla_\mathbf{v}\nabla_\mathbf{v}:(f\langle\Delta\mathbf{v}\Delta\mathbf{v}\rangle) \tag{A.4}$$

which describes convection and diffusion of the particle distribution in velocity (momentum) space under the action of velocity (momentum) fluctuation-induced advection and diffusion. The process described by this equation is called *diffusive acceleration* and is applied widely in space plasma physics to cover the acceleration of particles to high energies by collisionless shocks. This of course requires knowledge of the velocity space and real space diffusion coefficients for which in application to high energy particles various forms have been given, mainly based on pitch-angle scattering of energetic particles on Alfvén waves, ion-cyclotron waves for ions, and whistlers for electrons, including their spatial dependence. The favourites of such diffusive acceleration are the environments of shocks as we discussed in Chapter 13. All those processes are non-stationary as becomes clear from the above briefing.

In contrast Olbert's Kappa distribution is a thermodynamic stationary distribution. Attempts to base it in statistical mechanics go back to the nineties, when it was found that they actually describe a thermodynamic generalization of the Boltzmann distribution in application to stationary states far from thermal equilibrium. They are most easily constructed generalising the Boltzmann-Gibbs partition function which is given by

$$\mathscr{Z}(\mu, V, T) = \sum_i \exp\left(-\frac{\varepsilon_i - N_i \mu}{k_B T}\right) \tag{A.5}$$

where i numbers the micro-states, μ is the chemical potential, T temperature, N_i the number of particles per state, and $\varepsilon_i = p^2/2m$ particle energy. Now following Olbert's prescription one replaces the exponential by the Olbert function to obtain

$$\mathscr{Z}_\kappa(\mu, V, T) = \sum_i \left[1 + \frac{\varepsilon_i - \mu N_i}{\kappa T}\right]^{-\kappa - s} \tag{A.6}$$

where $\kappa \in \mathscr{R}$ is a real rational number, and s is a constant. This has to be extremized in the usual way to obtain the normalised average probability of finding particles of energy ε_i in state i

$$p_{i\kappa}(\varepsilon_i, \mu, T) = \mathscr{Z}_\kappa^{-1} \left[1 - \frac{\varepsilon_i - \mu N_i}{\kappa T}\right]^{-\kappa - s} \tag{A.7}$$

where κ can still be a function of energy and temperature but in general is taken constant. Since $\mathscr{Z}$ is just a factor here, the probability is a power law decaying with energy roughly as indicated. However, any thermodynamic quantities like the average energy, free energy, fluctuations, must be obtained from the above Olbert partition function $\mathscr{Z}$ by the thermodynamic prescriptions. The first and most important is that the application to an ideal gas must lead to the ideal gas equation of state.

For this to achieve one transforms the sum into a three-dimensional integral over momentum space, with **p** the particle momentum, and normalises to the known density $\langle n \rangle = \langle N \rangle / V$ and the quantum phase space element $\hbar^3$. Then calculating the energy moment $\int \varepsilon d\varepsilon f(\mathbf{p})$ of the distribution function $f(\mathbf{p})$ gives the average energy density (or pressure P) of the ideal gas and the equation of state

$$P = \frac{3}{2} \frac{\kappa}{\kappa + r - 5/2} nT \tag{A.8}$$

Three degrees of freedom require that $P = \frac{3}{2}nT$ which immediately determines the unknown constant as $r = 5/2$.

In this light Olbert's Kappa distribution turns out a valid thermodynamic quasi-equilibrium distribution function which can be used to describe stationary states far from equilibrium. Any thermodynamically relevant properties of such systems can then be obtained from the thermodynamic prescriptions.

The imposed stationarity is of course provided by internal interactions which however in the statistical mechanical description are hidden in the value of κ which for any relevant physical problems must be assumed $0 < \kappa \ll \infty$. It is easy to demonstrate from the distribution function that $\kappa \gg 1$ reproduces the classical Boltzmann distribution and thus ordinary thermodynamics. It is interesting to note at this place that a strict theory for electron interactions in pure electron plasma including spontaneous emission of Langmuir waves, Landau damping (absorption), and induced emission, in fact the three Einstein processes (Einstein coefficients) applied to this case just reproduces the above phase space distribution in the time-asymptotic limit including the approximate power of $r = 5/2$.

In general one would expect that $\kappa(\langle \varepsilon \rangle)$ may become a function at least of average energy $\langle \varepsilon \rangle$ and possibly of space and time as well. In this case the calculation of the distribution function will become rather involved and probably can only be performed by numerical computational methods even in case of stationarity. This is suggested by the above mentioned nonlinear theory of wave-electron interaction where stationarity is obtained in the long time limit only. The evolution of κ should then cause intermediate deformations of the distribution function and affect any statistical mechanical properties.

The above theory is not restricted to classical non-relativistic systems. It could be extended to relativistic energies of particles. It is easy to extend it to include a high energy cutoff ε_c through $\exp(-\varepsilon_i/\varepsilon_c)$ in order to exclude the divergence of higher moments of the distribution function, as this does not change anything on the theory. With its help a large number of moments can be formed from Olbert's Kappa distribution, including average velocities and higher order moments necessary in fluid theory.

Moreover, the theory has also been extended into the quantum domain where one derives a kappa-Fermi and kappa-Bose distribution including Bose-Einstein condensation. These can be applied to quantum interactions, a very important achievement in solid state physics and also high density quantum plasmas where skewed distributions may be expected for instance in the interaction between Fermions and phonons to produce or limit superfluidity and superconductivity.

Further Reading

In the bibliography at the end of this chapter we provide just a few references for the interested reader. Since the eighties of the past century the kappa distribution has received wide attention in the space plasma community. It has been applied to particle spectra as well as to obtain the various modifications of plasma wave dispersion relations, wave particle interactions, and also nonlinear interactions appearing in the so-called plasma turbulence (the perturbation theoretical approach to higher order wave-particle interaction the elements of which are contained in our companion monograph *Advanced Space Plasma Physics*). Many of those attempts of formally using the kappa distribution are purely academic calculations replacing the Maxwell-Boltzmann equilibrium distribution by the expression given in Chapter 6 in the hope to find some fundamentally new result. Since, as explained above, that form of the kappa distribution is just a fit but otherwise non-physical, little could be learned from it. Only when the relation of the kappa distribution to thermodynamics was ultimately clarified, its real value came to light. The number of publications on the kappa distribution developing various aspects of its theory and also applying it has since exploded. Of the vast amount of papers we just mention the earlier ones here. The reader may inform himself after having been guided by the small selection below.

The first public mentioning of the kappa distribution by Stan Olbert was on an International Conference at Boston College the proceedings of which were not published before 1968 [5]. It found its first application in [1], an unpublished PhD thesis, followed by its first journal appearance [11] which since has incorrectly been acknowledged as its very origin. Among the later important experimental applications in space plasma physics we just mention [2]. An important theoretical support the kappa distribution found when it was shown [3] that electrons in a photon bath do indeed assume a power law shape with power determined by the photon flux. Similarly in [12] it was demonstrated by a more elaborate theory that a complete interaction theory of electrons in plasma when including spontaneous emission (instability), induced emission (wave particle interaction), and absorption of/scattering by Langmuir waves indeed generates a kappa distribution with the correct thermodynamic properties in the long-term limit. Tentative arguments based on chaotic Lévy flight dynamics of particles [7] provided some further theoretical evidence for the formation of stationary power law distributions far from equilibrium. The first theory based on thermodynamic arguments provided [8] was based on Boltzmann statistics with a modified collision term guided by the inverse functional form of Boltzmann's entropy. By standard methods it resulted in a realistic kappa distribution but lacked application to the required thermodynamic constraints. These were given later [9, 10]. A fairly complete compilation of the properties and applications of the correct kappa distribution is contained in [4] including a complete list of references. The application to the construction of fluid moments and thus justification of the kappa distribution as basic to fluid theory far from equilibrium and the recipe for calculation of higher moments by adding an exponential cut-off is contained in [6]. Extension of the kappa distribution and the related statistical mechanics to relativistic energies is formally possible as well. Recently it also became possible to extend the theory into the quantum domain by deriving Fermi and Bose distributions thereby providing the functional forms of Olbert's entropy.

A.2   Mirror Modes

In Chapter 11 we briefly considered the magnetic mirror mode as one of the fundamental low-frequency instabilities which may, under certain conditions, arise in magnetised high temperature plasmas, like those of the magnetosheath, behind the Earth's bow shock wave. Mirror modes have also been observed in the solar wind, mostly as single magnetic drop-outs which are believed to be nonlinear so-called solitary structures or *solitons* (see our companion monograph *Advanced Space Plasma Physics*) arising in one dimension from self-interaction of the growing low frequency mode and entering a temporary balance between gradual steepening of the amplitude and dispersion of the Fourier superposition of a broad spectrum of waves of different wavelengths. These solitary structures differ substantially from the long chains of mirror modes much more frequently observed.

An example of such chains observed in the dayside magnetosheath is reproduced in Fig. A.1, which shows a ten minutes chain of magnetic undulations passing over the spacecraft transported by the fast ambient plasma flow velocity of few 100 km/s. Such structured chains of magnetic bubbles are practically stationary in the flow. The differences from bubble to bubble are simply caused by the geometrical arrangement of the bubbles to form the undulating magnetic texture. Not shown here is the necessary pressure balance between the magnetic and plasma pressure respectively plasma density (for the assumed grossly constant temperature). The density N or rather pressure $P = Nk_BT$ should be (and of course is) in approximate anti-correlation with the magnetic field $(\delta B)^2/2\mu_0 \sim \delta Nk_BT$, except for small temporary and transient fluctuations.

Figure A.1 shows in addition also the electric field spectrum in a wide spectral range of four decades with the magnetic field overlaid as white line in the form of the electron cyclotron frequency $f_{ce} = eB/m_e$ which oscillates around $\nu \sim 1$ kHz. The range above f_{ce} contains basically electrostatic signals (for explanation see Chapter 10) except for some signals which are caused by the Fourier trace of very steep magnetic and density gradients, mostly those weak signals which reach up to 100 kHz which map the steep boundaries of the electron mirror modes. The intense emission below f_{ce} are whistlers excited by the whistler instability of trapped electrons (described in Chapter 11). The observations of 30% magnetic depletions indicate that the mirror mode is firstly large amplitude and by no means an infinitesimal linear oscillation, secondly the large excursions into the high frequency spectral range indicate the presence of very steep magnetic and density gradients which in linear infinitesimal mirror modes should noch exist; they belong to an electron mirror branch as noted in passing in the mirror mode subsection in Chapter 10.

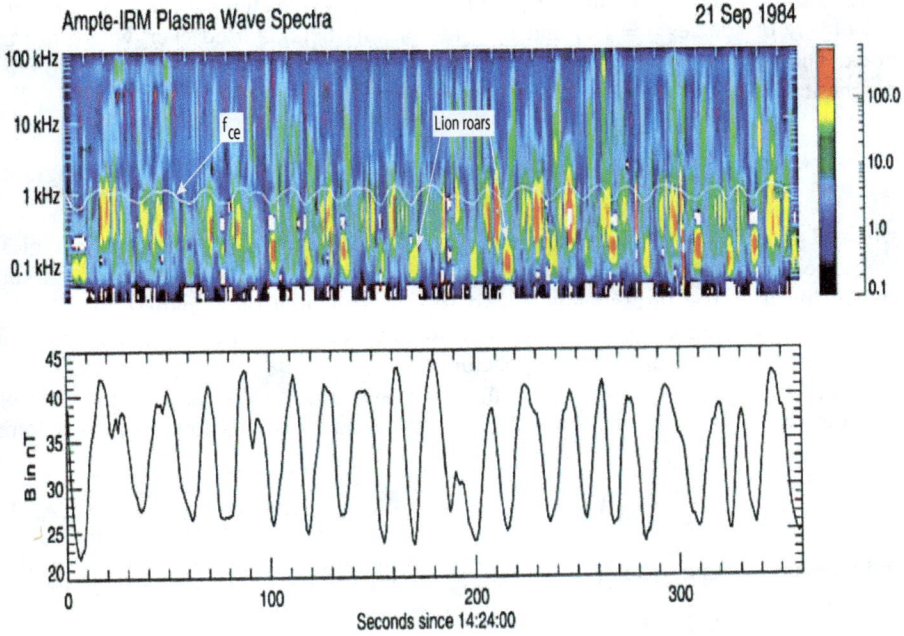

Figure A.1: Typical mirror mode oscillations in the magnetosheath as observed by the AMPTE-IRM spacecraft during a 10 min time interval. The lower panel shows the variation of the magnetic field exhibiting regular drop outs of the field for on the average roughly 5 s. The reduction of the magnetic amplitude amounts to $\sim 30\%$ at an external undisturbed field of $B \sim 40$ nT. These oscillations of the magnetic field are about stationary structures swept over the spacecraft by the plasma flow of speed of a few ~ 100 km/s. The irregular shape is caused by the oblique geometries and the magnetic texture of the mirror mode bubbles and the superimposed electronic mirror mode structures which by the moderate resolution of the instrument are not resolved. The upper panel shows simultaneous measurements of the spectrum of the electric field. The oscillating white line is the colour coded electron cyclotron frequency f_{ce} which exactly follows the magnetic field in the bottom panel. Above this line the waves are of electrostatic nature. Low frequency whistler modes in the bubbles are indicated as lion roars. The higher frequency lion roars near but below f_{ce} are caused by electron mirror modes and trapped electrons at the flanks of the mirror bubbles.

In order to understand the physics one needs considering the full kinetic dispersion relation of the mirror mode which yields the normalized to the ion cyclotron frequency growth rate

$$\frac{\gamma(\mathbf{k})}{\omega_{ci}} \approx \frac{k_\parallel \lambda_i}{A_i + 1} \sqrt{\frac{\beta_{i\parallel}}{\pi}} \left[A_i + \sqrt{\frac{T_{e\perp}}{T_{i\perp}}} A_e - \frac{k^2}{k_\perp^2 \beta_{i\perp}} \right] \tag{A.9}$$

Here λ_i is the ion inertial length, $A_j = (T_\perp/T_\parallel)_j - 1 > 0$ the temperature (or pressure) anisotropy of species $j = i, e$, and $\beta_{\parallel,\perp}$ the plasma beta as used everywhere in this book. The wave grows under the condition that $\gamma > 0$ which yields a condition on the magnetic field

$$B < B_{crit} \approx \sqrt{2\mu_0 N T_{i\perp}} \left(A_i + \frac{T_{e\perp}}{T_{i\perp}} \right)^{1/2} |\sin\theta| \tag{A.10}$$

where $\theta = \sin^{-1}(k_\perp/k_\parallel)$ is the angle of propagation (i.e. the direction of the bubble). Once the local magnetic field drops below this critical value, the mirror mode should evolve linearly. The threshold is largest for perpendicular and smallest for parallel propagation (inhibited by the last term in the growth rate) where the wave does not exist anyway. Mirrors are inclined structures.

If in the above growth rate neglecting the ion anisotropy, only the electron anisotropy remains, and one obtains the pure electron mirror mode, a much shorter scale structure

$$\frac{\gamma_e(\mathbf{k})}{\omega_{ci}} \approx k_\parallel \lambda_i \sqrt{\frac{\beta_{e\parallel}}{\pi}} \left[A_e - \frac{k^2}{k_\perp^2 \beta_{e\perp}} \right] \sqrt{A_e + 1} \tag{A.11}$$

and threshold magnetic field

$$B < B_{crit,e} \approx \sqrt{2\mu_0 T_{e\perp} A_e} |\sin\theta| \tag{A.12}$$

This electron mode is shorter scale because usually $\beta_e > \beta_i$. However, its threshold is higher than that for the pure ion mirror mode because usually the electron temperature in collisionless plasmas, because of the weak interaction between both particle species, exceeds the ion temperature $T_e > T_i$. In a given external magnetic field it can therefore start already at a higher field value than the ion mode but can be most easily grow inside ion mirror bubbles where the magnetic field is already reduced. On the other hand, due to the higher magnetic threshold value it can exist completely on its own if the threshold condition is satisfied. Electron mirror bubbles also cover a larger angular range in θ than the ion mode.

We show a high resolution Equator-S spacecraft observation of two adjacent mirror bubbles from a long sequence in the magnetosheath in Fig. A.2. The ion mirror mode in this figure is particularly large with a $\sim 50\%$ magnetic depletion $\Delta B/B$. This observation nicely resolves the bubbles including the modification of their walls. Since on the short time scale it is highly improbable that these modifications are caused by fluctuations in the streaming velocity, they are clear indications of the presence of chains of electron mirror modes inside the ion bubbles. These are much lower amplitude while deforming the ion bubble at its bottom and on the flanks in quasi-regular manner. The high time resolution moreover allows seeing the small amplitude magnetic modulation caused by the excitation of lion roar whistlers as seen in Fig. A.1 by those electrons that are trapped both in the ion mode as well as in the electron modes. This is seen in different places of the magnetic shape thus confirming the existence of trapped electrons in both modes.

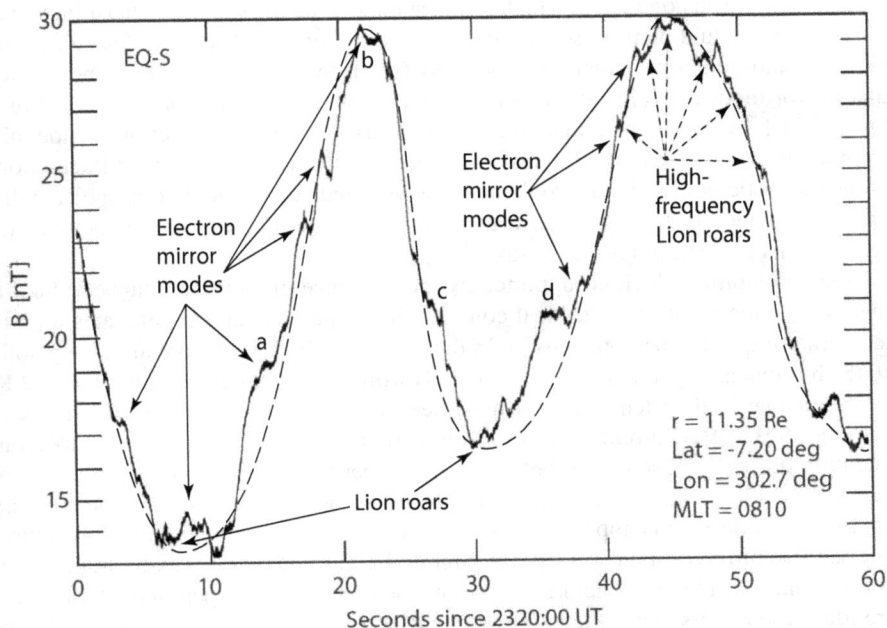

Figure A.2: High resolution mirror mode observation from Equator-S spacecraft. Shown are just two mirror bubbles. In this particular case the magnetic drop out amounts to $\sim 50\%$ of the external field amplitude. The resolution of the instrument is high enough to resolve the modulation of the magnetic profile caused by the electron mirror mode as indicated by the small letters, as well as the very small amplitude electron whistler waves which are attributed to the lion roar noise mostly trapped in the electron mirror oscillations.

The notorious observation of extended chains of such large-amplitude mirror-mode bubbles (shown in the figures) in high temperature plasmas poses a most interesting and also most difficult problem which so far has not been satisfactorily solved. Claims that, in order to maintain pressure balance, the magnetic depletions simply suck in particles from along the magnetic field are hardly valid, because bubbles occur in extended chains, as seen in Fig. A.1 where they cover more than 10 min time (in fact the observations extend over the much longer time of several hours). Even less believable are claims that nonlinear wave-wave interactions causing anomalous collisions and thus dissipation would heat the internal plasma to temperatures which become capable of balancing the pressure imbalance. The moderate whistlers and weak ion sound waves observed inside mirror bubbles could if at all hardly generate sufficient anomalous collisions in comparably short times, in particular when accounting for the incapability of whistlers to be involved in such processes anyway, as they affect just a minor resonant trapped electron population and not the bulk electrons. In addition such claims contradict thermodynamics. Plasmas on those scales are manifestly collisionless, ideal conductors which produce insufficient dissipation. Quasilinear calculation in the fluid picture (see our companion monograph *Advanced Space Plasma Physics*) and numerical kinetic calculations (see reference list below) demonstrated that mirror modes should saturate on extremely low quasilinear magnetic amplitude levels which by far underestimate the really observed magnetic depletions inside mirror bubbles. Quasilinear amplitudes would become drown in the thermal fluctuations of the magnetic field and thus would never be detected. Such calculations indicate that pressure balance is readily reduced to near zero by the quasilinear interaction thus requiring its fast continuous reconstruction.

One possibility which is supported by the existence of a critical magnetic field is that at high temperatures and ideal conductivity some mechanism generates a high-temperature quasi-superconductivity in the plasma (which is the dream of any solid state physicist as superconductivity in metals works below a temperature of $T \sim 2$ K, and moderate critical temperatures have been reached only using special alloys or extreme pressures. Currently achieved temperatures in high-temperature superconductivity are still substantially below room temperature which, compared to Fermi temperatures in metals of $T_F \sim 10^4 - 10^5$ K are in fact low). Plasmas at the contrary are by their nature high temperature with T in the range of metallic Fermi temperatures but because of their diluteness have negligible Fermi temperatures even though their about infinite conductivity naturally satisfies the condition of superconductivity: they are ideal conductors. Thus they are the opposite to metals, subject to the Boltzmann distribution and thus no superconductors because it is believed that no condensate can form under such conditions.

The problem of understanding the evolution of the mirror mode hints on the possible necessity of a revision of this notion. The generation of mirror bubbles might in fact arise from a Meissner-London-Landau-Ginzburg effect in high-temperature classical plasma in the course of which a localised global diamagnetism evolves as described by Ginzburg-Landau theory. Ideas in that direction came up recently and are based on the fact that particles trapped in any linearly excited very-low amplitude mirror bubble undergo bounce motions in the mirroring magnetic field. Close to their mirror points those particles have large perpendicular energy and very low parallel velocities. At those localities they can resonate with the permanently present thermal ion-acoustic background and for some short time become locked to the ion waves. All mirroring particles having their mirror points along the perpendicular wavelength of the ion sound wave are in this case simultaneously in resonance, locked to the same wave and forming a particle condensate which is correlated over the perpendicular wavelength of the ion sound wave, roughly one ion gyro-radius. Since this happens to a large fraction of trapped particles in the bubble at all times, a fraction of permanently resonant particles may form a stationary condensate all over the bubble with the exception of the region around the magnetic minimum in the bubble centre which is void of mirror points and where the particles have largest parallel speed.

The presence of such a condensate with large perpendicular energy involves a large anisotropy which is basically provided by the electrons and thus attributes to the above growth rate of the ion mirror mode as well as to an electron mirror mode, which destroys the linear and quasilinear state of the bubble evolution, letting it grow to much larger amplitudes than quasilinear. This process can approximately be described by Ginzburg-Landau fluid theory of superconductivity, because in the collisionless plasma the condensate behaves like a superfluid, which causes a Meissner effect and results in the formation of large magnetic bubble amplitudes. If these ideas are correct, a high temperature plasma of the kind of that of the magnetosheath downstream of the collisionless bow shock behaves like a very high-temperature super-conductor, and mirror modes are the result of the generation of macroscopic diamagnetism in close similarity to solid state superconductors, even though the very mechanism of condensate formation is different and classical.

A.3   Turbulence

When referring to turbulence we break the structure of this book which so far dealt with the basics of space plasma physics only. Turbulence is a much wider field. Mankind was certainly aware of it for millennia when observing the flow of water in rivers and creeks or the fast changes in the structure of clouds. It may have pondered about its causes attributing them to Ariel, the god of the winds who blew to move them around on the sky. The Greeks noted it when stating that the river remains the same and unchanged, while the flow changes continually when flowing down the river. One never steps into the same river was the wise observation of Heraclitus, but he or any of the Greek philosophers had not any idea about the structure and cause of turbulence. This remained so until recent time. Leonardo drew pictures and thought a life long about turbulence but was not in the possession of the instrumentation which is necessary for its understand.

Emergence of Turbulence

In principle turbulence has not primarily to do with plasma or even space plasma physics. It is a general property of flows at some large critical number called the Reynolds number in fluids or magnetic Reynolds number in electrically conducting magnetised fluids described in MHD. At large Reynolds numbers the fluids loose their viscosity respectively resistivity thus dissipation is suppressed, which allows any small disturbance to grow to large amplitudes and decay into other shorter amplitude disturbances by the intrinsic nonlinearity of the flow contained in the dissipationless convective derivative of the velocity

$$\frac{d\mathbf{v}}{dt} \equiv \left(\frac{\partial}{\partial t} + \mathbf{v} \cdot \nabla \right) \mathbf{v} = 0 \tag{A.13}$$

shows that the fluid velocity is undergoing a self-interaction whenever it changes in the course of the flow. For instance, for a simple sinusoidal disturbance of the flow in space $\mathbf{x} = z\hat{\mathbf{z}}$ a term $\sin z \cdot \cos z = \frac{1}{2} \sin 2z$ arises, which shows that the sinusoidal oscillation introduced into the flow generates a shorter wavelength disturbance which

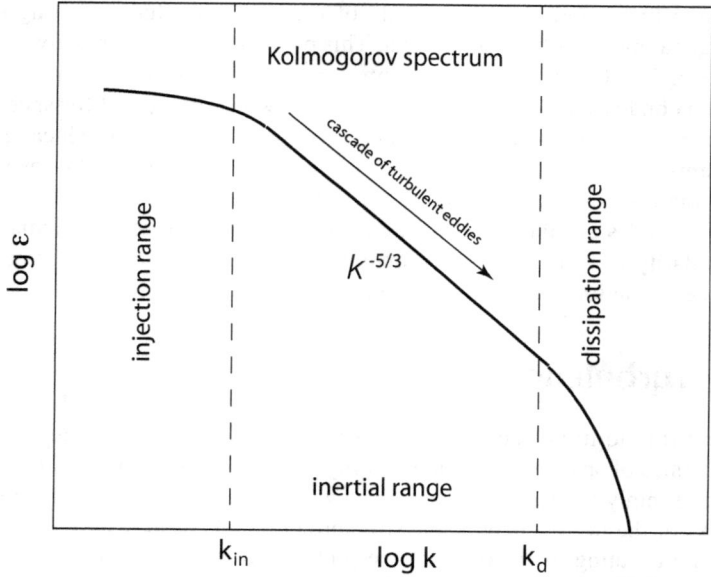

Figure A.3: A schematic of the Kolmogorov spectrum of turbulence. It shows the turbulent energy spectrum of the eddies as function of the inverse scale k (turbulent wave number). The eddies start at long wavelengths, the energy injection range $k < k_{in}$. Eddies at the injection wavenumber undergo turbulent cascading by causing energy flow down to smaller scales into the direction of increasing k which produces the power law shape of the spectrum with power $-5/3$. At the dissipation wavenumber k_d the energy flux enters the dissipation range $k > k_d$ where the eddies are destroyed by some unspecified mechanism and lose their energy to microscopic heating.

adds to it and thus steepens the sinusoidal oscillation. This well known behaviour of so-called *simple waves* is at the heart of generation of a whole spectrum of short waves which cause the emergence of rather steep turbulent vortices in the flow. Though this happens in uncharged media, it also happens in plasmas simply because the above equation is completely general and holds as well in electrically charged fluids. Neither the Greek, nor Leonardo knew about it and so could not grasp the mechanism of generation of turbulence, which, however, turns out much more involved than just given by the above simple total time derivative that just hints on why turbulence consists of a whole spectrum of steep eddies in all non-dissipative flows at sufficiently large flow velocity. Such flows are in general not laminar, a property which they have only at low speeds and sufficient viscosity which readily wipes out any small scale disturbance and transforms its free energy into friction and heat.

One way to turbulence is based on the theory of complexity. This is a purely mathematical approach which is considered to describe the emergence of turbulence in a recursive way by considering the fractal evolution of structure resulting from recursive relations in complex space, recursively mimicking the self-interaction contained in the convective derivative by transforming it into finite differences. It started with Poincaré's 1905 observation that particle motion in classical mechanics is rather vulnerable to small unpredictable disturbances causing the path of the particle to become erratic. This led Lyapunov to investigate stability on general mathematical grounds and formulate his stability criterion. These mathematical considerations are highly sophisticated and are believed to have wide application in physics. In particular it is believed by a large community that they pave the way to the microscopic understanding of emerging turbulence from thermal chaos.

Plasma Turbulence

Turbulence in plasmas is, however, different from turbulence in neutral fluids in several respects. It involves not just the hydrodynamic equations or, when including just the magnetic field **B**, the magnetohydrodynamic equations, but since any disturbance emerges from small kinetic scales, evolves to and through larger scales back again into small scales, it also involves the fully kinetic plasma theory. This had led to distinct between *kinetic plasma turbulence* and *turbulence in plasma*. The former is turbulence in collisionless high temperature plasmas where one remains on the level of kinetic theory. It is also called *weak turbulence* in the sense that it is subject to perturbation theory based on the number of waves which do indeed mutually interact. The lowest order in this theory is *quasi-linear theory* where the linear waves, which are excited by the available free energy of the particle population, just react back onto the particle distribution to deplete its unstable part and match it back again into the stable Maxwell-Boltzmann distribution. This generally leads to a rather low saturation level of the initial instability which had started growing from thermal noise. Nevertheless, in many cases this suffices to describe a large number of kinetic processes in collisionless plasmas like the stability of Earth's Radiation Belts in interaction between the trapped bouncing particle distribution and the whistler wave spectrum excited by it. Interestingly enough, this process of saturation and the wave spectrum can and have been observed in space in the magnetosphere and have for long time been the best documentation of wave particle interactions.

The next order-of-magnitude effect in kinetic plasma turbulence is *three-wave* interaction, where two waves interact to generate a third wave. From here one can step up to higher orders, include particles interacting with three waves and more waves. This theory readily becomes very complicated, though is in principle perturbatively manageable because it contains a small parameter, the energy ratio of the wave to plasma and magnetic energy which serves as the necessary expansion coefficient. It is the equivalent to the coupling constants in the field theories of particle physics. There exist quite good presentations of this kind of theory, its elementary state is also contained in our companion monograph *Advanced Space Plasma Physics*. There are, however a number of much more elaborate accounts available in the literature.

One readily realizes that another branch of kinetic turbulence exists when considering different scales. Short scale waves riding on long waves in such a *mean field* theory play the role of disturbances which can be averaged to account for their mean effect on the evolution of larger-scale waves. This is the main idea of mean field theory, in this case applied to turbulence. It causes the amplitude of the large-scale waves to slowly evolve in time and space under the average effect of the small-scale turbulence. This evolution causes a number of interesting effects known as *solitary waves* or *solitons* which are single wave packets with very little dispersion. They can propagate on the plasma over long distance, and can also appear in chains, if the plasma is dissipationless. In case the steepness of these waves reaches the dissipation scale, they however start dissipating at their fronts and evolve into shock waves or weak shocklets which are still single and of small scale and thus different from those large and strong shocks which we have considered in Chapter 13. In general these shock waves are microscopic still small scale and low amplitude. However when they occur in large numbers in high temperature plasma, then they become an important component of turbulence in the scale range of their wavelengths.

Space Plasma Turbulence

In the three past decades the space plasma community has dug into the problem of observation of turbulence in space, the solar wind in the first place, and the magnetosheath, the two regions which are known to be turbulent and being sufficiently accessible to measurement and observation with single and multi spacecraft missions. Meanwhile a wealth of observations has been accumulated which for lack of space we cannot discuss here. The advantage of the solar wind, for instance, is that it is over wide ranges a quasi-stationary plasma flow that allows the accumulation of long time series of the magnetic field, temperature and density fluctuations which can then be analysed to determine their Fourier spectra. Since the flow is quasi-stationary and comparably high speed of few Hundred km/s, it is also assumed that the spatial structures of the turbulence are very simply convected over the stationary spacecraft which is practically at rest against the solar wind, such that the time spectrum in fact mimics the spectrum of the turbulent spatial structures. This is the so-called *Taylor assumption* put forward in 1938 which in application turned out to be rather fruitful. Since magnetic fields rather than the fluctuations of velocity, density, temperature and the particle distribution are easiest to measure, mostly magnetic turbulence spectra have been analysed in the solar wind, based on the Taylor assumption. The surprising most frequently found result was that the turbulent magnetic spectra obtained obeyed a spectral power law shape of about precisely the kind predicted for fluid turbulence already in 1941 by Kolmogorov in his fundamental attempt to understand the formation of turbulence. It holds for stationary and isotropic turbulence. It is this Kolmogorov theory which still prevails even though a number of improvements have been constructed. For instance dropping the assumption of isotropy, which in a sufficiently strong magnetic field which imposes a preferred direction on the plasma on scales of the electron and ion gyroradii, is quite reasonable, does not make Kolmogorov theory invalid. It just let it become anisotropic with slightly different power laws parallel and perpendicular to the field. The turbulence becomes more perpendicular then which, however, in the solar wind is just a minor, actually negligible effect for the weak fields and high speeds encountered there. The turbulent velocity spectra are practically unaffected, and the density reacts to maintain pressure balance.

Kolmogorov theory resulted from the insight that turbulence in incompressible flows evolves from and into turbulent eddies organized according to their scales. Thus taking a Fourier spectrum of a long record of observations in the solar wind, the energy spectrum $S(\varepsilon)$ of the turbulent fluctuations should exhibit a shape that mirrors the distribution of turbulent energy contained in the eddies and vortices according to their inverse scales, respectively wave numbers k. Under stationary and homogeneous turbulence one expects a certain canonical functional shape of the energy spectrum. This shape in Kolmogorov turbulence turns out power law as shown in Fig. A.3. In developed Kolmogorov turbulence the k-dependence of the spectral energy density is found to be

$$S(k) \propto k^{-5/3}, \qquad k_{in} < k < k_d \tag{A.14}$$

with canonical power $\kappa = -\frac{5}{3}$. Kolmogorov obtained it from simple dimensional considerations. Remember that the simple wave equation suggests shorter wavelength generation by self-interaction of the fluctuations in velocity. These are in any realistic fluid balanced by pressure P and viscosity v and possibly another external driving force $\mathbf{f}(\mathbf{x}, \mathbf{v}, t)$

$$\frac{\partial \mathbf{v}}{\partial t} + \mathbf{v} \cdot \nabla \mathbf{v} = \frac{1}{\rho} \nabla P + v \nabla^2 \mathbf{v} + \mathbf{f}(\mathbf{v}, \mathbf{x}, t) \tag{A.15}$$

with $\rho = mn$ the mass density. The Fourier transform of the fluctuation is given by

$$\mathbf{v}(k) = \frac{1}{8\pi^3} \int d^3 x \, \mathbf{v}(\mathbf{x}) e^{-i\mathbf{k} \cdot \mathbf{x}} \tag{A.16}$$

The total turbulent energy per unit mass (note that the mass m which for all particles is the same here, plays no role) is given by

$$\frac{1}{2} \langle |\mathbf{v}|^2 \rangle = \int dk \, S(k) \tag{A.17}$$

for simplicity restricting to spherical eddies. This energy is dissipated through viscosity at rate

$$\varepsilon_d = 2v \int dk \, k^2 \, S(k) \tag{A.18}$$

which happens at large wave numbers respectively small scales! In stationary fully developed turbulence only two numbers determine the state of turbulence: the energy dissipation rate and the viscosity. Dimensionally this means that $[\varepsilon_d] \sim L^2/T^3$ (with L length and T time). The dimension of viscosity is $[v] \sim L^2/T$. Combination gives the eddy length scale and velocity for dissipation

$$\ell_d \sim (v^3/\varepsilon_d)^{1/4}, \qquad v_d = (v\varepsilon_d)^{1/4} \tag{A.19}$$

and consequently a dissipation time

$$\tau_d \sim \sqrt{v/\varepsilon_d} \tag{A.20}$$

Thus the power spectrum of the turbulence has the self-similar form

$$S(k\ell_d) \sim \ell_d v_d^2 F(k\ell_d) \sim v^{5/4} \varepsilon_d^{1/4} F(k\ell_d) \tag{A.21}$$

with F dimensionless. In the inertial range $k < k_d$ dissipation is negligible, therefore the power spectrum $S(k)$ is scale free and independent on v, and one has

$$F(k\ell_d) \sim (k\ell_d)^\kappa \tag{A.22}$$

Inserting the dimensions for v_d^2 and ℓ_d one has

$$S(k) \propto v^{1/2} \varepsilon_d^{1/2} v^{3/4} v^{3\kappa/4} \varepsilon_d^{-\kappa/4} k^\kappa \varepsilon_d^{-1/4} \tag{A.23}$$

The condition that $S(k)$ does not depend on v then gives immediately that $\kappa = -5/3$, which reproduces the Kolmogorov spectrum. The Kolmogorov constant of proportionality $C_K \approx 1.6$ has been determined by numerical simulations only, as there is no precise theory to fix it otherwise.

If applying the same reasoning to the MHD equations (which contain more terms) in strong magnetic fields where the whole problem becomes essentially nearly two-dimensional, then the Kolmogorov spectrum changes shape. The strong magnetic field leaves its trace on the spectrum and allows only for a power $\kappa \sim \frac{3}{2}$. The Kolmorovov spectrum becomes a somewhat flatter Iroshnikov-Kraichnan spectrum

$$S(k) \propto k^{-3/2} \tag{A.24}$$

Observations have shown that these power laws, in particular the Kolmogorov spectrum, fit the magnetic energy spectrum quite well in a certain spectral range called the *inertial range*, because there the eddies behave dissipationless, and the energy flows from large scales to ever smaller scales as predicted by the picture of simple waves. Large eddies cause small eddies to appear successively. Above this law has been suggested from simple dimensional considerations. It can in fact be derived by a more rigorous theory from the Navier-Stokes equations of hydrodynamic turbulence for the velocity fluctuations, assuming a stationary spectral energy flow, weak dissipation in the inertial range $k_{in} < k < k_d$, and onset of dissipation at $k > k_d$. It can also be obtained under the same assumptions from the equations of MHD, which is somewhat more involved. In higher order one obtains higher moments than energy flux. One then can determine so-called structure functions of the turbulence which allow the construction of the distribution function of fluctuations. This distribution turns out to be of the kind of Olbert's kappa distribution, a very interesting fact which has not yet found sufficient attention. It offers a bridge to the underlying statistical mechanics which is not contained in the fluid theory of turbulence on that all these investigations are based.

Comments on Dissipation in Turbulence

Turbulence remains one of the incomplete scientific fields. Hundred years of more or less intense research have added many different facets to its understanding but so far did not achieve a complete theory which goes substantially beyond the insights of Kolmogorov, Iroshnikov, and Kraichnan (cited below in the list of selected references). It has instead become on the one hand a highly mathematical enterprise in the theory of complex systems and in turbulent field theory, on the other hand it has become a playground for the numerical simulants and at the same time also for the observers of space plasmas who hope to obtain more precise indications of its intrinsic mechanism as well as to discover so far unknown effects. Though, in space we have an ubiquity of observations related to turbulence, in particular in the solar wind, plasma sheet, and magnetosheath, with the well established permanent observation of Kolmogorov spectra though however mostly in a fairly narrow range usually barely covering just one decade in wave number k.

Theory instead predicts that the inertial range $k_{in} < k < k_d$ between injection and dissipation should be of the order of $k_d/k_{in} \sim \text{Re}^{3/4}$. Reynolds numbers Re are very large in all regions of observation, which is due to the large electrical conductivity and thus absence of resistivity and viscosity. There are practically no molecular interactions present of any kind. Therefore one has $\text{Re} \sim 10^3$ or larger, theoretically implying a very extended inertial range of at least three or four decades in k. Observations do strongly contradict this. Why, is barely known. One can however, speculate that in collisionless plasmas nonlinear or anomalous dissipation processes set on much earlier at larger k than on the molecular level which implies that one needs referring to kinetic theory in order to determine the effective collision frequencies, anomalous viscosities and substantially reduced Reynolds numbers. This remains an unresolved problem but shows, where the problems lie. MHD simulations of plasma turbulence, for instance, as done in many of the publications, will not be capable to contribute to any realistic description of turbulence unless artificially large viscosities and resistivities will be included. Both efforts, theories and simulations of generation of those anomalous transport coefficients and simulations of turbulence will have to be combined.

Kinetic plasma theory offers a number of possibilities in that direction. Instabilities in weak kinetic plasma turbulence, the only form of interest in solar wind turbulence, for instance, lead to nonlinear saturation levels which enter the general expressions of anomalous transport coefficients which, when the wave number range of the instability is considered in the turbulent spectrum, should provide sufficient dissipation for causing a deviation from the Kolmogorov shape of the spectrum. This happens for instance in the ion inertial range $\lambda_i^{-1} < k < \lambda_e^{-1}$ which is between the ion and electron inertial lengths. There, kinetic Alfvén waves should naturally become included into the spectrum. If their nonlinear interaction and saturation contributes to the generation of any anomalous transport coefficient, deviations from the inertial range are expected here. Moreover, steepening of other waves in this range as well to become small-scale shocklets or current filaments will generate a mess of other kinetic instabilities inside the steep wave front with effects on the transport properties.

Such effects will naturally let the dissipation range start at much longer scales and smaller wave numbers than expected when it is deferred to the molecular scale range. What is, however, most interesting, is the fact that such small scale structures in the ion inertial range are mainly carried by electrons as the ions are non-magnetic. This implies that they result into small-scale current filaments. These are of the scale of the electron inertial length λ_e and will undergo reconnection. It may thus be suspected that reconnection is the ultimate state of dissipation in turbulent plasmas with locally weak dissipation only which, however, due to the extremely large number of current filaments adds up to become substantial and ultimately strong enough to cut the turbulence off at these scales to let it dip and merge into dissipation of its energy.

With this in mind one generally concludes that probably in all plasmas the ultimate dissipation of plasma turbulence occurs in the ion inertial range just above scales of the electron inertial length $k\lambda_e < 1$ where reconnection in myriads of eddies, which in fact are electron current filaments, generates the required dissipation in order to terminate the inertial range and enter the dissipation range, where the cascade and turbulent spectrum is cut off and the turbulent energy is transferred into heat and acceleration of particles. This may even be the general mechanism of termination of turbulence well above the molecular scale.

There is a final comment to make on dissipation. Plasma Turbulence itself is in the above sense a dissipative process even though in the inertial range it apparently does not cause any dissipation. It however causes the production of ever smaller scale eddies and vortices which ultimately end up in narrow current sheets or shocklets which undergo reconnection and cause the required real dissipation. This implies that in any fluid plasma theory dealing with scales on or larger than the injection scale the usually made ideal assumption is barely justified if only the fluid on smaller scales undergoes turbulence. In this case averaging over the turbulent scale range becomes appropriate to be included to modify the basic MHD equations. This implies that MHD becomes non-ideal and includes anomalous turbulent viscosities and resistivities. Turbulent plasmas on the large scales are then never ideal fluids.

In the turbulence reference list we just select some eclectically chosen citations on more basic work on turbulence.

A.4 Space Climatology and Space Weather

In Chapter 5 we briefly discussed the main magnetic disturbance of the geomagnetic field, the magnetic (or geomagnetic) storm which actually is a *magnetospheric storm* because it affects not just the geomagnetic field, its violent signature on the surface of Earth, but the entire space occupied by the geomagnetic field, the magnetosphere, including its plasma content and its bottom, the moderately resistive ionosphere. Even more generally, it is part of a much larger event, the *space storm* of which it is just the magnetospheric signature, caused by a huge violent disturbance of the *inner heliosphere*, that part of the heliosphere which lies inside a few astronomical units, i.e. the region of cosmic space occupied by the inner planets of the solar system up to the orbit of planet Mars.

Digging into its physics, it has become gradually clear in the past seven or eight decades that the magnetospheric storm itself is not the physically most interesting event. As in many other natural sciences and fields in physics, the most obvious and spectacular process turns out to consist of physically much more important elements. An example is the aurora which for millennia fascinated humans and mankind, at least in the North, but is nothing more than the tiny though beautiful signature of enormously complicated processes taking part in the lower magnetosphere and upper ionosphere. What concerns the geomagnetic and magnetospheric storm in general, its "element", so to say, is the *substorm*, or again the magnetospheric substorm, which itself is the accumulation of many rather complicated and interconnected plasma processes in Earth's environment including its upper atmosphere and the ionosphere. The alert reader, having read the various chapters of this book on plasma processes, will have gotten an impression of the complications arising, when accounting for all those processes together in their effects on Earth and civilisation.

All these processes are certainly very interesting to investigate. Within the past half century, however, mankind has become aware that they can have vast effects on mankind and civilisation in its highly developed modern state which by its high and sensitive technology has become rather vulnerable to any as well natural damages like earthquakes, tsunamis, all kinds of weather impacts, volcanic eruptions, seaquakes, impacts of large meteorites, the slowly increasing solar radiation intensity, catastrophic evens in the biosphere and, probably the most dangerous of all today, the change in Earth's climate caused by civilisation itself, the anthropogenic atmospheric greenhouse effect which has become today's main concern. Though mankind is currently confronted with all those hazards it is exposed to and it has to be prepared for them, awareness has also increasingly arisen concerning the dangers caused by violent events in space physics other than impacts of meteorites or larger objects which, though most dangerous indeed, are very rare with probability of huge impacts only within roughly hundreds of thousands or even millions of years. Civilisation with its advanced technology has meanwhile become vulnerable to much less violent weaker influences from space. This has become obvious when intense currents have been induced in the upper layers of Earth's crust during some violent magnetospheric storms which caused breakdown not just of communication by free space electromagnetic communication but of entire power stations leaving whole countries or even continents without electric power. This fact which happened a few times already has raised the necessity of prediction which because of its lesser occurrence is not as urgent as weather prediction in meteorology but by far not negligible. It is this insight which has brought in the needs for a more specialised applicational space activity which has the character of a *space meteorological service* to mankind and civilisation, called *Space Weather*, a necessary applied engineering rather than scientific activity concerned with and mostly focussed on the hazardous effects of space plasma events on human needs and activities. It may be noted at this place that meanwhile one journal has extended its name from *Space Weather* to *Space Weather and Space Climate*, having become aware of the much wider relevance of the physical (and chemical) processes which are continuously taking place in space.

Space Weather

Space weather, or more precise *Space Meteorology*, which in this sense is a rather narrow subfield of a much wider *Space Climatology*, has meanwhile integrated almost the entire space plasma community whose research activities contribute in one or the other way to it. Not that all space plasma physicist work on space weather. This by no means is the case. The majority of space plasma physicists leaves the activity to the space weather activists, while the results of their research directly or indirectly contribute to space weather. This is actually the road all useful science takes: from apparently useless research to engineering applications. Here we hint on some of its main achievements. Meanwhile a large number of publications including several monographs and conference proceedings has appeared. The reader who has become interested to devote her/his life to this kind of service might consult any of those elaborate collections.

From the above it should have become clear that space weather is not a genuine scientific field but more an engineering effort of being aware of the hazards of violent processes taking place in space. These, in the first place need to be identified and their consequences be explored. In the second place strategies should be developed of how to predict those hazardous events in order to prepare for their entry. And in the third place the question goes whether anything could be done to prevent them to occur or at least to mitigate their danger and effects on Earth's surface.

Which are the hazards expected from near-Earth space or more generally originating in space and relate in this or the other sense to space plasma physics? It is clear that the main threat from space does not come from the extraterrestrial plasma. It comes from large impacts resulting from disintegration of comets, larger fractions of solar system asteroids, collisions with extra-solar objects crossing the heliosphere (and in public phantasy also from the spurious unrealistic invasions of extraterrestrials). Because a large fraction of their mass is converted into radiation when impacting on Earth, they cause large nuclear explosions with all their devastating consequences. The probability of such events can be estimated. It is proportional to the size of the objects and depends on their origin. For all large object which really can become dangerous, this probability is very low, one in several hundred thousands years. Asteroids come from the asteroid belt between the orbits of Mars and Jupiter and some other smaller agglomerations of small bodies which are well under control. Smaller objects like the Tunguska comet fraction which hit Siberia in the early twentieth century are more frequent. They cause local damages but are as well quite rare. Space observations enable early warning.

The threats exerted by space plasma events are of different nature than the impacts. Impacts are dangerous because of their large mass and correspondingly high energy when with full free fall speed hitting the Earth. In comparison plasma effects are energetically negligible. There are essentially two mutually related categories of plasma effects, those primarily related to energetic particles and those primarily related to magnetic effects. However, within the past half century mankind has entered the electronic age with its main agent: information. It has in this time become extremely vulnerable to the tiniest damages which the information technology could suffer from. This electronic age started silently in the beginning of the twentieth century with the invention of wireless communication by using electromagnetic waves. Though this was an enormous achievement which survived the two World Wars, it suffered strongly from so called atmospheric disturbances which for long time were believed to be exceptionally caused by lightnings and thunderstorms until it was realised that they rather were correlated with geomagnetic disturbances. (The first hint came Hundred years earlier in September 1859 during the largest geomagnetic storm on record, the Carrington Event, when the young telegraph system was severely damaged.) This could be avoided replacing wireless communication by electromagnetic wave transport by cross-continental cables, a technology that seemed to be immune against any electromagnetic damaging.

This belief was suddenly destroyed on March 13, 1989 when during an extreme geomagnetic storm, the so far largest storm in the space age, the electric transmission system in Québec collapsed for nine hours. Investigation into the cause of the breakdown of the hydro station revealed that very strong currents had been induced in transmission lines by the violent geomagnetic disturbance. These currents had damaged the cables leading to short cuts, fire and caused the entire system to collapse. The storm was accompanied by aurora visible down to Florida and Texas and caused damages in Australia and Africa as well. Its cause was a solar *Coronal Mass Ejection* (CME) preceding the storm by a few days, the travel time from the Sun to Earth through the inner heliosphere. The CME was preceded by a large *solar flare* eruption emitting very high energy x-rays from a *solar active region* including sunspots. This sequence completed the *causal chain of events* such that it became quite clear that the damage on the transmission line was not caused by human inference but by a well established sequence of natural effects which led from the solar active region, through the precursor flare on the sun, to the CME and, as its consequence, the disturbance caused in the magnetosphere, the result of which was the *violent magnetospheric storm*, accompanied by *aurorae* that was visible down to subtropic latitudes, whose effect was the distortion of *signal transport* in the ionosphere and generation of *induction currents* in the Earth's subpolar crust which in Canada caused the breakdown of the electric transmission system leaving the whole province for nine hours without electricity.

With the explosive rise of the computer technology and the Internet the threat of civilisation has increased exponentially. That computer technology and internet communication is vulnerable to lightning strokes is well known. However, induction effects by violent, strong or even moderate external electromagnetic radiation and magnetic variations may cause damage to this increasingly important technology. It is therefore no surprise that after 1989 the community became alarmed deciding that forecasting of at least violent storms was absolutely needed in the interest of warranting the continuity of communication. A space weather forecasting system was required. It should include observation of the sun and its activity from ground and space, best from spacecraft located in the first *Lagrange point* L1 in solar direction to detect CMEs passing across them to hit the magnetosphere, the effects of interaction of CMEs with the bow shock, magnetopause, and the magnetosphere as a whole, and the transmission of those effects down to Earth's surface after crossing the ionosphere. It is clear that these requirements include a complete knowledge of the processes how CMEs are generated on the sun, their propagation across space, effects on the solar wind, and a full theory of the magnetospheric and ionospheric response. Of course, space weather does not require knowledge of all the details of these processes or interactions; rather, like weather forecasting in meteorology, they can be treated as black boxes which, when given a certain input, with some probability provide a certain output that predicts what and where is to be expected as effect on Earth's surface. The reference list at the end of this section collects the papers which take the final step in this chain of black boxes, the role the magnetosphere and to a lesser extent, the ionosphere play; their responses in all those cases when the input from the solar wind and its large disturbances like CMEs is given and what in that case is expected to happen at the surface of Earth. This defines the narrow domain of space weather.

Space weather in this sense is a statistical approach to predict the effects of violent processes in Earth's plasma environment on civilisation. It collects effects, compiles them according to their frequency and strength and estimates the probability of their occurrence. The sun, its atmosphere, corona and the solar wind are taken as input. Including them would belong an as well statistical science of *Space Climatology* which would extend out to the distant planets of the solar system and even include the entire heliosphere.

Elements of Space Weather

The core of space weather, when understood in the above explained sense, is the great magnetospheric storm. It has, however, become clear soon that the magnetospheric storm is a conglomerate of a number of physically much more important processes which form is building block, the so-called substorm. Many substorms in a row and simultaneously evolving and decaying in different though possibly adjacent regions of the magnetosphere, when caused by a large disturbance of the entire magnetosphere, are what is known under the notion of a magnetospheric storm. It is thus most important to know the physics of substorms.

Chapter 5 already discussed a number of properties of the magnetospheric substorm. It approached the dynamics of the inner magnetosphere from the point of view of a so-called ionospheric-magnetospheric current system, the heart of the coupling between magnetosphere and ionosphere, including transverse currents flowing in the magnetosphere across the magnetic field in the form of drift currents and being connected with the ionosphere through field-aligned currents flowing down into the ionosphere, where they close across the resistive ionosphere, and returning up to the magnetosphere. At the end of this book we are now in the position to list the more elaborate ingredients of a substorm.

Substorms

The substorm is, so to say, the *elementary response* of the entire magnetosphere-ionosphere system to an external disturbance imposed onto it. For it to happen, it is not required that the whole magnetosphere in all its regions becomes affected. In the spirit of space weather it is also not required to refer to all the microscopic processes involved to get the substorm started, grow and decay. Historically it was identified as a magnetic disturbance at Earth's surface called a *bay* (because of the nearly symmetric decrease and recovery of the magnetic field in the recordings, a diamagnetic effect attributed to a partial westward ion drift-*ring current* in the equatorial magnetosphere) being of duration of $\delta\tau \sim 500 - 2000$ s. However, this picture was too simplistic.

With the realisation that magnetic reconnection (see Chapter 12) is the trigger of a magnetospheric disturbance, it became clear that the exchange of magnetic flux between the magnetosheath and the polar magnetosphere, i.e. connection of the reconnected polar field lines to those in the magnetosheath which are convected over the magnetopause, causes a cross-magnetospheric convection, adds flux to the central current in the magnetosphere and causes reconnection there with extra flux released downtail and transported away, and flux added to the inner closed magnetosphere. The substorm has the following properties:

- Triggering a substorm requires by current knowledge an external energy input into the magnetosphere. This proceeds most efficiently through reconnection between a southward (in geocentric coordinates) component of the magnetic field in the magnetosheath and the northward low latitude magnetospheric field causing, under certain conditions forcing the two oppositely directed fields to merge. In this case extra magnetic flux from the external region, the magnetosheath, is added to the magnetosphere and, by the comparably fast — and further accelerating to supersonic and even super-alfvénic velocities — magnetosheath flow, is pulled over Earth's polar cap downstream in antisolar direction and added to the tail where it accumulates, strengthens the central-tail current, fills the plasma sheet, leads to the formation of a near-Earth neutral line and ultimately restores tail reconnection to occur relatively close to Earth.

- Pulling the field lines over the polar cap causes an intensification and deformation of the polar ionospheric current system through enhanced upper ionospheric convection.

- Reconnection in the tail causes dipolarization of the earthward tail magnetic field and earthward plasma motion, forming dipolarization fronts.

- It also forms the partial ring current and intense field aligned currents to close the magnetospheric-ionospheric current system, resulting in a so-called substorm current wedge in the substorm expansion phase and in the ionosphere the formation of a substorm electrojet.

- Aside of these macroscopic magnetic and current properties, reconnection in the tail leads to the generation of localized electric fields perpendicular to and along the magnetic field which accelerate the plasma into bursts of convection and injects electrons along the magnetic field into the auroral ionosphere to cause an auroral substorm, and aurorae. It also launches kinetic Alfvén waves along the magnetic field to flow from the tail into the ionosphere and cause further local acceleration by their field-aligned electric fields. Intense interaction between these waves and the plasma generates localized potential drops, double layers, electron and ion holes and violent acceleration of electrons into bursts which even may result in x-ray emission.

There is a large number of further properties of the processes taking place during substorm which are covered by the physics of reconnection at both locations in space, the magnetopause and the geomagnetic tail current sheet. These are important for the understanding of the very physics of substorms but are of lesser importance for the average characteristics of space weather.

Magnetospheric Storms

The most important is that each geomagnetic storm with its violent effects is composed of an overlapping sequence of substorms which occur in different places of the magnetosphere, thus affecting different locations on the Earth's surface, preferentially the subpolar regions on both hemispheres. From the physics of reconnection and the former explanation it becomes clear that reconnection at the magnetopause is by itself a local effect depending on the local direction of the magnetosheath magnetic field. It will hence be distributed in patches over the magnetopause thus affecting different stretches of the tail, causing several partial neutral lines and X points in the tail plasma sheet and current layer. The overlap of all those substorms arising in different places then causes the irregular appearance of the magnetospheric, geomagnetic and ionospheric storms. It thus becomes clear that in view of the prediction of the hazards related to space weather one has to refer to statistical methods, the collection of information about magnetospheric storms. This has meanwhile become an own scientific branch and industry which has little overlap with space plasma physics while being of enormous importance for its practical consequences. Space weather in this sense has brought space plasma physics to enhanced practical public attention. It has pulled it from the aesthetic delight when observing auroral phenomena down to the practical needs on Earth.

Substorm Triggering

Substorms are the reaction of the magnetosphere to external processes. These belong to the general field of Space Climatology which encompasses the generation of CMEs in the solar atmosphere and corona which we are not going to discuss here. They also cover the entire physics of the solar wind which on the one side continuously transports plasma and magnetic fields away from the sun and inner heliosphere to the other planets and the outer heliosphere until merging with the interstellar medium at distances beyond 100 AU. In general, as discussed elsewhere in this book, the solar wind is responsible for the comet-like form of the magnetosphere with its extended tail, inner plasma configuration, tail current sheet and distant neutral line. This means that already the solar wind and its mere existence signs responsible for the external process and the quiet time space weather on Earth with its daily and seasonal variations. It determines, so to say, the space climate in Earth's near cosmic environment. Substorms and their larger relative the magnetospheric storm are in this view the disturbances in space climate caused by the eruptions on the sun and in particular by those huge events like CMEs. They transport large amounts of mass, energy, and irregular fields towards the magnetosphere until interacting with the geomagnetic field and the plasma content of the magnetosphere.

Before a disturbance of the solar wind like a CME hits the magnetosphere, it has no information about its presence. The impact comes as a shock. Monitoring space weather thus requires observation of space climate away from Earth and its changes. This can be done by ground based observation of the solar activity and radiation. Optical radiation does not vary substantially. Therefore observations are required in the radio and UV bands as far as the atmosphere allows those radiations to penetrate to the ground. X rays can only be observed from space, and it is those which provide the relevant information about the start of a CME from the sun when an X ray flare occurs a few days before the arrival of a CME at Earth. But already in quiet space climate the solar wind, flowing at high speed causes a disturbance near Earth, the bow shock wave, an intrinsically non-stationary curved (hyperbolic) shield in front of the magnetopause with its small quasi-perpendicular and its extended quasi-parallel sections described in Chapter 13. Behind this shock and between it and the magnetopause we have the genuine region of contact of Earth with space, the magnetosheath with its generally sub-magnetosonic speed and turbulent magnetic field. Knowing its physics and monitoring its state is crucial for the science of space weather. Its local conditions along the magnetopause surface determine the conditions under which reconnection and in what form it occurs.

But reconnection, though being the main driver of substorms, is not the only process which adds plasma and fields to the magnetosphere. There are other streaming effects which cause the magnetopause to flatter at its flanks. The main effect is caused by the tangential flow of the solar wind along the magnetopause outside its nose. There instabilities of the Kelvin-Helmholtz family can be excited which produce large amplitude excursions of the flank magnetopause where local reconnection may occur which, however, have the important effect of mixing the solar wind field and plasma with that of the magnetosphere and injecting solar wind plasma into the magnetosphere while extracting magnetospheric plasma from it. The latter plasma is thus lost from the atmosphere, ionosphere and plasma sheet to free space. These effects do all contribute to substorm generation and mitigation and thus belong to its dynamics.

Once a substorm has been triggered by all those external effects, its internal and independent local dynamics determines its further fate. This dynamics consists of a wealth of complicated processes like bursts in the magnetospheric plasma flow caused by tail reconnection, jets ejected from reconnection, the mentioned dipolarization fronts, twisted magnetic flux ropes and bundles which all determine in which way information is transported toward the surface of Earth.

Space weather, to account for such anomalous behaviour of the magnetosphere, makes use of average models of the magnetosphere calculated numerically from the magnetohydrodynamic equations in various approximations. Determination of such models of the magnetosphere under different quiet and disturbed conditions has become a whole industry. It heavily depends on the grid resolution and the more sophisticated inclusion of ever more precisely defined transport parameters and spatial structures from observations in order to predict what could happen, how the magnetosphere and ionosphere would react if particular conditions are met in the solar wind and in the magnetosheath behind the bow shock. The incomplete reference list below provides some more information about those space weather activities.

References

References related to Olbert's Kappa Distribution

[1] J. H. Binsack, *Plasma studies with the Imp-2 Satellite*, Boston: PhD-Thesis MIT (1967), pp. 200.

[2] S. P. Christon *et al.*, Energy spectra of plasma sheet ions and electrons from ~ 50 eV/e to ~ 1 MeV during plasma temperature transitions, *J. Geophys. Res.* **93** 2562 (1988).

[3] A. Hasegawa *et al.*, Plasma distribution function in a superthermal radiation field, *Phys. Rev. Lett.* **54**, 2608 (1985).

[4] G. Livadiotis and D. J. McComas, Understanding kappa distributions: a toolbox for space science and astrophysics, *Space Science Rev.* **175**, 183 (2013).

[5] S. Olbert, in Carovillano, McClay (eds.) *Physics of the Magnetosphere, Proceeding of a Conference at Boston College, Astrophysics and Space Science Library* **40**, 641 (Dordrecht: Reidel, 1968).

[6] K. Scherer *et al.*, Toward a realistic macroscopic parametrization of space plasmas with regularized kappa-distribution, *Astron. Astrophys.* **643**, A20 (2020).

[7] M. F. Shlesinger *et al.*, Strange kinetics, *Nature* **363**, 31 (1992).

[8] R. A. Treumann, Kinetic theoretical foundation of Lorentzian statistical mechanics, *Phys. Scripta* **59**, 19 (1999).

[9] R. A. Treumann and C. H. Jaroschek, Gibbsian theory of power-law distributions, *Phys. Rev. Lett.* **100** 155 005 (2008).

[10] R. A. Treumann and W. Baumjohann, Lorentzian entropies and Olbert's kappa distribution, *Front. Phys.* **8**, 221 (2020).

[11] V. M. Vasyliunas, A survey of low-energy electrons in the evening sector of the magnetosphere with OGO 1 and OGO 3, *J. Geophys. Res.* **73**, 2839 (1968).

[12] P. H. Yoon *et al.*, Langmuir turbulence and suprathermal electrons, *Space Science Rev.* **173** 459 (2012).

References related to Mirror Modes

[1] W. Baumjohann *et al.*, Waveform and packet structure of lion roars, *Ann. Geophys.* **17**, 1528 (1999). [Contains the first resolution of the wave form of ion-mirror mode-trapped lion roars, showing that these are resonant whistlers generated by trapped electrons in the mirror magnetic field.]

[2] O. D. Constantinescu, Self-consistent model of mirror structures, *J. Atmos. Solar-Terr. Phys.* **64**, 645 (2002). [An attempt to model the 3d-geometric bottle form of a mirror structure in application to observations in the magnetosheath.]

[3] O. D. Constantinescu *et al.*, Magnetic mirror structures observed by Cluster in the magnetosheath, *Geophys. Res. Lett.* **30**, 1802 (2003). [Observations of mirror mode chains in the data of the Cluster spacecraft.]

[4] A. Hasegawa, Drift-mirror instability in the magnetosphere, *Phys. Fluids* **12**, 2642 (1969). [A first demonstration that fluid theory is insufficient to describe mirror modes. The kinetic theory leads to the appearance of a finite frequency of the linear mirror mode whenever density gradients are taken into account.]

[5] E. A. Lucek *et al.*, Identification of magnetosheath mirror mode structures in Equator-S magnetic field data, *Ann. Geophys.* **17**, 1560 (1999). [A thorough high resolution investigation of very long mirror mode chains in the magnetosheath listing their propagation direction, mean amplitudes and geometric parameters.]

[6] N. Noreen *et al.*, Electron contribution in mirror instability in quasilinear regime, *J. Geophys. Res.* **122**, 6978 (2017). [Inclusion of the electron contribution to the ion mirror mode and performing a zero order non-linear (quasilinear) calculation which shows that the mirror mode readily saturates on a very low amplitude level by approximately restoring pressure isotropy (which had with lesser rigour been shown in our companion monography *Advanced Space Plasma Theory* but here is demonstrated from kinetic theory thus raising doubts on any simple explanation of the formation of large amplitude mirror modes. Moreover, the simulations in this paper indicate the separate existence of the electron mirror mode however still lacking its importance beyond the quasilinear saturation for the lack of the theoretical mechanism. Such a semi-classical mechanism is found in the interplay between the bounce dynamics of particles trapped in mirror modes and their interaction with the ion-acoustic wave background which in this particular case leads to the otherwise unexpected condensate formation in high temperature plasma.]

[7] D. J. Southwood and M. G. Kivelson, Mirror instability: I. Physical mechanism of linear instability, *J. Geophys. Res.* **98**, 9181 (1993). [The rigorous derivation and discussion of the linear mirror mode theory from semi-kinetic theory in application to the magnetosheath.]

[8] R. M. Thorne and B. T. Tsurutani, The generation mechanism for magnetosheath lion roars, *Nature* **293**, 384 (1981). [Application of the theory of whistler instability to trapped electrons in mirror mode leading to excitation of trapped low frequency electron whistlers called lion roars.]

[9] R. A. Treumann and W. Baumjohann, Electron mirror branch: observational evidence from "historical" AMPTE-IRM and Equator-S measurements, *Ann.*

Geophys. **36**, 1563 (2018). [The first experimental proof of the existence of the electron mirror mode as a separate structure in spacecraft data in the magnetosheath.]

[10] R. A. Treumann and W. Baumjohann, Condensate formation in collisionless plasma, *Frontiers in Physics* **9**, 713551 (2021). [The proposed semi-classical mechanism of condensate formation under high temperature conditions based on particle trapped in a magnetic mirror configuration and near their mirror points being in resonance with the ion-acoustic thermal background.]

[11] T. L. Zhang *et al.*, Mirror mode structures in the solar wind at 0.72 AU, *J. Geophys. Res.* **114**, A10107 (2009). [An observation of mirror modes in the streaming high temperature solar wind.]

[12] Y. Zhang *et al.*, Lion roars in the magnetosheath: The Geotail observations, *J. Geophys. Res.* **103**, 4615 (1998). [A thorough investigation of the properties of lion roars observed in the magnetosheath.]

References related to Turbulence

[1] D. Biskamp, *Magnetohydrodynamic Turbulence* (Cambridge Univ. Press, Cambridge, 2009). [A readable general monograph on turbulence with main emphasis on numerical simulations. Not updated to the modern simulational and theoretical achievements.]

[2] R. C. Davidson, *Methods in Nonlinear Plasma Theory* (Academics Press, New York, 1972). [Develops the theoretical formalism of weak kinetic plasma turbulence up to the early seventies. Very lucid systematic presentation of the perturbational approach. Lacks however reference strong turbulence like plasma collapse and soliton turbulence.]

[3] P. Davidson, *Turbulence: An Introduction for Scientist and Engineers* (Oxford University Press, Oxford, 2015). [General textbook introduction to the whole field of turbulence on the undergraduate and graduate level. No space plasma turbulence included.]

[4] U. Frisch, *Turbulence* (Cambridge Univ. Press, Cambridge, 1995). [Standard representation of turbulence in the general sense. No mentioning of plasma and space plasmas. Main emphasis on laboratory measurements and in theory on the origin of turbulence in chaos theory.]

[5] T. S. Horbury *et al.*, Anisotropy in space plasma turbulence: Solar wind observations, *Space Sci. Rev.* **172**, 325 (2011). [Typical review of solar wind turbulence observations and measurements of magnetic spectra, no mentioning of its kinetic aspects. Indicates Kolmogorov spectra, anisotropies, observed dissipation scales.]

[6] P. S. Iroshnikov, Turbulence of a conducting fluid in a strong magnetic field, *Soviet Astron. - AJ* **7**, 566 (1964). [The original replacement of the Kolmogorov law in strong magnetic fields.]

[7] A. N. Kolmogorov, Dissipation of energy under locally isotropic turbulence, *Dokl. Akad. Nauk SSSR* **30**, 9 (1941). [The original work on Kolmogorov turbulence.]

[8] R. H. Kraichnan, On Kolmogorov's inertial-range theories, *J. Fluid Mech.* **62**, 305 (1974). [The original work on extension of Kolmogorov turbulence to magnetized fluids following the original work of Iroshnikov published in 1964.]

[9] R. A. Treumann and W. Baumjohann, *Advances Space Plasma Physics* (Imperial College Press, London, 1997). [Contains both the basics of weak turbulence and strong turbulence on the kinetic level in plasma.]

[10] M. Wilczek *et al.*, On the velocity distribution in homogeneous isotropic turbulence: Correlations and deviations from Gaussianity, *J. Fluid Mech.* **676**, 191 (2011). [Highly mathematical presentation of the expected particle distributions in turbulence.]

[11] H. Xu *et al.*, Multifractal dimension of Lagrangian turbulence, *Phys. Rev. Lett.* **96**, 114503 (2006). [The fractal approach to turbulence in the Lagrangian formulation.]

References related to Space Weather

[1] D. Baker *et al.* (Eds.), *The Scientific Foundations of Space Weather* Space Science Series of ISSI Vol. 67 (Springer Netherlands 2019).

[2] P. Cannon *et al.*, *Extreme Space Weather: Impacts on Engineered Systems and Infrastructure* (Royal Academy of Engineering, London, 2013).

[3] J. P. Eastwood *et al.*, The scientific foundations of forecasting magnetospheric space weather, *Space Sci. Rev.* **212**, 1221–1252 (2017).

[4] H. Koskinen, *Physics of Space Storms*, (Springer-Verlag, Berlin-Heidelberg, 2011).

[5] T. Lopez *et al.* (Eds.), *Geohazards and Risks Studied from Earth Observations* Space Science Series of ISSI Vol. 82 (Springer Netherlands 2020).

Additional selected references related to space weather

[6] V. Angelopoulos *et al.*, Tail reconnection triggering substorm onset, *Science* **321**, 931 (2008).

[7] V. Angelopoulos *et al.*, Electromagnetic energy conversion at reconnection fronts, *Science* **341**, 1478 (2013).

[8] D. N. Baker *et al.*, Timing of magnetic reconnection initiation during a global magnetospheric substorm onset, *Geophys. Res. Lett.* **29**, 43 (2002).

[9] W. Baumjohann *et al.*, Average electric wave spectra across the plasma sheet and their relation to ion bulk speed, *J. Geophys. Res.* **94**, 15221 (1989).

[10] M. Bodeau, Review of better space weather proxies for spacecraft surface charging, *IEEE Trans. Plasma Sci.* **43**, 3075 (2015).

[11] L. Bolduc, GIC observations and studies in the Hydro-Quebec power system, *J. Atmos. Sol.-Terr. Phys.* **64**, 1793 (2002).

[12] J. E. Borovsky *et al.*, Estimating the effects of ionospheric plasma on solar wind/ magnetosphere coupling via mass loading of dayside reconnection: ion-plasma-sheet oxygen, plasmaspheric drainage plumes, and the plasma cloak, *J. Geophys. Res.* **118**, 5695 (2013).

[13] M. D. Cash *et al.*, Validation of an operational product to determine L1 to Earth propagation time delays, *Space Weather* **14**, 93 (2016).

[14] L. B. N. Clausen *et al.*, On the influence of open magnetic flux on substorm intensity: ground- and space-based observations, *J. Geophys. Res.* **118**, 2958 (2013).

[15] E. W. Cliver and W. F. Dietrich, The 1859 space weather event revisited: limits of extreme activity, *J. Space Weather Space Climate* **3**, A 31 (2013).

[16] S. W. H. Cowley and C. J. Owen, A simple illustrative model of open flux tube motion over the dayside magnetopause, *Planet. Space Sci.* **37**, 1461 (1989).

[17] J. P. Eastwood *et al.*, What controls the structure and dynamics of Earth's magnetosphere? *Space Sci. Rev.* **188**, 251 (2015).

[18] J. P. Eastwood *et al.*, Sunjammer, *Weather* **70** 27 (2015).

[19] J. P. Eastwood *et al.*, The economic impact of space weather: where do we stand? *Risk Anal.* **37**, 206 (2017).

[20] I. A. Erinmez *et al.*, Management of the geomagnetically induced current risks on the national grid company's electric power transmission system, *J. Atmos. Sol.-Terr. Phys.* **64**, 743 (2002).

[21] W. D. Gonzalez *et al.*, What is geomagnetic storm? *J. Geophys. Res.* **99**, 5771 (1994).

[22] E. Gordeev *et al.*, The substorm cycle as reproduced by global MHD models, *Space Weather* **15**, 131 (2017).

[23] M. A. Hapgood, Towards a scientific understanding of the risk from extreme space weather, *Adv. Space Res.* **47**, 2059 (2011).

[24] M. A. Hapgood, Prepare for the coming space weather storm, *Nature* **484**, 311 (2012).

[25] Y. Kamide, What is an "Intense Geomagnetic Storm"? *Space Weather* **4**, S06008 (2006).

[26] L. Kepko *et al.*, Substorm current wedge revisited, *Space Sci. Rev.* **190**, 1 (2015).

[27] L. M. Kistler *et al.*, Ion composition and pressure changes in storm time and nonstorm substorms in the vicinity of the near-Earth neutral line, *J. Geophys. Res.* **111**, A11222 (2006).

[28] D. J. Knipp, Advances in space weather ensemble forecasting, *Space Weather* **14**, 52 (2016).

[29] M. Kubicka, Prediction of geomagnetic storm strength from inner heliospheric in situ observations, *Astrophys. J.* **833**, 255 (2016).

[30] J. J. Love, Credible occurrence probabilities for extreme geophysical events: earthquakes, volcanic eruptions, magnetic storms, *Geophys. Res. Lett.* **39**, L10301 (2012).

[31] S. E. Milan *et al.*, Overview of solar wind-magnetosphere-ionosphereatmosphere coupling and the generation of magnetospheric currents, *Space Sci. Rev.* **206**, 547 (2017).

[32] Y. Miyashita *et al.*, Geotail observations of signatures in the near-Earth magnetotail for the extremely intense substorms of the 30 October 2003 storm, *J. Geophys. Res.* **110**, A09S25 (2005).

[33] T. D. Phan *et al.*, The dependence of magnetic reconnection on plasma beta and magnetic shear: evidence from magnetopause observations, *Geophys. Res. Lett.* **40**, 11 (2013).

[34] A. Pulkkinen *et al.*, Solar shield: forecasting and mitigating space weather effects on high-voltage power transmission systems, *Nat. Hazards* **53**, 333 (2010).

[35] I. G. Richardson *et al.*, Sources of geomagnetic storms for solar minimum and maximum conditions during 1972–2000, *Geophys. Res. Lett.* **28**, 2569 (2001).

[36] P. Riley, On the probability of occurrence of extreme space weather events, *Space Weather* **10**, S02012 (2012).

[37] J. Safrankova *et al.*, Reliability of prediction of the magnetosheath BZ component from interplanetary magnetic field observations, *J. Geophys. Res.* **114**, A12213 (2009).

[38] C. J. Schrijver *et al.*, Understanding space weather to shield society: a global road map for 2015–2025 commissioned by COSPAR and ILWS, *Adv. Space Res.* **55**, 2745 (2015).

[39] V. A. Sergeev *et al.*, Recent advances in understanding substorm dynamics, *Geophys. Res. Lett.* **39**, L05101 (2012).

[40] M. I. Sitnov *et al.*, Great mysteries of the Earth?s magnetotail, *EOS Trans. AGU* p. 97 (2016).

[41] T. Terasawa *et al.*, Solar wind control of density and temperature in the near-Earth plasma sheet: WIND/GEOTAIL collaboration, *Geophys. Res. Lett.* **24**, 935 (1997).

[42] G. Toth *et al.*, SpaceWeather Modeling Framework: a new tool for the space science community, *J. Geophys. Res.* **110**, A12226 (2005).

[43] R. A. Treumann and W. Baumjohann, Collisionless magnetic reconnection in space plasmas, *Front. Physics* **1**, 31 (2013).

[44] L. Turc *et al.*, Statistical study of the alteration of the magnetic structure of magnetic clouds in the Earth's magnetosheath, *J. Geophys. Res.* **122**, 2956 (2017).

[45] V. M. Vasyliunas, The largest imaginable magnetic storm, *J. Atmos. Sol.-Terr. Phys.* **73**, 1444 (2011).

[46] R. Walker *et al.*, Source and loss processes in the magnetotail, *Space Sci. Rev.* **88**, 285 (1999).

[47] W. Wang *et al.*, Initial results from the coupled magnetosphere-ionosphere-thermosphere model: thermosphere-ionosphere responses, *J. Atmos. Sol.-Terr. Phys.* **66**, 1425 (2004).

[48] M. Yamada *et al.*, Magnetic Reconnection, *Rev. Modern Physics* **82**, 603–664 (2010).

[49] Y. Yu and A. J. Ridley, Response of the magnetosphere-ionosphere system to a sudden southward turning of interplanetary magnetic field, *J. Geophys. Res.* **114**, A03216 (2009).

B — Basic Relations

B.1 Useful Constants and Numbers (SI units)

c	velocity of light	$3.00 \times 10^8 \, \mathrm{m \, s^{-1}}$
μ_0	free space magnetic permeability	$4\pi \times 10^{-7} \, \mathrm{H \, m^{-1}}$
ε_0	vacuum dielectric constant	$8.85 \times 10^{-12} \, \mathrm{F \, m^{-1}}$
e	electron charge	$1.60 \times 10^{-19} \, \mathrm{C}$
m_e	electron mass	$9.11 \times 10^{-31} \, \mathrm{kg}$
m_p	proton mass	$1.67 \cdot 10^{-27} \, \mathrm{kg}$
e/m_e	charge to mass ratio	$1.76 \times 10^{11} \, \mathrm{C \, kg^{-1}}$
$\mu = m_p/m_e$	proton to electron mass ratio	1836
k_B	Boltzmann's constant	$1.38 \times 10^{-23} \, \mathrm{J \, K^{-1}}$
$\hbar$	Planck's constant	$1.05 \times 10^{-34} \, \mathrm{J \, s}$
R_0	ideal gas constant	$8.31 \, \mathrm{J \, K^{-1} \, mol^{-1}}$
$R_\odot$	solar radius	$6.96 \times 10^8 \, \mathrm{m}$
R_E	equatorial radius of Earth	$6.37 \times 10^6 \, \mathrm{m}$
AU	Sun-Earth distance	$1.50 \times 10^{11} \, \mathrm{m}$
$B_\odot$	Solar maximum surface field	$0.25 - 0.35 \, \mathrm{T}$
B_E	Earth's surface field	$3.11 \times 10^{-5} \, \mathrm{T}$
M_E	magnetic moment of Earth	$8.05 \times 10^{22} \, \mathrm{A \, m^2}$
g_E	Earth's gravitational surface acceleration	$9.81 \, \mathrm{m \, s^{-2}}$
$T_\odot$	Solar coronal temperature at 1.1 $R_\odot$	1.1–1.6 K
$k_B T_{e,sw}$	Solar wind thermal electron energy at 1 AU	$\sim 50 \, \mathrm{eV}$
n_{sw}	Solar wind density at 1 AU	$\mathrm{cm^{-3}}$
v_{sw}	Solar wind velocity at 1 AU	320–$1400 \, \mathrm{km \, s^{-1}}$
B_{sw}	Interplanetary magnetic field at 1 AU	$\sim 10 \, \mathrm{nT}$

B.2   Energy Units

Thermal energy and temperature are related by $W = k_B T$, with Joule (J) and Kelvin (K) used interchangeably. Particle energy is measured in eV. Overall energy budgets are often given in the cgs-unit erg. These units are related as follows:

1 J	—	7.24×10^{22} K	6.24×10^{18} eV	10^7 erg
1 K	1.38×10^{-23} J	—	8.62×10^{-5} eV	$1.38 \cdot 10^{-16}$ erg
1 eV	1.60×10^{-19} J	1.16×10^4 K	—	$1.60 \cdot 10^{-12}$ erg
1 erg	10^{-7} J	7.24×10^{15} K	6.24×10^{11} eV	—

B.3   Useful Formulas

The numerical values of all plasma parameters and other important quantities can, of course, be calculated using the formulas given in the main text. However, it is often useful to have a simpler formula, which already incorporates the numerical values of any constants included in the formula as well as possible unit conversion factors. The units are adapted to the typical range of values found in space plasmas.

f_{pe}	electron plasma frequency	$9.0 \times 10^3 \sqrt{n}$	in Hz
f_{pi}	proton plasma frequency	$2.1 \times 10^2 \sqrt{n}$	in Hz
f_{ge}	electron gyrofrequency	$2.8 \times 10^1 B$	in Hz
f_{gi}	proton gyrofrequency	$1.5 \times 10^{-2} B$	in Hz
f_{lh}^*	lower hybrid frequency	$8.8 \times 10^2 B$	in Hz
f_B	Buneman frequency	$2.7 \times 10^2 \sqrt{n}$	in Hz
f_{pe}/f_{ge}	frequency ratio	$3.2 \times 10^2 \sqrt{n}/B$	—
ν_e	electron collision frequency	$2.9 \times 10^{-9} n_e T_e^{-\frac{3}{2}} \ln \Lambda$	in Hz
r_{ge}	electron gyroradius	$3.1 \times 10^1 \sqrt{T_e}/B$	in km
		$1.1 \times 10^2 \sqrt{W_e}/B$	in km
r_{gi}	proton gyroradius	$1.3 \times 10^3 \sqrt{T_i}/B$	in km
		$4.6 \times 10^3 \sqrt{W_i}/B$	in km

*in strong magnetic fields

Continuation of Table

v_{the}[†]	electron thermal speed	$3.9 \times 10^3 \sqrt{T_e}$	in km/s
		$1.3 \times 10^4 \sqrt{W_e}$	in km/s
v_{thi}[†]	proton thermal speed	$9.1 \times 10^1 \sqrt{T_i}$	in km/s
		$3.1 \times 10^2 \sqrt{W_i}$	in km/s
v_E	E×B drift speed	$1.0 \times 10^3 \, E/B$	in km/s
v_A	Alfvén speed	$2.2 \times 10^1 \, B/\sqrt{n}$	in km/s
λ_D	Debye length	$6.9 \times 10^1 \sqrt{T_e/n}$	in m
		$2.4 \times 10^2 \sqrt{W_e/n}$	in m
$\lambda_e = c/\omega_{pe}$	electron inertial length	$5.31 \times 10^3/\sqrt{n}$	in m
$\lambda_i = c/\omega_{pi}$	ion inertial length	$2.28 \times 10^5/\sqrt{n}$	in m
$\beta = 2\mu_0 n k_B T/B^2$	plasma beta	$4.03 \times 10^{-7} nT/B^2$	—
n	plasma density		in cm^{-3}
$T_e \, \& \, T_i$	electron & proton temperature		in 10^6 K
$W_e \, \& \, W_i$	electron & proton thermal energy		in keV
B	magnetic field strength		in nT
E	electric field strength		in mV/m

[†]see Eq. (6.64) for definition of v_{th}

B.4   Vectors and Tensors

The vector and tensor relations given in the following sections are useful in the context of plasma physics. Here, we have denoted scalars by ϕ, vectors by $\mathbf{A}$, and tensors by $\mathbf{T}$.

Dyadic Notation

Second-rank tensors can be written in *dyadic notation* as

$$\mathbf{T} = \mathbf{AB} \tag{B.1}$$

Thus a second-rank tensor can be constructed from two vectors, i.e., first-rank tensors. In component notation the dyadic product reads

$$T_{ij} = A_i B_j \tag{B.2}$$

where i, j represent the components x, y, z. Hence, the dyadic product represents a two-dimensional matrix

$$\mathbf{T} = \begin{pmatrix} A_xB_x & A_xB_y & A_xB_z \\ A_yB_x & A_yB_y & A_yB_z \\ A_zB_x & A_zB_y & A_zB_z \end{pmatrix} \tag{B.3}$$

Exchanging the vectors in the dyadic product $\mathbf{T} = \mathbf{AB}$ results in the transposed dyad

$$\mathbf{T}^t = \mathbf{BA} \quad \text{with} \quad \mathbf{T} = \mathbf{AA} = \mathbf{T}^t \tag{B.4}$$

For symmetric tensors, dyad and transposed dyad are identical.

Second-rank tensors or dyadic products of two vectors are the ones most often used. However, at times it is useful to represent scalars and vectors as zero- and first-rank tensors, respectively, and occasionally third-rank tensors are used. The latter are defined as dyadic products of three vectors, like the heat tensor given in Section 6.5.

Dyads can be functions of complex quantities like frequency and wavenumber, in which case they become complex functions themselves. For complex tensors it is useful to introduce another kind of transposed tensor, the *Hermitean conjugate tensor*, $\mathbf{T}^\dagger$. This tensor is defined as the transposed of the conjugate complex tensor

$$\mathbf{T}^* = (\mathbf{AB})^*, \quad \text{so that} \quad \mathbf{T}^\dagger = \{(\mathbf{AB})^*\}^t \tag{B.5}$$

With this definition, one can show that

$$\mathbf{A}^* \cdot \mathbf{T} \cdot \mathbf{B} = \mathbf{B} \cdot \mathbf{T}^\dagger \cdot \mathbf{A}^* \tag{B.6}$$

Algebraic Relations

The algebraic relations, inner and cross-product, are governed by the following rules:

$$\mathbf{A} \cdot \mathbf{B} = \mathbf{B} \cdot \mathbf{A} \tag{B.7}$$

$$\mathbf{A} \times \mathbf{B} = -\mathbf{B} \times \mathbf{A} \tag{B.8}$$

$$\mathbf{A} \cdot (\mathbf{B} \times \mathbf{C}) = \mathbf{B} \cdot (\mathbf{C} \times \mathbf{A}) = \mathbf{C} \cdot (\mathbf{A} \times \mathbf{B}) \tag{B.9}$$

$$\mathbf{A} \times (\mathbf{B} \times \mathbf{C}) = (\mathbf{A} \cdot \mathbf{C})\mathbf{B} - (\mathbf{A} \cdot \mathbf{B})\mathbf{C} \tag{B.10}$$

$$(\mathbf{A} \times \mathbf{B}) \cdot (\mathbf{C} \times \mathbf{D}) = (\mathbf{A} \cdot \mathbf{C})(\mathbf{B} \cdot \mathbf{D}) - (\mathbf{A} \cdot \mathbf{D})(\mathbf{B} \cdot \mathbf{C}) \tag{B.11}$$

$$\phi\mathbf{T} = \mathbf{T}\phi \tag{B.12}$$

$$\mathbf{CT} = \mathbf{CAB} \neq \mathbf{ACB} \neq \mathbf{ABC} \neq \mathbf{TC} \tag{B.13}$$

$$\mathbf{C} \cdot \mathbf{T} = (\mathbf{C} \cdot \mathbf{A})\mathbf{B} = \mathbf{B}(\mathbf{C} \cdot \mathbf{A}) = \mathbf{T}^t \cdot \mathbf{C} \tag{B.14}$$

$$\mathbf{C} \times \mathbf{T} = (\mathbf{C} \times \mathbf{A})\mathbf{B} = -(\mathbf{A} \times \mathbf{C})\mathbf{B} \tag{B.15}$$

Differential Relations

The vector derivatives tensors obey the following rules:

$$\nabla(\phi\psi) = \phi(\nabla\psi) + \psi(\nabla\phi) \tag{B.16}$$

$$\nabla\cdot(\phi\mathbf{A}) = \mathbf{A}\cdot(\nabla\phi) + \phi(\nabla\cdot\mathbf{A}) \tag{B.17}$$

$$\nabla\times(\phi\mathbf{A}) = \phi(\nabla\times\mathbf{A}) - \mathbf{A}\times(\nabla\phi) \tag{B.18}$$

$$\nabla\cdot(\mathbf{A}\times\mathbf{B}) = \mathbf{B}\cdot(\nabla\times\mathbf{A}) - \mathbf{A}\cdot(\nabla\times\mathbf{B}) \tag{B.19}$$

$$\nabla\times(\mathbf{A}\times\mathbf{B}) = \mathbf{A}(\nabla\cdot\mathbf{B}) - \mathbf{B}(\nabla\cdot\mathbf{A}) + (\mathbf{B}\cdot\nabla)\mathbf{A} - (\mathbf{A}\cdot\nabla)\mathbf{B} \tag{B.20}$$

$$\nabla^2\phi = \nabla\cdot(\nabla\phi) \tag{B.21}$$

$$\nabla^2\mathbf{A} = \nabla(\nabla\cdot\mathbf{A}) - \nabla\times(\nabla\times\mathbf{A}) \tag{B.22}$$

$$\nabla\times(\nabla\phi) = 0 \tag{B.23}$$

$$\nabla\cdot(\nabla\times\mathbf{A}) = 0 \tag{B.24}$$

$$\nabla\cdot(\phi\mathbf{T}) = (\nabla\phi)\cdot\mathbf{T} + \phi(\nabla\cdot\mathbf{T}) \tag{B.25}$$

$$\nabla\cdot\mathbf{T} = \nabla\cdot(\mathbf{AB}) = (\mathbf{A}\cdot\nabla)\mathbf{B} + \mathbf{B}(\nabla\cdot\mathbf{A}) \tag{B.26}$$

$$\nabla\times\mathbf{T} = \nabla\times(\mathbf{AB}) = (\nabla\times\mathbf{A})\mathbf{B} - (\mathbf{A}\times\nabla)\mathbf{B} \tag{B.27}$$

Integral Relations

There are two sets of integral relations for vector derivatives. One set of equations prescribes a transformation from a volume to a surface integral while another set gives the relation between a surface integral and the integration over the path enclosing the surface. From the equations below, Eqs. (B.29) and (B.32) are of special importance in electrodynamics and plasma physics. After their inventors, they bear the names *Gauß' theorem* and *Stokes' theorem*, respectively.

Using the notation $dV = dx\,dy\,dz$ and $d\mathbf{A} = \mathbf{n}\,dA$ with $\mathbf{n}$ the normal vector of a surface element dA, the volume integrals of the vector derivatives can be rewritten as integrals over the closed surface, $\mathbf{A}$, bounding the volume, V,

$$\int_V (\nabla\phi)\,dV = \oint_A \phi\,d\mathbf{A} \tag{B.28}$$

$$\int_V (\nabla\cdot\mathbf{Q})\,dV = \oint_A \mathbf{Q}\cdot d\mathbf{A} \tag{B.29}$$

$$\int_V (\nabla\times\mathbf{Q})\,dV = -\oint_A \mathbf{Q}\times d\mathbf{A} \tag{B.30}$$

In a similar way, surface integrals of vector derivatives can be rewritten as integrals along the closed path, C, enclosing the surface, $\mathbf{A}$,

$$\int_A (\nabla\phi) \times d\mathbf{A} = -\oint_C \phi\, d\mathbf{s} \tag{B.31}$$

$$\int_A (\nabla \times \mathbf{Q}) \cdot d\mathbf{A} = \oint_C \mathbf{Q} \cdot d\mathbf{s} \tag{B.32}$$

where $d\mathbf{s} = \mathbf{t}\, ds$, i.e., the product of a line element, ds, with its tangent vector, $\mathbf{t}$.

B.5 Electrodynamics in a Nutshell

Plasma physics builds heavily on knowledge of electrodynamics. In the following we give a short derivation of Maxwell's equations and other useful formulae.

Maxwell's Equations

Here we give a derivation of Maxwell's equations of electrodynamics, starting from first principles. Conventionally Maxwell's equations are postulated without derivation. One can, however, show that very simple assumptions on the continuity of charges and currents suffice to provide a satisfactory basis from which they can be derived axiomatically. Assume that there exist net space charge, ρ, and current, $\mathbf{j}$, densities

$$\rho = \sum_s q_s n_s, \qquad \mathbf{j} = \sum_s q_s n_s \mathbf{v}_s \tag{B.33}$$

where the summation is over the different species (electrons, protons and ions) with different charges, q_s, and densities, n_s. The net space charge is the source of the electric fields. Currents are generated when the charges of a species move with an average velocity, $\mathbf{v}_s$. They will become the sources of the magnetic fields.

Since neither charges nor particles can be destroyed in electrodynamics, any local temporal change of the charge density will be due to the divergence of the electric current which transports charges in or out of the volume element. Hence, we have the fundamental charge and current conservation equation

$$\frac{\partial \rho}{\partial t} + \nabla \cdot \mathbf{j} = 0 \tag{B.34}$$

The second term in this equation is a divergence. Hence, to solve the equation we define the charge density as the divergence of some vector, $\mathbf{E}$,

$$\nabla \cdot \mathbf{E} = \rho/\varepsilon_0 \tag{B.35}$$

where ε_0 has been introduced as a dimensional factor. Inserting into Eq. (B.34) we get

$$\nabla \cdot \left(\mathbf{j} + \varepsilon_0 \frac{\partial \mathbf{E}}{\partial t} \right) = 0 \tag{B.36}$$

Since the divergence of the vector in the brackets vanishes, it must be the curl of another vector, **B**,

$$\nabla \times \mathbf{B} = \mu_0 \mathbf{j} + \varepsilon_0 \mu_0 \frac{\partial \mathbf{E}}{\partial t} \tag{B.37}$$

Here we have introduced another dimensional factor μ_0 to account for the possibly different dimensions of **E** and **B**. By taking the divergence of this expression it is easy to show that it just reproduces the original charge and current continuity equation.

The electric field, **E**, is generated by the charge density, but we must still ask for the charges generating the magnetic field, **B**. From observation one knows that such charges, called magnetic monopoles, do not exist. Thus **B** must be divergence-free

$$\nabla \cdot \mathbf{B} = 0 \tag{B.38}$$

Since the time derivative of Eq. (B.38) vanishes, too, we conclude that

$$\frac{\partial \mathbf{B}}{\partial t} = -\nabla \times \mathbf{E} \tag{B.39}$$

must be the curl of a vector. However, this vector can only be the electric field, since no other quantities are involved. The minus sign arises from the experimental *Lenz's rule*.

Poisson's Equation

Poisson's equation is Laplace's equation including a source term. We encounter it in the Poisson law of electrodynamics, i.e., the Maxwell equation (B.35)

$$\nabla \cdot \mathbf{E} = \rho/\varepsilon_0 \tag{B.40}$$

when replacing **E** by the gradient of the electric potential, $-\nabla\phi$. Then Eq. (B.40) reads

$$\nabla^2 \phi = -\rho/\varepsilon_0 \tag{B.41}$$

In vector form with a vector source **Q** the same equation becomes

$$\nabla^2 \mathbf{A} = -\mathbf{Q} \tag{B.42}$$

Using the notation $d^3x = d\mathbf{x} = dxdydz$, these equations have the following solutions

$$\phi(\mathbf{x}) = \frac{1}{4\pi\varepsilon_0} \int \frac{\rho(\mathbf{x}')}{|\mathbf{x}-\mathbf{x}'|} d^3x' \quad \text{and} \quad \mathbf{A}(\mathbf{x}) = \frac{1}{4\pi} \int \frac{\mathbf{Q}(\mathbf{x}')}{|\mathbf{x}-\mathbf{x}'|} d^3x' \tag{B.43}$$

expressing the simple truth that the fields are generated by their sources which are the singularities of the fields, and that the field value in space outside the sources can be obtained simply by integration over all the sources.

Biot-Savart's Law

Integrating Ampère's law, i.e., the Maxwell equation (B.37) for $\partial \mathbf{E}/\partial t = 0$,

$$\nabla \times \mathbf{B} = \mu_0 \mathbf{j} \tag{B.44}$$

with respect to the surface $\mathbf{A}$, which is enclosed by the path C and penetrated by the current $\mathbf{j}$, and using Stokes' theorem (see Appendix B.4) one obtains

$$\int_A (\nabla \times \mathbf{B}) \cdot d\mathbf{A} = \oint_C \mathbf{B} \cdot d\mathbf{s} = \mu_0 \int_A \mathbf{j} \cdot d\mathbf{A} = \mu_0 I \tag{B.45}$$

where $d\mathbf{s} = \mathbf{t}\,ds$ is the product of a line element, ds, with its tangent vector, $\mathbf{t}$, while $d\mathbf{A} = \mathbf{n}\,dA$ is the product of a surface element dA with its normal vector, $\mathbf{n}$. Hence, the line integral along the magnetic field around the current along the boundary C of the surface A is proportional to the total current I flowing through the surface.

For a line current this curve is a circle of radius r, and the above integral yields $2\pi r B_\phi$. Hence, the only non-vanishing component of the magnetic field, the azimuthal component, decays radially as $1/r$

$$B_\phi(r) = \frac{\mu_0}{2\pi} \frac{I}{r} \tag{B.46}$$

Taking the curl of Ampère's law and remembering $\nabla \cdot \mathbf{B} = 0$, one finds, with the magnetic disturbance caused by distributed currents

$$\nabla^2 \mathbf{B} = -\mu_0 \nabla \times \mathbf{j} \tag{B.47}$$

This is a vectorial Poisson equation, called *Biot-Savart's law*. According to Eq. (B.43), its solution is

$$\mathbf{B}(\mathbf{x}) = \frac{\mu_0}{4\pi} \nabla \times \int \frac{\mathbf{j}(\mathbf{x}')}{|\mathbf{x} - \mathbf{x}'|} d^3x' = \frac{\mu_0}{4\pi} \int \frac{\mathbf{j}(\mathbf{x}') \times (\mathbf{x} - \mathbf{x}')}{|\mathbf{x} - \mathbf{x}'|^3} d^3x' \tag{B.48}$$

Calculating the curl yields the second of these forms which is Biot-Savart's law in integral form.

Faraday's Law

When integrating the Maxwell equation (B.39) over the surface $\mathbf{A}$ perpendicular to the magnetic field, one obtains

$$\int_A (\nabla \times \mathbf{E}) \cdot d\mathbf{A} = \int_A \frac{\partial \mathbf{B}}{\partial t} \cdot d\mathbf{A} \tag{B.49}$$

The left-hand side can be transformed using Stokes' theorem, while the right-hand side gives the time variation of the magnetic flux

$$\Phi = \int_A \mathbf{B} \cdot d\mathbf{A} \tag{B.50}$$

through the surface. Hence, one finds

$$\frac{d\Phi}{dt} = -\oint_C \mathbf{E} \cdot d\mathbf{s} \tag{B.51}$$

where $d\mathbf{s}$ is a line element of the curve C surrounding the surface A. The change in the magnetic flux is equal to the induced potential difference along C.

B.6 Plasma Entropy

Knowing the ideal gas equation of state, it is possible to derive an expression for the entropy, $\mathscr{S}$, in an ideal gas which is useful to calculate $\mathscr{S}$ from measured quantities. The differential of the entropy is defined as the ratio of heat, $d\mathscr{Q}$, and temperature, T, where

$$d\mathscr{Q} = dW + pd\mathscr{V} \tag{B.52}$$

with W the internal energy and $\mathscr{V}$ the volume. For an ideal gas

$$p = Nk_BT/\mathscr{V} = (\gamma-1)W/\mathscr{V} \tag{B.53}$$

where $n = N/\mathscr{V}$ is the number of particle per volume or particle density yielding

$$pd\mathscr{V} = Nk_BTd\mathscr{V}/\mathscr{V} \tag{B.54}$$

or substituting into the above equation for the heat produced

$$d\mathscr{Q} = Nk_BT\left[d\ln\mathscr{V} + (\gamma-1)^{-1}d\ln T\right] \tag{B.55}$$

from which we immediately obtain the desired form of the differential entropy

$$d\mathscr{S} = Nk_Bd\ln\left[\mathscr{V}T^{1/(\gamma-1)}\right] \tag{B.56}$$

Performing the integration and remembering that for an ideal gas $Nk_B = R_0 = \text{const}$ (see Appendix B.1), one finds the expression for the entropy in the form

$$\mathscr{S} - \tilde{\mathscr{S}}_0 = R_0\ln\left[\mathscr{V}T^{1/(\gamma-1)}\right] \tag{B.57}$$

The quantity $\tilde{\mathscr{S}}_0$ is an integration constant which is of no importance because only differences in entropy are of interest. Because of the same reason one can replace the volume by the density. This adds only another constant $\mathscr{S}_0$, depending on the number of particles in the volume to the entropy. One then has finally

$$\mathscr{S} - \mathscr{S}_0 = R_0\ln\left[\frac{T^{1/(\gamma-1)}}{n}\right] \tag{B.58}$$

as a formula to calculate the entropy change in an ideal gas from the measurements of temperature and density in the gas. With $\gamma = 5/3$ the exponent of T becomes 3/2.

The above equations apply only to isotropic cases. For anisotropic plasmas the corresponding expression for the entropy is

$$\mathscr{S} - \mathscr{S}_0 = R_0 \ln \left[\frac{1}{n} \left(\frac{\gamma_\| T_\| + 2\gamma_\perp T_\perp}{3} \right)^{3/2} \right] \tag{B.59}$$

where $\gamma_\|, \gamma_\perp$ are the adiabatic indices parallel and perpendicular to the magnetic field (see p. 163).

B.7   Aspects of Analytic Theory

Plasma wave theory makes extensive use of complex variables, since frequency and wave number are complex quantities. Hence, the dispersion relation and all the integrals it contains become complex functions or integrals in the complex plane. Integrals in the complex plane are contour integrals. For their solution special methods have been developed already in the past century. Here, we review the basic methods as far as they apply to plasma theory.

Cauchy-Riemann Equations

A continuous complex function, $f(z) = u(x,y) + iv(x,y)$, in the complex plane, $z = x + iy$, must have a unique derivative with respect to z, $\partial f(z)/\partial z$. Because z itself consists of two variables, x, y, this uniqueness is not obvious. The condition of uniqueness is found by requiring that the derivatives from both sides x and y lead to the same result

$$\frac{\partial f(z)}{\partial z} = \frac{\partial u}{\partial x} + i \frac{\partial v}{\partial y} = \frac{\partial v}{\partial y} - i \frac{\partial u}{\partial y} \tag{B.60}$$

Comparing real and imaginary parts one finds

$$\frac{\partial u}{\partial x} = \frac{\partial v}{\partial y} \qquad \frac{\partial v}{\partial x} = -\frac{\partial u}{\partial y} \tag{B.61}$$

These equations are known as the *Cauchy-Riemann equations*. They constitute the necessary and sufficient conditions for the existence of a unique derivative of $f(z)$ and, hence, the condition for its continuous behavior in the complex plane. Such a function can be non-analytic only at a finite number of isolated points, which are called singular points.

Cauchy's Integral Formula

If $f(z)$ is continuous and analytic in a region, R, of the complex plane which is bounded by a curve, C, and if a is an interior point of R then *Cauchy's integral theorem* states that the integral

$$\oint_C \frac{f(z)dz}{z-a} = 2\pi i f(a) \tag{B.62}$$

is proportional to the value of the function $f(z)$ at the interior point a. Hence, in a more general form the *Cauchy integral formula*

$$F(z) = \frac{1}{2\pi i} \oint_C \frac{f(\zeta)}{\zeta - z} d\zeta \tag{B.63}$$

along a closed curve, C, over an continuous complex function, $f(\zeta)$, defines an analytic function, $F(z)$, in the interior of C. One can then easily define its nth derivative with respect to z as

$$\frac{\partial^n F(z)}{\partial z^n} = \frac{n!}{2\pi i} \oint_C \frac{f(\zeta)}{(\zeta - z)^{n+1}} d\zeta \tag{B.64}$$

so that $F(z)$ possesses all orders of derivatives and can thus be expanded into a Taylor series around any point inside C.

Contour Integration

With the help of these results the Cauchy integral can be used to calculate the value of line integrals which contain singular points. If $z = b$ is a singular point of the function $f(z)$ then in the neighborhood of $z = b$ the function $f(z)$ can be expanded as

$$f(z) = \sum_{n=0}^{\infty} \left[a_n(z-b)^n + \frac{a_{-n}}{(z-b)^n} \right] \tag{B.65}$$

This series is called *Laurent series*. When this expansion has an infinite number of singular terms, the singularity at $z = b$ is called essential and cannot be treated in a simple way. However, when there are only a finite number m of terms which are singular, which is the usual case in physical application, then the function is said to have a pole of order m. In particular, when $m = 1$ the function has a simple pole. In the case of a pole one can define a new function

$$\phi(z) = (z-b)^m f(z) \tag{B.66}$$

for $z \neq b$ so that $\phi(b) = a_{-m}$. This function is analytic at $z = b$. We can now expand this function into a Taylor series around b

$$\begin{aligned} \phi(z) &= (z-b)^m f(z) \\ &= a_{-m} + a_{-m+1}(z-b) + \cdots + a_{-1}(z-b)^{m-1} + a_0(z-b)^m + \cdots \end{aligned} \tag{B.67}$$

Therefore the term a_{-1} is the coefficient of the $(m-1)$st derivative of $\phi(z)$

$$a_{-1} = \left| \frac{1}{(m-1)!} \frac{d^{m-1}[(z-b)^m f(z)]}{dz^{m-1}} \right|_{z=b} \tag{B.68}$$

For the special case of a simple pole this residue reduces to the special case $m=1$

$$a_{-1} = \lim_{z \to b}(z-b)f(z) \tag{B.69}$$

These residua are very useful in calculating complex line integrals. When a function $F(z)$ has a series of poles the line integral over this function turns out to be the sum of all the residua, r_n, of this function at the poles inside the closed contour, C, of integration

$$\oint_C F(z)dz = 2\pi i \sum_n r_n \tag{B.70}$$

where the residua are calculated using the above rules. One first determines the poles of $F(z)$ and their order. Then one expands $F(z)$ around each pole into a Laurent series and determines the coefficient a_{-1} at this pole by simple use of the above formula. The value of the integral is just the sum over all r_n inside C.

If one knows that a function is analytic in one domain of the complex plane but one does not know its behavior at the outside one can use the method of analytic continuation to extend the region of analyticity of the function. Analytic continuation requires investigation of the behavior of the poles during the crossing of the contour C to the outside. If the poles all remain inside C, the continuation is trivial. But if the poles move out of C, one must deform the contour in such a way that the pole remains always on the same side of the new contour. The pole pushes the contour ahead of it. This process is called *analytic continuation*.

Plemelj-Dirac Formula

When the integrand in the complex contour integral has a simple pole but the pole approaches the real axis $\zeta \to \pm x$ either from above or from below that is from the half-plane where the integrand is analytic with the exception of the pole, then the Cauchy integral takes one particularly simple form discovered by Plemelj and Dirac

$$\lim_{\eta \to 0} \int_{-\infty}^{\infty} \frac{f(x)dx}{x-a\pm i\eta} = P \int_{-\infty}^{\infty} \frac{f(x)dx}{x-a} \mp i\pi f(a) \tag{B.71}$$

where a is real and the symbol P designates the principal value of the integral which is defined as the sum of two integrals

$$P \int_{-\infty}^{\infty} \frac{f(x)dx}{x-a} = \lim_{\varepsilon \to 0} \left[\int_{-\infty}^{a-\varepsilon} \frac{f(x)dx}{x-a} + \int_{a+\varepsilon}^{\infty} \frac{f(x)dx}{x-a} \right] \tag{B.72}$$

taken along the real x-axis. The above formula can symbolically be written as

$$\lim_{\eta \to 0} \frac{1}{x \pm i\eta} = \frac{P}{x} \mp i\pi\delta(x) \tag{B.73}$$

As one easily realizes either form of this formula arises from the integration along the real axis and the deformation of the contour into a half-circle around the pole at $x = a \pm i\eta$. Hence, the factor two in front of the imaginary part disappears, and the principal value integral must be included as a "boundary condition" during analytic continuation between the two parts of the complex plane.

Maxwellian Integrals

Maxwellian integrals are integrals over Gaussian functions multiplied by some power of the integrand. They appear frequently in plasma physics where velocity distribution functions are modeled by products or sums of Maxwellians. The basic integral is the definite integral along the full real axis over a Gaussian function

$$\int_{-\infty}^{\infty} \exp(-x^2)\, dx = \sqrt{\pi} \tag{B.74}$$

This integral is closely related to the error function

$$\mathrm{erf}(z) = \frac{2}{\sqrt{\pi}} \int_{0}^{z} \exp(-x^2)\, dx \tag{B.75}$$

The Gaussian function is twice the value of the error function at infinity, with the factor of two a consequence of the symmetry of the integrand. The integral can be calculated using the methods of complex path integration described below. Generalizations of the above integral needed in plasma physics are of the form

$$\int_{0}^{\infty} x^{\alpha} \exp(-ax^2)\, dx = \begin{cases} \Gamma(l+1/2)/2a^{l+1/2} & \text{for } \alpha = 2l \\ l!/2a^{l+1/2} & \text{for } \alpha = 2l+1 \end{cases} \tag{B.76}$$

where $\Gamma(l+1/2) = [(2l-1)!!/2^l]\sqrt{\pi/a}$, and a has a positive real part to make the integral converging. Because of the asymmetry of the integrand for odd α it is clear that in the second case the integrals from $-\infty$ to $+\infty$ vanish identically, while for even α they are twice the value given above. The above formula can be verified by observing that the basic Maxwellian integral can be reproduced by multiple differentiation with respect to a.

Plasma Dispersion Function

The *plasma dispersion function* $Z(\zeta)$, sometimes also called *Fried-Conte function* because Fried and Conte first tabulated it in 1961, is a special example of an analytic function which is instrumental in plasma physics and particularly in plasma wave theory. It arises from the singular integral over a Maxwellian distribution (see Section 6.3)

$$Z(\zeta) = \frac{1}{\pi^{1/2}} \int_{-\infty}^{\infty} \frac{\exp(-z^2)dz}{z - \zeta} = i\pi^{1/2}\mathrm{erf}(\zeta) \tag{B.77}$$

with $\mathrm{Im}\,\zeta > 0$. By differentiation with respect to ζ it is easy to show that $Z(\zeta)$ satisfies the first order differential equation

$$\frac{dZ(\zeta)}{d\zeta} = -2[1 + \zeta Z(\zeta)] \tag{B.78}$$

which can be used also as a recurrence relation to replace derivatives. By analytic continuation one finds some other useful relations

$$\begin{aligned} Z(-\zeta) &= -Z(\zeta) + 2\pi^{1/2}i\exp(-\zeta^2) & \text{for } \mathrm{Im}\,\zeta > 0 \\ \tilde{Z}(\zeta) &= Z(\zeta) - 2\pi^{1/2}\exp(-\zeta^2) & \text{for } \mathrm{Im}\,\zeta < 0 \end{aligned} \tag{B.79}$$

where $\tilde{Z}(\zeta)$ is the analytic continuation of $Z(\zeta)$. The complex conjugate of the plasma dispersion function is

$$[Z(\zeta)]^* = Z(\zeta^*) - 2\pi^{1/2}\exp(-\zeta^2) \tag{B.80}$$

The expansion of $Z(\zeta)$ for small argument $\zeta < 1$ is found by Taylor expansion of the above Cauchy integral and integrating term by term

$$\begin{aligned} Z(\zeta) &= i\pi^{1/2}e^{-\zeta^2} - 2\zeta\left(1 - \frac{2\zeta^2}{3} + \frac{4\zeta^4}{15} - \cdots\right) \\ &= \pi^{1/2}\sum_{n=0}^{\infty} \frac{i^{n+1}\zeta^n}{\Gamma(1 + n/2)} \end{aligned} \tag{B.81}$$

The asymptotic expansion for $\zeta \gg 1$ is given by

$$Z(\zeta) = -\frac{1}{\zeta}\left(1 + \frac{1}{2\zeta^2} + \frac{3}{4\zeta^4} + \cdots\right) + \sigma\pi^{1/2}i\exp(-\zeta^2) \tag{B.82}$$

where

$$\sigma = \begin{cases} 0, & \mathrm{Im}\,\zeta > 0 \\ 1, & \mathrm{Im}\,\zeta = 0 \\ 2, & \mathrm{Im}\,\zeta < 0 \end{cases} \tag{B.83}$$

The plasma dispersion function is closely related to a number of other functions. One of them is the error function of complex argument

$$\mathrm{erf}(\zeta) = \left(1 + \frac{2i}{\pi^{1/2}}\int_0^{\zeta} e^{t^2}dt\right)e^{-\zeta^2} \tag{B.84}$$

C — Extensions

C.1 Coulomb Logarithm

In Section 4.1 we derived the collision frequency in a plasma, where the particles undergo pure Coulomb collisions. Here we present a rigorous derivation.

Rutherford Scattering

Let a charge, q, be scattered from a much heavier charge, q', in such a way that the heavy charge can be considered to be at rest. The Coulomb force F acting on q is then

$$F = \frac{qq'}{4\pi\varepsilon_0 r^2} \tag{C.1}$$

with r the instantaneous distance between q and q'. The problem is symmetric (see Fig. C.1) and for equal sign of the charges the path described by the lighter charge (in the rest frame of the heavier charge) will be a hyperbola with its symmetry axis along x. Only the x component of the momentum of the scattered charge, $F\cos\phi/m$, is changed during the collision. With an initial velocity $v_0 \cos\phi_0$, we obtain

$$2mv_0 \cos\phi_0 = \frac{qq'}{4\pi\varepsilon_0} \int\limits_{-\sin\phi_0}^{\sin\phi_0} \frac{dt}{d\phi} \frac{d\sin\phi}{r^2} \tag{C.2}$$

for the total change in momentum. In a central force field the angular momentum

$$mr^2\frac{d\phi}{dt} = mbv_0 \tag{C.3}$$

is conserved (b is the collision or impact parameter defined in Fig. C.1; see also p. 61). Hence, the value of the integral is

$$2mv_0 \cos\phi_0 = \frac{qq'\sin\phi_0}{2\pi\varepsilon_0 bv_0} \tag{C.4}$$

Since the angle of deflection is $\theta = \pi - 2\phi_0$ and thus $\cot(\theta/2) = \tan\phi_0$, we can rewrite the above equation and find the well-known Rutherford scattering formula

$$\cot(\theta/2) = \frac{4\pi\varepsilon_0 mbv_0^2}{qq'} \tag{C.5}$$

486

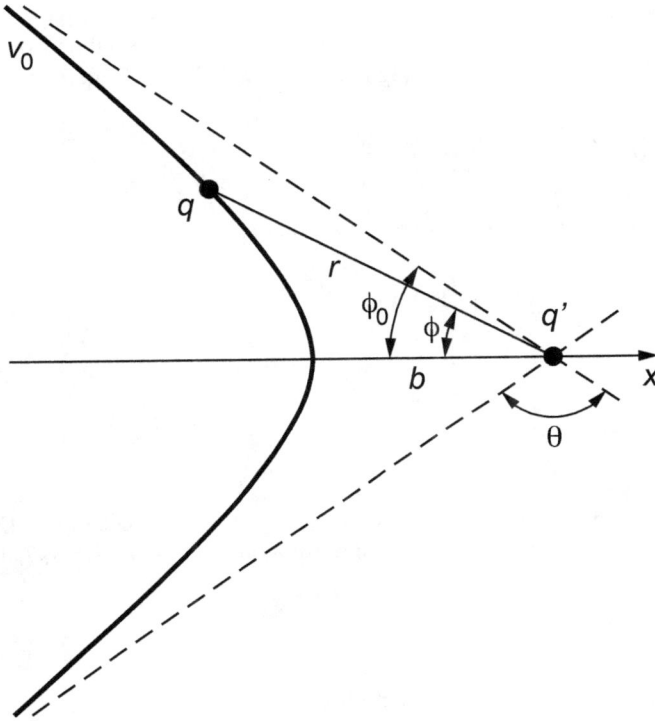

Figure C.1: Geometry of scattering a charge q at a heavier charge q'.

Mean Scattering Angle

To calculate the collision frequency one needs the average collisional cross-section with the average taken over all particles incident on q' with their different impact parameters, b. The differential cross-section is defined as

$$d\sigma = 2\pi b \, db \tag{C.6}$$

A stream of particles, all with initial velocity v_0 but different impact parameters, $b_{min} < b < b_{max}$, will be deflected by a mean scattering angle

$$\langle \theta^2(v_0) \rangle = \frac{\int \theta^2(v_0, b) \, d\sigma(b)}{\int d\sigma(b)} \tag{C.7}$$

where the integration in both integrals is over the interval $b_{min} < b < b_{max}$. Taking into account Debye screening, a good approximation for b_{max} is the Debye length. It is more difficult to find the lower limit of the impact parameter. The usual choices are

$$b_{max} = \lambda_D \tag{C.8}$$
$$b_{min} = qq'/4\pi\varepsilon_0 m v_0^2$$

with the latter formula resulting from the balance between initial particle energy and Coulomb energy. The main contribution to the integrals comes then from small-angle scattering of incident particles. Using the small-angle limit of Eq. (C.5)

$$\theta^2(v_0, b) = \frac{4}{v_0^4 b^2} \left(\frac{qq'}{4\pi\varepsilon_0 m} \right)^2 \tag{C.9}$$

and performing the integrations one obtains finally

$$\langle \theta^2(v_0) \rangle = \frac{4}{\pi} \frac{b_{min}^2}{b_{max}^2} \ln \left(\frac{b_{max}}{b_{min}} \right) \tag{C.10}$$

where

$$\frac{b_{max}}{b_{min}} = \frac{4\pi\varepsilon_0 \lambda_D m v_0^2}{qq'} \tag{C.11}$$

is the ratio of incident energy to Coulomb energy at the distance of one Debye length from the scattering center. For a thermal plasma one can use $mv_0^2 = k_B T_e$ to obtain

$$\frac{b_{max}}{b_{min}} = \frac{4\pi\varepsilon_0 \lambda_D k_B T_e}{qq'} \tag{C.12}$$

and setting $q = q' = e$ one can rewrite this expression as

$$\Lambda = \frac{b_{max}}{b_{min}} = 4\pi n \lambda_D^3 \tag{C.13}$$

Λ is within a factor 4π equal to the plasma parameter introduced in Eq. (1.5) which is proportional to the number of particles in a Debye sphere. The logarithm $\ln \Lambda$ of this quantity is called the Coulomb logarithm and has been used in the expression for Spitzer's plasma collision frequency.

C.2　Transport Coefficients

The Coulomb logarithm enters the collision frequencies. The two equations from where this is obvious are the retardation of a particle by collisions

$$\frac{d\mathbf{v}}{dt} = -v_c \mathbf{v} \tag{C.14}$$

and the equation describing energy losses during frictional motion

$$\frac{dW_{kins}}{dt} = \frac{m_s}{2} \frac{dv^2}{dt} = -v_W \frac{m_s}{2} v^2 \tag{C.15}$$

Under anisotropic conditions the collision frequencies have parallel and perpendicular components. For two species, α, β, the general expressions for the collision frequencies obtained from a Boltzmann collision integral formulation are

$$
\begin{aligned}
v_{c\alpha\beta} &= 2(1 + m_\alpha/m_\beta) v_{0\alpha\beta} x_\beta^2 \psi(x_\beta) \\
v_{w\alpha\beta} &= 2v_{c\alpha\beta} - v_{\perp\alpha\beta} - v_{\|\alpha\beta} \\
v_{\perp\alpha\beta} &= 2v_{0\alpha\beta}[\phi(x_\beta) - \psi(x_\beta)] \\
v_{\|\alpha\beta} &= 2v_{0\alpha\beta} \psi(x_\beta) \\
v_{0\alpha\beta} &= (q_\alpha^2 q_\beta^2 n_\beta/4\pi\varepsilon_0^2 m_\alpha^2 |v_\alpha - v_\beta|^3) \ln \Lambda_{\alpha\beta}
\end{aligned}
\tag{C.16}
$$

where the two functions of $x_\beta = |v_\alpha - v_\beta|/\sqrt{2} v_{\text{th}\beta}$ are given by

$$
\begin{aligned}
\phi(x) &= \frac{2}{\sqrt{\pi}} \int_0^x dt \exp(-t^2) \\
\psi(x) &= \frac{1}{2x^2}\left[\phi(x) - x\frac{d\phi(x)}{dx}\right]
\end{aligned}
\tag{C.17}
$$

The Coulomb logarithm for electron-electron collisions has the value

$$
\ln \Lambda_{ee} = \begin{cases} 16.0 - 0.5\ln n_e + 1.5\ln T_e & T_e \le 7\cdot 10^4 \,\text{K} \\ 21.6 - 0.5\ln n_e + \ln T_e & T_e \ge 7\cdot 10^4 \,\text{K} \end{cases}
\tag{C.18}
$$

while we have for electron-ion collisions $\ln \Lambda = \ln \Lambda_{ei} = \ln \Lambda_{ie}$ from Eq. (C.13) with the numerical values

$$
\ln \Lambda_{ei} = \begin{cases} 16.0 - 0.5\ln n_e + 1.5\ln T_e - \ln Z_i & T_e \le 1.4\cdot 10^5 \,\text{K} \\ 21.6 - 0.5\ln n_e + \ln T_e & T_e \ge 1.4\cdot 10^5 \,\text{K} \end{cases}
\tag{C.19}
$$

In both cases, the density is measured in cm^{-3} and the temperature in K. Temperature equilibrium between electrons and ions is reached in a time

$$
dT_e/dt = v_{\text{eq}ei}(T_i - T_e)
\tag{C.20}
$$

where the equilibrium collision frequency is determined as

$$
v_{\text{eq}ei} = \frac{e^2 q_i^2 n_e \ln \Lambda_{ei}}{3(2\pi)^{1/2}\pi m_e m_i \varepsilon_0^2 (v_{\text{the}}^2 + v_{\text{thi}}^2)^{3/2}}
\tag{C.21}
$$

Characteristic collision times for the electrons and ions are

$$
\begin{aligned}
\tau_e &\approx 2.8\cdot 10^5 \, T_e^{3/2}/(n_e \ln \Lambda_{ei}) \\
\tau_i &\approx 1.7\cdot 10^7 \, T_i^{3/2}/(n_e \ln \Lambda_{ei})
\end{aligned}
\tag{C.22}
$$

Again, the density is measured in cm^{-3}, and the temperature in K. The momentum transfer rate, $\mathbf{R}_{ei} = -\mathbf{R}_{ie}$, first used in Eq. (7.42) can be written as

$$\mathbf{R}_{ei} = en_e \left(\frac{\mathbf{j}_\parallel}{\sigma_\parallel} + \frac{\mathbf{j}_\perp}{\sigma_\perp} - \frac{0.7\nabla_\parallel T_e}{e} - \frac{3}{2} \frac{\mathbf{B} \times \nabla T_e}{eB\omega_{ge}\tau_e} \right) \tag{C.23}$$

where the parallel and perpendicular conductivities are

$$\begin{aligned} \sigma_\parallel &= 2\sigma_\perp \\ \sigma_\perp &= \varepsilon_0 \omega_{pe}^2 \tau_e \end{aligned} \tag{C.24}$$

The electron and ion heat fluxes are given by

$$\begin{aligned} \mathbf{q}_e &= \frac{0.7T_e}{e}\mathbf{j}_\parallel + \frac{3T_e}{2eB\omega_{ge}\tau_e}\mathbf{B} \times \mathbf{j}_\perp - \kappa_{e\parallel}\nabla_\parallel T_e - \kappa_{e\perp}\nabla_\perp T_e - \frac{5n_e T_e}{2m_e B\omega_{ge}}\mathbf{B} \times \nabla T_e \\ \mathbf{q}_i &= -\kappa_{i\parallel}\nabla_\parallel T_i - \kappa_{i\perp}\nabla_\perp T_i + \frac{5n_i T_i}{2m_i B\omega_{gi}}\mathbf{B} \times \nabla T_i \end{aligned} \tag{C.25}$$

and the thermal conductivities entering these expressions are for $\omega_{gs}\tau_s \gg 1$

$$\begin{aligned} \kappa_{e\parallel} &= 3.16 n_e T_e \tau_e / m_e \\ \kappa_{i\parallel} &= 3.90 n_i T_i \tau_i / m_i \\ \kappa_{e\perp} &= 4.66 n_e T_e / (m_e \omega_{ge}^2 \tau_e) \\ \kappa_{i\perp} &= 2.00 n_i T_i / (m_i \omega_{gi}^2 \tau_i) \end{aligned} \tag{C.26}$$

C.3   Geomagnetic Indices

Magnetic indices are derived from ground-based magnetograms and are meant to quantify disturbed states of the Earth's plasma environment. While planetary range indices like Kp describe only the overall disturbance level and will not be described here, two indices quantify the disturbance and the dissipation of energy of a certain element of the geo-plasma space. These are the *Dst index*, which was introduced in Section 3.6 and quantifies the ring current, and the *AE index*, which gives a measure of the auroral electrojets and the substorm activity (see Sections 5.5 through 5.7).

AE Index

The auroral electrojet indices *AE*, *AU*, and *AL* were introduced as a measure of global auroral electrojet activity. The present auroral indices are based on 1-min readings of the northward H component trace from twelve auroral zone observatories located between about $65°$ and $70°$ magnetic latitude with a longitudinal spacing of $10 - 40°$.

For each of the twelve observatories, the readings of the H component are referenced to a quiet day level, H_0. The base value H_0 for the month under consideration is calculated as the average over all the readings from the five most quiet days in that month. The data of all twelve observatories are then plotted as a function of universal time. The upper and lower envelopes are defined as AU and AL, while AE is defined as the separation between the upper and lower envelopes

$$
\begin{aligned}
AU(t) &= \max_{i=1,12} \left\{ H(t) - H_0 \right\}_i \\
AL(t) &= \min_{i=1,12} \left\{ H(t) - H_0 \right\}_i \quad\quad\quad\quad\quad\quad \text{(C.27)} \\
AE(t) &= AU(t) - AL(t)
\end{aligned}
$$

where t is universal time. AU and AL are thought to represent the maximum eastward and westward electrojet current, respectively. AE represents the total maximum electrojet current and is most often used.

The main uncertainties of the AE index stem form the use of the H component, from longitudinal gaps in the distribution of the twelve observatories, from the small latitudinal range covered by these magnetic stations, and from the effects of strong local field-aligned currents.

At some magnetic observatories the angle between the local magnetic H component and the global eccentric dipole north-south direction is greater than $30°$. Since the electrojets tend to flow along the auroral oval, i.e., perpendicular to the global eccentric dipole north-south direction rather than perpendicular to the local H direction, these observatories tend to underestimate the electrojet current.

The most severe longitudinal gaps in the AE observatory coverage are in Siberia, but also in western Canada and the Atlantic sector gaps spanning more than two hours of local time exist. Substorm current wedges associated with weak or moderate substorms may cover less than two hours of local time and can easily be missed by the twelve-station AE network.

Probably more severe is the small latitudinal range covered by the AE observatories. During times of very weak activity, when the interplanetary magnetic field is northward directed and convection ceases (see Section 5.2), the auroral oval contracts northward and the electrojets tend to flow poleward of $70°$ latitude. In this situation, the AE network, with all its stations south of $70°$, will not detect the maximum disturbances.

The first three uncertainties of the standard AE index can, in principle, be avoided by the use of eccentric dipole coordinates and by including more stations in the network. However, the last uncertainty cannot be overcome even by an ideal AE index. For regions east or west of strong local field-aligned currents, a significant part of the north-south component of the magnetic disturbance stems from the field-aligned currents. This effect is most pronounced behind the head of the westward traveling surge, where strong field-aligned currents flow upward (see Section 5.7). Here, the southward perturbation due to the westward flowing ionospheric current can be reduced by up to 30% by the northward magnetic field associated with the upward field-aligned current to the west.

The other two electrojet indices have the same uncertainties as the AE index, but in addition are influenced by azimuthally uniform non-electrojet fields like that of the ring current, which cancel out in AE.

Dst Index

The ring current index Dst was introduced as a measure of the ring current magnetic field and thus its total energy, as described in Eq. (3.37). Since the westward ring current causes a reduction of the terrestrial dipole field, Dst is typically negative. During a magnetic storm, a typical Dst trace looks like the magnetogram in Fig. 3.12. The present Dst index is based on hourly averages of the northward horizontal H component recorded at four low-latitude observatories, Honolulu, San Juan, Hermanus, and Kakioka. All four observatories are $20 - 30°$ away from the dipole equator to minimize equatorial electrojet effects (see Section 4.5) and are about evenly distributed in local time.

At each observatory a magnetic perturbation amplitude is calculated by subtracting from the hourly averaged measured H component a quiet time reference level, $H_0(t')$, and the Sq field, $H_{sq}(t')$ (see Section 4.5), which both vary with local time, t'. All four magnetic disturbances are then averaged to further reduce local time effects and multiplied with the averages of the cosines of the observatories' dipole latitudes, Λ_i, to obtain the value of the ring current field at the dipole equator

$$Dst(t) = \frac{1}{16} \left[\sum_{i=1}^{4} \cos \Lambda_i \right] \cdot \left[\sum_{i=1}^{4} \left\{ H(t) - H_0(t') - H_{sq}(t') \right\}_i \right] \tag{C.28}$$

where t is the universal time.

In contrast to the AE index, where most of the uncertainties lie in the uneven and too widely spaced station network, the Dst uncertainties are mainly caused by magnetic contributions of sources other than the ring current to the H component measured at the four Dst observatories, namely the magnetopause current, the partial ring current, and the substorm current wedge.

The magnetopause current (see Section 8.6.2) is regulated by the solar wind pressure. Its magnetic perturbation peaks around noon. A typical magnetopause contribution is included in the quiet time reference level, but dynamic variations of the solar wind pressure can yield positive or negative deviations from this average. In particular, sudden changes of the solar wind pressure associated with an interplanetary shock front can change the local time average of the magnetic perturbation due to the magnetopause current by typically $10 - 40$ nT.

The westward partial ring current (see Section 3.3) prevails in the afternoon sector and its closure via field-aligned currents causes a southward magnetic disturbance around the dusk meridian. The substorm current wedge (see Section 5.7) dominates in the midnight and early morning sector. Here, again mainly the field-aligned currents create a significant northward magnetic disturbance.

The combined effects of these three variable current systems yield an uncertainty of the quiet time reference level. Experimentally, one has found that the uncertainty maximizes around noon, where it may reach up to 50% of a typical *Dst* value of 50 nT. On the nightside, the uncertainty is relatively small, around $5 - 10$ nT. Hence, solar wind pressure changes are the dominant source of uncertainties in the *Dst* index.

C.4 Liouville Approach

The derivation of the Vlasov equation from the Klimontovich-Dupree equation given in Chapter 6 is only one way of finding the kinetic equations of a plasma. Historically, one has gone a different route, starting from the Liouville equation of statistical mechanics and descending from it along the construction of reduced distribution functions to the Vlasov equation. The two approaches are entirely equivalent. We have chosen the more simple treatment in the main text, but since the Liouville equation approach is slightly more rigorous we will, for completeness, sketch it here.

Liouville Equation

The Liouville approach starts from the assumption that each particle with index i spans its own six-dimensional phase space. Hence, the total phase space of all, say n, particles has $6n$ dimensions numbered by the index i of the particle to which the individual phase space belongs, $\mathbf{x}_1, \mathbf{v}_1, \ldots, \mathbf{x}_i, \mathbf{v}_i, \ldots, \mathbf{x}_n, \mathbf{v}_n$. The interaction between all these particles is contained in $\mathscr{H}_n(\mathbf{x}_1, \mathbf{v}_1, \ldots, \mathbf{x}_i, \mathbf{v}_i, \ldots, \mathbf{x}_n, \mathbf{v}_n, t)$, the total n-particle Hamiltonian which is a function of $6n + 1$ coordinates.

As done in Section 6.1, one can define an exact phase space distribution function of all coordinates and time, $\mathscr{F}_n(\mathbf{x}_1, \mathbf{v}_1, \ldots, \mathbf{x}_i, \mathbf{v}_i, \ldots, \mathbf{x}_n, \mathbf{v}_n, t)$, which is again conserved during the dynamic evolution of the plasma. This conservation can be expressed as the total time derivative of $\mathscr{F}_n$, taken with respect to all $6n$ coordinates which is the same as writing the Poisson bracket of $\mathscr{F}_n$ with the n-particle Hamiltonian $\mathscr{H}_n$

$$\frac{\partial \mathscr{F}_n}{\partial t} = [\mathscr{H}_n, \mathscr{F}_n] \tag{C.29}$$

This equation is the *Liouville equation* for the exact phase space distribution function $\mathscr{F}_n$. As with the Klimontovich-Dupree equation, this equation cannot be solved directly without knowing all the orbits of all the particles under all of their interactions. Instead solutions are found by taking moments of this equation.

Reduced Distribution Functions

The moments of the distribution function are defined as integrals over some of the microscopic individual phase spaces of particles contributing to $\mathscr{F}_n$. Such an integration, for instance, with respect to the whole individual phase space of particle i, gets rid of the coordinates of this particle and thus of its individuality. Its contribution to $\mathscr{F}_n$ and its dynamics is smeared out to all the remaining particles as an average effect on the distribution function. The loss of individuality is no problem in plasma physics because all electrons are indistinguishable as are all ions of the same kind.

Performing $6(n-1)$ such consecutive integrations, the individuality of all particles will be destroyed and one is left with a reduced distribution function $\mathscr{F}_1(\mathbf{x}_1, \mathbf{v}_1, t)$ which depends merely on the phase space coordinates of one particle, which can be any electron or ion in the plasma. The distribution function $\mathscr{F}_1$ describes all electrons or ions equally well, distinguishing them only with respect to their velocities and positions. Hence, at a given position $\mathbf{x}_1$ there can be many particles of same velocity $\mathbf{v}_1$ at time t, and the function $\mathscr{F}_1(\mathbf{x}_1, \mathbf{v}_1, t)$ gives the *probability density* of finding these indistinguishable particles at this place.

Accordingly, the equation describing the dynamic evolution of $\mathscr{F}_1$ is obtained by performing $6(n-1)$ integrations on Eq. (C.29). This procedure sounds easy, but it introduces a serious difficulty insofar as the Hamiltonian is a nonlinear function, and thus the integrations over subspaces of the phase space produce non-vanishing average terms in each step which appear on the right hand side of the Liouville equation, which thus becomes inhomogeneous. In the last step, the integration over the coordinates of index 2, one obtains an equation which contains a large number of correlation terms, on the right which depend on all the higher order reduced distribution functions.

BBGKY Hierarchy

Hence, the moment procedure does not close, but produces a whole hierarchy of evolution equations for the reduced distribution functions. To obtain the lowest one, one must solve that for the former and so on. This hierarchy is called *BBGKY hierarchy* after the initials of their inventors. It shows that the statistical character of a plasma is, in principle, very complicated and cannot be resolved entirely without some severe assumptions. These assumptions are that at some stage in the hierarchy, one simply neglects the higher order correlations and their contributions to the reduced distribution function.

At the lowest level, neglecting all the higher order correlations, the continuity or kinetic equation for $\mathscr{F}_1$ becomes

$$\frac{\partial \mathscr{F}_1}{\partial t} = [\mathscr{H}_1, \mathscr{F}_1] \tag{C.30}$$

Expanding the Poisson bracket and replacing $f(\mathbf{x}, \mathbf{v}, t) = \mathscr{F}_1(\mathbf{x}_1, \mathbf{v}_1, t)$, i.e., dropping the index 1, since the lowest level phase space distribution function depends only on the space and velocity coordinates of one particle, one obtains

$$\frac{\partial f}{\partial t} + \mathbf{v} \cdot \nabla_{\mathbf{x}} f + \frac{q}{m}(\mathbf{E} + \mathbf{v} \times \mathbf{B}) \cdot \nabla_{\mathbf{v}} f = 0 \tag{C.31}$$

This is exactly the same equation as the Vlasov equation (6.20). The method of reduced distribution functions and the Liouville equation thus lead to the same result for the

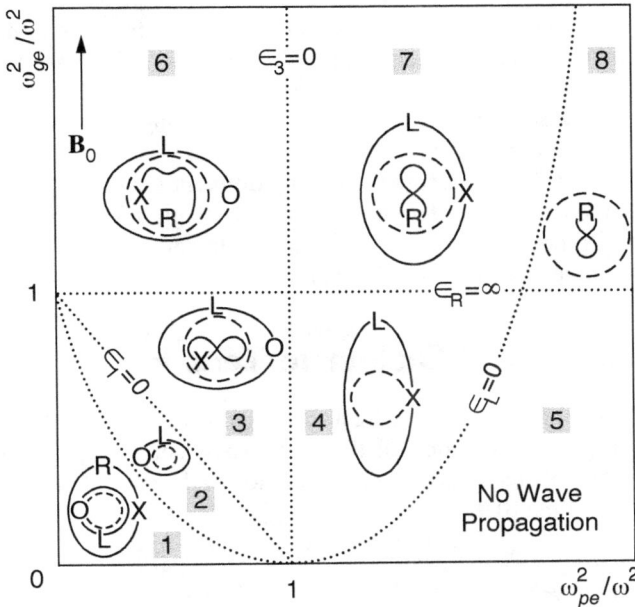

Figure C.2: CMA diagram for a cold electron plasma.

kinetic equation of a plasma if all the correlations between the particles as well as the collisions are neglected.

C.5 Clemmow-Mullaly-Allis Diagram

Investigation of wave propagation even in a cold plasma has turned out to be a complicated play with the dispersive properties of the different wave modes. There is, however, a simple way of classifying the different types of waves in dependence on the plasma parameters, the so-called CMA-diagram in Fig. C.2.

For this diagram one chooses the ratios of cyclotron, ω_{ge}^2/ω^2, and plasma frequencies, ω_{pe}^2/ω^2, to the wave frequency as rectangular axes in a plane. In this coordinate system the resonances and cut-offs are represented as particular curves. For example, the resonance frequency of the R-mode, $\omega_{R,res} = \omega_{ge}$ does not depend on the plasma density and in this system becomes a horizontal line valid for all ratios ω_{pe}^2/ω^2. Similarly, in a pure charge-compensated electron plasma the cut-off $\epsilon_3 = 0$ is a constant vertical line cutting the abscissa at $\omega_{pe}^2/\omega^2 = 1$ and valid for all ratios ω_{ge}^2/ω^2, while the resonance $\epsilon_1 = 0$ connects the point $\omega_{ge}^2/\omega^2 = 1$ on the ordinate with the point $\omega_{pe}^2/\omega^2 = 1$ on the abscissa. The cut-off $\epsilon_L = 0$ can be written as $1 - \omega_{pe}^2/\omega^2 = \omega_{ge}^2/\omega^2$ and describes a parabola with its minimum at $\omega_{pe}^2/\omega^2 = 1$.

The two straight lines, $\epsilon_R = \infty$ and $\epsilon_3 = 0$, divide the CMA-plane into four regions which are further subdivided by the remaining resonances and cut-offs at different angles. The total number of separate regions obtained is eight. Now, in each of these regions one can draw a polar plot of the phase velocity of the different waves normalized to the velocity of light, ω/kc, as a function of the angle θ between the magnetic field (parallel to the ordinate) and the direction of the wave vectors. The dashed circles denote $v_{ph} = \omega/k = c$. One observes the different connections between the L-, R-, O-, and X-modes in the different regions of the diagram as well as the changes of the polarization properties from region to region. In some regions no propagation is possible due to reflection from this domain.

C.6 Magnetized Dielectric Tensor

In a hot plasma of arbitrary temperature the calculation of the dielectric tensor, $\epsilon(\omega, \mathbf{k})$, entering the dispersion relation Eq. (9.57) demands solving the linearized Vlasov equation (10.84). This requires explicit calculation of the current integral over all particle species, s, in Eq. (10.92)

$$\delta\mathbf{j}(\mathbf{k}, \omega) = -\sum_s \frac{\epsilon_0 \omega_{ps}^2}{n_0 \omega} \int_0^\infty \int_0^{2\pi} \int_{-\infty}^\infty v_\perp dv_\perp d\psi dv_\parallel$$

$$\delta\mathbf{E}(\mathbf{k}, \omega) \cdot \int_0^\infty d\tau e^{-i\varphi(\tau)} \{\mathbf{v}\mathbf{k}(\tau) + [\omega - \mathbf{k} \cdot \mathbf{v}(\tau)]\mathbf{I}\} \cdot \frac{\partial f_0[\mathbf{v}(\tau)]}{\partial \mathbf{v}(\tau)} \qquad (C.32)$$

under the additional assumption of gyrotropy, i.e., that there is no difference between the two directions transverse to the ambient magnetic field, $\mathbf{B}_0$. In this case we can write the wave vector as

$$\mathbf{k} = k_\perp \hat{\mathbf{e}}_\perp + k_\parallel \hat{\mathbf{e}}_\parallel \qquad (C.33)$$

Similarly, we decompose the velocity into transverse and parallel components

$$\mathbf{v} = (v_\perp \cos\psi, v_\perp \sin\psi, v_\parallel) \qquad (C.34)$$

The motion of particles in uniform magnetic fields consists of a parallel translation with speed $v_\parallel = \text{const}$ and a gyration (see Chapter 2). A particle starting with initial phase angle ψ has at time τ the velocity

$$\mathbf{v}(\tau) = [v_\perp \cos(\omega_g \tau + \psi), v_\perp \sin(\omega_g \tau + \psi), v_\parallel] \qquad (C.35)$$

where $\tau = t' - t$ is the difference between observation time, t', and initial time, t. Correspondingly its position is, after integrating the velocity

$$\mathbf{x}(\tau) = \mathbf{x} + \omega_g^{-1}[v_\perp \sin(\omega_g \tau + \psi), -v_\perp \cos(\omega_g \tau + \psi), v_\parallel \tau] \qquad (C.36)$$

It is assumed that the wave has so small amplitude that the path of the particle is not distorted. With the help of these expressions we can now calculate the velocity-derivative vector of the distribution function appearing in Eq. (C.32)

$$\frac{\partial f_0[\mathbf{v}(\tau)]}{\partial \mathbf{v}(\tau)} = \left[\frac{\partial f_0}{\partial v_\perp}\cos(\omega_g\tau+\psi), \frac{\partial f_0}{\partial v_\perp}\sin(\omega_g\tau+\psi), \frac{\partial f_0}{\partial v_\parallel}\right] \tag{C.37}$$

Moreover, using the above representation of the position vector, the phase function of the plane wave, $\varphi(\tau) = -\omega\tau - \mathbf{k}\cdot[\mathbf{x}-\mathbf{x}(\tau)]$, can be rewritten as

$$\varphi(\tau) = (k_\parallel v_\parallel - \omega)\tau + \xi[\sin(\omega_g\tau+\psi) - \sin\psi] \tag{C.38}$$

where $\xi = k_\perp v_\perp/\omega_g$. Inserting for the exponent of the exponential factor in Eq. (C.32) lets the exponential depend on the two sin-functions. Fortunately we can make use of the following addition theorem of Bessel functions, $J_l(\xi)$

$$\sum_{l=-\infty}^{l=\infty} J_l(x)\exp(-il\phi) = \exp(-ix\sin\phi) \tag{C.39}$$

to decompose the exponential into a sum over the product of Bessel functions and simple exponentials which turn out to be trigonometric functions as

$$\sum_{l,l'=-\infty}^{l,l'=\infty} J_l J_{l'}\exp\{-i[(k_\parallel v_\parallel + l\omega_g - \omega)\tau + (l-l')\psi]\} = \exp[-i\varphi(\tau)] \tag{C.40}$$

Vice versa, we write the trigonometric functions in Eq. (C.37) as exponentials. In addition, observing that

$$-i[\mathbf{k}\cdot\mathbf{v}(\tau) - \omega]\exp[-i\varphi(\tau)] = \frac{d}{d\tau}\exp[-i\varphi(\tau)] \tag{C.41}$$

allows to simplify the scalar product in Eq. (C.32). We realize that the factor of $\delta\mathbf{E}(\omega,\mathbf{k})$ in Eq. (C.32) is the linear conductivity. Inserting it into Eq. (10.92), we obtain

$$\epsilon(\omega,\mathbf{k}) = \left(1-\sum_s\frac{\omega_{ps}^2}{\omega^2}\right)\mathbf{1} - \sum_s\frac{\omega_{ps}^2}{n_{0s}\omega^2}\int v_\perp dv_\perp dv_\parallel d\psi$$

$$\int_0^\infty d\tau\left(i\mathbf{k}\mathbf{v}(\tau) - \mathbf{1}\frac{\partial}{\partial\tau}\right)e^{-i\varphi(\tau)}\mathbf{v}\cdot\frac{\partial f_{0s}[\mathbf{v}(\tau)]}{\partial\mathbf{v}(\tau)} \tag{C.42}$$

The integration over τ can now be performed with the help of the above decompositions into sums. It splits into a number of separate integrations over exponentials of different l, l', which can subsequently be simplified using the orthogonality conditions of trigonometric functions, when performing the sum over l' and when ultimately performing the integration over the velocity phase angle, ψ, in the interval 0 to 2π. Since Eq. (C.42) contains a tensor, the final result also becomes a tensor

$$\in(\omega,\mathbf{k}) = \left(1 - \sum_s \frac{\omega_{ps}^2}{\omega^2}\right)\mathbf{I} - \sum_s \sum_{l=-\infty}^{l=+\infty} \frac{2\pi\omega_{ps}^2}{n_{0s}\omega^2}$$

$$\int v_\perp dv_\perp dv_\parallel \left(k_\parallel \frac{\partial f_{0s}}{\partial v_\parallel} + \frac{l\omega_{gs}}{v_\perp}\frac{\partial f_{0s}}{\partial v_\perp}\right) \frac{\mathbf{S}_{ls}(v_\parallel, v_\perp)}{k_\parallel v_\parallel + l\omega_{gs} - \omega} \qquad (C.43)$$

where we have introduced

$$\mathbf{S}_{ls} = \begin{bmatrix} \dfrac{l^2\omega_{gs}^2}{k_\perp^2}J_l^2 & \dfrac{ilv_\perp\omega_{gs}}{k_\perp}J_lJ_l' & \dfrac{lv_\parallel\omega_{gs}}{k_\perp}J_l^2 \\[2ex] -\dfrac{ilv_\perp\omega_{gs}}{k_\perp}J_lJ_l' & v_\perp^2 J_l'^2 & -iv_\parallel v_\perp J_lJ_l' \\[2ex] \dfrac{lv_\parallel\omega_{gs}}{k_\perp}J_l^2 & iv_\parallel v_\perp J_lJ_l' & v_\parallel^2 J_l^2 \end{bmatrix} \qquad (C.44)$$

with $J_l'(\xi) = dJ_l(\xi)/d\xi$. In the above calculation we have used

$$\int_0^{2\pi} d\psi \exp[i(l-l')\psi] = 2\pi\delta_{l,l'}$$

$$\int_0^{2\pi} d\psi \cos(\psi+\phi)\exp[i(l+l')\psi] = \pi\left[e^{i\phi}\delta_{l+1,l'} + e^{-i\phi}\delta_{l-1,l'}\right] \qquad (C.45)$$

and the following recursion formulas and sums

$$J_l(z) = \frac{z}{2l}[J_{l-1}(z) + J_{l+1}(z)]$$

$$J_l'(z) = \frac{1}{2}[J_{l-1}(z) - J_{l+1}(z)]$$

$$\sum_{l=-\infty}^{\infty} J_l^2(z) = 1 \qquad (C.46)$$

$$\sum_{l=-\infty}^{\infty} lJ_l^2(z) = 0$$

$$\sum_{l=-\infty}^{\infty} \frac{J_l(z)J_{l-r}(z)}{q-l} = \frac{(-1)^r\pi}{\sin\pi q}J_{r-q}(z)J_q(z)$$

Weber Integrals

The following so-called Weber integrals are used for calculating the magnetized plasma response function for longitudinal waves:

$$\int_0^\infty x\,dx\,J_0^2(px)e^{-q^2x^2} = (2q^2)^{-1}\exp\left(\frac{-p^2}{4q^2}\right)$$

$$\int_0^\infty x\,dx\,J_l(px)J_l(rx)e^{-q^2x^2} = (2q^2)^{-1}\exp\left[\frac{-(p^2+r^2)}{4q^2}\right]I_l\left(\frac{pr}{2q^2}\right)$$

(C.47)

Index

www.ingramcontent.com/pod-product-compliance
Lightning Source LLC
Chambersburg PA
CBHW070740220326
41598CB00026B/3710